In these two volumes Professor Nef relates the history of the coal industry from 1550 to 1700 and displays, with a wealth of detail, an impressive knowledge of his subject. The scope of the work is enormous for the two volumes are divided into five parts, each of which covers a complete aspect of the coal industry in England. Part I surveys the early history of the coal industry, the chief coal-producing areas, and the principal markets for coal, and in Part II the author elucidates the causes of the expansion of the industry, among them the transfer of property at the Reformation and the growth of industries which depended on coal for their fuel; while Part III is concerned with the ownership of natural resources and mineral rights. Part IV, an excellent study of 'coal and capitalism', is one which every student of economic history should read for the abundant information which it contains about the capitalist structure of the industry and the condition of labour, and in Part V Professor Nef concludes with an account of coal and public policy.

The scale of the work is further demonstrated by seventeen appendices, which include twenty pages of bibliography, statistics of production and consumption, statistics of shipments of coal, coal prices, extracts from the colliery accounts, and sundry illustrative documents. Yet despite the immensity of the work every statement is fully documented by the author who displays a notable degree of impartiality.

This is an extraordinarily readable work as Professor Nef possesses the gift of imagination and he manages to convey a sense of the romance of a great industry; notwithstanding this however he amply supports his main thesis that England's 'strength appears to have rested to some measure on coal, a hundred years before the Coal Age is supposed to have began', and these two volumes, which first appeared in 1932, constitute a substantial work of great value.

THE RISE OF THE
BRITISH COAL INDUSTRY

THE RISE OF THE
BRITISH COAL INDUSTRY

J. U. NEF

John

Volume One

FRANK CASS & CO. LTD.
1966

Published by Frank Cass & Co. Ltd.,
10 Woburn Walk, London W.C.1
by arrangement with George Routledge & Sons Ltd.

First Edition   1932
Second Impression   1966

Printed in Great Britain by
Thomas Nelson (Printers) Ltd., London and Edinburgh.

TO MY WIFE

CONTENTS

VOLUME I

CONTENTS

MAPS

ILLUSTRATIONS

KEY TO ABBREVIATIONS

Augm.	Augmentation Office.
Cal.	Calendar.
Cal. S.P. Colonial	*Calendar of State Papers, Colonial Series.*
Cal. S.P.D.	*Calendar of State Papers, Domestic Series.*
Cal. S.P. Irish	*Calendar of State Papers, Irish Series.*
Cal. of Treas. Bks.	*Calendar of Treasury Books.*
Cal. S.P. Venetian	*Calendar of State Papers, Venetian Series.*
Chanc. Deps.	Court of Chancery, *Depositions Taken by Commission.*
Chanc. Proc.	Court of Chancery, *Proceedings.*
Court of Augm. Proc.	Court of Augmentations, *Proceedings.*
Court of Requests Proc.	Court of Requests, *Proceedings.*
Deps.	Depositions.
Dict. Nat. Biog.	*Dictionary of National Biography.*
Duchy of Lancs.	Duchy of Lancaster.
Durham Curs. Recs.	Palatinate of Durham, *Cursitor's Records.*
Eccles. Com.	Ecclesiastical Commissioners.
Exch. Deps. by Com.	Court of Exchequer, *Depositions Taken by Commission.*
Exch. K.R.	Exchequer, King's Remembrancer.
Exch. L.T.R.	Exchequer, Lord Treasurer's Remembrancer.
Exch. Q.R.	Exchequer, Queen's Remembrancer.
Exch. Spec. Com.	Court of Exchequer, *Special Commissions of Inquiry.*
H.M.C. and *Hist. MSS. Com.*	*Historical Manuscripts Commission.*
Land Rev. Misc. Bks.	Exchequer, Land Revenue Office, *Miscellaneous Books.*
Min. Accts.	*Ministers' Accounts.*
P.C.R.	*Privy Council Register.*
Palat. of Durham	Palatinate of Durham.
Palat. of Lancs.	Palatinate of Lancaster.
Parl. Surveys	Augmentation Office, *Parliamentary Surveys.*
Pat. Rolls.	Chancery Enrolments, *Patent Rolls.*
Pat. Spec.	*Patent Specifications.*
Roy. Soc. Philos. Trans.	Royal Society, *Philosophical Transactions.*
S.P.D.	*State Papers, Domestic Series.*
Star Chamb. Proc.	Court of Star Chamber, *Proceedings.*
V.C.H.	*Victoria County History.*

PREFACE

THE material for a history of the British coal industry from 1550 to 1700 proved to be so abundant that I could make use of only a portion of it. The treatment had to be selective ; and the method of selection was determined partly by the nature of my interest in the subject. Even though the sources for a description of coal mining and the coal trade, in a narrow sense, were more than sufficient to occupy me, I was unable to confine myself to such a description. To study the history of this great industry is to become increasingly aware of its relation to most other industries, and, indeed, to the whole historical fabric of industrial civilization. I have been naturally drawn into such allied subjects as the general industrial development of the period, the progress of invention, and the history of mining law in its relation to the ownership of natural resources. But my interest in the wider implications of the rise of the coal industry has not, I believe, interfered to any great extent with the presentation of such information as might be expected in a monograph on coal mining and the coal trade. I can only hope that subsequent research among the manuscript sources, some of which are described in Appendix A, may fill in the gaps which I am conscious of having left.

Owing to delays, occasioned partly by illness and partly by the necessity of undertaking a thorough revision of the entire book, more than three years have now passed since portions of it were in type. I have tried to keep up with the important material which has been appearing during this period, but it has not been possible to incorporate everything of value. I have made much less use than I should have liked of Mr. Ashton and Mr. Syke's *Coal Industry of the Eighteenth Century*, and of the interesting section on coal in the recently published volumes of Mr. Lipson's *Economic History of England*.

The importance of giving a rough quantitative picture of the coal industry has led me, perhaps rashly, to attempt a number of statistical conclusions; but, as I have tried to indicate in each case, many of these are necessarily of a tentative and sometimes even of a conjectural nature. The maps and illustrations make no pretence at completeness. The former have been introduced mainly to help the reader in following the survey of the principal coalfields and markets in Part I. They are based on materials of a kind which make precision at all points impossible ; but I have tried to show in each case what is uncertain and what ambiguous. As is pointed out on page 23, the word " Midlands " has been used in Part I in a special sense, for the purpose of grouping certain coalfields which could conveniently be brought together in connection with the survey of production. Elsewhere the word is used in its ordinary sense. I have tried in so far as possible to avoid the use of special mining terms. In the cases in which they do appear I have attempted to define them

when they are first introduced. The whereabouts of these definitions
is usually given in the index.

I am acutely conscious of the many shortcomings of this book.
A seventeenth-century writer referred to a coal mine as " a Jack of
all Trades shop " ; and an historian of the coal industry ought to
be a master of many disciplines. His equipment should include a
wide knowledge of technology, of the economic history of Europe,
and even of political, constitutional, and legal history. When I set
out upon my task, I had almost no knowledge of any of these subjects,
and what I have since acquired is very imperfect.

Before referring to my major obligations, I must acknowledge
other assistance generously accorded me. Abroad, my researches
were facilitated by the help which I received from a number of persons,
particularly in Great Britain. I wish to thank especially Mr. William
Angus for helping me with sources in the Register House in Edinburgh ;
Mr. H. H. E. Craster for giving me information concerning material
on coal in the Bodleian Library ; the Dean of Durham for giving me
access to the documents in the Cathedral Library in Durham, and for
permitting me to print portions of one of them ; Mr. F. W. Dendy
for allowing me to draw upon his unrivalled knowledge of the Company
of Hostmen and of Newcastle history ; the Ecclesiastical Commissioners for permission to consult their documents in the New
Exchequer Buildings in Durham ; Dr. Hubert Hall for advising me
concerning collections of manuscripts in the north of England, and
for making it possible for me to examine several collections
which would otherwise have remained closed ; Professor Henri
Hauser for advising me concerning French and other continental
sources ; Mr. A. J. Hawkes for supplying me with information
concerning the exhibition of mining literature held at the Public
Library in Wigan ; Betriebsrat Willy Meyer, of Dortmund, for
helping me to understand some of the technical aspects of coal mining,
and for advice concerning authorities on mining in Germany; Mr. A. M.
Oliver for helping me with the records in the Town Hall and elsewhere in Newcastle, for supplying me with interesting information
about the history of the town, and for permitting me to use a transcript
which he has of a valuable document preserved in the Bodleian
Library ; Principal Rees for advising me concerning English sources ;
the late Lord Sackville and Lord De La Warr for permitting me to
consult the collection of manuscripts owned by the former ; Mr. A. H.
Thomas for calling my attention to various collections of documents
in the London Guildhall ; Mr. A. P. Wadsworth for calling my attention
to a number of documents among the records of the Palatinate of
Lancaster ; Dr. Marguerite Wood for assisting me in consulting the
manuscripts of the Duke of Hamilton, and for allowing me to make
use of some of her transcripts of these manuscripts. My work in the
Public Record Office, the British Museum, the National Library of
Scotland, and the Archives Nationales was made both more fruitful
and pleasanter than it would otherwise have been by the assistance
of members of the various staffs. In the Record Office I am especially
indebted to Mr. D. L. Evans and Mr. A. E. Stamp for their help.
Special thanks are due Mr. Ambrose Heal and the authorities of

Lincoln's Inn Library for permission to reproduce an illustration and a plan. I am indebted to Mr. H. F. Milne for drawing the maps. My English publishers have been both kind and patient in accepting my many delays and the continual changes and corrections which I have made in the proofs.

In my own country I acknowledge with thanks help of divers sorts. My general approach to history owes much, as will be apparent, to my former teacher at Harvard, the late Professor F. J. Turner. My friend Mr. Powers Hapgood enabled me to see at first hand something of the technique of mining and the nature of the miner's life to-day; he also guarded me against mistakes by reading over a part of the book which deals with the technical side of the industry. My colleagues Professors W. E. Dodd, F. H. Knight, Jacob Viner, and Helen Wright were all kind enough to read portions of the manuscript or the proof. While a student at the Robert Brookings Graduate School, I obtained useful suggestions from Professor Walton Hamilton, Dr. C. O. Hardy, Dr. L. S. Lyon, and Professor Walter Shepherd, each of whom read a part of the manuscript. Professor H. L. Gray helped me to understand certain documents in the Public Record Office. Professor E. R. A. Seligman kindly lent me a valuable tract on the coal trade which he had acquired for his own library. Miss Bertha Londeen helped me to prepare the index.

My principal scholastic obligations in the United States are to Professor E. F. Gay and Dr. Conyers Read. Professor Gay read the entire manuscript of Volume I, made many valuable criticisms and suggestions, encouraged me when encouragement was much needed, and put at my disposal on more than one occasion the wide knowledge and understanding of economic history of which he has given so generously to many students. I derived from him, among other things, some of my ideas about the influence of the contract to deliver upon industrial capitalism. To my friend Dr. Read, under whom I studied as a college freshman, I owe my first interest in economic and social history. Some years later, he suggested to me the possibility of attempting a work on a large scale in English history. He showed me that the *Port Books*, in the Public Record Office, might be used as the basis for a quantitative study of the coal industry, gave me much wise counsel concerning sources and methods of research, and read through in manuscript the whole book with the exception of the last two chapters of Part IV. I have profited greatly by his numerous criticisms and suggestions, and by his constant encouragement and sympathetic understanding.

This study was undertaken and carried through largely under the guidance of Professor R. H. Tawney, and I can never adequately acknowledge how much its composition owes to him. His interest in my work began nearly ten years ago, when I started gathering material. He suggested many fruitful lines of investigation, called my attention to a number of sources, helped me either directly or indirectly to obtain access to most of the private and semi-private collections of documents which I consulted, and undertook to publish my book in the series of which he and Professor Power are the editors.

He read through the major portion of the manuscript at various stages of its composition, giving far too generously of his time. I scarcely dare to think what portions of Volume I, and of Part I in particular, would be like but for the share which he took in their revision ; and it is a source of regret to me that the book itself is not more worthy of his interest in it. Apart from his personal assistance, he made it possible for me to obtain the invaluable help of Miss M. E. Bulkley. There is hardly a section or an appendix which has not been substantially improved by her criticism and advice. She has either verified directly, or had verified, a large portion of the references, and has saved me from making many errors. I am indebted to Dr. Power for assistance of various kinds, and especially for enabling me to obtain the very efficient help of Mr. F. J. Fisher in reading the proofs. He has read through the entire book with much zeal and patience, and there are few sections which do not bear the trace of his valuable suggestions.

My greatest debt is to my wife. She has done part of the research, has found most of the illustrations, and has helped me in all manner of ways. But for her unflagging support, interest, and sympathy, the book would never have been completed.

J.U.N. 1932

INTRODUCTION

DEPOSITS of mineral fuel have been present in the earth for thousands of years, but other civilizations rose and fell and left them undisturbed. Except in the case of the Chinese, who are said to have made use of it as long ago as 1000 B.C.,[1] there is no unchallengeable evidence that ancient peoples ever mined coal.

When, towards the end of the thirteenth century, the Venetian wanderer Marco Polo penetrated to the interior of Cathay, he was surprised at—what seemed to him—the strange practice of the natives, who burned for fuel a " black stone . . . dug out of the mountains, where it runs in veins."[2] They had probably done so for centuries. China, after North America, is apparently the richest territory in the world in coal resources ; and there are certain districts in which the seams, pushing upwards between rocky strata, pierce the surface of the earth at a great number of points, providing what the early Germans picturesquely called " day " (tage) coal.[3] It was from these outcrops—to give them their English name—that the Chinese dug their supplies. We have no evidence that they ever sank deep pits. Doubtless the demand for coal was not great enough to tempt diggers to follow the seams many feet underground, for it was confined to small areas within a few miles of the outcrops. In these areas, when wood and turf were scarce, coal was burned in the family hearth or the smithy forge. The situation of the more important Chinese coal measures, far from the sea and from navigable rivers, and, consequently, from the chief centres of population, often makes intensive exploitation unremunerative even to-day, when western capitalists have begun to push railway lines into the interior. How much less remunerative, then, must such exploitation have been before the age of steam transport, when the carriage of fuel ten miles from the outcrop at least tripled—if it did not quadruple—the selling price.[4] It was partly because of the unfavourable location of the seams that coal never became an important factor in the ancient civilization of the Chinese.

A belief that coal was known in the Graeco-Roman world is based, first of all, upon certain passages in the writings of contemporaries. The most frequently cited is one from Theophrastus, the

---

[1] T. J. Taylor, " Archæology of the Coal Trade," in *Proc. of the Arch. Inst. of Great Britain and Ireland at Newcastle*, 1858, vol. i, pp. 151–2 ; G. Decamps, *Mémoire historique sur l'origine de l'industrie houillère dans le bassin du couchant de Mons*, 1879, p. 14 ; Ludwig Beck, *Geschichte des Eisen*, 1884–1903, vol. i, pp. 297–8 ; Thomas Y. Hall, " On the Progress of the Coal Mining Industry in China," in *Trans. North of Eng. Inst. of Mining Engineers*, vol. xv, 1865–6, pp. 67–74.
[2] M. Komroff, *Travels of Marco Polo*, 1926, pp. 170–1.
[3] The word " Day-Coal " was used by a physician of Newcastle in a letter to the Royal Society in 1676 (*Roy. Soc., Philosophical Transactions*, vol. xi, 1676–7, pp. 762–6), and this usage became common in England at about that time.
[4] See below, p. 103.

1

Greek philosopher, pupil and disciple of Aristotle. Theophrastus, in a book about stones, mentiors as a curiosity a kind of black stone, probably lignite, which smiths burned occasionally in place of charcoal, and which was found in the province of Liguria in northern Italy, and in the province of Elis in Greece.[1] Students of classical literature have suggested a few other passages which may refer to mineral fuel,[2] though none is as convincing as the passage in Theophrastus. During the nineteenth century, the excavations of archæologists in various counties of England have revealed the presence of coal ashes amid Roman remains,[3] and have therefore established a strong presumption [4] in favour of the theory that coal was used to some extent in Britain during the Roman period. If, however, coal mining had been of any significance in the life of the Roman Empire, if it had even been as important an industry as in ancient China, we should hardly be without definite references to it in the very considerable number of classical treatises that have come down to us ; [5] Theophrastus could not well have neglected to mention the digging, as well as the burning, of his " black stones " ; and the Greeks and Romans could not have failed to find words, such as appear in early Chinese, and in several western European tongues during the Middle Ages, to distinguish mineral fuel from charcoal.[6]

Apart, then, from a restricted use of it by the Chinese, and possibly a casual use by the Greeks and Romans, there is no evidence that coal was ever burned by any of the ancient peoples.[7] It may have been put to other uses. This is suggested by the fact that, until the nineteenth century, in various districts of England, some grades, which are geologically closely allied to jet,[8] served as material for making buttons,

---

[1] R. L. Galloway, *Annals of Coal Mining and the Coal Trade*, 1898, p. 1.

[2] Sallust, for instance, mentions mineral fuel burned by the smiths at Casta Aelia in Spain (C. de Brosses, *Histoire de la république romaine . . . par Salluste*, 1777, vol. i, p. 581). See also J. P. Bünting, *Sylva Subterranea*, 1693, ch. ii.

[3] At Lanchester and Ebchester (Durham), Manchester, North Brierley (Yorks), Wroxeter and Oakengates (Shropshire). (See Galloway, *op. cit.*, p. 5 ; *V.C.H. Shropshire*, p. 455.) Similar evidence is said to have been discovered concerning the use of coal by the Romans in the neighbourhood of Liége (F. Henaux, *Histoire du pays de Liége*, 1856, vol. i, p. 19 ; R. Malherbe, " Historique de l'exploitation de la houille dans le pays de Liége," in *Mémoires de la soc. libre d'émulation de Liége*, new series, vol. ii, 1862, p. 276.)

[4] L. F. Salzman (*English Industries of the Middle Ages*, 2nd ed., 1923, p. 1) considers the evidence conclusive.

[5] Ancient writers, like Cæsar, who show an extensive knowledge of mining methods in connection with the extraction of lead, iron, and copper, never mention the mining of coal (Decamps, *op. cit.*, pp. 24–7).

[6] The Greek *anthrax* and the Latin *carbo* mean charcoal. In the Chinese language, as in early English, French, and German, mineral fuel was distinguished from charred wood by a qualifying prefix. The Chinese sometimes called the new fuel " ice-charcoal " (see the writings of Liu An—122 b.c.—referred to by Komroff, *op. cit.*, p. 170, note).

[7] Such evidence as has been advanced in favour of the use of coal by the early Britons has been met by Galloway (*op. cit.*, pp. 2–4) with what seem to be conclusive objections. The absence of satisfactory evidence is, of course, no proof that coal was unknown.

[8] Particularly the " cannel " coal of Wigan, also found in small quantities in Staffordshire and near Sheffield in Yorkshire, and the " peacock " coal of Staffordshire. (Cf. below, pp. 115 sqq.)

spoons, plates, dishes, snuff-boxes, ink-wells, candlesticks and various ornaments, for carving statues and for paving church floors.[1] At Haigh, near Wigan, it even served for the roof and walls of a summer house, built in the seventeenth century by the lord of the manor.[2] In medieval Germany women sometimes wore coal trinkets as jewellery ; [3] and coal tokens, used centuries ago either as amulets or money, have been found in Dorsetshire.[4] At a period before men thought of mining for fuel, this carboniferous mineral (small pieces of which were found loose along the seashore or near the outcrops) may have possessed a value due to its scarcity. Its use on a large scale as a combustible is clearly a peculiar feature of our own time. There is, perhaps, no other natural product which is so essential a part of our life, and which yet has been of so little importance in the lives of ancient peoples.

England is commonly believed to be the country in which coal resources were first intensively exploited ; and, although that belief may be qualified by certain reservations,[5] it is on the whole confirmed by historical research. Less can be said on behalf of the equally prevalent belief that in England, and Great Britain generally, the coal industry was of little or no importance before the nineteenth century. It is generally supposed that, before the end of the eighteenth century, except possibly within the counties of Durham and Northumberland, there was, properly speaking, no mining for coal by means of shafts, and that the small quantities of mineral fuel burned came from short trenches or from open-works dug like stone quarries into the outcropping seams. Anyone who had made a considerable study of early industrial history would, of course, reject such a view of early coal mining. But even historians have tended to minimize the development of the coal industry before the Industrial Revolution. A majority of them, if asked when a history of coal mining should begin, would probably say in the eighteenth century, adding, no doubt, that it might be well to summarize the earlier history in a fairly long introductory chapter.

But a long chapter would not provide enough space for the previous record of a fuel which was described already in 1597 as " one principall commoditie of this Realme ".[6] We are not obliged to exhaust all English literature from Shakespeare to Milton to find a substantial

---

[1] R. Plot, *Natural Hist. of Staffordshire*, 1686, pp. 125–6 ; Aléon Dulac, *Observations générales sur le charbon minéral*, 1786 (Bibliothèque nationale, *Manuscrits français*, no. 11857), ff. 4–5 ; *Hist. MSS. Com., Report on MSS. of the Duke of Portland*, vol. vi, p. 191 ; Archives nationales, 0'1293 (letter written by M. Ticquet from England in 1738) ; Defoe, *Tour through Great Britain*, 1769 ed., vol. iii, p. 105 ; Sir John Sinclair, *Statistical Account of Scotland*, 1791–9, vol. xv, pp. 4–5.

[2] Cf. Wigan Public Library, *Catalogue of the Jubilee Exhibition of Early Mining Literature*, 1928, p. 1.

[3] G. Agricola, *De Natura Fossilium*, cited Beck, *op. cit.*, vol. ii, pp. 104–5.

[4] It has been thought that coal pieces may have been used as money in England in the Anglo-Saxon period (Hutchins, *Dorsetshire*, 1st ed., vol. i, p. 197, cited Ruding, *Annals of the Coinage*, 1840, vol. i, p. 4, note).

[5] See below, pp. 7–8.

[6] *A Discourse Touching the Stapling of Newcastle Coles, Tyne, and Leade* (*S.P.D., Eliz.*, vol. cv, no. 30). Cf. below, pt. v, ch. i.

number of references to the coal trade, as " one of the greatest sorts
of traffic in the kingdom ", " one of the greatest home trades within
the Commonwealth of England." " Coale," said the Earl of Wemyss,
speaking on behalf of the Scottish colliery owners in 1658, " is one of the
greatest staple comodityes of the [Scottish] nation." [1] We cannot
overlook such explicit testimonials to the importance of this industry
in the national life long before the nineteenth, or even the eighteenth,
century. We cannot accept either as the starting point for a history
of coal.

How much farther back shall we carry this history ? No certain
records have been found to show that coal was used in western Europe
before, at any rate, the last decade of the twelfth century ; the earliest
reference to what is unmistakably coal occurs in the Chronicle of
a monk named Reinier, of the Priory of St. Jacques at Liége, who
wrote, under the year 1195 or 1213 (the date is uncertain), of the
discovery of " black earth very similar to charcoal ", which was being
used by smiths and metal workers.[2]

Until near the end of the nineteenth century, it was believed
that we had documentary proof of the use of these " black stones "
in England in the ninth century, and, even to-day, statements to that
effect are found in books on mining. But Galloway, who has done so
much to till the soil for future historians of the coal industry, has
raised what seem to be insuperable objections to translating as mineral
fuel the Anglo-Saxon word *graefa* in the passage upon which this
belief rested.[3] Other references hitherto cited as evidence that coal
was burned in England before the thirteenth century are based on
misunderstanding of the meaning of the word as it was used in medieval
documents. Until the seventeenth century, the English " coal " (in
its various spellings), like the Latin *carbo*, meant charcoal ; coal
was invariably distinguished by the addition of a qualifying prefix,[4]
such as " sea- ", " pit-," " earth-," " stone," " peacock-," or
" smithy-," designations derived from the place whence the new fuel
was taken, from its physical appearance, or from the use to which
it was put. By far the most common of these words was " seacoal ".
The origin of this term (no equivalent for which appears to be found
in the colliery districts of early France, Germany, or the Belgian
provinces), has caused considerable controversy and is still unsettled.[5]
During the seventeenth century the English first used the word " coal ",
or " coals ", as a term for all types of mineral coal (although the

---

[1] *S.P.D., James I*, vol. clxxx, no. 77 ; *S.P.D., Interreg.*, vol. clxxx, no. 12 ;
Lansdowne MSS., 213, no. 30 ; R. Welford, *History of Newcastle and Gateshead*,
1885-7, vol. iii, p. 265.

[2] Malherbe, *op. cit.*, pp. 274-5 ; Beck, *op. cit.*, vol. ii, pp. 101 sqq. The value
of Reinier's testimony is challenged by Malherbe (*op. cit.*, pp. 277-8), even if
the earlier date is the correct one, on the ground that a reference of 1202,
describing a coal *pit* at Liége, reveals a stage of mining technique which, he
thinks, could hardly have been reached within a decade of the " discovery "
of coal.

[3] Galloway, *op. cit.*, pp. 8-10.

[4] Bishop Fleetwood was one of the first to point this out (see his *Chronicon
Preciosum*, 1707 ed., p. 118). Cf. S. Dowell, *Sketch of the History of Taxes in
England*, 1876, p. 106.

[5] See Appendix P.

older designations were not immediately given up), and it is only since the Industrial Revolution that the word has lost its original meaning.[1] In early French, Flemish, and German, we also find coal distinguished from charred wood by the addition of prefixes or adjectives similar to the English.[2] Indeed in German and French the compound words *steinkohle* and *charbon de terre* have persisted, and are in use to-day. The great confusion over names is in itself evidence of the novelty of coal in the Middle Ages, each district tending to supply a word of its own for the new fuel.

Attempts to show that coal was burned on the Continent before the last decade of the twelfth century do not rest on a misinterpretation of the word " coal " ; yet they are, for the most part, hardly more convincing than the corresponding attempts of English antiquarians. Tradition has it that in the neighbourhood of Zwickau in Saxony,[3] and in the Mühlbachtal near Essen,[4] coal was worked in the tenth century. But scholars have been unable to discover any written evidence of its use in either district before the fourteenth century.[5] Judgment must be reserved until better grounds are found for believing these traditions. They are hardly better established, as yet, than the fantastic legend that the Roman legions, when first they marched up the Rhone valley, fled in terror from " black " men who emerged out of pits near St. Etienne.[6]

In one instance the case in favour of genuine records of the burning of coal before the time of Reinier deserves to be taken more seriously. In five separate ordinances promulgated by the Abbey of Rolduc between 1113 and 1125, reference is made to the exchange of a commodity called *Kalkulen* ; and, while there is nothing to indicate the use to which this commodity was put, two German antiquarians have assumed that the word is an early form of *Kolkulen*, a common term for coal at a later date in the district between Cologne and Aachen.[7] They have concluded that coal was mined at the beginning of the twelfth century in the Wurmrevier, or Aachen coalfield ; although Decamps, in his valuable monograph on mining near Mons, points out that the Abbey also owned land in the Liége and the Ruhr districts, and that there is no reason to think that the ordinances in question applied especially to the Wurmrevier.[8] It is not even

---

[1] Both meanings are given in Johnson's *Dictionary*.

[2] There are equivalents for most of the early English words for coal in French and German. Thus we have : *charbon de terre, de roche, de pierre, de forges, de fevres*, and *faux charbon* ; *Steinkohle, Pechkohle, Bergkohle, Tagekohle, schwarze Erde, Kohlerde, schwarze Kreide*, etc. The word *houille* is of Liége origin. *Charbon de mer* was never used in France, as far as I know, except to describe coal imported from England (see Appendix P). Probably most of the early words used to describe coal were invented by the inhabitants of the districts in which it was found. Later on, however, certain of the words came to be identified with coal of a particular grade. " Stone coal," for instance, was sometimes used to mean anthracite. Cf. below, pp. 115–6.

[3] R. F. Koettig, *Notizen über den Steinkohlen-Bergbau Sachsens*, 1861, p. 1.

[4] Otto Hue, *Die Bergarbeiter*, 1910, vol. i, p. 347.

[5] *Ibid.*, pp. 347–8.

[6] M. Rouff, *Les mines de charbon en France au XVIIIᵉ siècle*, 1922, p. xiii.

[7] Käntzeler and Michel, " Zur Geschichte der Kohlenbergwerke im Wurmrevier," in *Echo der Gegenwart* (Aachen daily newspaper), 1873, no. 126.

[8] Decamps, *op. cit.*, p. 30. It has been generally assumed that coal workings

certain that *Kalkulen* really refers to coal ; does not *Kalk* suggest lime, a commodity very common in medieval Germany ? [1] If there was any such early use of coal on the Continent, however, we might expect it to have been in the basin between Aachen and Mons, for, until the end of the eighteenth century, that area contained by far the most important colliery districts outside Great Britain.

To say that we have no certain records before the end of the twelfth century is far from saying that coal was not used at an earlier date. Many years might elapse between the first use of " black stones " as fuel and the first documentary evidence of such use. The earliest documents that mention coal, in nearly all the British and continental mining districts except Liége, describe " workings " of a type that would hardly be undertaken until the mineral had been dug for decades. Our first record—dated 1243—of an actual " working " in England shows that miners had already got beyond the stage of mere quarrying for coal, and had begun to dig trenches down into the dip of the seams.[2] A mining ordinance issued by ecclesiastics in 1228—our first reference to coal in the field south of Mons—proves that in some cases miners had given up open works altogether, and had sunk shallow pits.[3] While it is not until 1236 that we have so much as a reference to mineral fuel in any of the English mining fields, we know that by 1200 *charbon de roche* from England was one of the commodities regularly imported at Bruges, the busy centre of a precocious commercial district,[4] and that by 1226 " seacoal " was well enough known in London to give the name to a street.[5]

While it is reasonable to assume that coal may have been burned for some years, or decades, before we get any written records, it is equally reasonable to assume, from the absence of records, that coal must have been little used in the twelfth century, as compared with the thirteenth, when dozens of unassailable references establish the fact that it was being worked in almost every field of

---

in the Liége district " were older and more important than those in the Wurmrevier " ; and this view is strengthened by the discovery of records which show that the citizens of Aachen (situated in the centre of the Wurm mining field) purchased Liége coal, brought overland a distance of about 20 miles, as late as 1353–4 (Beck, *op. cit.*, vol. i, pp. 769–71). Unless it be assumed that the smiths of Aachen found the Liége fuel of a quality so much better suited to their operations than that of their own district (cf. below, pp. 119–20) that they preferred to pay the expense of the long carriage rather than to use the coal from pits near at hand, it is difficult to account for this bringing of fuel overland on any ground other than that the mines round Liége were exploited earlier than those round Aachen.

[1] I only mean to suggest that the interpretation of Käntzeler and Michel is debatable, not that my interpretation is better, or as good. Even if it could be proved that *Kalk* did mean lime, it might be argued that *Kalkulen* meant " lime-coal ", for one of the earliest uses for mineral fuel was in limeburning. The word " lime-coal " was very rarely used, however, though " smith-coal " was common in French and German, as well as in English.

[2] L. F. Salzman, *English Industries of the Middle Ages*, p. 4.

[3] Decamps, *op. cit.*, pp. 382–3.

[4] *Cartulaire de l'ancienne Estaple de Bruges*, 1200, cited N. S. B. Gras, *Evolution of the English Corn Market*, 1915, p. 110 note.

[5] Galloway, *op. cit.*, p. 29.

England and Scotland, and in a number of fields on the Continent.[1] From different places in the Midlands come reports that poor peasants or their cattle have stumbled into unfenced pits and lost their lives, either from the fall or from drowning. In South Wales tenants' services sometimes include the carrying of " seacoal " to supply fuel for burning lime.[2] Round the Firth of Forth, and in Durham and Northumberland, monasteries make grants of the right to take, or dig for, coal within ecclesiastical holdings.

Strictly speaking, therefore, a history of the coal industry should start with the thirteenth century. That was originally my intention. An examination of such records as may be found in printed sources convinced me, however, that if the aim be to study the influence of coal upon industry, invention, the ownership of natural resources, industrial and financial organization, political power and state policy, there is little to be learned from the history of British coal mining during the Middle Ages, for the reason that it had then little influence upon any of these matters. While further research among medieval sources will undoubtedly be repaid by the discovery of many new documents which mention coal, it will be surprising if these documents do not confirm the impression that this fuel was little used in England before the sixteenth century. We shall find that the coal diggers borrowed their mining technique almost entirely from the metallurgical miners, that their industrial organization simply reflected features of the early manorial and gild economy, that statesmen were hardly aware of the existence of coal mines, that mineral law would not have been very different in 1550 if no one had ever burned a block of coal. What is here said is not meant to discourage the undertaking, or to disparage the value, of a study of coal mining in the Middle Ages. Such a study can be most profitably begun in the archives at Liége,[3] rather than in any British depositories, and, if thoroughly carried

---

[1] The following is a list of the earliest references I have found to the use of coal in various districts in Britain and on the Continent :—
(?) 1195—Liége (see above) ; 1200 to 1219—Linlithgowshire and East Lothian (Galloway, *op. cit.*, p. 18 ; R. W. Cochran-Patrick, *Early Records of Mining in Scotland*, 1878, pp. xliii, 1) ; 1228—Mons (see above) ; 1236—near Blyth, Northumberland (Galloway, *op. cit.*, p. 21) ; 1240—Pontefract, Yorkshire (*V.C.H. Yorks*, vol. ii, p. 338) ; 1249—Neath, Glamorganshire (*Catalogue of MSS. relating to Wales*, ed. Edw. Owen, 1900, p. 558) ; 1255—Forest of Dean (*Pat. Roll*, 40 Henry III) ; 1257—Nottinghamshire (Galloway, *op. cit.*, pp. 26–7) ; 1260—Clee Hills, Shropshire (*V.C.H. Shropshire*, p. 449) ; 1264—Colne, Lancashire (*V.C.H. Lancs.*, vol. ii, p. 356) ; before 1270—Kingswood Chase (*V.C.H. Glouc.* vol. ii, p. 235) ; 1272—Sedgeley, Staffordshire (*Collections for a History of Staffordshire*, ed. Wm. Salt Arch. Soc., vol. ix, 1888, p. 29) ; 1276—Chilvers Coton, Warwickshire (*V.C.H. War.*, vol. ii, p. 219) ; 1290—Silkstone, Derbyshire (*V.C.H., Derby*, vol. ii, p. 350) ; 1291—Dunfermline, Fife (Cochran-Patrick, *op. cit.*, p. 1) ; 1297—Charleroi (G. Arnould, *Mémoire sur le bassin houiller du couchant de Mons*, 1877, p. 15) ; 1317—Essen (Hue, *Die Bergarbeiter*, vol. i, p. 347) ; 1321—Roche la Molière (L. J. Gras, *Histoire économique des mines de la Loire*, 1922, pp. 31, 42) ; early fourteenth century—Aachen (K. T. von Inama-Sternegg, *Deutsche Wirtschaftsgeschichte*, 1879–1901, vol. iii, pt. ii, p. 145) ; 1348—Zwickau, Saxony (Hue, *op. cit.*, vol. i, p. 348) ; 1366—Altwasser, Silesia (*ibid.*, p. 349).

[2] J. W. Willis Bund, *The Black Book of St. David's*, 1902, pp. 181, 303.

[3] Where the records of the local mining court, a body unique in the history of coal mining, are collected. (Cf. below, pp. 287–8.)

out, it may be relied upon to provide an interesting chapter in medieval economic history.

In the presence of a great many scattered references to coal pits, and to purchases of coal, and in the absence of any precise information concerning early output, it is easy to get an exaggerated impression of the importance of the medieval coal industry. It is certainly too much to say that during the fourteenth century coal provided as many promising openings for English capital as cloth,[1] which, after corn, might reasonably be regarded as the chief domestic commodity. Before the sixteenth century the cost of purchasing mining equipment, of digging a pit to the seam, and perhaps a short trench to drain off the water from the working place, rarely exceeded £15, while less than £5 often sufficed.[2] Even when we multiply these figures by twelve or fifteen, as we must to get their equivalent in modern money, the charge seems small. And it was borne, in a very large number of cases, by ecclesiastics rather than by lay landlords or merchants.[3]

Did merchants in the thriving commercial towns find an opportunity to employ their savings in the purchase and sale, if not in the mining, of coal ? Except for the Tyne valley there was no district from which, until after 1500, the new fuel was regularly carried in quantities of more than a few hundred tons per annum, for distances of more than a few miles from the outcrops.[4] In other districts coal, chiefly for local consumption, was dug by husbandmen, who worked either to supply themselves, or—in return for some kind of wage—to hand over the product of their labour to the lord of the manor.[5] From time to time a modern excavator presents us with

---

[1] As Miss Alice Law asserts ("The English Nouveaux-Riches in the Fourteenth Century," in *Trans. Roy. Hist. Soc.*, N.S., vol. ix, 1895, pp. 57–8).

[2] A pit sunk at Beaupark (Durham) in 1531 cost £1 9s. (J. B. Simpson, *Coal Mining by the Monks* reprinted from *Trans. North of Eng. Inst. of Mining Engineers*, vol. xxxix, 1910, p. 17) ; one at Rainton in 1443 cost £4 8s. 9d. (*ibid.*) ; one at Aldengrange £14 5s. 6d. (*ibid.*) ; one at Bilton (Northumberland) in 1479, £11 (*A History of Northumberland*, vol. ii, 1895, pp. 452–3) ; one at Kilvey (Glamorganshire) in 1400 £3 (see Appendix K (i)) ; one at Falkland (Fife) in 1500, 10s. (£5 Scots) (*Accounts of the Lord High Treasurer of Scotland*, vol. iii, p. 87). Occasionally larger sums were invested in coal mining, e.g. £60 for an adit (Galloway, *op. cit.*, p. 71), but this was exceptional.

[3] See below, pp. 134 sqq.

[4] Sometimes a few loads were carried on packhorses for considerable distances overground, as from the Mendip Hills to Glastonbury in 1456 (*V.C.H. Somerset*, vol. ii, p. 380), or from near Wakefield to York in 1499 (Galloway, *op. cit.*, p. 77).

[5] One finds old records of numerous diggings of the kind mentioned in the text in districts like the Wear Valley between Durham and Bishop Auckland (Simpson, *op. cit.*, p. 15, and *passim*; Greenwell MSS. (Newcastle Public Library), D 84 ; V.C.H. Durham, vol. ii, pp. 323–4 ; Galloway, *op. cit.*, pp. 51–4, 69–72 ; Mathias Dunn, *View of the Coal Trade*, 1844, p. 12) ; in the West Riding of Yorkshire round Leeds, Wakefield, and Halifax (V.C.H. Yorkshire, vol. ii, pp. 336–41 ; Galloway, *op. cit.*, pp. 60–1, 76–7 ; J. Lister, "Coal Mining in Halifax," in *Old Yorkshire*, 2nd Series, 1885, pp. 269–70) ; and in Lancashire all through the Forest of Pendle (*Duchy of Lancs. Deps.*, 22/L/31, 77/B/6 ; *Duchy of Lancs. Pleadings*, 109/A/9 ; V.C.H. Lancashire, vol. i, p. 13 ; vol. ii, pp. 346–7 ; vol. iv, p. 108 ; H. T. Crofton, "Lancashire and Cheshire Coal Mining Records," in *Trans. Lancs. and Chesh. Antiq. Soc.*, vol. vii, 1889, pp. 34–5 ; Galloway, *op. cit.*, pp. 28–9, 61, 77).

direct physical evidence of early workings, when he finds the soil beneath the surface of a plot of countryside honeycombed with shallow pits, widening out at the bottom and resembling in their shape a bell or an inverted cone.[1] The revenues obtained from such workings did not commonly exceed a few pounds a year,[2] except in a few manors like Whickham and Gateshead on the Tyne.[3] From the nature of the pits, as revealed by excavators, it may be concluded that work was usually abandoned as soon as the coal within a few feet of the surface had been taken ; from the size of the ordinary revenues it is clear that the annual output of a single mine usually amounted to a few score, or, at the most, a few hundred tons.[4] Generally the consumer himself brought a sack, a packhorse or a cart to the pits to carry away his own fuel, and, as a consequence, there was no place for a middleman.

The market for coal dug in the lower Tyne valley, it is true, was not confined, even in the Middle Ages, to the immediate vicinity of the outcrops. Ships which anchored before the town of Newcastle carried coal to London, to other English ports between Berwick and Dover, to Flemish, Dutch, and occasionally to French and German towns along the coast from Dieppe to Lübeck.[5] Yet, though the market comprised a coastline of nearly two thousand miles and included many of the most important early trading centres, it absorbed coal only in very small quantities. I have compiled, from the early rolls kept by the customs officers, a few statistics of exports from Newcastle to foreign countries. For four separate years between 1375 and 1515 the

---

[1] For the technical side of early coal mining see Galloway, *op. cit.*, pp. 32–4, 56–8 ; *A History of Northumberland*, vol. ii, p. 541 ; *V.C.H. Shropshire*, p. 451 ; *V.C.H. Lancashire*, vol. ii, p. 359 ; *V.C.H. Yorkshire*, vol. ii, pp. 335–8 ; *V.C.H. Gloucestershire*, vol. ii, p. 219 ; W. R. Scott, *Joint-Stock Companies*, 1910–12, vol. ii, p. 460 ; M. Rouff, *Les mines de charbon en France au XVIIIe siècle*, 1922, p. 39. For a hypothetical description of the common coal mining enterprise at the end of the fifteenth century see Salzman, *op. cit.*, pp. 10–11.

[2] Simpson, *op. cit.*, pp. 13 sqq. ; E. A. Lewis, " The Development of Industry and Commerce in Wales during the Middle Ages," in *Trans. Roy. Hist. Soc.*, N.S., vol. xvii, 1903, pp. 146–7.

[3] The principal seats of the medieval coal industry, outside the Tyne valley, appear to have been :—the lordship of Gower in Glamorganshire, where the profits were nearly £100 in 1400 (see Appendix K (i)), Wollaton and Strelley manors on the Trent (*Star Chamb. Proc., Henry VIII*, 18/115 ; 22/94 ; *Hist. MSS. Com., Report on the MSS. of Lord Middleton*, p. 307), Raby manor, to the south of Bishop Auckland (*Durham Curs. Recs.*, Bishop Dudley, roll 1, m. 7 ; *V.C.H. Durham*, vol. ii, pp. 322–4 ; *V.C.H. Yorkshire*, vol. ii, p. 340 ; Galloway, *op. cit.*, pp. 72–3) ; and Tranent on the coast of East Lothian (*Exchequer Rolls of Scotland*, vol. v, pp. 52–4, 64 ; P. M'Neill, *Tranent and its Surroundings*, 2nd ed., 1884, pp. 9–11). From Tranent some coal was shipped as early as the fourteenth century to Leith, for consumption in Edinburgh (*Exch. Rolls of Scot.*, vols. ix and x, *passim* ; *Accounts of Lord High Treas. of Scot.*, vols. i–v, *passim*) ; to Perth, Dundee, and Dunbar (*Exch. Rolls of Scot.*, vol. i, p. 159 ; vol. iv, p. 165 ; vol. vi, pp. 243, 606 ; vol. vii, pp. 4, 113, 179, 205, 233, 342, 395, 474, 533, 617 ; vol. viii, pp. 59, 167 ; vol. ix, pp. 116, 224 ; vol. xv, pp. 130–1, 256, 264 ; vol. xv, pp. 6–7) ; and even to Aberdeen (*ibid.*, vol. v, pp. 363, 658), Inverness (*ibid.*, vol. v, pp. 234, 659), Berwick (*ibid.*, vol. i, p. 64), Orkney and Shetland (*ibid.*, vol. xiii, p. 339),

[4] Excepting at the mines mentioned in the preceding footnote.

[5] *Exch. K.R. Customs Accounts*, 106/4, 107/57, 109/1 ; Salzman, *op. cit.*, p. 91; *Acts of the Parliaments of Scotland*, vol. i, p. 438a ; Galloway, *op. cit.*, p. 31 (cf. below, pp. 85–7).

statistics are complete. They show that exports ranged from 2,000 to 7,000 tons per annum, and that in the year 1377–8 the exports were probably as large as they were a century and a half later in 1513–14.[1] Nor is there any reason to think that the imports of Newcastle coal at London increased much during this period, for there was no attempt, between the accession of Edward II in 1307 and the second half of Elizabeth's reign, to increase the number of coal meters, who made it their business to measure coal unloaded on the Thames.[2] It is doubtful whether, before the sixteenth century, the annual shipments from the Tyne, both to foreign and to English ports, often exceeded 15,000 tons.[3]

Even such trade as there was in " seacoal " from Newcastle did not offer much of an opening for English capital. The ships which carried fuel were almost all owned and manned by Normans, Bretons, Gascons, and Flemings.[4] These foreigners came to the Tyne primarily to bring corn for the people of Durham and Northumberland (counties always short of grain supplies), in exchange for cloth, fish or grindstones, originally the chief trading assets of the district. They first took on board coal (found either in the Tyne valley or on the seacoast),[5] in place of sand, as ballast for the return trip ; and as ballast they continued to carry it throughout the Middle Ages. Unless coal was carried either as ballast or as supplementary cargo, the proceeds on its sale at English or foreign ports did not cover the charge of freight.[6] Shipmasters found an advantage in spending a little extra money at Newcastle to get coal (which could be had until the sixteenth century for about two shillings per ton at the ship-side) in the hope of finding a purchaser for it in the town where they planned to unload their more valuable cargo of cloth, grindstones or fish ; but they did not sail to Newcastle, as hundreds of shipmasters were destined to sail every year during the seventeenth century, with the sole object of buying " seacoal ". Nor do we find, in any of the coastal towns during the Middle Ages, traders who make it their chief business to deal in this new fuel.[7] In London small stocks, if they were to be had, were in the

---

[1] See Appendix D(i). I have reduced the figures, which are given in the customs accounts in chaldrons, to tons. The amount of coal contained in a chaldron increased considerably during the fourteenth and fifteenth centuries. I have attempted to allow for the changes in the size of this ancient measure (see Appendix C (i)).

[2] See below, vol. ii, pp. 253–4, 256. [3] Cf. below, pp. 25, 79.

[4] W. Cunningham, *Growth of English Industry and Commerce, Modern Times*, 1921 ed., pt. i, p. 247 ; and see below, vol. ii, pp. 23–4. " Ye shall se in peace time," wrote in 1552 Thomas Barnabe, an English merchant, "iii or iiii score ships of Normans or Brytons at ones, as soon as theyr fishing is don " (R. H. Tawney and E. Power, *Tudor Economic Documents*, 1924, vol. ii, pp. 99–100). See also *Cal. of Letters and Papers of Henry VIII*, vol. xx, pt. i, no. 53 ; *Acts of the Privy Council*, N.S., vol. iii, pp. 313–14 ; vol. x, pp. 89, 102 ; vol. xi, p. 268 ; *Exch. K.R. Customs Accounts*, 107/57. (Cf. below, vol. ii, pp. 23–5.)

[5] See Appendix P.

[6] R. Ehrenberg, *Hamburg und England im Zeitalter Elizabeths*, 1896, pp. 144, 295.

[7] S. Dowell (*Sketch of the History of Taxes in England*, 1876, p. 106) is mistaken in saying that there were " seacoal dealers " in Colchester in 1295. The taxing bill (*Rot. Parl.*, vol. i, p. 228b) refers, not to dealers, but to householders who had " seacoal " in their possession. At least one of them was a smith (Galloway, *op. cit.*, p. 31).

hands of smiths, ironmongers, limeburners, or of the clerk of the works at the Tower.[1] Even in Newcastle itself there was no town gild of traders in "seacoal"; the Hostmen, who eventually secured a monopoly of this traffic, were originally merchants who dealt in all kinds of exports and imports.[2] No one held large stocks of coal. When, in 1366, the clerk of the works wished to purchase 600 tons for fuel for burning the lime to be used as solder in construction work at Windsor Castle, he had not only to order it direct from the mine within the manor of Winlaton on the Tyne, and to hire the river barges, or "keels", needed for loading it at the wharf beside the manor, but also to arrange for hiring the labourers who manned the keels, the ships which brought the coal to London, and the agent who supervised the unloading there.[3] In the seventeenth century he could have had a consignment ten times as large, simply by placing an order with any one of a dozen metropolitan dealers.[4]

No one kept large stocks during the Middle Ages because they were so seldom in demand. Limeburners used the new fuel when they could get it ; but they wanted a large supply only when there was some great stone edifice to build—a castle, a fortress or a cathedral.[5] Their orders were uncertain, and came as windfalls to those who dug coal.

Smiths alone took continual supplies, to burn in shaping smelted metal into lances, "engines of war," horseshoes and falconry outfits, in forging ships' anchors and iron gates, or in casting church bells.[6] Perhaps a majority of all the early references to coal in the British Isles are associated with smiths. It supplied the smith with a fuel more economical to burn than wood, and provided him with a steady, hot flame suitable for the rougher kinds of iron work, though he found that

---

[1] *Exch. L.T.R. For. Accts.*, rolls 87, 89 (31 & 33 Henry VI). I am indebted to Professor Harold Gray of Bryn Mawr College for these two references. As late as 1551 ironmongers were apparently the chief London traders in seacoal (*City of London Repertories*, vol. viii, f. 87b ; vol. xii, f. 363 ; and cf. below, pp. 405-6). In Paris " marchands de fer " continued to be the chief coal traders of the city until the middle of the eighteenth century (Savary des Bruslons, *Dictionnaire universel de commerce*, 1741 ed., art. " charbon ").

[2] See Mr. Dendy's introduction to *Extracts from the Records of the Company of Hostmen of Newcastle upon Tyne* (*Publications of Surtees Soc.*, vol. cv, 1901) ; and cf. below, p. 405.

[3] R. L. Galloway, " Earliest Records of the working of Coal on the . . . Tyne," in *Arch. Aeliana*, N.S., vol. viii, 1880, pp. 194-6.

[4] When, in the middle of the sixteenth century, the Crown wanted consignments of 3,000 tons in connection with construction work at Boulogne, the King wrote to the Mayor of Newcastle, and Newcastle merchants managed the deliveries (*Cal. of Letters and Papers of Henry VIII*, vol. xxi, pt. ii, no. 312 ; *Acts of Privy Council*, N.S., vol. i, pp. 235, 550, 557, 558 ; vol. ii, p. 464).

[5] J. B. Simpson, *Coal Mining by the Monks*, p. 21 ; Galloway, *Hist. of Coal Mining*, 1882, pp. 10-11 ; Crofton, *op. cit.*, p. 36 ; *Cal. of Patent Rolls*, 1446-52, p. 330 ; E. A. Lewis, *op. cit.*, p. 146.

[6] *Cal. of Early Mayors' Rolls*, ed. A. H. Thomas, roll B, 5 ; *V.C.H. Yorkshire*, vol. ii, pp. 338, 341 ; Galloway, *op. cit.*, p. 12 ; *Cal. of Letters and Papers of Henry VIII*, vol. xxi, pt. i, p. 1172 ; Hector Boece, *The Boundis of Albioun* (1527), printed in P. Hume Brown, *Scotland before 1700*, 1893, p. 77 ; *Exch. Rolls of Scotland*, vol. i, p. 57 ; vol. iii, pp. 81, 653-60 ; vol. viii, pp. 51, 160, 189 ; *Accounts of Lord High Treas. of Scotland*, vol. i, pp. 250, 328, 330, 379 ; vol. ii, p. 150 ; vol. iv, pp. 276, 458-71, 510-11, 513-14, app. i ; vol. v, p. 157 ; vol. vi, pp. 160, 235 ; and *passim* ; *Cal. S. P. Venetian*, 1534-54, p. 543.

the greasy content might spoil his metal or make it brittle.[1] When he did fine work, he burned charcoal. Even for his cruder operations he added, at first, only a few pieces of seacoal to a fire made of wood and charcoal.[2] And, while he was ready to purchase a few hundredweight of the new fuel when he could buy it cheap near his house, he did not yet go out of his way to obtain supplies. In towns along the south coast of England, the smiths still depended entirely on fires of wood or charcoal at the beginning of Elizabeth's reign.[3]

Until the sixteenth century coal was hardly ever burned, either in the family hearth or the kitchen, at distances of more than a mile or two from the outcrops, and, even within the area thus circumscribed, it was used only by the poor who could not afford to buy wood. There are two principal exceptions. In the towns, villages, and hamlets bordering the Firth of Forth, in the Middle Ages, coal not only supplied artisans, mariners, and customs officials with fuel, but also furnished the castles of many a nobleman's family in the same district ; and was even ordered for use in the household of the Scottish kings.[4] And in Aachen, at least as early as the fourteenth century, it served to make fires in the Town Hall and in the Mayor's chambers.[5] These exceptions are easy to explain. Some of the seams near Aachen and round the Firth of Forth yielded fuel of a special quality, originally called in Scotland " great coal " and in England " pitcoal ",[6] which, though it burned more rapidly than the more common coal, gave off far fewer noxious fumes, and made in every way a pleasanter domestic fire.[7] Elsewhere, for the most part,[8] the bituminous mineral taken from the outcrops, though suitable (and indeed more economical than the better grades) for the operations of smiths and limeburners, was far inferior in quality to most of the fuel we use to-day, or to what

---

[1] Agricola, *De Natura Fossilium*, bk. 4, quoted Beck, *Geschichte des Eisen*, vol. ii, pp. 104–5. Agricola's remarks concerning the uses made of coal in his time have been said to apply particularly to the district round Zwickau in Saxony (J. P. Bünting, *Sylva Subterranea*, ch. xii).

[2] In casting an iron bell in south Staffordshire in 1371, 6 baskets of charcoal, 1,100 peats, and 2 chaldrons of " seacoal " were burned (Galloway, *Annals*, p. 61).

[3] W. Harrison, *Description of England*, 1577, bk. iii, ch. xvi.

[4] *Exch. Rolls of Scotland*, vol. iv, pp. 600, 603, 615 ; vol. vii, pp. 79, 179, 190 ; vol. ix, pp. 41–3 ; vol. xiv, pp. 36–7, 140–1, 162, 246, 264 ; vol. xv, pp. 6, 7 ; *Accts. of Lord High Treasurer of Scotland*, vol. iii, pp. 331, 333, 369 ; vol. iv, pp. 280–1 ; *Cal. of Letters and Papers of Henry VIII*, vol. xviii, pt. ii, p. 374 ; John Major, *On the Boundaries of Scotland* (1521), printed in P. Hume Brown, *Scotland before 1700*, p. 51.

[5] L. Beck, *op. cit.*, vol. i, pp. 770–1 ; vol. ii, p. 102.

[6] As distinguished from " seacoal ". " Pitcoal " was also a general term applied to many varieties of mineral fuel ; so we cannot assume that every reference to " pitcoal " in early times means the particular grade described in the text (cf. below, pp. 115 sqq.).

[7] G. Jars, *Voyages métallurgiques*, 1774–81, vol. i, pp. 307–8 ; and see below, p. 118.

[8] A coal similar to that found near Aachen and in the Firth of Forth district is said to have been mined in south Nottinghamshire (G. de Malynes, *Lex Mercatoria*, 1622, p. 271). This may partly explain the considerable market during the Middle Ages for coal from Wollaton and Strelley manors. The early Pembrokeshire anthracite, and the Wigan " cannel ", while they differed in chemical content from Scotch " great coal ", also made a much more pleasant fire than common bituminous coal. (Cf. below, pp. 115–7.)

could be had a little deeper in the earth than the early miners found it necessary to sink their pits.[1] As it burned, their coal gave off a continual cloud of choking, foul-smelling smoke, like that which sometimes emerges from a modern fire when it is first kindled, leaving behind a heavy deposit of thick black soot on the clothing and faces of all attending. With the common medieval habit of building the fire in the centre of the room, and allowing the fumes to escape through a flue in the roof, the consequences of substituting ordinary coal for wood were almost intolerable. The inhabitants of towns objected strongly to its use even by the smith, on the ground that his smoke polluted the air they breathed.[2]

So intense and so stubborn a prejudice as existed both in England and on the Continent against the new fuel was likely to be overcome only where the price of wood, and even of peat or turf, was very much higher than the price of coal. With the abundant supplies of timber available throughout Europe in the Middle Ages, such price conditions could prevail only where there was a considerable concentration of population on the very outcrops; for the cost of digging surface coal was small in comparison with the cost of carrying it eight miles overground. The charge for Newcastle coal in London, even though the shipmasters took it as ballast, was rarely less than three or four times the price on the Tyne. Probably nowhere in medieval Britain was there a district about the pits as densely inhabited as that round Liége, where miners actually dug coal within the town and all along the slope of the hills which rise on both sides of the river Meuse. It is at Liége, therefore, that we might expect the early coal industry to be more important than anywhere in Great Britain, and all the evidence tends to show that until Elizabeth's reign such was indeed the case. At Liége alone do we find a special jurisdiction for settling coal mining disputes; there alone are the coal miners organized into a *métier*, or gild.[3] Before any Englishman thought of treating raw coal to rid it of its noxious smell, the Liégeois made a kind of briquette out of it.[4] Liége coal, though it had to be carried in barges down the Meuse nearly two hundred miles to the sea, competed at French channel ports during the reign of Henry VIII with Newcastle " seacoal ",[5] unloaded from a ship that had sailed up almost to the pit's mouth. In 1515 a flooded mine at Liége caused the death of eighty-eight miners.[6] An accident of uch magnitude could hardly have occurred in England at this time; for, until the days of Elizabeth, the number of miners working together in a single pit or drift probably

---

[1] Cf. J. P. Bünting, *Sylva Subterranea*, ch. xi.
[2] See below, pp. 157–8.
[3] See below, pp. 287–8, 401.
[4] F. Henaux, *Histoire du pays de Liége*, 3rd ed., 1872, vol. ii, p. 35 ; Otto Hue, *Die Bergarbeiter*, vol. i, pp. 344, 351 ; Sir Hugh Platt, *A new, cheape and delicate Fire of Cole-Balles*, 1603. The Chinese are said to have been the first to make briquettes out of coal (Hue, *op. cit.*, vol. i, p. 342). See also below, pp. 246–8.
[5] *Cal. of Letters and Papers of Henry VIII*, vol. xiv, pt. ii, no. 244 ; vol. xvi, no. 98.
[6] H. Pirenne, *Histoire de Belgique*, 1900–20, vol. iii, p. 248.

did not exceed a dozen or so,[1] and the colliery owner who kept more than three pits in operation was quite exceptional.

A traveller, journeying through Great Britain before the Reformation, could mention the burning of mineral fuel only as a local curiosity. Pope Pius II, when still Æneas Sylvius, visited Scotland and found to his astonishment that " stones " were given to the poor as alms, and gladly received by them. " This stone," he observes, " whether by reason of sulphurous or some fatty matter which it contains, is burned instead of wood, of which the country is destitute." [2] His comments recall those of Theophrastus. He does not tell us how this strange mineral is procured. Apparently he knew less about the seams than Marco Polo had found out when he penetrated to far Cathay. Perhaps one might conclude that the coal industry of Great Britain in the Middle Ages was of no greater consequence in the national life than was that of ancient China. Even among the learned, the nature and properties of mineral fuel were by no means a matter of common knowledge before the sixteenth century.

But, by the beginning of the seventeenth century, men had already begun to count coal a national asset. The Venetian Ambassador to England could comment, during James I's reign, on the new method of baking bricks with what he called " mineral charcoal ", and could regret that Italians, having none of this fuel, had no prospect of becoming as proficient as the English in this new " art ".[3]

Wood was the fuel of Thomas More's Utopians.[4] Coal first steals into literature with the great Elizabethans. On more than one occasion Shakespeare gathers his characters about a " seacoal " fire. Webster, when he makes the brother of the Duchess of Malfi declaim, in his fury over her secret marriage with her steward,

> " I would have their bodies
> Burnt in a coal pit, with the ventage stopp'd
> That their curs'd smoke might not ascend to heaven,"

shows a familiarity with mining technique that would hardly have served as the material of poets in More's time.

Scholars are aware that Elizabeth's reign marks the beginning of an epoch in the history of British coal mining. Yet they rarely appear to appreciate how rapid was the expansion of the industry between the accession of Elizabeth and the Revolution of 1688. Nor

---

[1] There were eleven in one pit at Wollaton in 1526 (*Hist. MSS. Com., Report on MSS. of Lord Middleton*, pp. 429 sqq.) ; ten in one pit at Raby in south Durham in 1460 (*V.C.H. Durham*, vol. ii, p. 324). In a coal drift at Kilvey in Glamorganshire, in 1400, there were usually three hewers at work, besides a considerable number of porters, who cannot be properly considered as underground workers, since they carried coal not only from the working face to the entrance of the drift, but from the entrance to the waterside, where it was loaded into ships. (See Appendix K (i)).

[2] P. Hume Brown, *Early Travellers in Scotland*, 1891, p. 26.

[3] *Cal. S. P. Venetian*, 1617–19, pp. 320–1.

[4] " These Husbandmen plowe and til the ground, and breede up cattel, and provide and make ready woode, whyche they carrye to the citie either by lande, or by water, as they maye moste convenyently." In 1526, some years after he had written *Utopia*, More had occasion, when Chancellor of the Duchy of Lancaster, to sign his name to the decision in a dispute over rights to coal in Lancashire (*Duchy of Lancs. Deps.*, 19/T/3).

do they seem to realize what an extensive influence coal had upon the industrial, the commercial, the social, and even the political, development of England and Scotland before the Industrial Revolution. It is our object to show how important a part coal played in British history during this period. Since no number of quotations, however striking, can be of as much help as a few statistics, however imperfect, in persuading the sceptical that the output of coal in Great Britain doubled and redoubled and doubled again, within a century after 1580, an attempt is made in the following chapters to offer some quantitative indication of the growth of the coal industry.

J.U.N. 1932

PART I

THE EXPANSION OF THE COAL INDUSTRY
1550–1700

THE EXPANSION OF THE COAL INDUSTRY, 1550–1700

ANY attempt to present a quantitative picture of the increase in the output of coal before the nineteenth century must be subject to a wide margin of error. It is with considerable hesitation, therefore, that we venture to introduce the following table, with its estimates of the average annual production of the principal mining fields at four important stages in their history. The evidence on which the figures for the sixteenth and seventeenth centuries are based is of three kinds. We have, first, the actual records of coal shipments contained in the customs accounts, which vary in accuracy according to the ease or difficulty with which shippers could evade the customs officers. We have, next, a mass of heterogeneous data concerning the number and importance of collieries in particular districts. We have, finally, a few statements of seventeenth-century writers concerning the quantity of coal consumed in certain manufactures in some of these districts. From information of the second and third types, it is possible to gain an idea of the amount mined for local consumption, though these figures are inevitably of a conjectural nature. For the extent to which they enter into the different estimates in the table, and for a fuller discussion of the sources and ambiguities of the calculations which form the basis for all the tables in the two chapters which follow, the reader is referred to Appendix B.

Subject to these qualifications, the growth in the output of British coal during the last four centuries may be estimated to have been, perhaps, somewhat as follows :—

TABLE I

Estimated Annual Production of the Principal Mining Districts (in Tons).[1]

Coalfield.	1551–60	1681–90	1781–90	1901–10
Durham and Northumberland	65,000	1,225,000	3,000,000	50,000,000
Scotland	40,000	475,000	1,600,000	37,000,000
Wales	20,000	200,000	800,000	50,000,000
Midlands [2]	65,000	850,000	4,000,000	100,180,000
Cumberland. . . .	6,000	100,000	500,000	2,120,000

[1] In order to facilitate comparison, I have attempted, throughout this book, to give all figures relating to the output and shipment of coal in tons. But it must be remembered that in the sixteenth and seventeenth centuries coal was seldom accurately weighed. There was practically no uniformity in the measures used in the different mining districts and the different ports ; and the quantities of coal are always given in the early documents in the special measures of their district. I have reduced these quantities to tons, in each case, in accordance with the estimated content by weight of the particular measure, as it has been worked out in Appendix C. The accuracy of my figures, therefore, rests in part upon the accuracy of the estimates in that Appendix, and these estimates can hardly be altogether free from error.

[2] Including Yorkshire, Lancashire, Cheshire, Derbyshire, Shropshire, Staffordshire, Nottinghamshire, Warwickshire, Leicestershire, and Worcestershire.

Coalfield.		1551–60	1681–90	1781–90	1901–10
Kingswood Chase	. .	6,000 ⎫	100,000	140,000 ⎫	1,100,000
Somerset	. .	4,000 ⎭		140,000 ⎭	
Forest of Dean	. .	3,000	25,000	90,000	1,310,000
Devon and Ireland	. .	1,000	7,000	25,000	200,000
Total	. .	210,000	2,982,000	10,295,000	241,910,000
Approximate Increase [1]			14 fold	3 fold	23 fold

In making these calculations we have sought to avoid exaggerating the growth of production during the period with which we are specially concerned in this book. Even when due allowance is made for a possible over-estimate of the output at the end of the seventeenth century, and a possible under-estimate of the output in the middle of the sixteenth, these figures reveal a fact which must astonish anyone familiar with the traditional history of the industry. If we were to represent on a graph the growth in the output of coal in Great Britain from decade to decade, beginning with the twelfth century, we should have to draw, according to the traditional view, a more or less slowly rising line, climbing somewhat more steeply in some decades than in others, occasionally remaining level or even falling slightly, but on the whole representing a steady, moderate progression until the end of the eighteenth century, when the line would take a sudden upward bound, incomparably steeper and longer sustained than any previous climb it had made.[2]   But the estimates given in Table I do not confirm this traditional view. They show that in the hundred and thirty years following the accession of Elizabeth the production of coal, in comparison with previous production, increased scarcely less rapidly than during the hundred and twenty years following 1775, the period commonly thought of as ushering in the Coal Age. In the three centuries and a half which separate 1560 from the period in the Middle Ages when we get our first records of the burning of coal, progress in mining had been so slow that the annual output of England, Wales, and Scotland scarcely exceeded 200,000 tons. Before another century and a half had passed, the annual output had nearly reached 3,000,000 tons. It seems hardly an exaggeration to say that there appears to have been something like a revolution in the use of fuel in the period under review.

This impression is confirmed by an examination of actual statistics of coal shipments along the east coast. While Table I, which is partly based upon conjectural estimates, indicates a fourteen-fold increase in output during the hundred and thirty years between the accession of Elizabeth and that of William III, the following table, compiled from entries in the port books and other ledgers kept by tax collectors, shows that the actual rate of increase in the coal traffic at the two most important centres of the trade was much more rapid. In less

---

[1] It will be observed that these figures are not strictly comparable because the first increase covers a period of 130 years, the second a period of 100 years, and the third a period of 120 years.

[2] This is the view adopted by W. S. Jevons in *The Coal Question*, 1865. For a more detailed discussion of the bearing of my findings upon his argument, see below, pt. i, ch. iii.

than a century and a quarter, shipments from Newcastle multiplied nearly nineteen-fold, while imports at London multiplied more than thirty-fold.

TABLE II[1]

Shipments at Newcastle (in Tons).			Imports at London (in Tons).		
Mich.,	1563–Mich., 1564	32,951	Mar. 12,	1580–Sep. 28, 1580 [2]	10,785
,,	1574– ,, 1575	56,487	Mich.,	1591–Mich., 1592	34,757
,,	1591– ,, 1592	112,128	Xmas,	1605–Xmas, 1606	73,984
,,	1597– ,, 1598	162,552	,,	1614– ,, 1615	91,599
,,	1608– ,, 1609	239,271	,,	1637– ,, 1638	142,579 [3]
Xmas,	1633–Xmas, 1634	452,625	Midsum.	1667–Midsum. 1668	264,212
June,	1658– June, 1659	529,032	Mich.,	1680–Mich., 1681	393,453
Xmas,	1684–Xmas, 1685	616,016	,,	1697– ,, 1698	466,639

Even to the modern automobile manufacturer or oil magnate, accustomed as he has been to treble or quadruple his turnover every decade, this early expansion in the coal trade of the east coast cannot seem insignificant. To the contemporary English merchant, who had been accustomed to regard any considerable economic change as the affair of a lifetime, this expansion was something to marvel at.

> "England's a perfect World! has Indies too!
> Correct your Maps : New-castle is Peru,"

run some verses by the irrepressible author of one of the first poems on coal.[4] Rapidly as the Tyne trade increased, it could not keep pace with the figures as they leapt upwards in the popular mind. Before 1600 the Corporation of London, in a solemn document, informs the Privy Council that "there are yearelie shipped from the Towne of Newcastle unto severall places within the Realme and beyonde the seas, by computacon of such as observe the same, fyve or six hundred thousande chalders [or between 665,000 and 800,000

---

[1] For the actual sources from which the table has been derived, see Appendix D (i), (iv). As pointed out above, the figures of shipments are an understatement, since some coal escaped entry. They give, however, a fairly accurate picture of the *growth* of the trade. Those for London are the more accurate, because less coal escaped entry there than at Newcastle. In the case of London an addition of five per cent to the recorded imports should be made to allow for the extra chaldron in every twenty which was given free.

[2] Although the port book record covers only six and a half months of the year 1580, these are the months in which, owing to the seasonal nature of the trade, at least 75 per cent of the annual shipments were ordinarily received (cf. below, p. 396 ; vol. ii, pp. 388–9). If we allow on this basis for the imports during the remaining five and a half months, the traffic for the whole year would be 14,401 tons. The words "more than thirty-fold" in the text, are based on the assumption that the figure 14,400 tons may be taken as a reasonably accurate estimate of the imports in 1580.

[3] This is not a representative year, for the shipmasters declared a boycott of the colliery owners, and refused to take their regular ladings on the Tyne (see below, vol. ii, p. 280). It is reasonably certain that the imports at London in the 'thirties, under normal conditions, amounted to nearly 250,000 tons per annum.

[4] *News from Newcastle*, a poem "Upon the Coal-pits about New-castle upon Tine," 1651.

tons] of coale or thereabouts " ;[1] a statement which they are soon after at pains to explain they had made in ignorance of the facts.[2] At the time of the Civil War the figures first laid before the Council of State totalled twice the actual traffic.[3] Such estimates, which were frequently put forward, aroused in the mind of the general public a vivid picture of the great coal fleet entering the river, and the names of important mines like Stella and Blaydon, which bordered on the northern waterway, became household words to the citizens of London.

The remarkable growth in the coal industry revealed by the figures in the two preceding tables was undoubtedly most striking in the case of the east coast trade. But this growth was by no means confined to one section of the country. It occurred in every district of the British Isles in which coal could be mined or to which it could be brought. If the rate of increase in output appears to have slackened in the eastern parts of Great Britain after the reign of Charles I, in some of the western parts it was probably at least as rapid in the decades immediately following the Civil War as in the eighty years preceding. In order to make clear how complete was the change effected in the habits of the British people with respect to fuel during the century and a half following the accession of Elizabeth, we propose to consider, first, the growth in the production of coal, and then the development of the markets to which it was sent.

---

[1] *Lansdowne MSS.*, 65, no. 11. Another draft of this document is printed in Dendy, *Extracts from the Records of the Company of Hostmen of Newcastle-upon-Tyne* (*Publications of Surtees Soc.*, vol. cv, 1901), pp. 2–5.

[2] *Remembrancia MSS.*, vol. ii, no. 139.

[3] *Cal. S.P.D.*, 1644–5, p. 98.

CHAPTER I

THE PRINCIPAL COAL-PRODUCING REGIONS

In examining the expansion of the coal industry during the sixteenth and seventeenth centuries, it is convenient to divide the British Isles into a number of separate producing regions, corresponding roughly to the principal coalfields of our own time. The following regions have been selected. First, the counties of Durham and Northumberland, which are usually regarded as containing a single coalfield, but which for our purposes can be better subdivided into several smaller fields. Second, the whole of Scotland, which obviously includes several mining districts—Fife, the Lothians, Stirlingshire and Clackmannan, Lanarkshire and Ayrshire, to mention only the most important. Third, the whole of Wales, which includes two fields, that of the north and that of the south, united only in the sense that they lie mainly within the same political boundary, and which also includes within the southern field itself two areas yielding entirely different grades of coal, the one anthracite, the other bituminous.[1] Fourth, the Midlands, which includes a great number of small fields (many having no connection with one another until the age of steam railways) within the counties of Yorkshire, Lancashire, Cheshire, Derbyshire, Nottinghamshire, Shropshire, Staffordshire, Worcestershire, Warwickshire, Leicestershire, and Rutlandshire. Of the six other regions—Cumberland, Kingswood Chase, the Forest of Dean, Somerset, Devon, and Ireland—the second, third, and fourth could, on geographical grounds, be drawn within a single coal-producing region more easily than the mining districts of Shropshire, Leicestershire, and the West Riding, but, for the purpose of comparing output, they can better be kept separate. While such a division is obviously arbitrary, it has seemed on the whole the most satisfactory for this survey of production.

The expansion of the mining industry was not uniformly rapid in each of the selected regions. The following figures, being based on the estimates contained in Table I, must not be taken as precise percentages, but they probably give a fairly accurate picture of the relative position of the more important fields in the different periods.

TABLE III

Percentage of Total Output produced in the Different Coalfields

Coal-producing Region.	1551–60	1681–90	1781–90	1901–10
Durham and Northumberland	31	41	29	21
Scotland	19	16	15	15
Wales	10	7	8	21
Midlands	31	29	39	41
Cumberland	2	3	5	1
Other fields	7	4	4	1

[1] There is, of course, no sharp dividing line between the anthracite and the bituminous areas. The one shades into the other ; the steam coal of Glamorgan in the neighbourhood of Swansea and Neath ranking, in its chemical properties, somewhere about midway between hard and soft coal (cf. below, pp. 115–6).

To explain the changes in the percentage of the total output furnished by each region during the last two centuries would require a more detailed analysis of the recent history of the coal industry than is possible in this book.[1] We are concerned only with those changes which took place in the period from 1550 to 1700. It would appear that the most striking fact revealed by the tables concerning this period is the growth in the relative importance of Durham and Northumberland as coal-producing counties.[2] Before Elizabeth's reign, this region produced about thirty per cent of the total output. At the end of the seventeenth century, its share had risen to about forty per cent. It is natural, therefore, to begin our survey of production by a closer examination of conditions in this field.

(i) *The Durham-Northumberland Region*

Within the Durham-Northumberland coalfield it is possible to distinguish at least seven mining districts, four of which were in the seventeenth century without access to navigable water. These may be conveniently called " land-sale " districts. The three remaining districts—the lower Tyne valley, the lower Wear valley, and the Northumberland coast between Cullercoates and Blyth—were all so situated that the output of the mines could be loaded into colliers for shipment coastwise and to foreign countries. We may therefore call them " sea-sale " districts. In the survey which follows, an attempt is made to show the development of coal mining in each of these seven districts during the period from 1550 to 1700. It has not been possible, however, to follow any strict chronological plan in treating the material, except in cases where statistical data concerning the increase in output are available.

---

[1] It may be suggested, however, that the decline in the relative importance of coal production in Durham and Northumberland during the eighteenth century is to be explained, on the one hand, by the increasing taxes placed on coal shipped coastwise (cf. below, vol. ii, pp. 308–14), and, on the other hand, by the movement of industry into the Midlands and the new facilities for carrying inland coal along canals and streams made navigable for the first time, and by the increasing trade from the west coast of England and Wales, stimulated in the case of coal by the low taxes on shipments to Ireland (cf. below, vol. ii, p. 313 n.). The relative decline of this region in the nineteenth century, and the corresponding growth in the relative importance of the Midland coalfields, is less rapid, according to the figures in this table, than might be expected. The spread of railway lines gave the Midland fields a great advantage after the middle of the nineteenth century, in the disposal both of raw fuel and of manufactured articles produced with the aid of coal. But it must be remembered that, at just about the time when the railroad made possible the competition of Midland with Durham and Northumberland coal in the London market, the north of England mine owners had a new market opened to them by the removal of the duties on the export of coal (cf. below, vol. ii, p. 238). Between 1830 and the decade prior to the World War, the annual exports (including bunkers) increased from less than a million to more than eighty million tons ; and the lion's share of these exports fell to the mining fields of Durham and Northumberland and South Wales.

[2] By far the greater part of this coal came from Durham (cf. below, pp. 25–36, 361 n.).

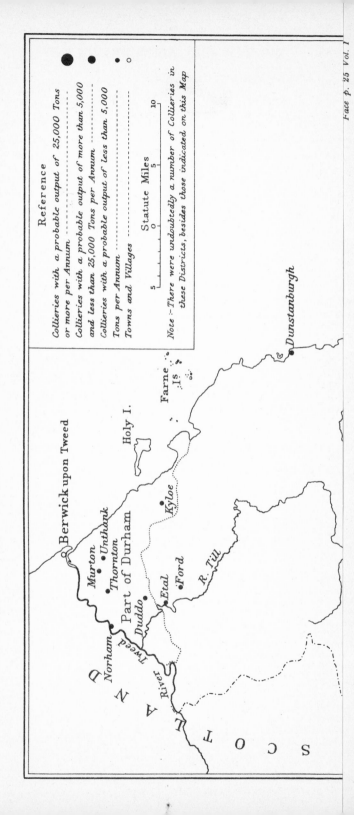

Reference

Collieries with a probable output of 25,000 Tons or more per Annum ---------

Collieries with a probable output of more than 5,000 and less than 25,000 Tons per Annum ---------

Collieries with a probable output of less than 5,000 Tons per Annum ---------

Towns and Villages ---------

Statute Miles

5   0   5   10

Note :- There were undoubtedly a number of Collieries in these Districts, besides those indicated on this Map

(a) Sea-sale Districts

The Lower Tyne Valley.—During the sixty years from 1565 to 1625, the shipments of coal from Newcastle probably increased at a more rapid rate than at any other period in their history. Records in the port books show that they jumped from about 35,000 tons per annum, in the sixties of the sixteenth century, to more than 50,000 tons in 1575, to more than 100,000 tons in 1592, to more than 200,000 tons in 1609, and to about 400,000 tons in 1625.[1] In two generations the coal trade from the Tyne had multiplied twelve-fold. Colliery owners looked forward as a matter of course to a greatly increased market in each succeeding year.

After the first quarter of the seventeenth century the pace slackened. The Tyne industry was partly thrown out of joint by the war with France and Spain, by the fierce raids of the Dunkirk pirates on east-coast shipping, especially between 1626 and 1630, and by the Dutch war of 1653, when the enemy warships attempted to prevent the passage of the colliers to the south of England. The industry was brought almost to a standstill by the Civil War. While Newcastle was in royalist hands, Parliament forbade all trade with the Tyne valley, and was so well able to enforce its will that only about 55,000 tons of coal were shipped in the two years ending at Michaelmas, 1644.[2] Even when there was no attempt to block traffic from the Tyne, the annual shipments in the decade preceding the Civil War did not normally exceed 450,000 tons.[3] Not until 1659 is there an official record of an annual shipment of more than 500,000 tons; not until after 1710 is the 600,000 ton mark regularly surpassed;[4] and not until near the end of the eighteenth century do we first get as much as a million tons shipped in any year.[5] Between 1565 and 1625 the annual shipments doubled every fifteen years, but it was almost a century and a half after 1625 before they doubled again.

Further evidence of the early expansion in the Tyne coal trade is afforded by the fact that new collieries were opened nearly every year after the accession of Elizabeth, and that the area from which the miners dug coal was greatly extended. At the end of the reign of Henry VIII almost all the " ship's coal " came from pits sunk within the manors of Whickham and Gateshead, close to the river and directly

[1] See Appendix D (i) for the actual figures. We have no record of overseas shipments for 1625, but a record in the State Papers shows that about 350,000 tons were shipped coastwise. The annual exports at this time usually amounted to from 30,000 to 50,000 tons (cf. *S.P.D., James I*, vol. cxvi, no. 79). In 1623 and 1624 the coastwise traffic appears to have been nearly as great as in 1625 (cf. Appendix D (i), note 7), showing that figures approaching the one for 1625 were not exceptional for the closing years of the reign of James I.

[2] See Appendix D (i), and below, vol. ii, pp. 285–8.

[3] See *Cal. S.P.D.*, 1644–5, pp. 98–9.

[4] Appendix D (i). As we have already pointed out, the actual amounts shipped would be somewhat in excess of these official records.

[5] The highest annual shipment recorded before 1766 was 924,000 tons. For the six years preceding 1776, the average annual shipments from the Tyne amounted to 984,350 tons, i.e. 379,000 Newcastle chaldrons. Not until 1784 did they regularly exceed 1,000,000 tons. Not until after 1820 was the two million ton mark passed (Dendy, *op. cit.*, p. 261 ; Dunn, *View of the Coal Trade*, pp. 25, 72).

opposite Newcastle. In Elizabethan times, besides the development of new mines at Whickham and Gateshead, other large " sea-sale " collieries were set up farther west, at Winlaton, Stella, and Ryton. Winlaton colliery alone produced more than 20,000 tons for shipment in the year 1581–2, when a commission appointed by the Court of Exchequer examined the account books of the enterprise.[1] The first decade of the next century saw the development of other important collieries on the north bank of the Tyne, at Newburn, Denton, Elswick, and Benwell.

Two factors determined where a pit should be sunk to extract coal for shipment, the depth of the seam below the surface, and the distance of the shaft head from a navigable part of the river. Inventions which facilitated sinking operations, drainage, and the prevention of noxious gases tended to increase the economic advantages of mines near the water ; inventions which diminished the cost of transport made it profitable to open pits farther inland. Where there was no choice between two mines as regards depth or distance from the water, that nearest the mouth of the river would be opened first. The first shipments of coal in medieval times came from pits close to the river bank in Whickham, which, of all manors in the Tyne valley with abundant outcropping seams, was the one nearest the mouth of the river. The exhaustion of surface coal from land along the water front had led to the opening of a few pits somewhat farther inland, probably before the reign of Henry VIII, but the introduction of improved drainage machinery, such as had long been used by the German metal miners,[2] resulted in the further exploitation at greater depths of those seams beside the river.

Most of the development under Elizabeth and James I took the form of deeper mining. It was found that the seams in the manors of Elswick and Benwell, practically abandoned after the surface coal had been exhausted in the middle of the sixteenth century, could now be reached easily with the help of the new pumps. At the beginning of the seventeenth century, almost all the large collieries—Elswick, Benwell, Denton, Newburn, Gateshead, Whickham, Winlaton, Blaydon, and Ryton—stood little more than a stone's throw from the Tyne. A line of new pits, most of them more than fifteen fathoms deep, stretched along both sides of the river for nearly eight miles west of Newcastle. Even as far up as Wylam and Crawcrook, collieries had begun to contribute to the supply of coal for shipment. On the other hand, the mine at Chopwell, which was no farther from Tynemouth than Wylam or Crawcrook, and afforded a good thick coal at a depth of only seven fathoms, was adjudged practically valueless in 1611 " by reason . . . carriage is far from the water ".[3] It was thought worth while, because of its proximity to the river bed, to attempt to reach the high main seam, which dips steeply to the east of Newcastle. Mining was carried on at Jarrow as early as 1617, but whether for

---

[1] *Exch. Deps. by Com.*, 29 Eliz., East. 4. The transcript made from the account books is printed below, Appendix K (ii).

[2] See below, pp. 241–2.

[3] *Exch. Spec. Com.*, no. 5037.

" sea-sale " or not is uncertain.[1] In 1654 coal from Walker colliery, also situated to the east of Newcastle (and destined to become, after the invention of the steam pump, one of the leading eighteenth-century mines), was being brought to the wharf at the junction of the Tyne with the Ouseburn.[2] As late as 1636 the three chief collieries to the south of the Tyne were still in three manors bordering the river—Whickham, Blaydon, and Stella.[3]

But before 1636 there can be traced a distinct movement of mining away from the water front, for the seams at depths made accessible by German technique were being rapidly exhausted.   In a document written about 1610, the owners complain of additional expenses "in regard that the greatest quantitye of coles are now wrought at further pytts then they were the last yeare ", a movement away from the river which, they say, is likely to continue.[4] On the manor of Benwell, by means of an ancient map drawn for commissioners appointed in the reign of Charles I,[5] it is possible to follow the mining operations financed by a group of adventurers, known as Sir Peter Riddell and Partners. Their first pits were opened between 1618 and 1620 in a field known as Stumple Wood, which extended only 300 yards inland. Later, in 1627 and 1628, the partners were obliged to sink pits in Meadow Fields, about 600 yards from the river, and to carry the coal in wagons through various freeholds to the staiths built in Stumple Wood.[6] Another water-front colliery to the west of Benwell, in the manor of Denton, had yielded a profit of at least £1,000 annually during the early years of the seventeenth century ; [7] but returns fell off steadily, and in 1649 the colliery was unoccupied, and, in the opinion of parliamentary commissioners " worth no more, as the uppermost coale is quite wrought out ".[8] The mine owners had begun to turn from the manor which bordered the Tyne to those next behind them, to Throckley in 1634, and in 1651 to Fenham, whence a colliery company leased a wayleave through the Newcastle town moor.[9]

On the south bank of the river the same movement is noticeable. While the manors flanking the Tyne maintained their pre-eminent position in the matter of coal production as late as 1638, newer pits were already being dug some distance inland. After 1610 Greenlaw freehold, situated beyond the manor of Whickham, but nominally a part of it, became the source of great quantities of " ship coal ", and the productive pits within the manor itself often lay a mile or two from the water.[10] Suggestive also is the growing importance of the

---

[1] *Hunter MSS.*, 11, no. 16.
[2] *MS. Journal of the Common Council of Newcastle*, vol. for 1650–60, ff. 203–5. Cf. G. Jars, *Voyages métallurgiques*, vol. i, pp. 193–6.
[3] See Appendix I, and below, p. 361.
[4] Dendy, *op. cit.*, p. 59.
[5] See map facing p. 305.
[6] *Exch. Spec. Com.*, no. 5567.
[7] See below, p. 147.
[8] *Parl. Surveys, Northumberland*, no. 5.
[9] *P.C.R.*, vol. xliv, pp. 544–5; *MS. Journal of the Common Council of Newcastle*, 1650–60, ff. 458, 460, 463.
[10] *Palat. of Durham, Bills and Answers*, bdl. 9 (Harding *v.* Barlowe) ; *Exch. Deps. by Com.*, 3 Chas. I, East. 11.

mines at Ravensworth, whence it was found profitable to bring coal
to the river, even though, in addition to the expense of transport, the
adventurers had to pay wayleave to the copyholders of Whickham
for the right of carriage through their lands.[1]   To judge from an
account of rents paid to the Bishop of Durham in 1635, Blackburn
colliery, situated, like Chopwell, some distance up the Derwent valley,
already produced coal for shipment.[2] Most of the more advantageously
situated seams about Newcastle had been worked to as great a depth
as was practicable by the period of the Commonwealth, and recourse
was had, not only to seams farther inland, but to seams farther up
the river.   Crawcrook colliery—to give one example—took its place
among the chief concerns of the north.   " The Coal-Pits nearest the
Water," a writer could remark in 1689, " are almost quite exhausted
and decayed." [3]

The movement of mining away from the banks of the Tyne was
facilitated by a reduction in the cost of carrying coal overland from
the pits to the water, effected by the introduction of railed wagonways,
along which horses could pull cart-loads of coal with about one-sixth
of the labour necessary along ordinary paths.[4]   By 1700 coal for
shipment was frequently carried eight or ten miles to the river, from
collieries at Tanfield Moor, Marley Hill, Blackburn Fell, Orbside, and
Pontop.   Not until the nineteenth century, when the steam railway
again reduced the costs of carriage, did the colliery adventurers find
it worth while to sink pits much farther inland.   In the meantime the
discovery of the steam engine for pumping out water had made it
possible to reach seams at greater depths nearer the river.

During the closing years of the seventeenth century, about
800,000 tons of coal were produced annually from a great number of
pits dotting the Tyne valley for several miles on either side of the river.[5]
The traffic in coal must have absorbed the attention of a spectator
seated on one of the hills behind Newcastle, almost to the exclusion
of any other sight.   Picture the mouth of the muddy, narrow river
Tyne, jammed with four or five hundred keels and two or three hundred
ships, all specially constructed to carry coal ; think of the hilly slopes
to the north and south covered with hundreds of small carts and
wagons, leaving behind them trails of black refuse on the green
countryside ; and then think of a time when this same countryside
was at rest except for the occasional movement of some husbandman
driving his mule to Newcastle for corn, when the only evidence of the
coal industry was a few pits at the water's edge by the hamlet of
Gateshead, and a few ships, whose masters took on board a lading of

---

[1] *Chanc. Proc., James I*, A 9/24 ; *Palat. of Durham, Decrees and Orders*,
vol. i, ff. 107, 152, 228.

[2] *Ecclesiastical Commissioners, Ministers Accounts*, no. 190357.

[3] *Reasons offered to Parliament for the not laying of any further imposition
upon Coals imported to London*, 1689.   See also W. Green, "Chronicles and
Records of the Coal Trade in Durham and Northumberland ", in *Trans. North of
Eng. Inst. of Mining Engineers*, vol. xv, 1865-6, pp. 277-8 ; *Palat. of Durham,
Decrees and Orders*, vol. ii, ff. 193 sqq.

[4] See Galloway, *Annals of Coal Mining*, pp. 154–7 ; and below, p. 385.

[5] Nearly 600,000 tons per annum were shipped from the Tyne, and probably
200,000 tons more were consumed locally (see below, p. 36).

coal to ballast their cargoes of fish, or grindstones, or jersey cloths. In this comparison you have the contrast between an output of about 40,000 tons and one of about 800,000 tons, and a view of the change wrought round the town of Newcastle in the century following the accession of Elizabeth. When we see the colliers crowding into the Thames, writes Defoe, " we wonder how it is possible for them to be supplied, and that they do not bring the whole Coal Country away ; yet when in this Country we see the prodigious Heaps, I might say Mountains of coals, which are dug up at every Pit, and how many of those Pits there are, we are filled with equal Wonder to consider where the People should live who consume them." [1]

The Lower Wear Valley.—About two-thirds of all the coal produced in Durham and Northumberland at the end of the seventeenth century came from the Newcastle district. Yet for more than seventy-five years the importance of Sunderland as a coal-shipping town had been increasing so rapidly that the traders of the more ancient port were not a little jealous. Sunderland did not possess as many natural advantages for shipping coal as did Newcastle. The seams of Durham county, in their course towards the east, dip sharply some distance before they reach the sea, and consequently the nearest mine which could be worked by methods available before the eighteenth century lay some eight or ten miles up-stream from Sunderland. The river was difficult to navigate, for the shallow Wear is an insignificant rivulet in comparison with the Tyne, and the harbour at Sunderland could not accommodate large vessels, such as could load freely inside Tyne-mouth bar. Although Wear valley coal is said to have been carried to sea as far back as the fourteenth century, it is scarcely possible to speak seriously of a trade from Sunderland before 1600. A full examination of the very complete series of port books for Hull, Scarborough, Bridlington, and Grimsby does not reveal a single entry of Sunderland coal between 1580 and 1600, nor is there any reason to suppose that many entries could be found in the books for London, Yarmouth, or King's Lynn during the same period. [2] Such early diggings as had existed north of the town of Durham furnished fuel for local consumption, and to some extent for the manufacture of salt. Indeed, it may have been the establishment on the Wear, probably in the eighties of the sixteenth century, of a considerable salt industry, which first called the attention of English merchants to the possibility of shipping large quantities of coal. Some time before 1591 Robert Bowes, treasurer of Berwick, erected pans at Sunderland for boiling down sea-water, and, with a view to supplying them with fuel, invested large sums of capital, including £2,000 for an adit, in a colliery at Offerton, whence coal was brought down the Wear in keels to Sunderland. [3]

It was only after the growing demand for coal in the south of

---

[1] Defoe, *Tour*, 1769 ed., vol. iv, p. 230.

[2] See Appendix D (v). Time has permitted me to examine only three books for London and one for King's Lynn for the latter half of the sixteenth century. None of these appears to contain entries of Sunderland coal.

[3] *Cal. S.P.D. Addenda*, 1580–1625, p. 327.

England had begun to exhaust the supply from the best situated, relatively shallow mines in the Tyne valley, that a regular trade developed at Sunderland. Early in the seventeenth century the port books for Hull, Yarmouth, King's Lynn, and other east-coast ports north of London begin to show imports of Wear valley coal, and before the Civil War Sunderland had become, next to Newcastle, the principal coal-shipping centre in the British Isles. The coal in the manor of Harraton seems to have been first exploited for shipment in the nineties of the sixteenth century,[1] and between the years 1629 and 1638 the shipments of Harraton coal amounted to from 6,000 to 10,000 tons per annum.[2] Although temporarily unworked, when the parliamentary forces captured Durham and Northumberland from the Royalists, the mine at Harraton, according to Sir Lionel Maddison, was certain to prove " wonderfull benificiall ", because " the coles may be even from the pitt allmost put into the keeles for a very small matter leadinge ".[3]

Long before the Civil War, other important collieries had been opened on the opposite bank of the Wear, on Sir William Lambton's estate, and in the manor of Lumley, where were found the "best coals in the county . . . known for their goodness at London as well as here ".[4] The colliery in Lambton is said to have been rented for £800 per annum on the eve of the Civil War, and it may therefore be assumed that the annual output exceeded 30,000 tons.[5] Coal from that part of the manor of Chester which was near the river and farther up, on the Harraton side, was brought in keels to the ships at Sunderland in 1617, and, at least as early as 1634, pits were worked for ship-sale at Fugerhouse.[6]

Those writers who have argued that it is not possible to speak seriously of a coal trade from Sunderland before the Civil War [7] should examine the evidence to be found in the port books. It is improbable that the shipments coastwise and overseas from Sunderland, in the last decade of the sixteenth century, amounted to more than 2,000 or 3,000 tons a year. By 1609 they had almost reached 12,000 tons.[8] In the following year the traffic benefited greatly by the withdrawal,

---

[1] *Palat. of Durham, Bills and Answers*, bdl. 2 (case of Sir John Hedworth, October 3rd, 1605). At about the same time another colliery was opened on the manor of West Herrington (*ibid.*, bdl. 5; Pinshon *v.* Todd).

[2] *Exch. Deps. by Com.*, 9 Chas. 1, Mich. 11 ; *Palat. of Durham, Decrees and Orders*, vol. i, f. 533 (Bainbridge *v.* Hedworth and Conyers).

[3] *S.P.D., Charles I*, vol. Dvi, no. 59.

[4] *Ibid.*, vol. Dii, nos. 77–8; Welford, *History of Newcastle and Gateshead*, vol. iii, p. 289 ; Galloway, *Annals of Coal Mining*, pp. 166–7. Lambton and Harraton were the first two important wharves at which coal was loaded into keels for carriage to the ocean-going ships at Sunderland, and, according to a Harraton contract made about 1628, fuel was to be delivered at Sunderland in keels of " Lambton measure " (*Exch. Deps. by Com.*, 9 Chas. I, Mich. 11).

[5] Welford, *Records of the Committees for Compounding* (Surtees Soc., vol. cxi, 1905), p. 262. For the relation between rent and output, cf. below, pp. 322–7.

[6] *Palat. of Durham, Bills and Answers*, bdl. 24 (Parkinson *v.* Lee) ; *ibid.*, *Decrees and Orders*, vol. i, f. 130.

[7] Bell, *MS. Collection for a General History of Coal*, vol. iii, p. 60 ; Galloway, *op. cit.*, p. 165.

[8] See Appendix D (i).

as far as Sunderland and Blyth were concerned, of the tax of a shilling per chaldron, imposed on Tyne coal, which had been hitherto illegally collected at these ports as well as at Newcastle.[1] This gave an advantage to the Sunderland trade, the growth of which had already been sufficient to alarm Newcastle colliery owners, as is proved by their protest against the withdrawal of the tax.[2] Sunderland shipments increased to 35,000 tons per annum in the 'twenties and reached 70,000 tons in 1634.[3] In 1636 the King succeeded in reviving the shilling tax at Sunderland for a brief period which ended with the outbreak of the Civil War, but Charles II did not try to impose it again after the Restoration, by which time the shipments from the Wear exceeded 110,000 tons per annum.[4] They amounted to approximately 180,000 tons before the Revolution of 1688.[5]

The rate of increase in the Sunderland trade, unlike that in the trade of Newcastle, was scarcely diminished after 1625. In 1609 the shipments of coal from the Wear were less than a twentieth of those from the Tyne, in 1626 they were about a tenth, in 1634 more than an eighth, in 1660 about a fourth, and in 1680 nearly a third. Had it not been for " the ill condition of the harbour of Sunderland ",[6] which could not conveniently accommodate ships of more than seventy tons burden, the progress of the mining industry in the Wear valley might have been even more rapid. As late as 1685, the average coal cargo entered in the port books for Sunderland was 23 chaldrons (or about 60 tons), the largest cargo 50 chaldrons.[7] In the Newcastle trade the average cargo carried coastwise was 58 chaldrons as early as 1634.[8] The bar at the mouth of the Wear was choked with sand and rubbish from hundreds of ships, for the sailors recklessly cast their ballast into the already shallow harbour. In the opinion of a local authority on the coal mines, who wrote in 1708, " if any Storm arises at Sea, there is no safety in offering to go into Sunderland, that wants a Pier, as at Whitby and Burlington." [9]

Notwithstanding this handicap, the shipments from Sunderland grew so rapidly that the better situated seams, close to the river, proved inadequate to meet the demand. As early as 1624 it became necessary for two mining adventurers to petition the Privy Council for a wayleave from their colliery through an adjoining manor to the river.[10] By 1700 some of the principal mines in the Wear valley—Fat-

---

[1] See below, vol. ii, pp. 268–9.
[2] *Lansdowne MSS.*, 169, no. 17.
[3] See Appendix D (i).
[4] *Exchequer Decrees*, Series 4, vol. iv, ff. 244, 292 ; and see below, p. 36.
[5] See Appendix D (i). The figure there given is somewhat smaller than that in the text. In reaching the latter, allowance has been made for the fact that, at the end of the seventeenth century, the Sunderland chaldron was larger than the Newcastle chaldron (see appendix B, vol. ii, p. 370).
[6] J. C., *The Compleat Collier*, p. 52.
[7] *Exch. K.R. Port Books*, 201/6, 13.
[8] See below, p. 391. One shipment of 204 chaldrons is recorded in the Newcastle port book for 1706 (*ibid.*, bdl. 215). At the same time we find one shipment of 80 chaldrons from Sunderland ; but so large a cargo must have been quite exceptional. Even when we allow for the greater size of the Sunderland chaldron, the advantage of Newcastle remains considerable.
[9] *The Compleat Collier*, pp. 52–3.
[10] *P.C.R.*, vol. xxxii, p. 499.

field, Allan Flatts, and Birtley—stood at a considerable distance from the water.[1] Before 1725 coal was brought five miles overland from Chester-le-Street on wooden rails, at a great expense to the under-takers, not only in maintenance of the way, but in numerous rents paid for wayleaves.[2] On both sides of the river in the district round Lumley Castle were dozens of pits, each with its supply of horses and wagons for hauling away the produce, and the activity caused by the coal industry must have made the country resemble that round Newcastle.

The Northumberland Coast.—Until the nineteenth century, Durham coal found its only outlet by sea either through the Tyne or the Wear valley. Costs of transport overground remained so high that it was useless to attempt to compete either with the Newcastle or the Sunder-land trade by bringing coal from the numerous small collieries of the Bishop Auckland district fifteen or twenty miles to Stockton for shipment ; [3] and mining technique had not developed sufficiently to tap the deep seams in the neighbourhood of Seaham, which have recently furnished so excellent a coal for the export market. But, in Northumberland, outcrops were found along the strip of coastline between the rivers Tyne and Coquet, and during the reigns of Elizabeth and the first two Stuarts a number of ambitious attempts were made to work coal at various points in this district for shipment from Amble, Blyth, Hartley, and Whitley.

They were all failures. Until after the Restoration the shipments of coal from the Northumberland coast never appear to have exceeded 5,000 tons in any one year. A contract made in 1618, with a view to the development of the coal and salt industries near Amble, was terminated in 1620 because of a dispute between the contracting parties over the costs of financing the scheme ; [4] and there is no evidence that either coal or salt was shipped from this port. At Blyth enterprise was less easily discouraged. After the dissolution of the monasteries, coal pits and salt pans at Bebside and Cowpen were frequently leased by the Crown, and early in Elizabeth's reign certain lessees received encouragement from the Percys.[5] In 1595, nine salt pans are said to have been in working order ; but, by the end of the sixteenth century, the operation of pits and pans alike had been practically discontinued.[6] Not more than 800 tons of coal appear to have been shipped from Blyth in the year 1609,[7] although for four

---

[1] Dunn, *View of the Coal Trade,* p. 20.

[2] Lord Harley, " Journeys in England," 1725, in *Hist. MSS. Com., Report on MSS. of the Duke of Portland,* vol. vi, p. 104.

[3] Occasional records of coal shipments are found in the seventeenth-century port books for Stockton (*Exch. K.R. Port Bks.,* 195/13 ; 196/1, 2, 7, 8 ; 197/2, 7 ; 198/13 ; 201/16). They never exceeded 300 tons in a year, and it is probable that they were, like similar entries in the port books for Whitby and Hartlepool, reshipments of coal originally brought from Newcastle or Sunderland.

[4] *Chanc. Proc., James I,* D 9/10.

[5] See below, p. 175, note 8, and authorities there cited.

[6] *Exch. Spec. Com.,* no. 4347 ; *A History of Northumberland,* vol. ix, pp. 223–5.

[7] See Appendix D (i).

years there had been on foot a project at Cowpen and Bebside for opening a great new colliery which was to compete with the mines about Newcastle. Between 1605 and 1618 more than £6,000 was sunk in the venture by a group of London merchants and midland gentry ;[1] but all this capital only succeeded in increasing the shipments from Blyth to about 3,000 tons in 1616. Soon after, the enterprise collapsed altogether, and the survey of Crown Commissioners for 1621 contains the significant comment : " No coal pits wrought at Cowpen."[2] In 1636 a lease was granted by the Crown to David Errington at an annual rent of £16 14s., and the port book records indicate that Blyth carried on a small trade in coal during the 'thirties ;[3] but in 1649 commissioners found the upper coal at Cowpen quite worked out, the lower coal " overflowed " with water, and the colliery apparently yielding no return, since no rent had been paid by the lessee for many years.[4]

At Hartley the story of the industry prior to the Restoration is similar, though we have no evidence of a capital outlay comparable to that at Blyth. Salt-making was pushed towards the end of the sixteenth century by the Delaval family, which owned the manor of Hartley and a great deal of land in the neighbourhood ; but, owing to the lack of a suitable harbour, the produce of the pans had to be taken to Blyth for shipment.[5] Letters written by Sir Ralph Delaval in 1612 leave an impression that the main difficulty was to induce shipmasters to risk bringing their vessels to Hartley in order to carry away the salt and coal. It appears that a little coal was shipped thence, nevertheless, for in a lease of mines within the manor of Seaton Delaval, granted by Sir Ralph in 1611 to Sir William Slingsby, it was provided that the lessor might confiscate coals on the wharf (" staith ") for non-payment of rent.[6] Neither Slingsby's efforts, nor those of Delaval himself, bore fruit, and in 1628 the coal mines were described as yielding no benefit to the owner.

Adventurers in mines along the Northumberland coast were handicapped in competing with their rivals in the Tyne and Wear valleys by other disadvantages besides the lack of a suitable landing place for ships. At Newcastle and at Sunderland one harbour served all the mines of the valley, and the produce from a number of collieries could be brought to a single wharf. Thus the share in the total cost of building wharves, and of dredging the harbour, borne by a single enterprise was greatly reduced. On the Tyne, at the end of the seventeenth century, there were only about six staiths for some twenty-five collieries. The salt pans at South Shields received their fuel from workings throughout the valley. At Harraton colliery, on the Wear, in 1645, it was unnecessary to sink capital in drainage devices, as the pits were kept dry by the operations of a pumping engine on the

---

[1] For the history of this enterprise, see below, vol. ii, pp. 16, 34–5, 79.
[2] *A History of Northumberland*, vol. ix, p. 230. See also p. 355.
[3] See Appendix D (i).
[4] *Parl. Surveys, Northumberland*, no. 2.
[5] *A History of Northumberland*, vol. viii, pp. 22–3.
[6] *Waterford MSS.* See esp. no. 16 J : " A Booke of severall things touching the Estate of Seaton Delaval," 1612.

adjacent manor of Lambton.[1] But machinery, buildings, and wharves erected for coal mining and salt manufacture at Amble or Blyth, at Hartley or Whitley, could serve only the two or three collieries of the immediate hinterland. Each of these tiny mining districts had to have its own harbour, its own salt pans, and its own pumping engines, which meant that the capital required to start a large enterprise was greater per unit of coal produced than in the Tyne and Wear valleys.

But, with the increasing costs of mining in the Tyne and Wear valleys, occasioned by the working of deeper seams at longer distances from the river, it became possible after the Restoration for adventurers along the Northumberland coast, by a more carefully directed outlay of capital, to establish several successful collieries in competition with those above Newcastle and Sunderland. Mining at Whitley was stimulated by the erection of seventeen salt pans before 1677. The construction of a harbour and a pier at Cullercoates in that year insured the future of a new colliery started at Whitley between 1673 and 1675.[2] The pier protected the waiting coal ships from the fierce storms of the North Sea, which had formerly torn them from their moorings, and the entrance to the harbour was deepened so that larger vessels could load at Cullercoates than had ever loaded at Sunderland.[3] In 1679, shipments of 14,000 tons of coal were recorded in the port book for Cullercoates. In 1684 the shipments exceeded 23,000 tons,[4] and by 1690 the new industries had brought so many inhabitants to the district that Cullercoates was made into a township separate from Tynemouth. Progress continued during the first decade of the eighteenth century, but none of the enterprises long survived the disaster of 1710, when the outworks of the pier were washed away by a heavy sea.

Encouraged perhaps by the growth of Cullercoates, Sir Ralph Delaval, the admiral, a great-grandson of the pioneering coalowner of Hartley, undertook the building of a pier and the deepening of the harbour at Hartley pans.[5] This operation, which later excited Defoe's admiration,[6] was apparently completed before the death of Charles II. By 1691 a colliery at Seaton Delaval was already producing coal for sale to the ships and for salt manufacture, and, soon after, glass, lime, and copper works were established in connection with the mining industry, which continued to prosper throughout the eighteenth century.[7]

The growth of the town of Blyth, which " as well as those before described derives its origin from the coal-trade ",[8] also dates from the end of the seventeenth century. Joint-stock companies succeeded where so many ambitious adventurers had buried their savings earlier in the century ; pits were sunk as far inland as Plessey, and their

---

[1] *S.P.D., Charles I*, vol. Dvi, no. 59.
[2] *A History of Northumberland*, vol. viii, pp. 281–3, 397.
[3] *Exch. K.R. Port Bks.*, bdls. 198, 201.
[4] See Appendix D (i).
[5] *A History of Northumberland*, vol. viii, pp. 22–3.
[6] Defoe, *Tour*, 1769 ed., vol. iii, p. 240 ; and see below, p. 378 n.
[7] *Waterford MSS.*, no. 34 (O).
[8] Defoe, *op. cit.*, vol. iii, p. 241.

produce was carried five miles overland to the ships.[1] Blyth and Hartley lagged behind Cullercoates as coal-shipping centres until the eighteenth century, but the shipments increased from about 1,400 tons in 1674 to about 5,000 tons in 1679, and to about 9,000 tons in 1685.[2] It seems likely that the shipments of coal from all the ports along the Northumberland coast approached 35,000 tons per annum in the eighties of the seventeenth century.[3] Newcastle and Sunderland were faced with new competitors.[4]

In the following table an attempt is made to estimate, in round numbers, the growth of coal shipments from north-eastern England. Entries in the port books, upon which this table is based, fall somewhat short of the reality. This is especially the case with shipments overseas. The heavy impositions upon the overseas shipment of English coal led the traders to invent all kinds of devices for evading payment. Sellers of coal in the north of England invariably connived with the exporting buyers, by giving over-measure ; and we have evidence that in some cases less than three-fourths of the actual cargo was entered in the books.[5] Again, some coal was entered at north of England ports for shipment coastwise, only to be carried overseas. Although ship-masters were required to return certificates to prove that their cargoes had been unloaded in English ports, we know that certificates were sometimes forged. We cannot estimate with any confidence the quantity of coal fraudulently exported ; but we do know that only a very small portion of that carried coastwise from the north of England was reshipped for foreign parts.[6] If a record for six months of the year 1663 may be taken as a fair sample, it appears that approximately a ninth of all the north of England coal legally exported was entered, not in the port books for Newcastle, Sunderland, or Blyth, but in the port books for London, King's Lynn, Hull, Yarmouth, and a dozen other towns.[7] We have endeavoured to make allowance for reship-ments abroad of coal carried coastwise, and for fraudulent entries in the case of coal carried overseas. We have also, in computing the shipments from Sunderland for the decade 1681–90, allowed for the fact that the chaldron used in measuring coal on the Wear was then somewhat larger than that used on the Tyne.[8] But it has not been found feasible to make allowance for fraudulent entries in the case of

---

[1] *A History of Northumberland*, vol. ix, pp. 230–1 ; W. R. Scott, *Joint-Stock Companies*, vol. ii, p. 462 ; and cf. below, vol. ii, p. 46.

[2] Appendix D (i).

[3] In addition to the shipments from Cullercoates, Hartley, and Blyth, some fuel was probably shipped before the end of the seventeenth century from Newbiggin, where one landlord received the large rent of £100 per annum for leasing a coal mine (*Cal. S.P.D.*, 1663–4, p. 76), and from Amble. Defoe reported a half century later that the island at the mouth of the Coquet " is said to yield sea coal in great quantities " (*Tour*, 1748 ed., vol. iii, p. 227).

[4] See below, vol ii, p. 118.

[5] See below, vol. ii, p. 379.

[6] The port books for London, King's Lynn, Yarmouth, Hull, Whitby, Hartlepool, Stockton, etc., all contain entries of such reshipments.

[7] *Sackville MSS.*, " An Account for . . . Lord Buckhurst upon Coales exported . . . 3rd June to 29th Sept., 1663." Out of approximately 25,000 tons exported, about 3,000 tons were reshipments.

[8] See Appendix C (i).

coastwise shipments.[1] As the incentive to frauds probably increased in the seventeenth century, owing to the increasing taxation of the coal trade,[2] the rate of growth in shipments shown by this table is more likely to be an under- than an over-estimate.

TABLE IV

Estimated Annual Shipments from Durham and Northumberland (in Tons)

	1551–60	1595–1600	1631–40	1656–65	1681–90
The Tyne Valley—					
Shipments coastwise .	25,000	130,000	400,000	440,000	525,000
,, overseas .	10,000	30,000	50,000	40,000	62,000
Total	35,000	160,000	450,000	480,000	587,000
Per cent of grand total	97·2	97·5	87	80	73·5
The Wear Valley—					
Shipments coastwise .	500	1,500	60,000	100,000	160,000
,, overseas .	nil	1,500	10,000	15,000	20,000
Total	500	3,000	70,000	115,000	180,000
Per cent of grand total	1·4	1·8	12·6	19	22·5
The Northumberland Coast—					
Shipments coastwise .	500	1,000	2,000	4,500	30,000
,, overseas .	nil	nil	nil	500	3,000
Total	500	1,000	2,000	5,000	83,000
Per cent of grand total	1·4	·7	·4	1·0	4
Grand total . .	36,000	164,000	522,000	600,000	800,000

To these shipments we must add the coal mined for local consumption, in manufactures and as domestic fuel. The salt pans provided the chief markets, but other industries were also growing in importance during the seventeenth century—glassmaking, limeburning, shipbuilding, and the making of hardware.[3] At the same time coal mining and these industries, which were dependent upon it, were attracting an increasing labouring population,[4] all of whom used coal for their firing. The total amount burned annually in the lower Tyne valley at the end of the century probably exceeded 200,000 tons, while 50,000 tons may have been consumed in the lower Wear valley, and 40,000 tons along the Northumberland coast—or perhaps about 300,000 tons in all.[5] The quantity consumed in all three districts before the accession of Elizabeth did not perhaps exceed 15,000 tons.

(b) " Land-sale " Districts

The records kept by customs officials, which enable us to compile fairly accurate statistics of the quantity of coal carried by sea, and therefore to estimate with some confidence the rate of growth in production in the sea-sale districts, are lacking for the inland coal fields. We must fall back upon such evidence as is obtainable concerning the growth in the number of collieries in each land-sale district, and the

---

[1] For a fuller discussion of the table, see Appendix B.
[2] See below, pt. v, especially chs. iii and iv.
[3] See below, pt. II, ch. ii, passim, and p. 230.
[4] Cf. below, pt. IV, ch. iv, sec. (i).
[5] For figures concerning the amounts consumed in certain industries, see below, pp. 206–8, 219–20.

increase in the output of particular collieries. A considerable amount of evidence on the former point has been found in contemporary legal proceedings, in the accounts of travellers, and in local histories. In tracing the progress of particular mines, we are helped occasionally by reports from which it is possible to calculate the annual output, and frequently by reports giving the rent paid by the mine owners, the number of pits sunk, the number of men employed, or the annual " value " as assessed for purposes of taxation. The evidence provided by all these sources taken together is unsatisfactory as a guide to the actual production of coal in the various inland fields, but it is sufficiently impressive when the object is simply to show that the industry grew greatly in importance in the century and a half after 1550.[1] While we are on much less firm ground than in the case of our estimates for sea-sale districts, we are convinced that in saying, as we do, that the output multiplied six-, eight-, ten-, and in some cases even fifteen-fold or more, we are still reasonably near the truth.

South Durham.—It must not be supposed that the development of coal mining in Durham and Northumberland during the sixteenth and seventeenth centuries was confined to those districts having an outlet by sea. At least as far back as the fourteenth century mines, parcel of the Bishopric of Durham, to the south of Bishop Auckland, had been worked, and during the fifteenth century revenues flowing into the episcopal treasury from these mines almost equalled the revenues obtained from the colliery at Whickham.[2] A decline in the revenues received by the Bishops after 1550 from collieries round Raby has led one investigator to assume that a corresponding decline in the output of coal occurred.[3] But that is to be doubted. The decline in revenue probably resulted from a weakening of the Bishop's power to exact high rents from the lessees of his mines.[4] There is no good reason to believe that the demand for coal in south Durham and Yorkshire diminished at just the time when, in every part of the British Isles for which we have records, this demand was increasing.[5] Compared with the rapid expansion in the coal industry near the mouth of the Tyne and of the Wear, the growth of colliery enterprise in south Durham was doubtless slow enough ; but that there was a marked growth is incontestable.

Early in Elizabeth's reign an attempt was made to open new mines at Cotherstone and Lartington, up the Tees valley beyond Barnard Castle,[6] and, by the last decade of the sixteenth century, colliery

[1] In order to give complete figures for the approximate output of coal in Great Britain, it was necessary nevertheless to attempt the rough estimates of the production in land-sale districts which will be found in this chapter. For the procedure which was followed in making these estimates, see Appendix B (viii).

[2] Cf. below, pp. 137, 154.

[3] Galloway, *Annals of Coal Mining*, pp. 105–6.

[4] See below, pp. 144–55.

[5] See below, part I, ch. ii.

[6] *Exch. Spec. Com.*, no. 2617 (Certificate of " workers of cole mynes ", etc., as to the coal within and under the manors of Cotherstone and Lartington, Yorks).

ventures just south of Bishop Auckland had begun to attract Yorkshire and London capital.[1]   Already in 1568 the mines within the lordship of Raby had come under the control of a certain Henry Smith, who remained the principal entrepreneur in this district until near the end of Elizabeth's reign.[2]   The progress of his colliery ventures was rapid, at least after 1580.[3]   In the year 1587–8, when the revenues of the Bishop of Durham were in the hands of the Crown, and when rents generally were exceedingly low in proportion to output, Smith paid the substantial sum of £107 16s. 8d. per annum for the right to work pits at Raby, Hargill, and Grewborn.   His principal achievement was the opening of a new mine at Carterthorne within the lordship of Raby. In 1598 his collieries yielded him a clear profit of at least £100 a year, a return which at that time was exceeded only by the larger enterprises on the Tyne.   Most of the mines in which Smith had had an interest continued throughout the seventeenth century to produce, under successive entrepreneurs, substantial supplies of coal.   In 1635 one lessee paid a rent of £70 for Carterthorne ; and the Corporation of Durham paid £38 for Grewborn and Hargill, collieries bequeathed them by Smith.[4]   The rents paid to the Bishop must have been far short of the true value of the collieries, for in 1646 Carterthorne was subleased for £350 a year over and above the Bishop's rent.[5]

Another colliery, in Cockfield moor, not apparently included in Smith's concession, had come to be of some importance by 1595, when it was taken over by Henry Rosse, a merchant of London.   Rosse subleased it a few years later to two Newcastle merchants, obtaining a handsome yearly profit over and above the Bishop's rent.[6]

As on the Tyne and Wear during the seventeenth century, mining in south Durham developed so rapidly, and the seams situated nearest to the market were dug to such considerable depths, that it became profitable to sink new pits to shallower seams somewhat farther away, since the saving in mining costs more than made up for the additional charges for transport.   By the Civil War, a definite move-ment of capital had begun towards a second line of collieries farther north, at Shildon, Coundon, Auckland Park, Softly, Hamsterley, and Etherley.

The Crown mine at Softly, which had lain practically unworked since the beginning of the sixteenth century, still proved an unprofitable venture in 1632, " because the Bishop of Durham's pits [at Raby and Hamsterley] lie between Softly and the country, so that the country

---

[1] Cf. below, vol. ii, pp. 11, 37.

[2] Galloway, op. cit., p. 106.   The rent paid by Smith was not enough to reimburse the Bishop for the wayleave of £22 which he had to pay for carriage of coal from these mines through land within the lordship of Raby, formerly owned by the Earl of Westmorland (Exch. Spec. Com., no. 752).

[3] Exch. Deps. by Com., 22 and 23 Eliz., Mich. 5 ; 23 Eliz., East. 6 ; Ministers' Accounts, Eliz., no. 661 (Receipts from the See of Durham, 1587–8).   In the year 1588–9 Smith paid £35 8s. 4d. rent for his pits at Raby alone (ibid., no. 662).

[4] Ecclesiastical Commissioners, Ministers' Accounts, no. 190357 (Accounts of " Minera Carbonum, Tegalum, Ferrei ", etc., in the Bishopric of Durham, for 1635).

[5] Ibid., no. 190356.   See also no. 190333 ; and Galloway, op. cit., p. 170.

[6] Star Chamb. Proc., James I, 265/3.   See also Ministers' Accounts, Eliz., no. 661, for information concerning coal pits at Cockfield and Evenwood.

furnishes itself elsewhere." [1] Seventeen years later, however, parliamentary commissioners found a marked improvement in this colliery (at that time known as the " Kinges Collyrie in Lanton Hills "), and estimated that the mine lay under four hundred acres of surface land.[2]

More impressive still was the development of a new enterprise at Etherley. Of little value before the Civil War, the mine there was leased by the Bishop of Durham, in 1667, for the considerable fine of £1,100.[3] As coal sold at the pits for two or three shillings a ton, it does not seem likely that adventurers would have paid a fine of more than £1,000, or rent at a rate of £300 and more per annum, unless the colliery would produce upwards of 10,000 tons a year.[4] There appear to have been at least two collieries—Carterthorne and Etherley—producing as much as this at the end of the seventeenth century; and it is not unreasonable, therefore, to assume that the yearly output in south Durham had already reached some tens of thousands of tons.

The Upper Wear Valley.—For our purposes, the Wear valley field may be said to include the entire river basin between the sharp fork at Lambton Castle and that at Bishop Auckland. This area included the most populous part of the county between Durham and Bishop Auckland, a district already in the Middle Ages of some industrial importance, because of its lead mines. The coal consumed in the Wear Valley came neither from the pits sunk below Chester-le-Street to supply the ships at Sunderland, nor from the pits in South Durham, but from a number of small workings sprinkled all through the valley itself. By the middle of the seventeenth century there must have been at least twenty or thirty collieries within an area of about a hundred and fifty square miles. Every manor of any size had its own pits.[5]

The actual output in the Wear valley remains even more a matter for conjecture than the output in south Durham. It is probable that coal was already the common domestic fuel before the sixteenth century, for there are many early references to workings in the records of the monasteries at Finchale and Durham. Progress in mining during the century and a half following the accession of Elizabeth was probably somewhat less rapid than in the district to the south of

---

[1] *Exch. Spec. Com.*, no. 5276 (Survey of Coal Mines in the parish of Softly, October 24th, 1632). See also *Exch. Spec. Com.*, no. 749, on conditions at Softly in 1574.

[2] *Parl. Surveys, Durham*, no. 6 (November, 1649).

[3] Green, *Chronicles of the Coal Trade of Durham and Northumberland,* pp. 268–9. See also Galloway, *op. cit.*, pp. 169–70.

[4] Cf. below, pp. 325–6.

[5] We have records of these small enterprises near the town of Durham at Rainton and Southstreet; up the river Browney at Consett, Knitsley, and Lanchester; along the Wear at Plawsworth, Kimblesworth, Durham, Brandon, Brancepeth, Tudhoe, Spennymoor, Willington, Findon Hill, Sherburn, Hummock Moor, Morton, Framwellgate Moor, and perhaps also at Houghton-le-Spring, Coxhoe, and Thornley (*Eccles. Com. Min. Accts.*, no. 190357; *Hunter MSS.*, 22, no. 18; *Min. Accts., Eliz.*, nos. 661, 662; *Augm. Partics. for Leases*, 39/78; *Palat. of Durham, Decrees and Orders*, vol. i, ff. 226, 380 sqq.; vol. ii, ff. 26, 195–6; *Chanc. Proc., James I*, B 16/50; *S.P.D., Eliz.*, vol. ccxxxviii, no. 148).

Bishop Auckland. But it would be a mistake to assume that there
was no notable increase in production. The abundance of outcrops,
which made coal everywhere cheap, must have encouraged the local
population to use it freely, not only for heating houses, but in many
small local industries, and possibly also, at the end of the seventeenth
century, for such smelting of lead ore as was carried on in the county.[1]

Northern Northumberland.—It is not possible to consider the area
between Morpeth and the river Tweed as one coal-mining district in the
same sense that south Durham or the upper Wear valley was one
district, for the distances between clusters of pits were much greater.
Mining developed in the valleys of the Wansbeck, the Coquet, the Aln,
and the Tweed, all districts separated from one another by rough,
hilly country, and joined only by the highway from Newcastle to
Edinburgh. The problem of finding fuel to the south of Berwick,
on both sides of the ancient Devil's Causeway, led, during the reign
of Elizabeth, to the discovery of seams at shallow depths near Kyloe
and Etal.[2] During the reign of James I, mines were also worked at
Ford, Duddo, Norham, and Thornton.[3]

At about the same period a search for coal within a few miles
of Berwick bore fruit ; and for a time it appeared that this town
(which had received occasional supplies by sea during the Middle
Ages from the Tyne and the Firth of Forth) might develop a shipping
trade in coal. The colliery within the manor of Unthank must have been
an impressive sight for the local inhabitants, with its " howeses and
covers . . . windows . . . timberworke, ropes, frames and buildings
thereon made ". At one time the undertakers had as many as sixty
chaldrons of coal stacked at the shaft head awaiting sale. They were
faced with fierce competition from the proprietors of a rival enterprise on the adjoining manor of Murton.[4] These two enterprises,
between them, not only furnished the town of Berwick with its necessary
firing, but enabled its merchants to load a few vessels with coal for
carriage by sea.[5] But shipments soon ceased ;[6] and it is therefore
probable that the mines nearest the mouth of the Tweed declined
in importance after the middle of the seventeenth century.

Any falling off in the output of coal at Unthank and Murton
was more than compensated by the activity of Francis Blake, who
paid £450 in 1673 to get control of the colliery at Ford, and four
years later purchased for £500 another colliery in Gatherick.[7]
Blake was successful, not only in building up an extensive local sale
of household fuel, but in the development of several small manu-

---

[1] See below, pt. ii, ch. iii, passim, and pp. 245–6.
[2] *Augm. Partics. for Leases*, 112/12, 15, 27, 30 ; *Star Chamb. Proc., James I*,
154/12.
[3] Cf. *A History of Northumberland*, vol. xi, pp. 397, 470–1.
[4] *Star Chamb. Proc., James I*, 224/19–22 (Ralph Ord, gent., of West Ord, *v.*
Geo. Morton, Esq., of Murton, May, Nov., etc., 1609).
[5] *Exchequer Decrees*, Series 4, vol. ii, f. 114 ; vol. iii, f. 230 ; and see
Appendix D (i) (b) note.
[6] *Exch. K.R. Port Bks.*, bdls. 161, 162, 165.
[7] *Waterford MSS.*, Craster's *Index*, vol. i, p. 81 ; vol. iii, pp. 131–2.

factures, notable among which was a brewery, to consume such small coal as was deemed unfit for the domestic hearth. Coal from his pits came to be sold up Tweeddale as far as Melrose, and up Teviotdale as far as Minto.

Other mines were developed during the sixteenth century farther south, round the Earl of Northumberland's seat at Alnwick. Shafts had been sunk in the manor of Bilton to a depth of eleven fathoms as early as 1479; and, in 1533, when the mine was leased to an Alnwick trader named George Clarkson, a project was afoot for shipping coal by sea from Alnmouth. Under the terms of his lease, Clarkson had " staythe lies and place bothe at the . . . colefield and also at the water syde at Aylnemouth for uttering and carrying away the said coles for his most advantage ". This project could not have met with success, however, for, in a survey made in 1567 of the Earl's estates, there is no mention of a market by sea ; the exploitation of the seams now being advocated on the ground " that it shuld be no lesse commoditye to the countrye there about then the preservation of his lordship's woods ".[1] Henceforth every effort was made to exploit coal for a local market. Farther to the north, pits had been sunk on the Crown manor of Dunstanburgh as early as the reign of Edward VI.[2] Farther to the south, coal was being won at Amble and up Coquet Dale. Before the end of the seventeenth century the tenants of nearly every manor within fifteen miles of Alnwick had easy access to one or another of these small mining enterprises.

Still further mines were developed about Morpeth. During the reign of Elizabeth coal is said to have been discovered within the manor of Morpeth itself by John Gray, a Newcastle merchant. Owing to the great charge " in keeping labourers to drawe the water from the . . . mine ", the venture did not prove an immediate success, and, as late as 1602, it had not been possible for Gray to pay any rent.[3] We hear also of small workings at Butterlaw,[4] Ulgham, Ellington, and Horton Grange.

The Upper Tyne Valley.—West of Wylam coal was dug at least as early as the fifteenth century in the liberty of Tyndale and at Codden, near Acomb.[5] During the reigns of Elizabeth and James I there is evidence of a much more serious attempt to develop mining all through Hexhamshire.[6] " The mannor of Hexam," writes a landlord, who is negotiating in 1622 a lease of coal and lead mines, " lyith so far in the land from all vent by sea, or water, or in anie other kinde,

---

[1] *A History of Northumberland*, vol. ii, pp. 453–5. Defoe (*Tour*, 1769 ed., vol. iii, p. 246) found Alnmouth a sea port for shipping large supplies of corn, but he made no mention of coal.

[2] *Duchy of Lancs. Pleadings*, 51/B/10 (Edward Bradforth *v.* Sir Thomas Gray, receiver of Dunstanburgh manor, 1561).

[3] *Exch. Spec. Com.*, no. 1760 (Inquisition as to a coal mine at Morpeth, 1602).

[4] *Augm. Misc. Books*, vol. ccxi, f. 59*d*.

[5] *A History of Northumberland*, vol. iii, pp. 54, 150 ; vol. iv, 173, note.

[6] See, for instance, *Exch. Spec. Com.*, no. 1740 (Inquisition as to coal mines near Dilston Park, 1581).

saving to a small companie of poore villages, that coales being gotten, theare can be no great proffit expected." This landlord already has two collieries, one to the west of Hexham, the other about five miles to the east, but he wants to discover coal in Hexham itself. " My end in this business is to gett fyre near me, though it cost me dear, to save carriage." [1] That was also the " end " of thousands of manorial lords, freeholders, and copyholders all through England at the beginning of the seventeenth century ; and, with the abundant surface seams which the country possessed, it is no wonder that in hundreds of cases their desires were gratified. In Hexhamshire, within twenty years of the writing of this letter, several new pits appear to have been dug ; and in the hilly parish of Allendale, south of Hexham, the inhabitants made some use of an inferior kind of mineral known as " crow " coal, which they dug out of the mountain side.[2]

This general survey of the " land-sale " mining districts of Durham and Northumberland suggests that the output of coal at the end of the seventeenth century must have been several times what it was a hundred and fifty years earlier. As population was scattered in country villages, rather than assembled in growing towns, and as the cost of transporting coal overland was great, there was seldom any opening for a single large enterprise, except in south Durham. But the number of small collieries is impressive, especially when it is remembered that many of them were found in places which had no mines before the middle of the sixteenth century, and that everywhere the seams were being worked at greater depths than in the Middle Ages.[3] Even a small colliery now produced from 1,000 to 5,000 tons a year.[4] It is probable that the annual output from the four " land-sale " districts had multiplied some eight-or ten-fold since the accession of Elizabeth, and that it amounted to 100,000 or 150,000 tons at the end of the seventeenth century.

(ii) *Scotland*

The growth of the coal industry north of the Tweed seems to have been less rapid than in Durham and Northumberland. Owing to the lack of satisfactory statistical material concerning coal shipments, and because by far the greater part of the coal produced was not shipped at all, it is impossible to do more than hazard a guess concerning the output in Scotland. But the evidence of the rise of the industry, during the period from 1550 to 1700, to a place of great importance in the economic life of the country is scarcely less impressive for Scotland than for the north of England.

---

[1] An unsigned document among the *Sackville MSS*. The mine on the west was probably at Codden, where an effort had been made to work coal more intensively in 1590 ; that on the east was probably at Bywell (*A History of Northumberland*, vol. iii, p. 59 ; vol. iv, p. 102).

[2] *Ibid.*, vol. iii, p. 54 ; vol. iv, pp. 173, 249.

[3] See below, pp. 350–52.

[4] Cf. below, pp. 359–60.

(a) *The Firth of Forth District*

Contemporary Englishmen called Durham and Northumberland the Black Indies, and the term might have been applied with almost equal force to that part of Scotland bordering on the Firth of Forth. " Ane of the chief benefites," wrote in 1623 John Nicholson, the owner of a colliery at Lasswade, " quair with it has pleasit God to inritch the boundis of Lotheane "—and he might have added Fife, Clackmannan, and Stirlingshire—" is the commoditie of the coill, quhair with the countrey is so plentifullie servit and without the quhilk it could not long subsist." [1] At the end of the sixteenth century, the district was dotted with coal pits and saltpans. Standing on a height near Falkland, shortly before he was called to the English throne, James VI watched the small grey puffs of coal smoke from the saltpans curl together into a tinted horseshoe above the water, and compared his kingdom to a " beggar's mantle with a fringe of gold ".[2] Before the end of the seventeenth century the fringe had widened. For here, as on the Tyne and Wear, the increasing demand for coal soon exhausted the surface seams near the water, and the miners dug a second, and perhaps even a third, line of pits farther inland.

When did the coal industry of Scotland begin rapidly to expand ? In a sense, of course, it had been growing for centuries. Probably we should find, if the data existed, that the production of coal was greater in the fourteenth than in the thirteenth century, and greater in the fifteenth than in the fourteenth. But there is no reason to think that, on the Forth any more than on the Tyne, the coal industry underwent a period of rapid development until the last forty years of the sixteenth century. It is during these years that we first hear of considerable sums of money spent in sinking pits in Scotland. Sir George Bruce's famous colliery at Culross, which must have cost its owner several thousand pounds sterling, was started about 1575. No such ambitious industrial undertaking had hitherto been thought of in Scotland, and it was the marvel, not only of King James and his Scottish subjects, but also of travellers from less backward England. " I did never see, read or heare of any worke of man," wrote Taylor, the Water-Poet, in 1618, " that might parallell or bee equivalent with this unfellowed and unmatchable worke." [3]

At about the time when Bruce was constructing this masterwork, we hear coal mining spoken of for the first time as an industry. Previous references to coal in Scotland relate to isolated diggings, or deal with the casual use of " black stones " as fuel. Now the mines are referred to as a great national asset, and an adequate supply of cheap coal for the inhabitants becomes a national problem. The " masters of coal and salt " first appear in the records of meetings of the Scottish Privy Council. The closing years of the sixteenth century are accompanied by a remarkable expansion in the coal industry, and the following hundred years mark in Scotland, as in Durham and Northumberland, a period of continually increasing output.

[1] *Privy Council Register of Scotland*, 1st Series, vol. xiii, p. 368.
[2] A. S. Cunningham, *Mining in the Kingdom of Fife*, p. 25.
[3] J. Taylor, *The Pennyless Pilgrimage*, 1618, printed in P. Hume Brown, *Early Travellers in Scotland*, 1891, p. 116.

An estimate of the growth between 1550 and 1700 must be based, first, upon such evidence as can be obtained concerning the shipment of coal by sea and the amount consumed in salt works ; secondly, upon the records of the increase in the number and size of collieries in the Lothians, in Fife, Clackmannan, and eastern Stirlingshire. It is much more difficult to estimate the shipments made from ports along the Firth of Forth than it is to estimate shipments from ports in Durham and Northumberland. While it is possible to say with confidence that towns like Dysart, Culross, Methil, and Easter Cockenzie owed their existence to the development of a coal trade after 1550, no less than did the ports of Sunderland, Blyth, Seaton Sluice, and Cullercoates, there are, among the papers of the customs officers, no records of the actual shipments from Scottish ports which can be regarded as even roughly accurate.

The machinery set up by the Scottish Privy Council was altogether inadequate for collecting the small duties on coal.[1] Of the eight head ports at which regular offices were established, only three—Leith, Bo'ness, and Burntisland—were on the Firth of Forth.[2] Shipmasters called for coal at a score of other ports.[3] At some, like Elphinston or Stirling, there was before the Civil War no customs officer at all ; at others, like Wemyss, Kirkcaldy, or Dysart (already called " Little Holland " because of its trade with the Low Countries),[4] there was only one officer for several ports. And, even in towns where there was a regular office, the records were miserably kept ; " in some places there were noe bills of entrye taken ; . . . in others . . . noe accompt but what noe was made and drawne from these bills at the end of the quarter, which had beene too loosely flung about, without any fileing up of the same." [5] Under such circumstances it is plain that we cannot expect much help from the records of the collectors in estimating the early commerce of Scotland.

But enough information is obtainable in the customs records for the last half of the sixteenth century to prove that in these fifty years coal shipments from the Firth of Forth increased rapidly. Before 1540 there is no record of a duty collected on coal except at Leith, the port of Edinburgh ; and there, if the sum received in respect of customs be taken as a true measure of the shipments, these seldom exceeded 100 chaldrons.[6] But before 1600 duty was paid in some

[1] Cf. below, vol. ii, p. 226.
[2] Thomas Tucker, *Report upon . . . Excise and Customs in Scotland*, 1656 (ed. Bannatyne Club, 1824), p. 24.
[3] Cf. *Privy Council Register of Scotland*, 1st Series, vol. vi, pp. 373–4 (Act of April 27th, 1602, concerning the shipment of coal from Haymouth, Coldingham, Aberlady, Newhaven, Cokeny, Queensferry, Bo'ness, and Airth on the south side of the Forth, and Elie, St. Monance, Levensmouth, Wester Wemyss, North Queensferry, Limekilns, and Whalehaven on the north side).
[4] A. S. Cunningham, *op. cit.*, p. 38.
[5] Tucker, *op. cit.*, p. 43.
[6] *Exchequer Rolls of Scotland*, vol. iv, pp. 270, 362, 438, 515 ; vol. xvi, pp. 35, 65, 152, 360, 374 ; vol. xvii, pp. 48, 60, 185, 305, 458, 459, 464 ; vol. xviii, pp. 49, 89, 119. The shipments recorded in the years between 1515 and 1550 are as follows (the total in each case being given in chaldrons, for the approximate weight of which see Appendix C (i)) :—
1515, 4 ; 1516, 14 ; 1517, 24 ; 1518, 26 ; 1522, 221 ; 1523, 6 ; 1525, 24 ;

years at ports along the Firth of Forth, on as much as 1,600 chaldrons. Between 1570 and 1600, we get the first evidence of shipments from Burntisland, Bo'ness, Inverkeithing, Dysart, Crail, Pittenweem, Culross, Anstruther, and Kinghorn.[1] Shipments of from 300 to 400 chaldrons per annum are recorded in connection with Sir George Bruce's colliery at Culross. By 1575 salt, and, no doubt, coal as well, was shipped from a number of other ports, from Preston, Kirkcaldy, Granton, Musselburgh, and Crummy.[2]

In 1614, more than a decade after the imposition of the heavy tax on English coal had caused foreigners to go farther north for their ladings,[3] the recorded exports of coal are said to have reached 6,308 chaldrons.[4] Owing to the uncertain size of the chaldron, it is impossible to say how much coal this figure represents, but if the chaldron used was that later adopted as the standard in Scotland,[5] then the recorded exports must have already reached 30,000 tons per annum. There can be no serious doubt that, after the Restoration, the exports normally approached 50,000 tons a year,[6] and, knowing as we do that the Scottish export trade suffered during the Inter-regnum, it seems probable that, before the Civil War, this figure may have been frequently exceeded.[7]

Although the shipments coastwise from the Firth of Forth did not equal the shipments overseas at the end of the seventeenth century, they were probably sufficient to bring the total of water-borne coal well up towards the 100,000 ton mark.[8] By far the greater portion of the coal produced was, however, for local consumption.

A larger quantity of coal appears to have been consumed on the shores of the Forth, in the manufacture of salt for export, than was shipped by sea.[9] It is perhaps not too much to assume that, on the eve of the Civil War, more than 150,000 tons per annum was either carried overseas directly or burned to produce salt to be carried overseas. Before 1550 it may be doubted whether more than 10,000

1526, 24 ; 1527, 92 ; 1528, 103 ; 1529, 27½ ; 1530, 32½ ; 1531, 51 ; 1532, 67 ; 1534, 94 ; 1535, 147 ; 1537, 53 ; 1538, 103 ; 1539, 89 ; 1540, 101 ; 1541, 457 (233 from Kinghorn and Burntisland, 224 from Edinburgh) ; 1543, 57 ; 1546, 87½ ; 1550, 102.

[1] *Ibid.*, vols. xx–xxiii, *passim* (see index). The entries for Edinburgh increased rapidly during the fifties of the sixteenth century, reaching 829 chaldrons in 1561–2. In the 'seventies and 'eighties the entries, which begin to include other towns on the Firth of Forth, besides Edinburgh, regularly exceed 1,200 chaldrons, reaching 1,649 in 1581–2. During the 'nineties, in years for which there are no entries for Edinburgh, more than 1,200 chaldrons per annum were entered for Culross, Crail, Anstruther, Pittenweem, Kinghorn, Burntisland, and Inverkeithing.

[2] *Privy Council Register of Scotland*, 1st Series, vol. ii, pp. 442–3.

[3] See below, vol. ii, pp. 223 sqq.

[4] *Hist. MSS. Com., Report on the MSS. of the Earl of Mar and Kellie*, pp. 70–4.

[5] See Appendix C.

[6] See the discussion of Scottish exports in Appendix B.

[7] For the condition of the Scottish export trade during and after the Inter-regnum, see below, vol. ii, pp. 228 sqq.

[8] See below, p. 92.

[9] For a discussion of the quantity of coal burned in the salt pans, see below, pp. 207–8.

tons per annum was mined for these purposes. During the remainder of the sixteenth century, the local market for coal[1] grew considerably in relative importance, both because of the increase in the quantity burned in housekeeping, and because of the development of fuel-consuming industries other than salt making. A survey of mining enterprise in the Firth of Forth district reveals the development, after 1575, of an ever increasing number of important collieries along the coast on both sides of the Forth, and, especially on the south side, inland to a distance of five or ten miles.

The country between Edinburgh and Haddington undoubtedly remained the chief centre of the Scottish mining industry. All through the valley of the Esk, inland beyond Dalkeith and Lasswade, hamlets that had once consisted almost entirely of hovels built with the hide of oxen[2] became busy little mining villages, like Tranent, whose population in 1619 is said to have been made up of butchers, bakers, " coal-hewers," and agricultural labourers.[3] The backward condition of mining in this area before the middle of the sixteenth century is revealed by the fact that in 1548 a special messenger had to be sent to Lasswade, Dalkeith, Musselburgh, Newbattle, Inveresk, Edmonstone, Whitehill, and Cockpen to order that coal be dug to supply the French army.[4] Fifty years later there was a regular trade in coal all through East- and Mid-Lothian.[5] Along the shores between Easter Cockenzie and the mouth of the river Esk, the chief collieries for supplying the sailing ships and salt pans were found. As late as 1621 many of the pits lay near the water's edge, at Seton, Elphinston, Mekle and Little Fawside,[6] and Whitehill; but farther inland, at Smeaton and Pencaitland, there were at least two exporting collieries. Most of the coal from these mines, and most of the salt produced with the coal, was loaded for export at the villages of Newhaven, Musselburgh, and Fisherrow. The capital invested in 1621 was said to exceed £10,000 Scots at Little Fawside, £20,000 Scots at Smeaton, 20,000 marks Scots at Elphinston. The colliery at Elphinston employed at one time " fourtie families of men, wyffis and children " in drainage operations alone.[7]

The importance of the mines at Seton, in 1679, led the Earl of Winton to undertake the construction of a new harbour at Easter Cockenzie, to enable ships of 300 tons to drop anchor for ladings of coal.[8] By this time it is probable that, as in the Tyne and Wear

[1] See below, pp. 107–8.

[2] Æneas Sylvius, *Commentarii Rerum Memorabilium*, p. 4, cited in P. Hume Brown, *Early Travellers in Scotland*, pp. 26–7.

[3] *Privy Council Register of Scotland*, 1st Series, vol. xi, p. 495.

[4] *Accounts of the Lord High Treasurer of Scotland*, vol. ix, p. 204.

[5] *Privy Council Register of Scotland*, 1st Series, vol. v, p. 227 ; vol. xii, pp. xx, 387–8, 418–19, 433–5, 466–7, 474.

[6] In 1671 the seam at Fawside was devoured by one of those underground fires common to nearly all ages of recorded coal mining, and colliery operations were possibly suspended (*ibid.*, 3rd Series, vol. iii, pp. 319–20). Work was being carried on there, however, as late as November 26th, 1668 (*ibid.*, 3rd Series, vol. ii, pp. 563–4).

[7] *Ibid.*, 1st Series, vol. xii, p. 434 ; vol. xiii, pp. 207–8.

[8] *Ibid.*, 3rd Series, vol. vi, pp. 383–5. From a number of ports along the coast of East- and Mid-Lothian, regular shipments of coal, and of salt

valleys, pits were being dug farther inland. Meantime the growth of the market at Edinburgh, where the demand for coal presumably increased in each succeeding decade of the seventeenth century, had led to the development of a dozen or more " land-sale " collieries on the western side of the Esk valley.[1] To judge from the capital, 36,000 marks Scots, which is said to have been invested in one of these mines, some of them must have been no less important than those operating for sea-sale.[2]

In Linlithgowshire, where coal is heard of as early as 1200, there were pits at Bonnington, Kinneil, Bo'ness and Linlithgow in the fifteenth and sixteenth centuries, [3] but the first large colliery was apparently started about the beginning of the seventeenth century, at Carriden. The capital invested there in 1609 is said to have exceeded 50,000 marks Scots (or nearly £3,000 sterling).[4] By 1622 a colliery at Bo'ness owned by Roger Dunkieson, a merchant of Edinburgh, and his co-partners appears to have been equally important.[5] According to a representative sent by Cromwell to investigate the state of Scottish trade, " Burrowsto'ness . . . next to Leith, hath of late beene the chiefe port . . . in Scotland, as well because it is not farre from Edinburgh, as because of the greatt quantity of coale and salt that is made and digged here, and afterward carryed hence by the Dutch and others." [6] After the Restoration, Grange and Bonhard, villages more than a mile inland, inhabited entirely by " coal-hewers, sa[l]tors and uncostmen ", seem to have outstripped Carriden, Kinneil, and Bo'ness, at least in their output for overseas export.[7] There were no less than seven important collieries within six miles of Linlithgow,[8] not to mention smaller diggings farther inland, at Torphichen and Uphall.

In Fife, where the seams were accessible to seventeenth-century mining technique along a shore line of seventy miles, the first important exporting collieries were started during the second half of the sixteenth

manufactured with the help of coal, were made during the seventeenth century. For example, an account of the customs officials at Preston Pans from February 12th to November 1st, 1641, contains a record of the export of 555 chaldrons of coal and 177 chaldrons of salt (*Customs and Excise, Misc. Accts. and Papers*, 1617–41, General Register House, Edinburgh).

[1] *E.g.* at Ormiston, Carberry, Southside, Cockpen, Clerkington, Lasswade, Wolmet, and Arniston (*Privy Council Register of Scotland*, 1st Series, vol. vi, p. 236 ; vol. xiii, pp. 368–9, 668 ; 2nd Series, vol. iii, p. 563 ; vol. v, pp. 277–8 ; vol. vii, p. 389 ; vol. viii, pp. 96–7 ; 3rd Series, vol. i, p. 388 ; *Hist. MSS. Com., 3rd Report*, appx. p. 414b).

[2] *Privy Council Register of Scotland*, 2nd Series, vol. v, p. 278.

[3] *MSS. of the Duke of Hamilton, Charters*, no. 154, ii ; *Accounts of the Lord High Treasurer of Scotland*, vol. ix, pp. 107, 132, 145, 241.

[4] *Privy Council Register of Scotland*, 1st Series, vol. viii, pp. 568–9.

[5] *MSS. of the Duke of Hamilton, Charters*, no. 154 ; *Privy Council Register of Scotland*, 1st Series, vol. xiii, pp. 751–2. In 1643 the four exporting collieries of Linlithgowshire were Carriden, Grange, Bonhard, and Bo'ness (*ibid.*, 2nd Series, vol. viii, pp. 15–17).

[6] Thomas Tucker, *Report upon . . . Excise and Customs*, p. 28.

[7] *Privy Council Register of Scotland*, 3rd Series, vol. ii, pp. 155–6 ; vol. iii, p. 98 ; vol. v, p. 653 ; *State Papers* (General Register House, Edinburgh), no. 252 (printed below, Appendix F).

[8] *Privy Council Register of Scotland*, 3rd Series, vol. iii, p. 98.

century at Culross, Dysart, and Wemyss, where the works of Lord
Sinclair, like those of Bruce at Culross, extended under the sea.[1]
Coal must have been mined also farther east, for, in the customs
records, we find a tax collected on exports from Crail, Anstruther,
Pittenweem, and St. Monance.[2] After 1600, a dozen new enterprises
were started all along the shore, and inland as far as Fordel and
Cleish, in the county of Kinross.[3] Tucker found all the ports east of
Inverkeithing " pittiful small towns . . . inhabited by seamen, colliers
and salt-makers." [4] Except for Dysart, Wemyss, and Methil, where,
about 1660, the Earl of Wemyss, had laid dry the " Happie Mine " by an
adit driven through the rock for two miles from the shore, most of
the important exporting collieries after the Interregnum were situated
in the narrows of the Forth, west of the islet of Inch Garvie. Coal
from Fordel was shipped from St. David's Harbour.[5]

Not until the beginning of the seventeenth century were there
any important collieries along the river Forth ; but, by 1609, Airth
in Stirlingshire, on the south bank, and Alloa and Sauchy in
Clackmannanshire, on the north bank, already ranked, along with
Carriden and Culross, as the largest enterprises west of Queensferry.[6]
Before the Civil War pits were also dug at Quarrell and Falkirk on
the Stirling side of the river Forth. Farther south in Stirlingshire,
in the neighbourhood of Bannockburn, Larbert, and Falkirk, several
collieries were developed after the Restoration, principally to serve
the local market.[7]

From all the coalfields about the Firth of Forth comes abundant
evidence of the activity of miners, in sinking pits and exploiting
whatever seams they could find, both for a local market and for shipment by water. It appears that, before the end of the seventeenth
century, there must have been more than fifty collieries of some
importance, besides innumerable small pits, dug to serve those landlords who had not thought it worth while to compete either for the
export market or for the market provided by domestic consumers
in the towns and by the growing manufactures. Tulliallan colliery,
on the northern shore of the Forth, had an output of about 15,000

---

[1] *Acts of the Parliaments of Scotland*, vol. iii, p. 322b ; A. S. Cunningham,
Mining in the Kingdom of Fife, p. 8. The colliery at Dysart was owned by the
Sinclair family, and was of importance throughout the seventeenth century.

[2] *Privy Council Register of Scotland*, 1st Series, vol. iii, p. 217. The colliery
at Pittenweem appears to have ranked among the principal exporting mines
at the time of the Civil War (*Supplementary Parliamentary Papers* (General Register
House, Edinburgh), vol. iv, no. 7), and it is said to have been confiscated and
worked by Cromwell under the Protectorate (Sir John Sinclair, *Statistical Account
of Scotland*, 1791–9, vol. iv, pp. 369–70).

[3] The principal collieries near the coast, besides those already mentioned
in the text, were at Kincardine, Tulliallan, Vallifield, Pittencrief, Newmilne,
Torry, Crombie, Limekilns, Inverkeithing, Burntisland, and Kinglassie. There
were other collieries at Dury, Markinch, Lochgelly, Dunfermline, and Crossgates.

[4] Tucker, *op. cit.*, p. 31.

[5] Sinclair, *op. cit.*, vol. xvi, p. 519 ; A. S. Cunningham, *op. cit.*, pp. 9, 38 ;
State Papers (General Register House, Edinburgh), no. 252.

[6] *Privy Council Register of Scotland*, 1st Series, vol. viii, pp. 568–9 ; Tucker,
op. cit., pp. 29–30 ; *State Papers* (General Register House, Edinburgh), no. 252.

[7] *Privy Council Register of Scotland*, 3rd Series, vol. v, pp. 153–4.

tons in 1679. Torry colliery, which adjoined it, produced more than 10,000 tons per annum at about the same time.[1] If these two enterprises can be taken as fair samples of the more important Scottish collieries, the output of coal in the Firth of Forth district, at the end of the seventeenth century, may well have amounted to 300,000 or 400,000 tons per annum.

(b) The Lanarkshire District

The basin of the river Clyde from Douglas and Lanark down to Dumbarton was a much more fertile, and potentially a much more valuable industrial, region than the country round the Firth of Forth. In 1628, William Lithgow, observing the abundance of arable and pasture land and the varieties of what he called " coal and earth fuel ", wrote that Lanarkshire and the adjacent parts of Stirlingshire, Dumbarton, and Renfrewshire " may justly be surnamed the paradise of Scotland ".[2] But, in point of population, wealth, and coal produced, Lanarkshire did not yet rival the Lothians or even the kingdom of Fife. Until the deepening of the river Clyde in the eighteenth century, the growth of mining and of all industrial enterprise was handicapped by the impossibility of carrying bulky goods by water to Glasgow, except in very small boats. Coal was first shipped, not from Glasgow, but from Dumbarton,[3] some thirty miles nearer the sea. By 1620 a few tiny boats, none of them of more than ten tons, came to Glasgow for ladings of coal and salt.[4] It required thirteen ships to carry the eighty-five tons of coal sent to northern Ireland in the year 1621, and almost as many to transport an even smaller quantity of salt.[5] Even as late as 1672 the annual shipments of coal did not apparently exceed a few hundred tons.[6]

In spite of the handicap to trade from Glasgow, there is substantial evidence of a development of mining enterprise in the Clyde valley during the last three-quarters of the seventeenth century. In 1642 we hear of a colliery and a saltwork employing several " coilyears and salters " on the lands of Palmertoun within the barony of Douglas, and by this time adits for draining the mines were probably cut here

[1] *Accounts of the Kincardine and Tulliallan Coal and Salt Works*, 1679–80 ; *Accounts of the Torry Coal and Salt Works*, 1679–80 (both in the General Register House, Edinburgh).

[2] *Description of Scotland*, printed in P. Hume Brown, *Scotland before 1700*, p. 298.

[3] The first recorded export of coal from Dumbarton is in 1556, when 7 chaldrons and 12 barrels were shipped out, presumably to Ireland. Further exports from Dumbarton were recorded in 1571–2, 1575–6, 1576–7, 1577–80, 1580–1, 1581–2, indicating that shipments became regular at about this time (*Exchequer Rolls of Scotland*, vol. xviii, p. 338, vol. xx, *passim*). There is, however, no record of an early export of more than 30 chaldrons in a single year.

[4] Boats were no larger at the time of Tucker's investigation in 1656. According to him, some coal was shipped to France, together with plaiding and herrings (*Report on Excise and Customs*, p. 38).

[5] *Customs and Excise, Misc. Accts. and Papers* (General Register House, Edinburgh), for year ending October 30th, 1621.

[6] *Customs and Excise, Foreign Collectors' Accounts*, 1665–72 (General Register House, Edinburgh).

and in the adjoining parish of Carmichael.[1] In Stirlingshire, along what is now the Clyde-Forth Canal, coal had been worked in the parish of Campsie from early times, by driving a so-called " creeping heugh " into the sides of the hill, and later (probably before the end of the seventeenth century) by means of perpendicular pits—called " windlass heughs "—drained by adits.[2]

But most of the coal produced in the Clyde valley came from the immediate neighbourhood of Glasgow, from mines situated either within the estates of the barony of Glasgow, under the control of the town, or within the holdings of the Duke of Hamilton. At about the time of the Restoration we hear of two considerable coal works— one in the barony, " most usefull to the toune " of Glasgow, and another on certain lands called Gorballs adjacent to those of the town.[3] Within the Hamilton estates, bordering the river, were two collieries on the south bank, in Rutherglen and Cambuslang parishes,[4] and a third across the river, in Monkland parish, none of them more than five miles from Glasgow by land or water. Before 1700 a great wharf, capable of holding several thousand tons, was built to store the coal from the Hamilton collieries.[5] Most of the other parishes in the neighbourhood of Glasgow had their own pits. The " coalworks " of Blackburn, in the barony of Kilbride, produced fuel for the town of Hamilton throughout the seventeenth century. Those of Corseford, near Renfrew, supplied the surrounding villages.[6] The parish of Paisley had a tiny pit, which required only five miners to work it in 1634.[7] Across the Clyde, we hear of rivalry between two coal pits at Stevinson.[8] It is apparent that the district round Glasgow had already become an important mining centre before the end of the seventeenth century, and it does not seem unreasonable to assume that the annual output of coal in Lanarkshire amounted to from 70,000 to 100,000 tons. We may doubt whether, before the reign of Elizabeth, the yearly production was more than a tenth as much.

(c) The Ayrshire District

By virtue of their position on the coast, the ports of Irvine and Ayr were accessible to larger vessels than could reach Glasgow, and

---

[1] *Privy Council Register of Scotland*, 2nd Series, vol. vii, p. 556 ; Sinclair *Statistical Account of Scotland*, vol. xiii, p. 371.

[2] *Ibid.*, vol. xv, p. 331 ; vol. xviii, p. 241.

[3] *Privy Council Register of Scotland*, 3rd Series, vol. i, pp. xliv, 258 ; *Acts of the Parliaments of Scotland*, vol. vii, appx., p. 31.

[4] Sinclair, *op. cit.*, vol. v, p. 256 ; vol. ix, p. 8.

[5] *Hamilton MSS.* (Charter of November 16th, 1661 ; also MS. " Account of the Duchess of Hamilton with Wm. Lawson, mason in Glasgow, for building a new Reeve . . . for holding Coalls, 1716 " ; and a pamphlet entitled *Answers for John Fairie, tactsman of the Duke of Hamilton's Coalworks in Cambuslang . . . to the petition of John Lawson, Robert Finlay and the magistrates of Glasgow*, September 14th, 1764).

[6] *Hamilton MSS.*, no. 183, and *Charters*, no. 429. Coal was worked in the barony of Kilbride as early as 1603 and no doubt before.

[7] Sinclair, *op. cit.*, vol. vii, pp. 79–80.

[8] *Privy Council Register of Scotland*, 1st Series, vol. xiii, p. 47.

shipments of coal from these ports to Ireland were made at least as early as the middle of the sixteenth century.[1] At that time the salt manufacture near Ayr must have been more important than that of Lanarkshire, for Ayrshire salt competed with salt from Glasgow and Dumbarton even along the Renfrewshire bank of the river Clyde.[2] The early shipments of coal from Ayr, though larger than those from Glasgow, were irregular, and do not appear to have much exceeded a hundred tons in a single year.[3] There is no reason to believe that much progress in mining was made during the seventeenth century, when the chief pits were at Craigie and Stair, several miles from the sea.[4] There must have been a small local market for coal about the town of Ayr, where a cloth manufacture was begun before 1650,[5] but in this district there was no concentration of population such as made possible the early expansion of the coal industry near Glasgow. At Irvine, where seams were worked nearer the coast than at Ayr, only 66 tons were sent out in 1623.[6]

Not until 1678 was a real effort made to develop mineral resources in Ayrshire. In that year Robert Cunningham of Auchinharvie, a son of the physician to Charles II, began intensive mining activities. He spared no expense in testing the nature and depth of the seams, in driving an adit a mile and a half to lay them dry, and in building a harbour and salt pans at Saltcoats.[7] Within a few years he must have established a considerable trade with Ireland, for two decades later the new port had from fifty to sixty " small Barks and Ships ", which sailed into the Irish Sea " mainly with Coals ".[8] Besides the mining along the coast, there was some shallow digging for coal farther inland in Ayrshire, at Sorn and Stewardton;[9] but, as long as population was sparse and commerce slight, systematic exploitation of the seams was neglected, and it is doubtful if the total output of all Ayrshire during the last decades of the seventeenth century much exceeded 25,000 tons a year.

[1] The first exports recorded were 200 loads (about 20 tons) shipped from Irvine in 1541–2 (*Exchequer Rolls of Scotland*, vol. xvii, p. 458).

[2] *Privy Council Register of Scotland*, 1st Series, vol. iii, pp. 300–2.

[3] Only 401 chaldrons of coal are entered in the customs accounts for Ayr for a five-year period, 1574–9, and only 246 chaldrons for the eight years, 1564–72. The trade appears to have fallen off somewhat during the last decade of the sixteenth century, only 112 chaldrons being exported during the seven years, 1591–8. There are no entries for Irvine during the latter half of the century (*Exch. Rolls of Scotland*, vols. xix–xxiii, *passim*). Even if we make a substantial allowance for the coal which may have escaped entry in the customs accounts (see above, p. 44), it is obvious that the trade from Ayrshire was of little consequence before 1600.

[4] Sinclair, *op. cit.*, vol. v, pp. 369–70 ; vol. vi, p. 114.

[5] W. R. Scott, *Joint-Stock Companies*, vol. i, p. 244.

[6] See C. Innes's Preface to *Ledger of Andrew Halyburton, 1492–1503*, 1867, p. c.

[7] Sinclair, *op. cit.*, vol. vii, p. 173.

[8] J. Spruel, *An Accompt Current betwixt Scotland and England Ballanced*, 1705, p. 17.

[9] Sinclair, *op. cit.*, vol. xx, p. 154.

(d) *Other Scottish Mining Districts*

Outside the great Scottish coal belt stretching across the country from Saltcoats and Girvan on the Irish Sea to Fife Ness and North Berwick on the North Sea, early mining enterprise succeeded in reaching two relatively unimportant seams, one in Sutherland, the other in Berwick county. The discovery of the thin coal vein between Brora and Loth on the Moray Forth was made by John, tenth Earl of Sutherland, just before 1567, but it was first successfully worked some years later by Lady Jane Gordon, primarily to supply cheap fuel for the manufacture of salt, which " served not only Sutherland and the neighbouring provinces, but was also transported into England and elsewhere ".[1] Probably some coal was also used during the seventeenth century in casting great guns with iron extracted near Brora.[2] For domestic purposes, Sutherland coal, in spite of the tradition that it destroyed rats,[3] was not popular except among the local population ; and attempts to ship it coastwise to the adjacent towns did not apparently meet with much success, notwithstanding the high cost of bringing coal to the north of Scotland from Newcastle or the Firth of Forth. About a century after the discovery of the mines in Sutherland, coal was also found at Cockburnspath, and at Ayton in Berwickshire, but the sinkings made at Ayton did not yield enough return to relieve the neighbouring town of Eyemouth from its dependence on imported coal.[4] The total output from all these parts during the seventeenth century probably did not exceed a year's output from the collieries of Fife or of Midlothian at the end of the century. The student of the Scottish coal industry could hardly be criticized if he ignored them.

(iii) *Wales*

In estimating the output of the Welsh collieries, we again have the advantage of fairly complete records of the shipment of coal by sea. These records show that the development of the trade from Pembrokeshire, Glamorganshire, and Carmarthenshire, while on a smaller scale, of course, was hardly less remarkable than the development of that from Durham and Northumberland. The rate of increase was even more rapid than in north-eastern England, as may be seen by comparing the following table with Table IV.

---

[1] R. Chambers, *Domestic Annals of Scotland*, 1858, vol. i, p. 302.
[2] D. W. Kemp, *Notes on Early Iron-Smelting in Sutherland*, 1887, vol. i, p. 302.
[3] Thos. Pennant, *Tour in Scotland*, 1771, pp. 148–9.
[4] Sinclair, *op. cit.*, vol. i, p. 86 ; vol. xiii, p. 226 ; *Acts of the Parliaments of Scotland*, vol. viii, pp. 142–3.

TABLE V[1]

Estimated Annual Shipments of Coal from the Welsh Fields (in Tons)

			1551–60	1591–1600	1631–40	1681–90
South Wales—						
From	Milford Haven	. .	{300	{600	{3,000	20,000
,,	Tenby	. . .				8,000
,,	Swansea	. . .	1,800	3,000	{12,000	{40,000
,,	Neath	. . .	600	1,000		
,,	Llanelly and Burry Port	.	300	400	2,000	7,000
North Wales—						
From	Flintshire ports	. .	little or none	1,000	9,000	12,000
	Total	. . . .	3,000	6,000	26,000	87,000

Between 1560 and 1690, the shipments had multiplied nearly thirty-fold. Although the rate of increase in the coal produced for local consumption must have been less rapid, owing to the relative lack of manufactures, like saltboiling, which required large supplies of the new fuel,[2] the expansion of the mining industry in Wales was clearly comparable to that in the coalfields of the north of England and of the region bordering the Firth of Forth. Let us examine such further evidence of this growth as is provided in direct accounts of conditions in the Welsh counties.

(a) South Wales

Coal mining first became an important occupation of the inhabitants of South Wales during the sixteenth and seventeenth centuries. Although some coal was carried to sea at least as early as the fourteenth century,[3] it is clear that until after Elizabeth's reign the shipments were insignificant. There are " vaines and pitts growing in Wales ", wrote an anonymous authority on the coal trade, in the nineties of the sixteenth century, " but they are no more than needful for that countrye." [4] And a few years later Owen, the early Welsh chronicler, places " sea coales " eighth among the exports of Pembrokeshire, after corn, cattle, wool, butter and cheese, sheep and swine, herrings and oysters.[5] It is equally clear that a century later coal had become the principal article of shipment both from Pembrokeshire and Glamorganshire. It was no uncommon sight, early in the eighteenth century, to see a hundred sail of ships loading coal within Swansea Bay, and Defoe tells us that, chiefly as a result of its trade in fuel, Milford became the " largest, richest, and . . . most flourishing

---

[1] For an explanation of the sources from which this table is derived, see Appendix B.

[2] See below, pp. 207–8.

[3] See below, vol. ii, p. 422.

[4] *Hist. MSS. Com.*, *Report on the MSS. of the Marquis of Salisbury*, vol. xiv, pp. 330–1.

[5] George Owen, *Description of Pembrokeshire* (1603), ed. H. Owen, 1892, pt. i, pp. 54–8, 87.

town of South Wales except Carmarthen ".[1] Neither Tenby nor Llanelly, which is described about this time as a " tolerably good town ", whose " inhabitants are principally traders in sea-coal ",[2] were far behind Milford Haven or Swansea Bay, in point of ships loading with coal. A new activity had been instilled into all these ports as a result of the new trade.

There is good reason to suppose that this increase in the " sea-sale " of coal was preceded and accompanied by a substantial, if somewhat less rapid, increase in the " land-sale ". Owen comments upon the spread, which has occurred within his own memory, of the use of coal among the inhabitants of South Wales ; and, although he places coal eighth on his list of exports, he says that it " may be numbered as one of the chieffe comodities of this countrie ".[3] By the end of the seventeenth century it had doubtless become almost the universal fuel all through the mining districts ; [4] and, as there appear to have been about 100,000 persons living in the counties of Glamorgan, Pembroke, and Carmarthen, and those parts of Monmouth and Brecknock within ten or fifteen miles of the pits, we may perhaps put the annual consumption of coal in this area at about 80,000 tons.[5]

Mining in the parish of Llanelly was being pushed in 1613 to " the great reliefe of the country adjoyninge ".[6] During Elizabeth's reign, we hear of pits at Talbenny, Johnston, Freston, Picton, Roch, Jeffreston, Begelly, and in the Forest of Coedraeth in Pembrokeshire ; at Millwood, Llanharry, Gower, and Margam in Glamorganshire ; and at Dynas in Brecknockshire.[7]

There is evidence during the seventeenth century of more intensive mining, especially in the country behind Burry Port and round Swansea Bay. An adventurer sank more than £1,000 in a colliery at Llanelly in 1613.[8] Around Swansea new enterprises were launched at Llansamlet and Llangafelach.[9] By the latter part of the seventeenth century mines at Neath are said to have yielded one year the

---

[1] *A New Present State of England*, 1727 (?), vol. i, p. 312 ; Defoe, *Tour*, 1769 ed., vol. ii, pp. 363, 368.

[2] *A New Present State of England*, vol. i, p. 309.

[3] Owen, *op. cit.*, pt. i, pp. 54–8, 86–7, 92.

[4] For its possible use in connection with smelting metals, and its certain use in working these metals into finished products, cf. below, pt. ii, chs. iii, iv. The author of *A New Present State of England* speaks of the use of wood and turf, besides coal, " for daily fewell " (vol. i, p. 288) ; but he is writing of Wales generally rather than of the strip of mining country in South Wales.

[5] John Houghton, *A Collection for the Improvement of Husbandry and Trade*, ed. R. Bradley, 1727, vol. i, p. 75. For the basis on which Houghton founded his calculation of the population of England and Wales in 1693, and for my estimate of the consumption of coal per head in the coal-producing districts, see below, pp. 81 n., 83 n., 104.

[6] *Star Chamb. Proc., James I*, 288/5.

[7] Owen, *op. cit.*, p. 88 ; C. B. Wilkins, *A History of the South Wales Coal Trade*, pp. 21 sqq. ; *Cal. of Letters and Papers of Henry VIII*, vol. xv, no. 831 (62) ; *Augm. Partics. for Leases*, 217/14 ; 223/20 ; 225/98 ; *Exch. Spec. Com.*, nos. 3325, 3441 ; *Exch. Deps. by Com.*, 12 Eliz., East. 1.

[8] *Star Chamb. Proc., James I*, 155/5 ; 288/5.

[9] *Augm. Partics. for Leases*, 217/14 ; *Exch. Deps. by Com.* 12 Eliz., East. 1 ; *Chanc. Proc. Eliz.*, U.u. 1/50.

considerable profit of £600.[1] Neath river was accessible to small vessels as far up as the large town bridge, so that ships were able to take their lading within a short distance of the colliery. The statistics of shipments from Neath and Swansea are sufficient evidence of the growing importance of the Glamorgan industry ; and, although little coal was shipped from Cardiff during the seventeenth century,[2] it is certain that the Rhondda valley was not without its local " land-sale " pits. In the eighteenth century, Arthur Young writes that, in order to obtain manure, farmers from Herefordshire sometimes come twenty miles to Monmouthshire to get a supply of lime produced there in kilns heated with coal.[3]

(b) North Wales

Although references to profits from coal in the fourteenth and fifteenth centuries have been found in connection with the manors of Ewloe, Hopedale, and Mostyn in Flintshire and Brymbo in Denbighshire,[4] it is not until 1593 that the port books contain a record of coal shipments from the Dee estuary.[5] During the next two decades the trade developed rapidly. In 1616, when commissioners investigated the measures in use for " seacoles, stone coles . . . [etc.] shipped from Chester and member ports ",[6] the principal collieries were situated along the coast at Weppre, Bagillt, Leaderbrook, Uphfytton, Englefield, and Mostyn. At Mostyn there were three pits, according to Edward Ithell, who was appointed by Sir Thomas Mostyn " to see . . . coles measured by the barrell " ; and the monthly ladings taken by the shipmasters amounted to 100 tons, or nearly as much as the total annual shipments from all the mines in this district twenty-five years earlier.[7] The rapid expansion in the shipping trade probably continued

[1] Eliz. Phillips, *Pioneers of the Welsh Coalfield*, 1925, p. 15. Sir Humphrey Mackworth's statement that " the coal trade in Neath had been totally lost for 30 years and upwards, until he, in 1695, began to adventure great sums of money in finding and recovering the coal of that neighbourhood " (*loc. cit.*), is probably an exaggeration. On Mackworth, see below, vol. ii, p. 11.

[2] The port books for Cardiff contain almost no entries of coal during the seventeenth century, but from other sources it appears that about 500 tons were shipped from Cardiff in the year ending at Michaelmas, 1605 (*Additional MSS.*, 34318, f. 29), and about 300 tons in 1616 (*Exch. Decrees*, Series 4, vol. ii, f. 114d).

[3] Arthur Young, *Tour through the Southern Counties of England and Wales*, 1768, p. 118. And see below, pp. 205, 237, where it is shown that, already in the seventeenth century, a supply of cheap coal was indispensable for lime burning.

[4] E. A. Lewis, " The Development of Industry and Commerce in Wales during the Middle Ages," in *Trans. Roy. Hist. Soc.*, N.S., vol. xvii, 1903, pp. 121–73.

[5] See Appendix D (iii).

[6] The small " barrel havens " of Flintshire were always considered by the customs authorities as belonging to Chester.

[7] *Exch. Spec Com.*, no. 3648. See also concerning Englefield, *Parliamentary Surveys, Flint*, no. 3 ; Bainbridge, *Law of Mines*, 1878 ed., pp. 173–4 ; Galloway, *Annals*, p. 219. An account of the colliery at Mostyn in 1677 is given in a paper prepared for the Royal Society by Roger Mostyn (*Philos. Trans.*, vol. xii, 1677–8, pp. 895–9, described at length by Galloway, *op. cit.*, pp. 220–3 ; see also Birch, *Hist. of the Royal Society*, vol. iii, p. 339). From this it appears that a great deal of coal must have been extracted there. The pits were all sunk to a depth of from 40 to 60 yards, and the exhaustion of a seam 5 yards in

until the Civil War, and during the Interregnum most of the orders for coal to supply the English army in Ireland were placed at Chester,[1] the coal being, no doubt, obtained from the collieries in Flintshire ; but, after the Restoration, progress in the coal trade was slow, owing mainly to the growing competition from other mining fields for the Irish market.[2] The loss of the channel of the Dee, which changed from the Flintshire to the Cheshire shore, seriously handicapped the Flintshire trade in the eighteenth century.[3]

Farther from the sea, in the neighbourhood of Ewloe, a group of small " land-sale " collieries had been of some importance at least as early as the reign of Henry VIII.[4] During and after Elizabeth's reign mines were also worked up the valley of the Dee, near Wrexham, and in various parts of the lordship of Bromfield and Yale in Denbighshire, and also on the island of Anglesea.[5] We know very little concerning the history of these enterprises during the seventeenth century, but there is no reason to suppose that they suffered a loss of market such as hindered the rapid expansion of the mining industry along the Flintshire coast. Altogether, we may perhaps put the annual production of coal for local consumption in North and South Wales at the end of the seventeenth century at something approaching 120,000 tons, as compared with less than a sixth of this amount at the accession of Elizabeth.

(iv) *The Midlands*

The inland fields, which did not enjoy in the seventeenth century any outlet by water, are to-day among the most important in England. The seams lie under a hilly and, on the whole, unfertile country, with few navigable rivers, between Leeds and Preston on the north, and Coventry and Ludlow on the south. During the Middle Ages it was a sparsely settled region, of relatively little importance in the economic life of the country, but already in Elizabeth's reign it showed signs of awakening industry. This region included not only the north and south Staffordshire, the Leicestershire and the Warwickshire fields, all inland fields in a strict sense, but also the Lancashire-Cheshire,

---

thickness led to an attack upon another " Roach [seam] of Cole ", $3\frac{1}{2}$ yards thick, lying beneath the 5 yard seam. In the seventeenth century coal was also worked along the Flintshire coast in Prestatyn Hundred (*Calendar of Treasury Books*, vol. i, p. 118, January 22nd, 1660–1).

[1] *Hist. MSS. Com.*, 5th *Report*, appx., p. 350 ; *Cal. S.P.D.*, 1649–50, pp. 228, 284.

[2] Cf. below, pp. 109–10. Of the importance of the Irish market to the Flintshire coal owners in 1678 there is evidence in *Cal. of Treas. Bks.*, vol. v, pp. 991–2, 1002.

[3] Galloway, *Annals*, p. 358.

[4] *Star Chamb. Proc.*, *Henry VIII*, vol. v, f. 138. See also *Court of Requests Proc.*, 25/103, for the mining of coal at Ewloe in the reign of Philip and Mary.

[5] Galloway, *op. cit.*, pp. 78, 121, 219–20, 358 ; *Court of Augm. Proc.*, 6/36 ; *Parl. Surveys, Flint*, no. 1 ; *Harleian MSS.*, 2002, no. 7. On coal mining in the lordship of Bromfield and Yale see also *Court of Augm. Proc.*, 18/95 ; *Exch. Deps. by Com.*, 16 Chas. I, Mich. 22. On Anglesea see *Hist. MSS. Com.*, 5th *Report*, appx., pp. 415b, 416a.

the Yorkshire-Midland, and the Shropshire fields, accessible to water transport only at certain points, such as Liverpool on the Irish Sea, Nottingham on the Trent, and Broseley on the Severn. Behind these favoured positions lay vast quantities of unworked coal, which could as yet find only a small local sale.

While we are handicapped in estimating the output of these inland fields by the lack of any statistics of trade, such as can be obtained for districts near the coasts, we have sufficient information to show that the development of coal mining between 1550 and 1700 was everywhere remarkable, and that in some counties the rate of increase in production probably approached that in Durham and Northumberland. Before the canal and, later, the railway provided inland fields with wider markets, the countryside was already punctured at hundreds of places with small shafts, the coal heaped high about them, waiting to be carted away in wagons or on horse back. For no other inland districts have we more convincing evidence of the growth of colliery enterprise during the sixteenth and seventeenth centuries, than that revealed by the innumerable records concerning the mines sunk all through the Midlands.

(a) Yorkshire and North Midland Coalfield

East of a line drawn on the map from Nottingham through Worksop and Pontefract to Barwick-in-Elmet, the seams of the great Midland field were, because of their depth, inaccessible to early mining operations. West of this line, seams were to be found in all parts of the field within working distance of the surface. Here was a region of nearly a thousand square miles, in which one could scarcely walk for two hours without passing over a coal seam.

It is impossible to determine the average quantity of coal burned annually by the inhabitants of the region, but we know that in the reign of William III between 5,000 and 6,000 tons of coal were carried annually a distance of from four to six miles overland to be used in the town of Derby, a place of perhaps 3,500 souls.[1] And there were by this time at least a dozen towns as large as Derby, almost entirely dependent on coal for their fuel, besides scores of smaller towns and villages scattered through the mining field. In Yorkshire, in 1714, according to Thoresby, the pits were " now without number ".[2]

By the beginning of James I's reign there must already have been scores of them, for between 1550 and 1615 we find references to nearly a hundred different collieries, most of which appear to have been recently started, in the West Riding, southern Yorkshire, and Derbyshire.[3] There can be no doubt that there, as in the north of England,

---

[1] Houghton, *Husbandry and Trade*, vol. i, pp. 105, 108.

[2] *History of Leeds*, 1714, p. vi, quoted Galloway, *Annals*, p. 320.

[3] In the Leeds neighbourhood there were several coal mines in the manor of that name (*Duchy of Lancs. Pleadings*, 144/A/9, 149/A/35, 152/A/6, 171/A/12), in the manor of Rothwell (*ibid.*, 87/M/4, 87/M/7, 193/J/1 ; *Duchy of Lancs. Spec. Com.*, no. 317 ; *Duchy of Lancs. Rentals and Surveys*, 9/14), at Kippax, Allerton, Barwick-in-Elmet, Middleton (*Duchy of Lancs. Pleadings*, 116/F/2, 146/F/20, 106/A/1 ; *Chanc. Proc. Eliz.*, 4/5/42), and other places in the vicinity (*Duchy of*

coal mining first became an important industry during this period. The development appears to have continued throughout the seventeenth century, though it is improbable that the rate of increase in output was as rapid after 1625 as in the reigns of Elizabeth and James I.

There was much variation in the output of the various mines. At " certaine cole delphes lying in Wales Wood in the hethermost parte of Yorkshire " in 1599, 2,000 wainloads (or nearly as many tons) were raised yearly " towards furnishinge of the countries next adjoynynge ", though the " delphes " must have been too far from Sheffield or Rotherham to have competed for the markets of those towns.[1] At about the same time two other small enterprises in the same neighbourhood, at Woodsetts and Eckington, had sometimes as many as four or five hundred wainloads awaiting sale at the " bank ", or shaft head, and their annual production must have been three or four times that quantity.[2] From references to rents paid and profits drawn elsewhere, it is clear that there were other mines in this region with a much larger output. Christopher Anderson, of Lostock in Lancashire, paid £613 6s. 8d. in 1588 for the remaining thirty-four years of a crown lease of coal mines in two parcels of Leeds manor

---

Lancs. Pleadings, 73/M/10, 193/J/1 ; *V.C.H. Yorks*, vol. ii, p. 356). There were several coal mines in the manor of Wakefield (*Duchy of Lancs. Pleadings*, 42/F,7, 162/A/35 ; Galloway, *Annals of Coal Mining*, p. 113), at Ossett, Horbury, and Stanley (*Rentals and Surveys*, Gen. Ser., no. 991), at Castleford, Crofton (*Exch. Spec. Com.*, nos. 2621, 2655), Houghton (*Augm. Partics for Leases*, 159/12 ; *Exch. Deps. by Com.*, 21 Eliz., Hil. 8), and also at Lofthouse, Altofts, Featherstone, Pontefract, Crigglestone and other neighbouring places (*Duchy of Lancs. Pleadings*, 106/A/1, 116/F/2, 144/A/9, 146/F/20, 153/F/3, 208/A/27 ; *Duchy of Lancs. Deps.*,73/R/15 ; *Duchy of Lancs. Decrees and Orders*, vol. xx, ff. 196, 427 ; *Court of Augm. Proc.*, 9/99). Further west in the West Riding, there were mines at Halifax, Bradford, Kirkburton, Gomersal (Galloway, *op. cit.*, p. 114 ; *Duchy of Lancs. Pleadings*, 156/A/10, 208/A/27 ; *Star Chamb. Proc.*, *James I.*, 309/46), Hipperholme, Elland, Marsden and other places. In southern Yorkshire there were mines at Barnsley, Cudworth, Monk Bretton (*Duchy of Lancs. Spec. Com.*, no. 433 ; *Augm. Partics. for Leases*, 171/46 ; *Exch. Spec. Com.*, nos. 2777, 2592), Ardsley, Silkstone, and Thurnscoe ; still further south at Greasbrough and Rotherham (*ibid.*, no. 2759 ; *Cal. of Letters and Papers of Henry VIII*,vol. xviii, pt. i, no. 476 (12)), at Brampton-en-le-Morthen, Wales, Woodsetts (*Augm. Partics. for Leases*, 178/1 ; *Acts of the Privy Council*, N.S., vol. xxix, pp. 657–8 ; *Chanc. Proc., Eliz.*, O/2/46), and, to judge from the considerable amount of coal burnt in Sheffield (Galloway, *op. cit.*, p. 112), at a number of points in Hallamshire. Across the Yorkshire border, in Derbyshire and Nottinghamshire, there were mines at Bolsover, Eckington (Bainbridge, *Law of Mines and Minerals*, 1878 ed., pp. 56–8 ; *Exch. Deps. by Com.*, 11 James I, Trin. 6), Wingerworth, Tibshelf (*V.C.H. Derbyshire*, vol. ii, p. 353), Sutton-in-Ashfield, Totley, Chesterfield, and Shuttlewood. Farther south in Derbyshire, there were mines at Heanor (*Exch. Deps. by Com.*, 22–23 Eliz., Mich. 30 ; *Chanc. Proc., James I*, J/3/41), Langley, Stanley (*Star Chamb. Proc., James I*, 311/31 ; *Augm. Misc. Bks.*, vol. ccxii, f. 120), Butterley, Duffield Chase (*V.C.H. Derbyshire*, vol. ii, p. 352), Ripley (*Additional MSS.*, 6687, f. 183 ; *Cal. of Letters and Papers of Henry VIII*, vol. xvii, no. 220 (81)), and many other places (*Augm. Partics. for Leases*, 115/68, 116/28, 52 ; *Court of Augm. Proc.*, 31/46 ; *Court of Requests Proc.*,40/106). Besides all these mines in the Yorkshire coalfield proper, there were others farther north, at Kirkby Malzeard (*Star Chamb. Proc., James I*, 4/3, 227/1–3), Halton (*Court of Requests Proc.*, 28/56), Middleham, Masham, and Grewelthorpe. And the list of mines contained in this note is by no means exhaustive.

¹ *Acts of the Privy Council*, N.S., vol. xxix, pp. 657–8.
² *Chanc. Proc., Eliz.*, O.o. 2/46 ; *Exch. Deps. by Com.*, 11 James I, Trinity 6.

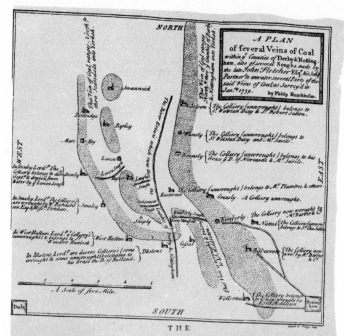

THE

CASE of the Petitioner *John Fletcher*, and his Co-partners;

AGAINST

The BILL for further and more effectually preventing the wilful and malicious Destruction of COLLIERIES and COAL-WORKS.

THE Reasons given for the bringing in of this Bill, are; That the Petitioner, and his Partners, have drowned several Collieries adjoining to theirs, particularly one of Mr. *Richardson*'s, with an Intent to enhance the Price of Coals, and procure to themselves a Monopoly.

As to the drowning of Collieries; the Fact is, That the Petitioner, and his Partners, have several Collieries in *Derbyshire* and *Nottinghamshire*, as appears by the Plan; and they drove, or headed, in their Colliery at *Smalley*, in *Derbyshire*, into the Old-Level to let Wind into the Work to prevent the Damp; which is an usual and a necessary Practice in the Working of Collieries. They also sunk Pits upon their own Land to get their own Coal, and did get, and might have got, great Quantities of Coal out of them. They also stopped up one old Sough, upon their own Ground; which could not have drowned Mr. *Richardson*'s Work at *Smalley*, had his Fire-Engine been in Repair, the Water not being above twenty Yards deep in the Engine-Pit, when the Engine was set down; which if it had been in any tollerable Repair would have drawn all the Water. However, all that the Partners did was upon their own Ground, and for the better carrying on their own Work.

Mr. *Richardson*, indeed, did bring an Action against the Partners for letting in the Water, and recovered a Verdict for 200 l. Damages: which Verdict could never have been obtained, had not the principal Witnesses on the Partners behalf been designedly made Defendants. However, this Verdict shews Mr. *Richardson* has a legal Remedy for any Damages he may sustain.

The Water whereby Mr. *Richardson* pretends he received this Damage is not a continued Water-Course, but is only what is called a Land-Flood, and only runs when it rains; and there is no Colliery whatsoever but the Ground will break and let in the Land-Flood, in some measure, after great Rains: To prevent which, as much as possible, the Petitioner, and his Partners, both before and since this Verdict have made Ditches upon fresh Ground to carry off all the Water; so that except Mr. *Richardson* has himself let in the Water on purpose to lay a Foundation for this Bill, it is impossible he can be drowned by the Land-Flood.

Another Complaint against the Petitioner, and his Partners, is from Mr. *Lowe*, who is possessed of *Denby* Colliery: And the Fact is, That the Partners rented that Colliery of his Father, and had a Lease for his Life; and during the Time they were possessed of this Colliery, they drove a Level from their own, called *Roby*'s-Colliery (which was before unwatered by *Loscoe* Sough) to Mr. *Lowe*'s Colliery, which unwatered it. After Mr. *Lowe* the Father's Death, the Petitioner, and his Partners, stopped up this Sough thus made by themselves, at their own Expence, upon their own Ground; by which they left Mr. *Lowe* in no worse but a better Condition, with respect to Water, than he would have been in case that Sough had never been made.

As to the Enhancing the Price of Coals: It will be time enough to make an Application to Parliament when the Petitioner, and his Partners, are guilty of that Offence, which is so far from the Case at present, that notwithstanding the Expence of getting Coal in Wages, and otherwise is much larger than ever, yet Coals are now as cheap as they have been for twenty Years last past to the Consumers. And by the Number of Collieries in the Plan, it may appear, that the Petitioner, and his Partners, have not a fourth Part of the Collieries in *Derby* and *Nottinghamshire*, worked and capable of being work'd, and therefore there is no Danger of a Monopoly.

That the Petitioner, and his Partners, and those under whom they claim, have been at near 20,000 l. Expence, in making Soughs and other expensive Work, to drain and work their Collieries; and now are under Leases, and stand liable to pay very large Rents to Persons thro' whose Grounds those Soughs are cut, and for their Coal-Mines. And if they are not to be at Liberty to stop up those Soughs upon their own Ground, then the Consequence will be, That other Persons who have contributed to no part of the Expence will reap all the Benefit, by having their Collieries laid dry at the Expence of the Partners; who will not be able pay their Rents. And if other Persons are to have the Benefit of the Soughs without having contributed to any part of the Expence, or being liable to pay any part of the Rents, there will be a greater Danger of a Monopoly in them than in the Partners.

If this Bill passes, it will be a total Discouragement to driving of Soughs; for who will be at such an Expence, when upon the Expiration, or some other Determination of a Lease, he can have no Power to stop them up, but some other Person must have the Benefit of them. And if no body will drive Soughs then Coals must be got by Fire-Engines; which being a much more expensive Way of draining, will considerably increase the Price of Coals, and leave Mines fuller of Water, and in a more ruinous Condition than they can be by stopping of Soughs.

Lord *Middleton*'s Pits, near *Nottingham*, are contiguous to the Petitioners Pits at *Kimberly* and *Bilborrow*, and his Complaint is, That the Petitioners have attempted to drown him, by stopping their Sough in *Kimberly*.

As to this; The Petitioner, and his Partners, are also possessed of a Colliery adjoining to *Kimberly*, runs from their Colliery at *Kimberly*, a large Annual Rent, and the Sough, which Lord *Middleton* complains they have attempted to stop up, runs from their Colliery at *Kimberly* into that which they hold of Lord *Stamford*, which fills that Colliery full of Water. The Petitioner, and his Partners, therefore apprehend they have a Right to stop up *Kimberly* Sough, merely to enable them to work Lord *Stamford*'s Colliery.

A PLAN OF THE COAL SEAMS NORTH-WEST OF NOTTINGHAM, 1739

(Lincoln's Inn Tracts, M.P. 102, f. 92)

Reproduced through the courtesy of the authorities of Lincoln's Inn Library

[To face page 59, V

and in the waste of " Wynmore ", in the Honor of Pontefract.[1] To be willing to pay so high a price, when other mining concessions near by could be had for a few shillings in rent, he must have expected a very considerable annual yield in coal. A contract to pay £80 a year to work a coal mine at Wakefield was entered into in 1620.[2] In 1641 Ralph Cook was said to have received more than £1,260 clear profit in a short period from working a colliery, probably in the West Riding, owned by Dr. John Scott, dean of York, and his brother.[3] Two collieries at Smalley and Heanor, in Derbyshire, appear to have produced, between them, several thousand tons per annum.[4]

When we come to that part of the great Midland field from which coal could find an outlet by water down the Trent, we find colliery enterprise conducted on a much larger scale than in other districts. By virtue of its proximity to the river, and the abundance of its easily accessible coal seams, the manor of Wollaton, which had been owned by the Willoughby family since the thirteenth century, offered the most favourable site for the development of mining in Nottinghamshire. Already in 1493 Sir Henry Willoughby had found it profitable to keep " goyng yerely five cole-pittes beside the levell pitt [i.e. adit] in the lordship ".[5] From information contained in the account books for six of the years in the period from 1526 to 1547, it seems probable that the annual output during these years averaged from 6,000 to 10,000 tons.[6] In the year ending Michaelmas, 1598, the actual produce from Wollaton amounted to 13,264¼ " rooks ",[7] or about 20,000 tons, and it is probable that this colliery continued to increase in importance throughout the seventeenth century.

Already in the reign of Henry VIII a colliery in the neighbouring manor of Strelley competed with that at Wollaton.[8] Mines in both manors were wrought with great energy throughout the sixteenth century,[9] and, in the reign of James I, the one in Strelley is known to have produced almost as much coal as Wollaton. A third colliery at Bilborough was only less important than these.[10] Early in the eighteenth century, mining enterprise was being carried on with even more intensity, as we learn from a plan exhibited to Parliament in 1739 by a group of colliery owners.[11] Supplies were now reaching the Trent from a number of pits, five and six miles away from the river,

---

[1] *Duchy of Lancs. Pleadings*, 171/A/12.
[2] *Chanc. Proc., James I*, G. 1/76.
[3] *Journals of the House of Lords*, vol. iv, pp. 391, 393.
[4] Houghton, *op. cit.*, vol. i, p. 105.
[5] *Hist. MSS. Com., Report on the MSS. of Lord Middleton*, p. 123.
[6] *Ibid.*, p. 492.
[7] *Ibid.*, p. 169.
[8] *Star Chamb. Proc., Henry VIII*, 22/94.
[9] *V.C.H. Notts.*, vol., ii, p. 326.
[10] *Hist. MSS. Com., Report on the MSS. of Lord Middleton*, pp. 148-9 ; *Chanc. Proc., James I*, B. 17/61, B. 40/70 ; and cf. below, vol. ii, pp. 15-6. North of Nottingham there was a substantial development of coal mining during the sixteenth century between Newstead and Kimberley ; and, though the pits of this neighbourhood were sunk too far from the Trent to compete with Wollaton and Strelley for the river market, they served the " inhabitants of the hole country thereabouts " (*Star Chamb. Proc., Edward VI*, 6/99).
[11] See plan facing this page.

and one group of co-partners is said to have invested what was then the large sum of nearly £20,000 in certain of these mines. There were at least five collieries contributing large supplies of coal for shipment on the river, and for carriage to the neighbouring towns.[1] From this evidence, we may estimate very tentatively that the Trent valley collieries in Nottinghamshire, which probably produced from 10,000 to 15,000 tons a year in the middle of the sixteenth century, were producing from 30,000 to 50,000 tons a year at the beginning of the seventeenth, and from 100,000 to 150,000 tons a year at the beginning of the eighteenth. The whole district to the west and north of Nottingham was given over to coal mining. It must have resembled in appearance the Tyne valley about Gateshead, or the Wear valley about Lumley Castle ; and in the Trent basin, as in Durham and in Scotland, it is possible to observe a tendency for pits to be sunk at greater distances from the water, so heavy was the drain upon supplies of readily accessible coal.

It is, of course, out of the question to make a satisfactory estimate of the output from the hundreds of small pits scattered through Derbyshire, northern Nottinghamshire, and Yorkshire, but with the growth of industry in the West Riding and the Sheffield district,[2] the production in the Yorkshire-Midland field as a whole can scarcely have increased less rapidly than in that particular section at the extreme southern tip of the field, and probably amounted to 300,000 or 400,000 tons per annum at the end of the seventeenth century. Before the accession of Elizabeth the output had not perhaps exceeded 30,000 or 40,000 tons.

(b) *The Lancashire-Cheshire Field*

Until about the middle of the sixteenth century the great majority of the inhabitants of southern Lancashire and northern Cheshire burned turf, or " fat earth ", as they called it, if timber was not available. For the most part they lived in complete oblivion of the coal seams which ran beneath their holdings. The expansion of the mining industry in Lancashire began somewhat later than in south Nottinghamshire, probably at about the same time as in the West Riding. After 1550 the records of the Duchy and the Palatinate of Lancaster begin to reveal the same activity in surveying for coal in Lancashire as in Yorkshire,[3] and the same eagerness, among local men of different stations and outsiders from other counties, to invest their own or other people's capital in collieries.[4] For Lancashire, indeed, the evidence of a rapid expansion is even more conclusive than for Yorkshire. It is possible that the number of pits dug and the

---

[1] *The Case of the Petitioners John Fletcher and his Co-Partners*, 1739.
[2] See below, pp. 226–7, 232.
[3] *Duchy of Lancs. Spec. Com.*, nos. 418, 648 ; *Duchy of Lancs. Deps.*, 61/R/2; *Exch. Spec. Com.*, no. 1208.
[4] *Duchy of Lancs. Pleadings*, 100/G/4, 102/T/10, 104/H/8, 109/A/5, 131/T/4, 185/T/12, 204/O/1, 211/G/6 ; *Star Chamb. Proc., James I*, 310/33 ; Crofton, *Lancashire and Cheshire Coal Mining Records*, pp. 40, 42, 56–7 ; *V.C.H. Lancs.*, vol. ii, p. 357.

capital invested was greater in proportion to the size and the population of the mining area.

Coal measures are found almost solidly under the whole of the Yorkshire-Midland field, but they are broken at many points under the soil of Lancashire. Nearly half the former field was inaccessible to seventeenth-century mining technique, but in Lancashire coal outcropped in almost all those districts where it is mined to-day. As a result of these geological conditions, even though the seams of the Lancashire-Cheshire field are actually found under an area only about one-fourth as large as that above the seams of the Yorkshire-Midland field, it was possible to market the coal of Lancashire by land-sale in an area about half the size of that accessible by land-sale to the coal of Yorkshire, Derbyshire, and Nottinghamshire.

It is possible to distinguish at least five separate colliery districts : first, the district about St. Helens and Prescot, five to ten miles west of Liverpool ; second, the district round Wigan ; third, the Forests of Trawden and Pendle, and the valleys of the Ribble and Calder from Barnoldswick and Colne down to Clitheroe and Blackburn ; fourth, the district round Manchester, Bolton, Bury, Rochdale, and Oldham ; fifth, the district where the Lancashire field extends into Cheshire and northern Derbyshire.

There is little evidence of any marked development in the output of coal in this last district. Parliamentary commissioners, who surveyed the manor and borough of Macclesfield in 1652, did not find any improvement in the value of the three " pitts or delfs " which their predecessors had surveyed in 1611 ; and, although a dispute brought before the Court of Star Chamber in 1606 reveals the presence of a mine at Fernilee, in Derbyshire, more important than that at Macclesfield, there could have been no very considerable demand for coal in the hilly, sparsely settled district along the Cheshire-Derbyshire county line.[1] But, for all the mining districts in Lancashire, the evidence of a rapid increase in output is abundant.

Mining in the south-western corner of the county appears to have developed extensively at the very end of Elizabeth's reign, with the coming of ships to the Mersey to load coal for the Irish market. Shipments at Liverpool increased from about 300 tons per annum during the period from 1563 to 1599, to about 1,200 tons between 1611 and 1621, and to more than 4,000 tons before the Civil War.[2] Yet, in spite of a growth in the general commerce of the town between 1660 and 1700, a growth that placed Liverpool inferior only to London and Bristol as a shipping port at the beginning of the eighteenth century, shipments of coal did not increase after the Civil War.[3] It would be a mistake to assume that the progress of coal ship-

---

[1] *Parl. Surveys, Cheshire*, no. 8 ; *Land Revenue Misc. Bks.*, vol. cc, f. 344 ; *Star Chamb. Proc., James I*, 50/23.

[2] See Appendix D (iii) and, for the shipment of coal from Liverpool at the beginning of Elizabeth's reign, J. A. Twemlow, *Liverpool Town Books*, 1918, vol. i, pp. 246, 400.

[3] Defoe, *Tour*, 1769 ed., vol. iii, p. 257 ; J. A. Picton, *Memorials of Liverpool*, 1873, vol. i, p. 131; T. Heywood, *The Moore Rental* (*Publications of Chetham Society*, vol. xii, 1847), pp. 80, 84 (the last two as cited by Crofton, *Lancashire and Cheshire Coal Mining Records*, pp. 61–2).

ments from Liverpool can serve as an adequate guide to the expansion of output in the St. Helens district. The indications are that the fuel which actually found its way to sea represented only a small fraction of that produced in south-western Lancashire. During the three years, 1594 to 1596, from 7,000 to 8,000 tons were dug in Prescot alone; and Prescot was at that time only one of at least half-a-dozen manors with working collieries.[1] It is certain that the mining industry developed rapidly after the Restoration, notwithstanding the stationary condition of the market for sea-borne coal. The collieries at Prescot and Hulton are heard of at the end of the seventeenth century, and they were undoubtedly much more important then than at the beginning.[2] We have a record of the output from two collieries: one in a " close " of the manor of Sutton, the other in the common lands of Widnes. At Sutton, in 1711, upwards of 5,000 tons were produced; at Widnes, in the four years 1717 to 1721, about 10,000 tons per annum.[3] There appear to have been, at the beginning of the eighteenth century, half-a-dozen or more mines as important as Sutton or even Widnes, and numerous other smaller enterprises, in the district from Windle down to the Mersey.

The cannel coal of Wigan is said to have been discovered in the fourteenth century,[4] but it was first seriously worked by Roger Bradshaw, two hundred years later, in Haigh manor, two miles north of Wigan.[5] By 1600 the Bradshaws already had competition from a mine at Aspull, and cannel was burned by " diverse tradesmen and other handycraft men . . . in Wigan ".[6] When Lord Guilford visited Sir Roger Bradshaw in 1676, he was impressed by the " vast piles " of cannel round the pits,[7] piles which were no doubt duplicated at the rival collieries of the Gerrards at Aspull. There were, in addition, before the middle of the seventeenth century, at least a dozen collieries, within a five-mile radius of the town of Wigan, all producing ordinary bituminous coal.[8] These and others are again heard of during the

---

[1] *Duchy of Lancs. Pleadings*, 183/L/5.

[2] *Palat. of Lancs. Bills*, 38/107, 56/49. The colliery at Prescot was worked under lease from King's College, Cambridge.

[3] *Ibid.*, 55/7, 61/11. The colliery at Sutton is said to have produced more than £500 worth of coal in the year; and as the pit head price at the beginning of the eighteenth century appears to have been between 2s. and 3s. per ton (see Appendix E), it is reasonable to assume that the output was in the neighbourhood of 5,000 tons. For Widnes we have a definite figure of 11,340 " works " produced in the four-year period. The " work " appears to have been a measure containing between three and four tons of coal (see Appendix C).

[4] David Sinclair, *History of Wigan*, 1882, vol. i, p. 89.

[5] *Duchy of Lancs. Pleadings*, 77/B/6 (Nicholas Butler *v.* Roger Bradshaw, Esq., and others); Leland, *Itinerary*, 1769 ed., vol. vii, pp. 47, 49.

[6] *Duchy of Lancs. Pleadings*, 211/G/6; *Duchy of Lancs., Decrees and Orders*, vol. xxiii, ff. 114, 132 (Myles Gerrard, Esq., *v.* Roger Bradshaw, Esq., and others).

[7] Cited Galloway, *Annals*, p. 188. See also *Duchy of Lancs. Deps.*, 75/13; J. H. Stanning, *Lancashire Royalist Composition Papers (Rec. Soc. Lancs. and Cheshire*, vol. xxix, 1896), pp. 34–51.

[8] At Hindley, Dalton, Winstanley, Shevington, Atherton, Adlington, Ashton in Makerfield, and in several holdings in the manors of Orrell and Up-holland and the town of Wigan (*V.C.H. Lancs.*, vol. ii, p. 358, and vol. iv, p. 108; Crofton, *op. cit.*, pp. 40, 55, 58; *Duchy of Lancs. Pleadings*, 73/O/5, 23/C/1, 183/L/13, 195/S/10, 204/O/1, 205/S/10).

latter half of the century.[1] The output of one such enterprise, at Shevington, is said to have exceeded 1,500 tons in the summer of 1663 or 1664,[2] and it is possible that the production of this more prosaic fuel may have been greater than that of cannel.

When, in 1650 and 1654, parliamentary commissioners surveyed the Crown manors in the Forests of Trawden and Pendle, with a view to increasing the rents paid by the tenants, they found the mining industry in a most healthy condition.[3] Small collieries had been started within more than a dozen manors since the accession of Elizabeth.[4] Pits were worth many times the value at which they had been let less than thirty years before. A mine in Haslingden manor, rented for 5s. a year, was estimated to yield an annual profit, over all charges, of at least £60; one at Padiham, rented for £2 13s. 4d. was now worth £40; another, in Ightenhill manor, rented at 6s., was now worth £20; another, in Trawden, rented at 11s., was now worth £60. The improvement in values apparently did not extend to agricultural land, for the commissioners recommended that no rents be raised except those of the coal mines. Not many years later, in 1672, a colliery at Clayton-le-Moors, near Blackburn, was said to be worth the clear yearly profit of £200,[5] which suggests that the output could not have fallen short of 6,000 tons a year.[6] If five or six of the score or more enterprises in this district produced as much as this colliery, it is clear that the output here, at the end of the seventeenth century, must have rivalled that in south-west Lancashire, and in the Wigan neighbourhood.

The opinion that Wigan was the leading coal-mining district of Lancashire in the seventeenth century [7] is due to the fame of its cannel, the occasion of frequent references. As a matter of fact it is likely that the output from mines near St. Helens or in the north was at least as great, and in the district round Manchester the output was almost certainly greater. Within the large manor of Rochdale more than a score of the freeholders appear to have mined coal within their holdings.[8] Little Hulton colliery appears to have been worked continuously from 1575 until after 1700,[9] and a seventeenth-

[1] *Palat. of Lancs. Bills*, 30/108; 38/27; 39/53, 84; 55/32.

[2] *Ibid.*, 30/108. The reference is to 30,000 "baskets". For an explanation of the amount of coal probably contained in a "basket", see Appendix C.

[3] *Parl. Surveys, Lancs.*, nos. 8, 19.

[4] At Colne, Burnley, Great and Little Marsden, Padiham, Read, Haslingden, Darwen, Blackburn, Greenfield, Clayton-le-Moors, Whalley, Clitheroe, Barnoldswick, Newton, and Halton West. See *Palat. of Lancs. Bills*, 36/159, 43/82, 66/14; *Duchy of Lancs. Pleadings*, 32/K/2, 78/A/7, 87/H/4, 96/T/15, 100/G/4, 102/T/13, 109/A/5, 109/A/9, 109/A/11, 131/T/4, 156/B/5, 159/T/7, 185/T/13, 189/L/7; *Duchy of Lancs. Deps.*, 61/R/2; *Duchy of Lancs. Spec. Coms.*, nos. 418, 648; *Exch. Deps. by Com.*, 1659, East. 20; *Star Chamb. Proc., James I*, 310/33; *V.C.H. Lancs.*, vol. ii, p. 357, vol. vi, pp. 272 n., 528 n.; Crofton, *op. cit.*, pp. 39, 41.

[5] *Palat. of Lancs. Bills*, 33/29.

[6] Cf. below, vol. ii, p. 366.

[7] *V.C.H. Lancs.*, vol. ii, p. 357.

[8] *Palat. of Lancs. Bills*, 32/44, 36/128, 41/144.

[9] A certain amount of cannel was obtained at Hulton, but for the most part the seams yielded an ordinary bituminous coal (Crofton, *op. cit.*, pp. 42–7).

century traveller along the road from Bolton to Manchester could perhaps have counted a dozen active mines to the south.[1] In the reign of James I the supply for Manchester, which already amounted to more than 10,000 tons per annum, came chiefly from the manor of Bradford,[2] east of the town and to-day a part of the city; but before 1700 the output of mines in the immediate vicinity had already proved inadequate to meet the heavy demand, and coal was brought with much labour over ground seven or eight miles from Worsley and Clifton, where, in 1689, it was estimated that the coal and cannel mines were "worth to be sold" £4,000 and upwards.[3]

With such abundant evidence of ambitious colliery enterprises all through Lancashire, it does not seem too much to assume that, by 1700, the output of coal in each of the four main producing districts was somewhere between 25,000 and 50,000 tons per annum, and that the yearly output from all the fields of Lancashire and Cheshire may easily have exceeded 150,000 tons. Before Elizabeth's reign coal had been dug casually in a few scattered manors, and it is unlikely that the output in any of these exceeded a few score tons per annum. It is probably not an exaggeration to estimate the increase in production during the succeeding century and a half at not less than fifteen-fold.

(c) Shropshire Coalfields

A survey of coal production in seventeenth-century Shropshire can be confined to the single district of Coalbrookdale, for that district probably furnished fully 95 per cent of all the coal raised within the county during the Stuart period. Shropshire, of course, possesses four other more or less isolated fields,[4] and some coal was dug from seams in the Clee Hills and in Leebotwood; but in those sparsely settled uplands no considerable "land-sale" market existed, and the districts were too far from the Severn to make it practicable to carry coal overland to the river. Wyre Forest, though cut by the Severn between Bridgnorth and Bewdley, contained no seams near enough to the surface to be accessible to the early miners.[5] The "discovery" of coal in the Shrewsbury field was hailed with joy by the citizens of that town in 1571,[6] but the high hopes of an adequate local supply of fuel raised by this discovery do not appear to have been fulfilled. During the Civil

---

[1] E.g. at Bolton, Farnworth, Pilkington, Little and Middle Hulton, Tyldesley, Kearsley, Walkden, Clifton, and Worsley (Duchy of Lancs. Pleadings, 189/H/8. 189/H/15; V.C.H. Lancs., vol. ii, pp. 357, 359; Crofton, op. cit., pp. 60–1), The list of collieries here given is of course far from exhaustive. There were also mines to the north and east of Bolton, at Tottington, Bury, Royton, and Oldham, as well as at Rochdale (Duchy of Lancs. Pleadings, 154/L/7, 182/H/3; Duchy of Lancs., Deps., 75/11; Crofton, op. cit., pp. 56–7, 61).

[2] Star Chamb. Proc., James I, 106/7.

[3] Exch. Deps. by Com., 4 James II, East. 21, cited Crofton, op. cit., pp. 63–6; Folkard, Industries of Wigan, 1889, pp. 11–13, cited ibid., pp. 49–53.

[4] Five, if we count the projection of the Denbighshire field into the northern part of the county.

[5] Little has been known of this field until recent times V.C.H. Worcestershire, vol. ii, p. 264).

[6] V.C.H. Shropshire, p. 450.

War, when the traffic up the Severn from Coalbrookdale was blockaded, the inhabitants of Shrewsbury, along with those of other neighbouring towns and villages, were reported to be " in great distress . . . for want of coles ".[1]

Coalbrookdale had a complete monopoly in supplying what became in the seventeenth century the principal river market for coal in the British Isles.[2] Natural conditions, combined with a skilful technique, enabled the miners to obtain a good quality of coal at a low cost. Not until the accession of James I, however, do we find any evidence of a considerable colliery on the Severn, comparable to Wollaton on the Trent. At that time mining enterprise in the manor of Broseley took on a new importance, and countless were the quarrels between competing adventurers.[3] Soon after, enterprise was pushed within the adjoining manor of Benthall. On the left bank at Madeley, Shirlett, Rowton, and adjacent places, much capital was sunk in new mines after the Civil War, and considerable supplies of coal must have been loaded into barges ; [4] but, apart from Madeley, which apparently ranked in 1661 in the same class as Broseley and Benthall,[5] none of these mines seems to have achieved as much success or prominence as those on the right bank. At the end of the seventeenth century nearly all the coal shipped up and down the river came from the three great collieries at Broseley, Benthall, and Barr,[6] which undoubtedly ranked among the principal mines in the whole country.[7] Their combined output may well have approached 100,000 tons per annum, and that of the district as a whole almost certainly exceeded 130,000 tons. The migration of Shropshire colliers to other fields during the eighteenth century [8] suggests that Coalbrookdale was probably more developed than other Midland mining districts during the seventeenth. The output of the Shropshire fields, taken together, had perhaps increased from about 6,000 tons before the accession of Elizabeth to about 150,000 tons at the end of the seventeenth century.

(d) Staffordshire Coalfields

Information concerning mining in Staffordshire before the middle of the sixteenth century is difficult to find, and the contrast between this paucity of records and the abundance of records during the Elizabethan age suggests that the output of coal increased rapidly in the second half of the sixteenth century. Coal from mines in the north, round Newcastle-under-Lyme, not only served the local

---

[1] *Ibid.*, p. 454. See below, vol. ii, pp. 288-9.
[2] See below, pp. 96-7.
[3] See, for instance, *Star Chamb. Proc.*, *James I*, 86/18, 294/25, 310/16. For an account of these quarrels, see below, pp. 308-9.
[4] *V.C.H. Shropshire*, pp. 455, 458-9.
[5] *Hist. MSS. Com.*, *15th Report*, appx., pt. x (*Report on the MSS. of the Shrewsbury Corporation*), p. 65.
[6] *Treasury Board Papers*, 34/51.
[7] See below, p. 360.
[8] T. S. Ashton, " The Coal-Miners of the Eighteenth Century," in *Econ. Hist. Supplement* of the *Economic Journal*, no. 3, pp. 329-30.

inhabitants, but was carried across the county line into Cheshire. Coal from mines all through the district lying between Cannock Chase and Birmingham was carried in some quantity overland into Worcestershire and Warwickshire ;[1] and some cannel from Beaudesert found its way, because of its special qualities,[2] as far east as the town of Leicester, in competition with coal from Warwickshire and Leicestershire itself.[3] We also have references early in the seventeenth century to small pits at Halesowen, to the west of Birmingham, in Worcestershire,[4] where the mines generally yielded a coal inferior to that found in Staffordshire itself.[5]

After the Restoration the industry expanded rapidly. The salt makers at Droitwich in Worcestershire, and at Nantwich, Middlewich, and Northwich in Cheshire, all took to burning Staffordshire coal.[6] The famous "Thick Coal" seam (reaching, according to Plot,[7] a breadth of fourteen yards at Sedgley), attracted many early mining adventurers to the neighbourhood of Dudley, Sedgley, and Wednesbury.[8] This district was at once the chief centre of population and of coal mining. All the townsmen of Dudley were either smiths or miners.[9] Dud Dudley, who, in advocating the further use of mineral fuel, was not interested in exaggerating the output from the neighbourhood, estimated the number of active collieries in south Staffordshire at twelve or fourteen, each with an output of from 2,000 to 5,000 tons per annum, besides twice as many collieries unable to work because of the lack of a market.[10] And he was writing before the salt makers at Droitwich had begun to use coal. In Cannock Chase, mining also was pushed at Apedale and in Beaudesert Park for a considerable "land-sale" market. The colliery at Beaudesert assumed a new importance when Lord Paget helped to finance it,[11] and another colliery at Talke yielded £400 a year clear profit in 1674 and 1675.[12] Still farther north we find an expansion of coal mining in the Cheadle district

---

[1] When officials wanted a supply of coal for Mary, Queen of Scots, at Tutbury, in 1584, it was to Staffordshire, not to Derby or Leicestershire, that they turned (Galloway, *op. cit.*, p. 116 ; *Hist. MSS. Com.*, *4th Report*, appx., p. 329). In spite of the failure to render navigable the river Stour, it seems that a little Staffordshire coal was brought to the Severn for shipment (*ibid.*, *5th Report*, appx., p. 160).

[2] For a discussion of the different grades of coal, and the purposes for which they were burned, see below, pp. 109 sqq. For the development of mining at Beaudesert in the late sixteenth century, when the manor was in the hands of the Crown, see *Exch. Q.R. Accounts*, 632/17 ; *Augm. Partics. for Leases*, 133/11.

[3] H. Stocks, *Records of the Borough of Leicester*, vol. iv, 1923, p. 241.

[4] *Exch. Deps. by Com.*, 5 James I, Hilary 17 (a part of which is quoted *V.C.H. Shropshire*, pp. 459–60).

[5] H. S. Jevons, *British Coal Trade*, p. 89.

[6] The consumption of coal in these salt-works at the end of the seventeenth century was probably nearly 20,000 tons. See below, p. 208.

[7] Plot, *Natural History of Staffordshire*, 1686, p. 127.

[8] Galloway, *Annals*, pp. 115–16, 191–2 ; *Chanc. Proc., Eliz.*, C.c. 6/40.

[9] T. Habington, *Survey of Worcestershire*, ed. J. Amphlett, 1895–9, vol. i, p. 195.

[10] *Mettallum Martis*, 1665, p. 35. Cf. Plot, *op. cit.*, p. 132, for a reference to pits at Amblecote, in Worcestershire, after the Restoration.

[11] *Ibid.*, pp. 125, 129–30.

[12] Galloway, *op. cit.*, p. 193.

as well as at Tunstall, Hanley, and Newcastle-under-Lyme,[1] where, as in south Staffordshire, the adoption of coal by the salt makers greatly increased the demand for the new fuel. It may be assumed that the annual output of coal in Staffordshire reached 100,000 or 150,000 tons at the end of the seventeenth century, possibly ten or fifteen times the output before Elizabeth's reign.

(e) Warwickshire Coalfield

Production in the small mining district between Tamworth and Coventry appears to have increased during the period from 1550 to 1700, at a rate not exceeded in any other coalfield. Before 1550 scarcely any coal had been mined in Warwickshire. The small quantity burned probably came almost entirely from Nuneaton, six miles north of Coventry, where, in 1544, " proffyts of collpitts . . . every year " were only £2 13s. 4d.[2] Crown commissioners appointed to survey the recently discovered mines at Bedworth in 1570 reported the coal there worth only 20s. a year if let, and a lease was actually made on those terms in 1574 to one Edmund Anderson, upon payment of a fine of £2.[3] About this time the Crown was making surveys farther to the south. In 1591 commissioners met with a mine of " sleck or basset ", i.e. small surface coal, and, as it had been found at the extreme south of the manor, north of which was the Bedworth mine, they concluded that " the myne or vaine " of coal must pass through the whole manor.[4] Coal had also been proved on the manor of Sowe, and at Foleshill, hard by Coventry, but hitherto diggings had been on a medieval scale.[5] During the last decade of Elizabeth's reign, adventurers from other counties, realizing the possibility of serving a large market to the south,[6] began to pour their capital into this district ; and, in place of twopenny leases, such as had been granted hitherto, we hear within ten years of debts running into thousands of pounds, contracted for sinking pits, draining seams, and purchasing mining equipment.[7] A period when the Crown officials beg for lessees is succeeded almost overnight by a period in which adventurers enter

---

[1] For the history of the coal industry in north Staffordshire during the whole period from 1550 to 1700, see Galloway, *History of Coal Mining*, pp. 68 sqq. ; *Duchy of Lancs. Pleadings*, 90/B/41, 144/B/11 ; *Duchy of Lancs. Rentals and Surveys*, 8/32 ; *Chanc. Proc., Eliz.*, U.u. 1/7 ; *Star Chamb. Proc., James I*, 92/6, 228/13 ; *Parl. Surveys, Staffordshire*, no. 38. Cf. Jars, *Voyages métallurgiques*, vol. i, pp. 253–4.

[2] *V.C.H. Warwickshire*, vol. ii, p. 220.

[3] *Ibid.* ; *Exch. Spec. Com.*, no. 2,326 ; *Augm. Partics. for Leases*, 151/47, 76. No coal had been worked at Bedworth before this date, to judge from Dugdale's reference to the village in Queen Elizabeth's time, " before the mines of coal were found " (*Antiquities of Warwickshire*, vol. i, p. 122).

[4] *Augm. Partics. for Leases*, 149/36.

[5] *Ibid.*, 149/18, 34, 50 ; 151/29.

[6] See below, pp. 101–2.

[7] *Exch. Spec. Com.*, no. 4689 (Inquisition as to the possessions of Sir Henry Beaumont, 20 James I). Cf. below, p. 355, and vol. ii, p. 15.

into a mad scramble for concessions, which are sold and resold at ever higher prices.[1]

The chief mines in the seventeenth century were at Bedworth and Griff, only a few miles north of the busy town of Coventry. Bedworth colliery, which is said to have been worth £300 per annum in clear profits as early as 1602, had already attained a yearly output of 20,000 tons before 1631.[2] Notwithstanding the activity of mining adventurers at Bedworth and at Griff, Coventry was nearly always short of coal. Diggings were pushed farther to the north and west, as there was a tendency to exhaust the readily accessible seams nearest the market.[3] By 1725 there are said to have been more than fifty collieries in this Midland field, most of them within sight of the traveller along Watling Street.[4] The output had perhaps increased from about 2,000 or 3,000 tons at the accession of Elizabeth to more than 70,000 tons at the end of the seventeenth century.

(f) Leicestershire Coalfields

Most of the townspeople of Leicestershire were driven to burn coal during the seventeenth century,[5] and several mines were worked in the field about Ashby-de-la-Zouch, but they were not sufficiently important to satisfy the thirst of local adventurers for investment,[6] nor to supply the county with all the fuel it needed. The inhabitants had to supplement the local stock of coal by overland importations from Warwickshire, Derbyshire, Nottinghamshire, and even Staffordshire. Leicestershire cannot, therefore, be counted among the important Midland fields. In Oakthorpe manor " certen mynes of cooles called see cooles " were found about 1575.[7] In the eighties of the sixteenth century, Camden commented upon the coal pits at Coleorton which " supply the neighbouring country, all about, with firing ".[8] This colliery at the original seat of the Beaumont family continued, no doubt, to be worked throughout the seventeenth century. In 1611 one of the Beaumonts took a lease of an ambitious enterprise at Measham, for which he paid the very high rent of

---

[1] *Chanc. Proc., Eliz.*, F.f. 5/31 ; *ibid., James I*, F. 4/53 ; *V.C.H. Warwick*, vol. ii, pp. 220-1 ; *Hist. MSS. Com., Report on the MSS. of Lord Middleton*, p. 320. Cf. below, vol. ii, pp. 13, 15.

[2] *Chanc. Proc., James I*, F. 4/53 ; *V.C.H. Warwick*, vol. ii, pp. 221-2. Cf. below, p. 360.

[3] *P.C.R.*, vol. xlix, p. 355 ; *S.P.D., Charles I*, vol. cccclxxx, no. 50 ; *V.C.H. Warwick*, vol. ii, pp. 223-4 ; *Chanc. Proc.*, Series II, 425/47 ; *Exch. Deps. by Com.*, 36 Charles II, Mich. 43.

[4] Galloway, *Annals of Coal Mining*, p. 335. It seems probable that the reference is to pits, rather than collieries. Cf. below, vol. ii, p. 139.

[5] *Ibid.*, p. 200 ; *Records of the Borough of Leicester*, ed. M. Bateson, vol. iii, pp. 128, 149, 152-3, 160, 172-3, 178-9, 201, 278, 414-15, 441, 448 ; *ibid.*, vol. iv, ed. H. Stocks, pp. 2, 168, 241-2 ; *A New Present State of England*, vol. i, p. 145.

[6] Cf. below, vol. ii, pp. 14 sqq.

[7] *Duchy of Lancs. Pleadings*, 108/W/12 (Humphrey Wolverton *v.* George Hastings and Edward Dykes, February 9th, 1579).

[8] *Britannia*, 1772 ed., vol. i, p. 414. The mines at Coleorton had apparently been worked much earlier.

£500 a year.[1] He found at the pithead about 3,000 tons remaining from the operations of his predecessor ; and it is probable that the annual output had already reached several thousands of tons. But this enterprise failed, and we have no evidence of any other large collieries in seventeenth-century Leicestershire. The mine at Oakthorpe, which had been worked in Elizabeth's reign in conjunction with another at Swadlincote,[2] was found, when it was surveyed by parliamentary commissioners in 1652, not to have produced any coal since the " beginning of the late wars ".[3]

Some " pitt coles " were apparently dug in eastern Leicestershire and Rutlandshire. In 1611 we hear of a mine in " Sturley Park ", where the Earl of Rutland's " yearly proportion " of coal was about 200 tons.[4] Twenty years earlier, witnesses informed Crown commissioners that, to their own knowledge, " there are under the earth [in Leighfield Forest, in Rutlandshire] certain mines of coal." These were let in 1590 at a rent of £5 per annum,[5] but little is heard of them in the seventeenth century. They could not have been of much importance, for the Earl of Rutland got some of the coal for his castle at Belvoir by land carriage from Nottinghamshire.[6] It is improbable that the output of coal in Leicestershire, which had possibly amounted to 1,000 to 2,000 tons per annum at the accession of Elizabeth, much exceeded 10,000 tons at the end of the seventeenth century.

(v) *Other British Coalfields*

At the end of the seventeenth century the entire coal production of all the remaining mines of the British Isles probably did not amount to much more than a sixth of the production of Durham and Northumberland. But in four of the small scattered fields, in Cumberland, Kingswood Chase, Somerset, and the Forest of Dean, mining was carried on with some intensity, and an examination of the records shows that in all of them there was taking place between 1550 and 1700 a development parallel to that in the greater coalfields ; a development, in short, which adds colour to the general picture of the expanding British coal industry.

(a) *Cumberland*

In western Cumberland the development of mining near the coast was so rapid during the seventeenth century that Whitehaven appears

[1] *S.P.D., James I*, vol. clvii, no. 51 (Statement of part of the losses sustained by the then Lord Beaumont on Measham coal mine " collected out of the Coale pitt bookes "). Cf. below, vol. ii, pp. 16–17.
[2] *Chanc. Proc., Eliz.*, O.o. 1/40 (John Osbaston v. John Smyth, February 10th, 1587). Sir Francis Willoughby had a connection with coal sales at " Okemore " (Oakthorpe) from 1605 to 1608 (*Hist. MSS. Com., Report on the MSS. of Lord Middleton*, p. 497).
[3] *Parl. Surveys, Derby.*, no. 26.
[4] *Hist. MSS. Com., Report on the MSS. of the Duke of Rutland*, vol. iv, pp. 480–4 ; see also vol. i, p. 415.　　　　[5] *Augm. Partics. for Leases*, 121/18.
[6] *Hist. MSS. Com., Report on the MSS. of the Duke of Rutland*, vol. iv, p. 544.

to have attained a position as a coal-shipping port inferior only to that of Newcastle, Sunderland, and, perhaps, Swansea. Generally it has been supposed that there was little or no trade from the coast of Cumberland before 1640, but it is not strictly true that no coal was shipped before that time. Already in the year 1605, 2,641 tons were carried to Scotland and Ireland, and 92 tons to the Isle of Man. In 1616, approximately 3,000 tons were carried to Ireland, 600 tons to Scotland, and 150 tons to the Isle of Man.[1]

These shipments have been overlooked, because it has been assumed that the earliest attempt to carry coal from Cumberland must have been made at Whitehaven, whence, it is true, little or no coal found its way overseas until the Civil War. The earlier shipments came almost entirely from Workington.[2] Seams, which cropped out at various spots along the sea-coast between Barrowmouth and Maryport,[3] were wrought in a number of independent freeholds in the manor of Clifton behind Workington, " to which place," according to a petition sent to the Privy Council in 1635 by Dorothy Salkeith, one of the freeholders, " coles there gotten have bene usually carryed, and from thence transported . . . for Ireland, [and] diverse parts both of England and Scotlande." [4]

This Workington trade may have had a very early origin. An ancient token, such as was given in lieu of money to the carters who brought coal to the wharves, has been found bearing a date difficult to decipher, but which is either " 1596 " or " 1526 ".[5] However far back it may be possible to trace shipments of coal, there is no doubt that, at the end of the sixteenth century, these shipments were greatly stimulated by the new demand in Dublin.[6] During the period from 1600 to 1640 the coal industry of Workington does not appear to have been so successful in meeting this demand as that of Flintshire and Lancashire.[7] Mutual jealousy between the numerous freeholders who worked pits at Clifton, and the excessive rents demanded by the lord of the manor, Sir Patricius Curwen, for wayleave through his lands,[8] prevented systematic exploitation. Sir John Lowther, who succeeded his father in 1644, as owner of the manor of Whitehaven, explained that " the county adjacent afforded coals sufficient for a staple export, but a great part of them were in the hands of small freeholders, and could not be wrought without great and expensive levels, which must go through several people's lands and draining all upon the Rise would enable such as have none of the charge to undersell and ruin those who did, so that the working of them under these

---

[1] See Appendix D (iii).

[2] I. Fletcher, " The Archæology of the West Cumberland Coal Field " in *Trans. Cumberland and Westmorland Antiq. Soc.*, vol. iii, 1878, pp. 296–7; *Exch. Spec. Com.*, no. 3678 (Inquisition as to measuring of coal shipped from the port of Carlisle, 1616).

[3] *V.C.H. Cumberland*, vol. ii, p. 348.

[4] *P.C.R.*, vol. xlv, p. 248.

[5] *Trans. Cumberland and Westmorland Antiq. Soc.*, vol. xv, 1899, pp. 393, 409.

[6] See below, pp. 90–1.

[7] See above, pp. 55–6, 61.

[8] *P.C.R.*, vol. xlv, pp. 248, 487–8.

circumstances was impracticable, and they were lost as well to the owners as to the county ".[1]

As a result of Lowther's own skill in getting control of the important coal-bearing lands, and in applying new and more efficient mining methods in his collieries round Whitehaven, Cumberland coal gained a virtual monopoly in the Dublin market.[2] Lowther apparently addressed himself to the problem of mining some time during the Interregnum. He remained the dominating force in the Cumberland coal industry for over half a century, until his death in 1706. From his father, Sir Christopher Lowther, he inherited the lands of the dissolved monastery of St. Bees at White-haven, and these served as a nucleus for his operations.[3] Wherever possible, he acquired by purchase coal-bearing property bordering on his original concession, and, by grant from the Crown in 1669,[4] he secured 150 acres between the high and low water mark near White-haven, a grant which enabled him to control the loading of ships. Further land was conceded him by the Crown in 1678, at which time he was already working several collieries. In 1675 he acquired a lease from Henry Fletcher of the Fearon mines in Distington parish, four miles north of Whitehaven, a property which was destined to remain permanently in the Lowther family.[5]

By this time, Lowther had succeeded in underselling nearly all his competitors in the Dublin market.[6] In the year 1688, when the Irish trade was doubly depressed, owing to restrictions imposed on trade by the Privy Council, and to the threat of a rebellion, exports from Whitehaven to Irish ports appear to have amounted to 25,000 tons.[7] It is probable that Walter Harris, who estimated the annual exports from Whitehaven at 38,400 tons in 1685, did not greatly exaggerate.[8] The output at Howgill colliery alone amounted to 15,196 tons in 1695, and to 25,703 tons in 1700.[9] We hear of eight other local mines worked by Lowther between 1668 and 1700.[10] But they did not

---

[1] Lincoln's Inn Broadsheet, "Case of Sir John Lowther, Bart., and the inhabitants of . . . Whitehaven" (probably in the year 1705), printed in Fletcher, *op. cit.*, pp. 271–4.

[2] Cf. below, pp. 109–10.

[3] Fletcher, *op. cit.*, pp. 270–2, 274–5 ; *V.C.H. Cumberland*, vol. ii, p. 359. On the early history of mining within the manor of St. Bees, see *Exch. Deps. by Com.*, 44 Eliz., East. 6 ; and below, pp. 134, 143–4.

[4] *V.C.H. Cumberland*, vol. ii, p. 359.

[5] *Ibid.*, vol. ii, p. 370.

[6] *P.C.R.*, vol. lxiv, p. 480 (Privy Council to the Earl of Essex, Lord Lieutenant of Ireland, August 4th, 1675) ; and see below, p. 110.

[7] About 12,000 tons were exported in the half-year ending June 25th, 1688 (see Appendix D (iii)).

[8] W. Harris, *Remarks on the Affairs and Trade of England and Ireland*, 1691, p. 19. (This estimate included the small trade from Workington.) A rapid expansion in the Whitehaven trade with Dublin began after 1720. Up to that time the annual shipments from Whitehaven rarely, if ever, exceeded 40,000 tons. [9] *V.C.H. Cumberland*, vol. ii, p. 312.

[10] A list of these other collieries, with the years in which it is known that they were producing coal, is as follows :—

Three Quarters Band	.	.	1675	
Greenbank	.	.	.	1675–95
Hensingham	.	.	1680	
Corbuckle	.	.	.	1680

all operate at once, and probably none of them was as large as Howgill.[1] Presumably the annual output of coal in the immediate neighbourhood of Whitehaven stood at between 50,000 and 90,000 tons at the end of the seventeenth century.

The production of coal in other parts of Cumberland was not large. At Workington, mining suffered from the competition of Whitehaven. The port was described in 1676 as " a fair haven, but not so much now frequented with ships, the colliery being decayed thereabout ".[2] Not until after 1700 did the local landlords succeed in exploiting the seams with enough efficiency to make Workington into a coal port of some importance.

Between Whitehaven and Workington, the Fletcher family of Moresby, lords of the manor of Distington, had attempted to develop their mineral resources as early as 1621, but their projects were hindered by the rights of freeholders,[3] just as at Clifton, near Workington, the projects of the freeholders had been hindered by the lord of the manor, Sir Patricius Curwen. By 1680, however, the Fletchers had achieved some success with their colliery, and shipments from Parton harbour came into competition with Whitehaven coal in Dublin.[4]

North of Workington, some coal was mined for land-sale in the country towards Carlisle.[5] Another mining district, about which very little is known, existed south of Penrith in western Cumberland, and over the county boundary in Westmorland. Small coal diggings are heard of in the Lake District in medieval times,[6] and during Elizabeth's reign pits were dug in connection with the new copper works set up by the Society of Mines Royal near Keswick.[7] South of the Cumberland mountains, scattered pits in the valley of the Lune, north of the Lancashire county line, probably contributed a few hundred tons to the yearly output from north-western England.[8]

(b) *The Kingswood Chase Field*

In the mining district that extends from Brislington parish, on the outskirts of Bristol, northwards along the eastern side of the

Whingill	.	.	.	1680
Priestgill	.	.	.	1682
Bransty	.	.	.	1668–96
Lattera	.	.	.	1693–1700

(*V.C.H. Cumberland*, vol. ii, pp. 352–3, 359–61, 371.)

[1] In 1695 the output at Greenbank colliery was only 2,321 tons, at Lattera 1,387 tons (*ibid.*, vol. ii, p. 361). [2] Fletcher, *op. cit.*, p. 297.

[3] *Chanc. Proc., James I*, C. 24/5 (Oswald Crofthatt and five other husbandmen, freeholders in Distington, *v*. Henry Fletcher, November 10th, 1621).

[4] *V.C.H. Cumberland*, vol. ii, p. 360.

[5] At Dearham, Flimby, Broughton Moor, Ewarigg, and Ellenborough (Fletcher, *op. cit.*, p. 301).

[6] In *Calendar of Inquisitions, Henry VII*, vol. i, p. 157, there is a reference to " Coal mines in Tynyefell, valueless on account of the Scots ".

[7] *V.C.H. Cumberland*, vol. ii, p. 359. See also *ibid.*, p. 380 ; *S.P.D., Eliz.*, vol. xlviii, no. 13 ; *Hist. MSS. Com., Report on MSS. of the Marquis of Salisbury*, vol. x, p. 217.

[8] See reference in 1661 to a coal mine in the manor of Casterton (*Calendar of Treasury Bks.*, vol. i, p. 119).

Cotswolds, colliery enterprises existed mainly for the purpose of supplying Bristol with fuel. With the general adoption of coal fires by the citizens, during the reigns of Elizabeth and James I,[1] Kingswood Chase became the "officina carbonum" of the town.[2] After 1600 the outcrops all the way from Brislington to Westerleigh were ruthlessly exploited. New pits, too numerous to count, were dug, before the old ones had been filled up.[3] Travellers to Bristol from the north and west comment frequently upon the "great store of coals"; [4] and by 1623 the coal under one square mile in Mangotsfield could be let for £300 per annum.[5] So intensively was mining enterprise carried on during the first half of the seventeenth century, that commissioners who surveyed Kingswood Chase in 1652 were pessimistic as to the future of the mining industry there.[6] But their fears were groundless. There is abundant evidence of a revival in mining activity after 1660 ; and in 1675 another group of commissioners found 156 "cole pitts that are wrought and left open " in the liberty of the Earl of Rochester alone.[7] There must have been hundreds of others. The countryside was covered with cottages built for the coal-miners and for the drivers who brought horses and cattle laden with the fuel to the town of Bristol.

(c) *The Somerset Field*

Except in its western half, which was supplied with fuel from Wales, Somersetshire was abundantly stocked with coal of its own, easily accessible to early mining operations. Coal was sold in the county in the fourteenth century, and pits were worked during the Middle Ages on the manor of Stratton-on-the-Fosse, and possibly elsewhere, though casually and intermittently. When a lease of the demesne lands of Stratton came, in 1544, into the hands of John Horner, whose descendants were active in mining enterprise there, the coal pits were of secondary importance.[8] In 1586, Camden referred to them as of use principally to smiths.[9]

Soon after this time began a scramble for concessions, accompanied by the opening of new enterprises and a rapid increase in output. Before the end of the reign of James I, a fine of £300 was paid by Edward and Thomas Leviband, for a lease of a colliery called

---

[1] See below, p. 108.
[2] *Gents Magazine*, vol. lix, 1789, p. 1098.
[3] At Stapleton, Brislington, Mangotsfield, Hanham, St. Phillip, Bitton, Barton Woods, Seston, Frameton, Cottred, Yate, Westerleigh, Rangeworthy, etc. (*Chanc. Proc., James I*, C. 9/69, C. 13/25 ; *Chanc. Proc.*, Series II, 371/4 ; *Chanc. Deps.*, P. 7/T (Player v. Trotman, Oct., 1623), P. 18/H (Player v. Hobbes, Oct., 1630) ; *Exch. Decrees*, Series IV, vol. i, ff. 190, 223 ; *S.P.D., James I*, vol. liv, no. 8.
[4] See e.g. G[uy] M[eige], *The New State of England*, 1691, vol. i, p. 88.
[5] *Chanc. Deps.*, P. 7/T.
[6] *Parl. Surveys, Gloucester.*, no. 12.
[7] *Early Treasury Warrants*, vol. 38, ff. 117–22 ; *Exch. Deps. by Com.*, 27 Charles II, Mich. 29. See also *V.C.H. Gloucester*, vol. ii, p. 236 ; Thos. Baskerville, *Journeys in England*, in *Hist. MSS. Com., Report on the MSS. of the Duke of Portland*, vol. ii, p. 298.
[8] *Cal. of Letters and Papers of Henry VIII*, vol. xix, pt. i, no. 1035 (42).
[9] *Britannia*, 1772 ed., p. 187.

" Les Holmes " in Stratton.[1]   Another colliery in the same manor—
the " Barrowe "—which was being worked at this time, was represented
to parliamentary commissioners, in 1651, by " persons well experienced
in coleworks ", as " well nigh spent ", and " will be wrought out in
less than ten years ".[2]   In spite of this pessimistic report, mining was
carried on even more intensively at Stratton after the Restoration.[3]

There is no mention of mines in either Midsomer Norton or
Farrington Gurney in a survey made of these manors early in
James I's reign ;[4] but a partnership set up a colliery in Midsomer
Norton about 1615,[5] and already in 1610 the output at two pits,
" newly entered into," near the parish of Clutton was 600 horse-
loads (or about 70 tons) a week.[6]   Soon after the Restoration we hear
of a coal mine in Farrington Gurney, while in 1678–9 the seams of
Mendip Forest between Binegar and Mells were worked not only at
Stratton, but in the manors of Ashwick, Kilmersdon, Holcombe,
Babington, Luckington, and Mells.[7]   It was at about this time that
a certain John Beaumont, who lived near Stoke Lane, counted " six
distinct coalworks " within five miles to the north of his house.[8]
Further north, round Clutton and Timsbury, was another cluster of
collieries worked mainly to supply Bath.   Judging from the evidence
of the increase in mining enterprise, both in Somerset and in Kingswood
Chase, it seems probable that the annual output in the two districts,
at the end of the seventeenth century, was in the neighbourhood of
100,000 tons.

(d) The Forest of Dean Field

Before the seventeenth century the district west of the Wye river
had been known above all for its iron mines, coal being won incidentally
and in small quantities, where it was found in connection with seams
of ore.   After 1600 coal mining was pursued for its own sake as a
separate industry, destined soon to supplant iron as the chief
business of the Forest.   The first proposal to farm the coal mines
as a separate venture was made in 1635 by Edward Terringham.[9]
At the same time Sir John Winter had various coal pits at
a place called " Norchards ", in his own manor of Lydney, at the

---

[1] *Land Rev. Misc. Bks.*, vol. ccvii, ff. 9–10 (Survey of the manor of Stratton-on-the-Fosse).

[2] *Parl. Surveys, Somerset,* no. 39.

[3] *Cal. Treas. Bks.,* vol. i, pp. 37, 117 ; vol. ii, p. 385 ; vol. iii, pp. 171, 254, 521 ; vol. iv, pp. 743, 808 ; vol. v, 89, 90, 92, 275, 499, 810, 1166 ; vol. vi, p. 746 ; *Early Treasury Warrants,* vol. 37, f. 256 ; *Exch. Deps. by Com.,* 30 Charles II, Mich. 11, summarized *V.C.H. Somerset,* vol. ii, pp. 380–1 ; *Land Rev. Misc. Bks.,* vol. 207, ff. 138–46.

[4] *Land Rev. Misc. Bks.,* vol. ccvii, ff., 11–12, 13–14.

[5] *Chanc. Proc., James I,* B. 20/72.

[6] *Hist. MSS. Com., 12th Report,* appx. i, p. 71.

[7] *Early Treasury Warrants,* vol. 37, ff. 225–7, 229 ; *Cal. Treas. Bks.,* vol. ii, p. 586 ; *Exch. Deps. by Com.,* 30 Charles II, Mich. 11, as summarized *V.C.H. Somerset,* vol. ii, p. 381.

[8] *Royal Soc., Philos. Transactions,* vol. xiv, 1679–81, pp. 6–8 (" Observations of the Fiery Damps in Mines ").

[9] *S.P.D., Charles I,* vol. ccciii, no. 61.

southern extremity of the Forest.[1] Neither appears to have been
very successful. Terringham was not able permanently to sustain his
position as farmer of Crown mines against the immemorial rights of
the free miners.[2] But the desire to obtain concessions calls attention
to the growing importance of coal in the economy of the Forest. In
1675, a " coale-miner ", named Henry Mathon, who had worked for
Terringham until that adventurer left the Forest in 1640, estimated
that Terringham's mines " might now be worth, if enjoyed without interruption, £500 per annum, besides all charges of managing
the same ".[3] Coal pits were dug throughout the Forest, and the
area of the minimum concession was extended in 1678 and again in
1692, 1728, and 1754, a fact which shows that mining operations were
developing on a larger scale.[4]

(e) The Devonshire Field

Anyone not familiar with the lignite deposits of south-western
England would be puzzled by the note in a document of about 1610, to
the effect that the inhabitants of certain parts of Devon and Cornwall
are sometimes " driven to goe 20 miles to the pits for coales at deere
rates ".[5] He might think that " coales " was written by mistake
instead of some other mineral, perhaps tin, or he might assume that
the writer had under-estimated the distance from Devon to the Mendip
Hills. The explanation is that lignite was being dug near Bovey Tracey
in Devonshire. In fact there is a record of payments for digging
" coal " in Dartmoor Forest as early as 1568.[6] But these pits, like
the Scottish mines at Brora and Ayton, are of no serious interest
to the student of the early coal industry, for their output was negligible.

(f) The Irish Fields

From the Irish point of view, which was rarely considered, it
was desirable to become independent of British coal supplies, and,
in spite of opposition from Englishmen interested in their own west-
coast fields,[7] much prospecting for mineral fuel was undertaken in the
seventeenth century in Ireland.[8] Before 1689, a colliery company
at Castlecomer in Kilkenny County had succeeded in extracting coal,
which was carried down the river Nore to be sold at St. Molin and
Ross for transport to Dublin.[9] Two decades later, we hear of coal pits

[1] *Exch. Deps. by Com.*, 13 Charles I, Mich. 42 ; 13 and 14 Charles I,
Hil. 16.
[2] See below, pp. 279–80.
[3] *Exch. Deps. by Com.*, 27 Charles II, Mich. 28 ; 35 Charles II, Mich. 40.
[4] Galloway, *Annals*, pp. 209, 340.
[5] *Harleian MSS.*, 6838, no. 34 (Exceptions against the bill for sands on
the sea coasts of Devon and Cornwall).
[6] *Rentals and Surveys*, General Series, Portfolio 6, no. 57.
[7] See Dean Swift, *A Proposal for the Universal Use of Irish Manufacture*,
1720, p. 10.
[8] *Sloane MSS.*, 4812, f. 16 ; R. R. Steele, *Bibliography of Tudor and Stuart
Proclamations*, vol. ii, no. 1440.
[9] See Proclamation of Nov. 29th, 1689 (Brit. Mus. C. 21, f. 12 (82)).

in Munster and Limerick.[1]   In the twenties of the eighteenth century, projectors were advancing schemes for raising in Dublin a joint stock with which to dig a canal for conveying fuel from Lough Neagh, by way of Newry and Carlingford Lough, to Dublin. [2] The great distance of seams from the coast prevented the success of these enterprises in both north and south.   Coal could be mined almost as cheaply at Kilkenny as at Whitehaven, but in Ireland the journey from the pithead to Waterford harbour more than tripled the price. We have found no evidence that Irish coal was shipped by sea until the middle of the eighteenth century, when a boat-load reached Havre.[3] As late as 1800, the total output of all Irish mines did not exceed 20,000 tons a year.[4]

Apart, then, from a few districts, none of which have ever contributed considerable supplies of coal, the history of production in every mining field during the century and a half following the accession of Elizabeth is one of revolutionary growth. From all the coalfields comes overwhelming evidence between 1550 and 1610 of a feverish interest in the discovery and development of mineral resources ; evidence so abundant as to make such previous records as can be found of mining enterprise in the same districts shrink to insignificance. For every lawsuit dealing with coal mines, found in the proceedings of the Star Chamber, the Court of Requests, the Court of Chancery, and the Court of the Duchy Chamber of Lancaster during the period of sixty-two years covered by the reigns of Henry VII and Henry VIII, there are at least forty suits during the sixty years following the accession of Elizabeth.   Everywhere surveys are made to determine the presence of coal seams.[5] Crown commissioners go from manor to manor consulting with such men of the district as are " skilled and expert . . . in the trades of getting Coales ", as to the possibility of finding mineral fuel, of inducing persons to search for it, or as to the terms on which it may be leased, if already found. A yeoman, tired of raising crops for small profit, a nobleman or gentleman, sometimes from another county, finding his revenues diminished by the falling value of money, perhaps combines his resources with others like himself, or borrows capital from town merchants, in order to enable him to sink pits in the hope of finding fuel to sell to the country people and the villagers round about.[6] If these efforts are frustrated again and

---

[1] *Sloane MSS.*, 4812, f. 16.

[2] *Additional MSS.*, 21134, f. 32. See below, vol. ii, p. 47.

[3] Archives nationales, F.[12]95, f. 427 (*Registre du conseil de commerce*). An examination of similar papers for the years 1700 to 1748 (*ibid.*, F.[12]51–94) reveals no imports of Irish coal at Havre.

[4] 58,000 barrels of $6\frac{1}{2}$ cwt. (M. Dunn, *View of the Coal Trade*, p. 143).

[5] *Augm. Misc. Bks.*, vol. cccxxx, f. 67 ; *Duchy of Lancs. Pleadings*, 144/A/9 ; *Exch. Spec. Com.*, nos. 1208, 2592, 2621, 2655, 2759, 2777 ; *Exch. Deps. by Com.*, 22 and 23 Eliz., Mich. 30 ; *Duchy of Lancs. Deps.*, 61/R/2, 73/R/15 ; *Duchy of Lancs. Spec. Com.*, nos. 317, 418, 433, 648 ; *Augm. Partics. for Leases*, 25/25 ; 112/12, 15, 27, 30 ; 115/68 ; 116/28, 52 ; 159/12 ; 171/46 ; 178/1 ; *Star Chamb. Proc., James I*, 154/12.

[6] *Exch. Deps. by Com.*, 11 James I, Trinity 6 ; *Star Chamb. Proc., James I*, 4/3, 309/46, 310/33 ; *Chanc. Proc., Eliz.*, L. 5/42 ; *ibid., James I*, J. 3/41 ;

again by the discovery of seams drowned out by water, eventually a more adventurous soul is induced to risk his or some creditor's savings in setting up a crude sort of draining apparatus. A period in which coal was found by accident, when the plough of the husbandman or the spade of the quarrier struck against a black rock, is now succeeded by a period in which conscious effort is directed to the search for mineral fuel. A period in which some such unsought-for outcrop was occasionally dug by the farmer to supply his own hearth is succeeded by a period in which coal-getting becomes a business in itself, employing a considerable population in England, Wales, and Scotland.[1]

Everywhere this change in the importance of coal mining occurs in the century and a half following the accession of Elizabeth. In the Tyne valley the expansion gains its full momentum during her reign. In the Midlands, eastern Scotland, and most other colliery districts, the period of maximum rapidity of growth appears to be the reign of James I. In a few districts on the west coast—South Wales, Cumberland, Lanarkshire, and Ayrshire—that period does not seem to be reached until the Restoration. But before 1690 there was hardly an important coalfield in which production had not multiplied ten-fold in the course of a century. It is obvious that this remarkable expansion of mining could have been brought about only by a remarkable development in the demand for coal. We turn therefore from the subject of production to that of markets.

Duchy of Lancs. Pleadings, 100/G/4, 102/T/10, 104/H/8, 109/A/5, 116/F/2, 131/T/4, 146/F/20, 153/F/3, 162/A/35, 171/A/12, 185/T/12, 204/O/1, 212/G/6 ; _Court of Requests Proc._, 28/56 ; Crofton, _Lancashire and Cheshire Coal Mining Records_, pp. 40, 42, 56–7 ; _V.C.H. Lancashire_, vol. ii, p. 357.
[1] See below, vol. ii, pp. 135–43.

CHAPTER II

THE PRINCIPAL MARKETS FOR BRITISH COAL

COAL is, of all commodities in general use, the one least valuable in proportion to its bulk. Costs of transport are inevitably a primary problem in its marketing. There is not, therefore, and can never be, a world market for coal in the sense that there is a world market for wheat and even for automobiles. But, while the separate markets for coal to-day are relatively numerous if they are compared with those for most other commodities, they are relatively few if they are compared with the separate markets which existed in the western Europe of the sixteenth and seventeenth centuries, when the expense of carriage was much greater, and when a considerable trade in goods of any kind between different countries, and even between different districts of the same country, had only begun to develop. Within Great Britain, we can count by the dozen the individual consuming areas, each dependent upon the output of a different group of mines ; and, if we include the Continent, we can count these areas by the score.

As is the case with all bulky goods, the transport of coal by water is vastly cheaper than by land. Water carriage has made it possible for the produce of the Welsh, Scottish, and Durham-Northumberland mines to compete with the produce of all other mines along the coasts of the North Sea, the Baltic, the Mediterranean, the eastern and even the south-western Atlantic. Water carriage has made it possible for the coal of West Virginia to compete with that of Utah and Washington in southern California.

Carriage in steam boats is, of course, much less expensive than in the sailing colliers which alone were available to the early traders, but the use of steam has not resulted in as large a reduction of the cost of sea, as of land, transport. In other words, the relative advantages of water over land carriage were greater in the sixteenth and seventeenth centuries than at the present time. Vessels laden with Newcastle coal sailed north as far as Inverness in Scotland, east as far as Stockholm and Danzig, south and west as far as Southampton and the Channel Islands. Welsh coal was sometimes freighted to Amsterdam, to Tangier, and, in small quantities, even to the American colonies. But the output of mines without access to water transport was seldom sold at a distance of more than fifteen miles from the pits.

The growth of the principal markets for water-borne coal, in the order of their importance, is shown by the following table [1] :—

---

[1] For a discussion of the sources from which this table has been derived, see Appendix B.

TABLE VI

Estimated Annual Trade in Water-borne Coal (in Tons).

To		1541–50	1681–90
Shipped by Sea	The East and South-East Coasts of England	22,000	690,000
	Foreign Countries and the Colonies	12,000	150,000
	The West and South-West Coasts of England (including Wales)	4,000	80,000
	Ireland		60,000
	Scotland	3,000	50,000
Shipped by River	River Valleys	10,000	250,000
	Total	51,000	1,280,000

(i) *Markets Reached by Water Carriage*

(a) *The English East and South-East-Coast Market*

At the end of the seventeenth century nearly 1,300,000 tons, or more than forty per cent of the estimated British production of 3,000,000 tons per annum, appears to have been carried by water ; and more than half the water-borne coal appears to have been consumed in eastern and south-eastern England. As is indicated on the map,[1] we include under this head the area from the mouth of the Tees along the shore line, within about twelve miles of the sea and of navigable rivers, like the Thames and the Ouse, as far south and west as Dorset.[2] Although some of the coal shipped to the south-east was finally distributed to inland towns and even to inland hamlets, nearly all the cargoes were first unloaded at a few principal ports, such as Hull, King's Lynn (which was described in 1653 as " a magazine of coal and salt for nine counties adjacent "),[3] Yarmouth, Ipswich, Colchester, and, above all, London.[4] The greater part of the London imports was sold within the City and suburbs ; but some quantity was emptied from the colliers into lighters, to be ferried to various points up-stream. So prodigious a number of sailing vessels, with their hulls full of coal, crowded the Thames during the summer months, and so spectacular became the trade between Newcastle and London, that many a citizen who drew his information from the gossip of the inns and taverns, where the shady manœuvres of merchants to raise the price of coal formed a favourite theme, assumed that the only reason why masters of colliers ever put into port at any point between the Tyne and the Thames was to forestall the city market.[5] It was

[1] Facing p. 19.

[2] I have also included in this east-coast market the town of Hartlepool, where several thousand tons of coal is said to have been burned annually in the alum works before the Civil War (*Cal. S.P.D.*, 1644–5, pp. 98–9 ; and cf. below, p. 209).

[3] *Cal. S.P.D.*, 1653–4, p. 476.

[4] See Appendices D (iv), (v), and G ; R. B. Westerfield, " Middlemen in English Business," in *Trans. Connecticut Acad. Arts and Sciences*, vol. xix, 1915, p. 221.

[5] Cf. below, vol. ii, pp. 104–5.

scarcely less common to under-estimate the trade of " other English " importing towns, than to over-estimate the trade of London.[1] There was a growing market for coal in several counties on the route between Newcastle and the capital. In 1696 officials in Norfolk declared that the inhabitants of that county used very little of any other fuel.[2] Imports at King's Lynn increased from 3,427 tons in 1575 to more than 22,800 tons in 1651 ; imports at Yarmouth amounted to more than 22,477 tons in 1682 ; and other coastal towns imported substantial quantities.[3] But these figures are small enough compared with those for London. The consumption of water-borne coal increased much more rapidly in the Thames valley than in other parts of the east-coast market, as is shown by the following table, which is based on statistics contained in the port books and in the records of the City of London.[4]

TABLE VII

Estimated Annual Consumption of Water-borne Coal in the East and South-East Market.

	In the Thames Valley.		In other parts.	
	Tons.	% of Total.	Tons.	% of Total.
1575–80	12,000	29	30,000	71
1651–60	275,000	65	150,000	35
1685–1699	455,000	66	235,000	34

These two sets of figures are not strictly comparable. While sea-borne coal possessed a virtual monopoly of the entire market in the Thames valley and to the south,[5] north of the Thames it came into increasing competition with coal from Nottinghamshire carried in barges down the Trent and the Humber as far east as Hull, with coal from the West Riding hauled sometimes as far as York, and even with coal from Warwickshire carried overland into Northamptonshire, Bedfordshire, and Huntingdonshire. Thus the area served by imports at towns north of Yarmouth shrank continually throughout the seventeenth and eighteenth centuries, while the area served by imports at London remained virtually intact.[6]

At the same time population in the Thames valley increased far more rapidly than in any other part of south-eastern England.[7]

[1] In 1628, for instance, the " London vent " was said to be about 360,000 tons per annum, and the imports at all other towns on the east and south coasts only 60,000 tons (*Stowe MSS.*, 326, f. 20). To get something approaching the truth we must halve the first figure and double the second.

[2] E. R. Turner, " The English Coal Industry in the Seventeenth and Eighteenth Centuries," in *Amer. Hist. Rev.*, vol. xxvii, 1921, p. 6.

[3] See Appendices D (v) and G.

[4] For a discussion of this Table, see Appendix B.

[5] Occasional cartloads of Midland coal, presumably from Staffordshire (cf. below, p. 102 n.), reached Oxford, and may have entered into competition with Newcastle coal brought up the Thames ; but the costs of bringing coal to Oxford, whether by land or water, were so high, that the amount consumed in and about the town, at the end of the seventeenth century, was not enough to concern us here. As late as the time of Adam Smith, the consumption of coal in Oxfordshire was still small, owing to the great price (*Wealth of Nations*, 1920 ed., bk. 1, p. 167).

[6] The question of competition for markets between coal-producing areas is dealt with more fully below, pp. 109 sqq.

[7] For the growth in the population of London, see below, p. 163 n.

The terms " London " and " other English ports ", frequently employed by contemporary writers in connection with the coastwise coal trade, give an altogether misleading idea of the number of buyers included under these two heads. " London " is easily contracted in the mind to include only the city proper—which, it is sometimes forgotten, had become a town of about half a million inhabitants by the end of the seventeenth century—while the term " other English ports " is sufficiently vague to bear indefinite expansion. But, if we compare the area actually served, at that time, by colliers unloading at London, with the area served by colliers unloading at other ports between Hartlepool and Southampton, we find the latter only slightly larger than the former in point of population. Coal deposited on the wharves at London, or farther up the river at Eton, Henley or Reading, was sometimes hauled in carts twelve miles or more overland. In Hertfordshire and other counties north of the river, Defoe writes in 1727, " there are many farmers, and others that are not farmers, who keep teams of horses on purpose to let them out, for the bringing of corn and meal, and malt, to London ; and generally these carriages go back laden with coale, grocery, wine, salt, oil, iron, cheese, and other heavy goods, for shopkeepers and tradesmen of the country ; and it is a very great return they make for the mere expense of this carriage." [1] Many years before Defoe wrote, coal, landed at London, can be traced as far north as Theydon Garnon, in Essex.[2] We may perhaps assume that the district provided with coal through the port of London should be expanded to include all the country (or at least all the towns and large villages [3]) within twelve miles of the river from Gravesend up to Oxford, to which point sea-borne fuel regularly found its way in small quantities after the middle of the seventeenth century.[4] This district comprised the counties of Middlesex, Surrey, Hertfordshire, and Berkshire, together with parts of the counties of Kent, Essex, Buckinghamshire, and Oxfordshire. It is said to have contained in the nineties of the seventeenth century about 200,000 houses, or more than a sixth of the population of England and Wales, according to the estimate of John Houghton, that busy collector of information concerning husbandry, manufactures and trade, whose weekly discourses were scarcely less famous in their time than the more polished writings of the *Spectator* and the *Tatler*.[5]

---

[1] Defoe, *The Complete English Tradesman*, 3rd ed., 1727, vol. ii, pt. ii, p. 33.

[2] Thorold Rogers, *History of Agriculture and Prices*, vol. v, p. 388.

[3] Coal replaced wood much more slowly in the country districts than in the towns. (Cf. below, p. 88.)

[4] Rogers, *op. cit.*, vol. v, pp. 387, 403, 405.

[5] John Houghton, *A Collection for the Improvement of Husbandry and Trade*, ed. R. Bradley, 1727, vol. i, pp. 73-5. This well-known estimate, made in 1693, is based on the hearth-tax returns of the number of houses in England and Wales, county by county. I assume that the Thames valley market included the whole of Middlesex, Surrey, Hertfordshire, and Berkshire, half of Buckinghamshire, one-third of Essex and Oxfordshire, and one-fourth of Kent. According to Houghton's estimate, these counties and parts of counties contained 203,332 houses. It is customary to assume that there were, on the average, about five persons living in each house. As is well known, the population of England and Wales at the close of the seventeenth century is estimated at between five and five and a half millions.

When Houghton wrote, the other section of the east and south-east coast market, in which Newcastle coal had a virtual monopoly, included the counties of Norfolk, Suffolk, Cambridgeshire, Sussex, and Hampshire, those parts of Essex and Kent not supplied through the port of London, and those parts of Yorkshire, Lincolnshire, and Huntingdonshire not yet reached by the output from Midland collieries. This area contained, according to Houghton's calculation, approximately 245,000 houses.

Even after we have allowed for the more rapid growth of population in the neighbourhood of the capital, and have given the words " London " and " other English ports " a rather more precise meaning, in terms of inhabitants, than could be deduced from contemporary usage, the figures of Table VII still indicate that the use of coal increased far more rapidly in the Thames valley than elsewhere in eastern and south-eastern England. Various factors were responsible for this. The concentration of population within the city encouraged large imports, both because the demand for coal was likely to be greatest where the largest number of persons settled together,[1] and because the city itself was easily accessible to sailing colliers, or lighters, through the excellent harbour and channel provided by the broad, slow-flowing Thames. In other parts of the east and south-east population remained scattered ; and not all the towns and villages had access to navigable water, upon which cargoes of coal could conveniently be unloaded. Except for the Hull, which was navigable from Hull to Beverley, the Ouse and the Cam, navigable from King's Lynn to St. Neots and Cambridge, the Yare from Yarmouth to Norwich, and the Medway winding into the heart of Kent, there was, along the whole of the east coast south of Durham, hardly a stream of any use to the trader in carrying coal inland.[2] Nor were suitable harbours always to be found along the coast. Although the mariner had plenty of landing places from which to choose after he had sailed south past Yarmouth, there is scarcely a record of a coal shipment to any place along the Lincolnshire coast.

Besides the natural obstacles to the more extensive distribution of coal in those parts of south-eastern England outside the Thames valley, the powerful corporations of the chief ports often imposed artificial obstacles, in the shape of dues or regulations, upon the carriage of coal inland. As early as the reign of Henry VIII, Richard Blaxton, a freeman of Cambridge, brought an action against the Mayor and Burgesses of King's Lynn, for having refused him permission to unload his own cargo out of a collier from Newcastle.[3] The town authorities took the position that their charter gave them the exclusive right to purchase, tax, and sell all coal brought into the river Ouse, which provided the sole channel for marketing this bulky commodity in

---

[1] This point is discussed in more detail below, see pp. 88–9.
[2] The Humber and the Yorkshire Ouse were easily navigable, of course ; but, by the end of the seventeenth century, sea-borne coal had been forced out of the greater part of the Humber valley by the competition from collieries in the West Riding and near the town of Nottingham (cf. below, p. 110).
[3] *Star Chamb. Proc., Henry VIII*, vol. v, f. 21.

Cambridgeshire and eastern Huntingdonshire. Nearly a century later, the Corporation of Yarmouth forced merchants to acknowledge that all coal entering the Yare, even though destined for sale up the river at Norwich, must be landed at the port of Great Yarmouth, " to be vented from thence *by land and not otherwise.*" [1] As the inhabitants of that part of Essex along the river Blackwater pointed out, in opposing a similar ruling of the Corporation of Maldon, such a provision " much increased " the expense of carriage, and enabled the authorities in the principal ports to engross and monopolize the whole of the coal, and to sell it again at any price they pleased.[2] Apparently the town of Maldon had a vested interest in the landings, which enabled it to unload all coal brought into the Blackwater channel even when this coal had to be carried back overland to be burned in places nearer the sea.

In spite of these obstacles, and in spite of the competition afforded by the Midland collieries, the rate at which the market for sea-borne coal expanded, not only in the Thames valley, but in counties all along the eastern and south-eastern coasts, is astonishing. Except for the Scottish highlands, there was, in the seventeenth century, no large area where coal was so dear. These counties had no collieries of their own, and the coal imported never sold for less than four or five times the pithead price, even under the most favourable conditions, and might bring from ten to twenty-five times the pithead price, when it had to be carted some distance inland, or when the supply was temporarily cut off by a naval blockade, or an illicit combination of middlemen.[3] Yet in London itself the consumption seems to have approached 16 cwt. per head,[4] as compared with an average consumption of perhaps 9 cwt. per head for the country as a whole.[5] If we except districts like the Tyne valley, or the Lothians, where the price of coal was usually not a fourth as much as in the Thames valley, the demand in and about London was greater, in proportion to the number of inhabitants, than in any other part of the country.

---

[1] *P.C.R.*, vol. xxxii, p. 261 (23rd Feb., 1624). The italics are mine.

[2] *Hist. MSS. Com.*, *4th Report*, appx., p. 79*b* (Petition to the House of Lords, 25th June, 1641). The Corporation of Bridgwater placed restrictions upon the carriage of coal inland by way of the river Parrett (*Exch. Deps. by Com.*, 23 and 24 Charles II, Hil. 18 ; 24 Charles II, East. 4).

[3] See Appendix E ; and below, vol. ii, pp. 81–2.

[4] The proportion of the coal arriving in the Thames which was actually consumed in London and Southwark is a matter for conjecture. A good deal depends on how we circumscribe the London area. If we include the whole metropolitan area, with its population of half a million, it seems likely that only an eighth of all the coal imported was consumed outside the city. As the imports at London exceeded 450,000 tons per annum at the end of the seventeenth century (Table VII), it is probable that the metropolitan population consumed annually about 400,000 tons, or about 16 cwt. per head.

[5] We have estimated the output of coal in Great Britain at the end of the seventeenth century at approximately 3,000,000 tons per annum (Table I). Probably 200,000 tons or more were shipped to Ireland and to foreign countries (Table VI), and rather more than 400,000 tons were consumed in Scotland (see above, pp. 45, 49–51). There remains nearly 2,400,000 tons consumed in England and Wales, with their population of five to five and a half million souls, or 9 cwt. per head.

(b) The Foreign Market

We have already given, in Table IV,[1] an estimate of the approximate amount of coal exported from the north-east coast of England. To this we must add the exports from Scotland,[2] which grew enormously in importance during the seventeenth century, but for which we have no such series of records as for the coal from the Tyne and Wear valleys. We must also add the exports from South Wales and other west-coast ports, where customs frauds could probably be practised with more impunity than on the east coast.[3]

Even if reasonably satisfactory allowances can be made for Scotch and Welsh coal, and for the quantity of English coal which escaped entry in the records, it is more difficult to present in tabular form a view of the average size of the foreign market, than it is to present a similar view of the south-eastern market, because the shipments overseas varied much more from year to year than the shipments coastwise.   During the wars with Louis XIV at the beginning of the eighteenth century, hardly any British coal reached France ; and during the three Dutch Wars the export trade was cut off almost altogether.[4]   The following tabular view, which is subject to a considerable margin of error, must be understood to represent the state of the foreign market for British coal when that market was not disturbed by the ever recurring sea warfare of the seventeenth century.

TABLE VIII [5]

Estimated Annual Exports of British Coal (in Tons)

From	1551–60.	1591–1600.	1631–40.	1661–70.	1681–90.
Durham, Northumberland	10,000	30,000	60,000	55,000	85,000
Scotland. .	1,000	7,000	60,000	50,000	60,000
South Wales (and Bristol)	1,000	2,000	4,000	4,000	5,000
Total . . .	12,000	39,000	124,000	109,000	150,000

What is striking in a comparison between the south-eastern market and the foreign market for English coal is the much slower growth in the latter during the period from 1550 to 1700.   In the middle of the sixteenth century foreign towns appear to have imported about half as much coal as the ports of south-eastern England, and during the Middle Ages exports probably made an even greater proportion of all shipments on the Tyne.   But, at the end of the seventeenth century, the entire foreign market appears to have absorbed only about a fifth as much coal as was shipped coastwise to be burned in

---

[1] See above, p. 36.
[2] See above, pp. 44–5.
[3] See below, vol. ii, p. 237.
[4] See Appendix D (i).   In 1666 and 1667 the exports from Newcastle and Sunderland were only about a sixth as great as in normal years.   See also the table of exports from English ports for the years 1711–13 (Appendix D (vi)).
[5] For an explanation of the sources, see Appendix B.

eastern and south-eastern England. During the reigns of Elizabeth and her first two successors, the foreign market increased less rapidly than the English. During the last half of the seventeenth century it hardly increased at all.

In order to explain the relatively slow growth of the foreign market, it is necessary to point out both that the demand for coal abroad increased less rapidly than in England, and that the restrictions imposed upon the foreign trade in coal were much greater than those imposed on the coastwise trade. Neither explanation by itself would be sufficient. In Elizabeth's reign, when few restrictions were imposed on the export trade either by the English or by foreign governments, exports increased less rapidly than coastwise shipments ; and it is certain that, even if free trade had continued, the foreign market would have been less important, at the end of the seventeenth century, than the domestic. But it would have grown much more rapidly than it did, particularly during the fifty years following the Civil War. The almost stationary condition of the export market for a long period following Charles I's reign is only to be understood after a discussion, on the one hand, of the English policy of taxing coal at the port of shipment, and, on the other hand, of the policy of foreign states in taxing it again at the port where it was unloaded.[1]

In attempting to discover what proportion of the coal exports reached each of the foreign countries, we are faced with even greater difficulties than in estimating the total quantity carried overseas. The English port books contain the port of origin of the ships loaded with coal, and frequently also the supposed destination of the cargo. But it is obvious that foreign ships did not always unload the coal at their home ports. And recent research in connection with the trade passing through the Sound has shown that even an entry giving the destination of the cargo cannot be regarded as conclusive proof that it was in fact consumed in the place indicated.[2] In some years a considerable proportion of the British coal imported at towns in the Baltic was originally unloaded in Holland and reshipped.[3]

While the records contained in the port books do not permit us to present a table giving the proportion of British coal imported by the various foreign countries, they are perhaps sufficient to enable us to indicate what were, broadly speaking, the principal markets. It is clear enough that overwhelmingly the greater portion of British coal exports during the sixteenth and seventeenth centuries were destined for continental ports between the Cherbourg and the Danish peninsulas.[4] While the shipments to Baltic ports, especially to Danzig, Lübeck, Stralsund, Stettin, Copenhagen, and Stockholm, increased rapidly during the early years of the seventeenth century,

---

[1] See below, pt. v, ch. ii.
[2] N. E. Bang, *Tables de la navigation et du transport des marchandises passant par le Sund*, 1906–22, vol. ii, Introduction.
[3] *Ibid.*, vol. ii, *passim*.
[4] This information is taken from the port books for Newcastle, Sunderland, and Blyth (for the references see Appendix D (i)).

and may have exceeded 15,000 tons per annum on one or two occasions,[1] these shipments probably did not average more than five or six thousand tons a year until the eighteenth century. And the overseas market west of the Cherbourg peninsula, although it probably grew more rapidly after the Restoration than the Baltic market, was, if anything, less important. Coal cargoes were actually carried to New England and Virginia, to ports in the West Indies and the Caribbean, to Tangier in Africa, to Spain and Portugal, and to Venice, but there was no regular trade with any of these places. Bordeaux and La Rochelle alone may be said to have carried on a commerce in coal ; and the records of French imports in the *Archives nationales* indicate that the maximum quantity unloaded at either of these towns in a single year did not exceed a thousand tons.[2]

From eighty to ninety-five per cent of the shipments of British coal overseas were imported at ports in northern France, western Germany, and the Low Countries, the great centres of the trade during the seventeenth century being Amsterdam and Hamburg.[3] It is probable that, during the reign of Charles II, imports at Hamburg amounted to more than 15,000 tons, and almost certain that at Amsterdam, in favourable years for the trade, they approached 50,000 tons.[4] Rotterdam may have been a coal port of scarcely less importance than Amsterdam ; but most of its coal apparently came by way of the Meuse from Liége.[5]

Our data are insufficient to enable us to apportion, even roughly, the coal exported to France, to Holland, and to western Germany, during the first half of the seventeenth century. There appears to have been much variation from decade to decade in the proportion going to each of these three countries. From the entries in the port books for Newcastle and Sunderland, which probably do not cover much more than half of all the British coal exported, it seems likely that, in the last decade of Elizabeth's reign, France and Holland each received a somewhat larger quantity of fuel from the north of England

---

[1] In 1612, 1,262 "lasts" of coal passed through the Sound—685 lasts from Scotland, 547 from England, 30 from the Low Countries (Bang, *op. cit.*). The figure for 1629 is 1,868. This appears to have been a record. Usually the amount was much smaller, e.g. 325 lasts in 1627, 332 in 1628, 112 in 1630, 920 in 1631, 93 lasts and 414 chaldrons in 1646, 95 lasts and 35 chaldrons in 1655. A last of coal appears to have been equivalent to 12 or 14 tons (see Appendix C (ii)).

[2] See Appendix D (viii).

[3] On the development of the coal trade at Hamburg, cf. Ehrenberg, *Hamburg und England*, 1896, pp. 295–6.

[4] The figure for Hamburg is based on entries in the port books for Newcastle (for the references see Appendix D (i)), which suggest that the annual shipments from the Tyne to that port sometimes reached 10,000 tons. It is perhaps legitimate to assume that nearly as much again was shipped to Hamburg from Scotland and from other English ports. The Newcastle port books for the years following the Restoration show that more than 20,000 tons per annum were frequently shipped for Amsterdam. In 1675, the port books for Newcastle and Sunderland combined show exports of more than 30,000 tons destined for that port (*Exch. K.R. Port Books*, 196/2). To this we must add the shipments from other English ports and from Scotland. While we cannot feel certain that all this coal actually reached its destination, the evidence is sufficient to prove that Amsterdam was the principal continental port for British coal.

[5] Cf. Pirenne, *Histoire de Belgique*, vol. v, p. 358.

than did western Germany, although Emden was at this time a coal port hardly less important than Amsterdam. Two decades later Holland appears to be much the most important foreign market ; but in the decade preceding the Civil War the German ports, and particularly Hamburg, appear to have received almost as much English coal as the Dutch.[1]

Whatever may have been the condition during the first half of the seventeenth century, it is certain that by the time of the Restoration Holland had become the great market for British exports. An account of the money received from a tax on north of England coal during a period of six months in 1663 shows that, out of about 25,000 tons shipped overseas, almost 16,000 tons, or more than sixty per cent, was destined for Dutch ports, about 11,000 tons being for Amsterdam.[2] A document explaining the condition of the Scottish export trade in 1670 states that the overseas shipment of coal " is most for Holland ".[3] There is abundant evidence that this continued to be the condition of the Scottish and the English export trade during the remainder of the seventeenth century, in spite of the wars which the country had waged against its principal customer. France appears to have received in the neighbourhood of 15,000 or 20,000 tons of British coal per annum,[4] western German ports may have received slightly more, and perhaps as much again went to the Spanish Nether-lands, to the Baltic countries, to Spain and Portugal and the English colonies. The remainder, from 60,000 to 100,000 tons in favourable years, went to Holland, particularly to the parts round the Zuider Zee. If, as seems probable, southern Holland received almost as much again from continental mines, by way of the Meuse, the Scheldt, and the Rhine, it follows that the Dutch provinces had become, next to south-eastern England, the principal market for water-borne coal in the world.[5]

(c) The South-Western Market [6]

Two generalizations may be made concerning the demand for coal in Great Britain during the sixteenth and seventeenth centuries.

---

[1] Exch. K.R. Port Books, 185/6 ; 188/6, 8 ; 190/5, 9 ; 191/9 ; 192/4, 5.

[2] Sackville MSS., as cited below, vol. ii, p. 307, n. 3.

[3] An Estimate of Scots Coal, 1670 (see Appendix F).

[4] See the statistics given in Appendix D (viii), which show that nearly 18,000 tons of coal were imported by France in 1671, nearly 13,000 tons in 1672, and more than 20,000 tons in 1683. Probably at least ninety per cent was British. It is probable, moreover, that substantial quantities of sea-borne coal, which were subject to the payment of a duty (see below, vol. ii, pp. 230–1), were imported without being entered in the records.

[5] Cf. below, pp. 127–9.

[6] It will be observed that, in this discussion of the markets for sea-borne coal, nothing is said of north-western England or of Wales. The amount shipped to these districts was unimportant, owing to the presence of collieries near the principal seaside towns—Workington, Whitehaven, Preston, Liverpool, Chester, Milford, Swansea, Neath, and Bristol. The total amount shipped along these coasts probably did not exceed 5,000 tons per annum, on the average, at the end of the seventeenth century (see Appendices D (iii) and G). For the coal imports at towns in North and South Wales, see Appendix G, and

The first is that, given an equal selling price, the demand was many times greater per head in and about cities like London, Dublin, Bristol, or Edinburgh, than in small towns and villages. The second is that, again given an equal selling price, the demand always tended to be greater in the eastern than in the western counties. Both these generalizations are illustrated by a comparison of the coal market in the south-west with that in the south-east.[1] If we define the Cornish peninsula as the area west of a line drawn on the map from the mouth of the river Parrett, on the Bristol Channel, to the port of Weymouth, on the English Channel, then (using Houghton's estimate of the number of houses once more) we may say that, at the end of the seventeenth century, it contained roughly the same population (500,000 persons) as the city of London.[2] But the consumption of coal in the Cornish peninsula does not appear to have exceeded 70,000 tons a year at this time,[3] or rather less than 3 cwt. to each inhabitant, as compared with nearly 16 cwt. in London, and about 9 cwt. in the whole of England and Wales.

Comparatively little coal was burned in the south-west, although it was a district in which coal prices were generally lower than in the south-east. To the chief ports on the northern side of the peninsula —Bridgwater, Minehead, and Barnstaple—coal could be brought from South Wales in half the time required to sail a laden collier from Newcastle to London,[4] and with much less risk of shipwreck, or of capture by the enemy in time of war or by pirates in time of peace. Freight charges on Welsh coal unloaded at any point along the Bristol Channel tended to be somewhat lower, therefore, than on Durham or Northumberland coal unloaded at King's Lynn or Yarmouth or London. And the coal could be distributed through the south-western market at somewhat less expense than through the south-eastern. The narrow peninsula, with its five hundred miles of broken coastline, scarcely contained a village which was not within a day's walk of the sea ; there were few sites more than ten miles away from a suitable harbour (such as was rarely found on the east coast north of Yarmouth), or a navigable river, on the banks of which coal could be unloaded either directly out of a sea-going vessel, or out of a lighter making the trip up and down stream, to and from the laden colliers.[5] During Charles II's reign, if not before, coal was regularly brought to villages in the heart of the peninsula at its widest point. Ships laden with Welsh coal sailed ten miles up the river Parrett to Bridgwater, " only to be unladen there into small boats, such as can

---

Exch. K.R. Port Bks., 1310/10. Occasional shipments were made from Whitehaven to Lancashire and also to ports on the Solway Firth (*ibid.*, 1448/6, 8).

[1] See note 6, p. 87.

[2] 56,310 houses in Devonshire and 25,374 houses in Cornwall (Houghton, *op. cit.*, p. 74). It is assumed that those parts of Somerset and Dorset, west of the line we have drawn from the mouth of the river Parrett to Weymouth, contained about 20,000 more houses.

[3] See below, p. 89.

[4] See below, p. 394.

[5] See Appendix G for a list of ports along the Cornish and Devon coasts at which coal was imported.

pass under the bridge and up to Ham Mills, where they are exposed to sale for the whole country." An aged husbandman, called as a witness in an action brought in 1672 by the Corporation of Bridgwater to establish its right to impede this traffic, described how he had been employed several times to drive pack-horses, bearing sacks of coal, from Ham Mills to the neighbouring " market towns " of Taunton, North Curry, Langport, and Wellington, all places " ordinarily supplied " with fuel in this manner. He " also has carried many thousand sacks of coal from Ham Mills to Stoke St. Mary ", two miles beyond Taunton, and has made trips to Yarcombe and Chard, only ten miles north of the English Channel, and west into Devonshire to Holcombe Rogus and Tiverton, thirteen miles north of Exeter.[1] A distribution of coal similar to that from Bridgwater was probably made from Barnstaple and Plymouth, and, on a small scale, from a dozen or more tiny ports like Padstow, St. Ives, Penzance, Truro, Penryn, Fowey, Looe, and Falmouth ; so it was through no want of facilities for reaching the inhabitants that the demand for coal was less intense in the south-west than in the south-east. The explanation is probably to be found, as has been suggested, in the different situation of the two districts, and in the fact that, in the former, population was divided between a great number of small towns and villages, while, in the latter, it was coming to be concentrated in the Thames valley and in London.

Imports of coal at towns in the south-west increased very rapidly, nevertheless, during the seventeenth century. It is doubtful whether the Cornish peninsula absorbed more than 3,000 or 4,000 tons annually before 1600.[2] But during the six warm months of 1651, by no means an unusually prosperous year, the imports approached 16,000 tons ;[3] and, in the last two decades of the seventeenth century, the annual imports probably reached 70,000 tons.[4] The rapid expansion in the

---

[1] *Exch. Deps. by Com.*, 23 and 24 Charles II, Hilary 18, and 24 Charles II, East. 24 (Mayor, aldermen, and burgesses of Bridgwater *v.* Richard and Robert Bobbett and Joseph Baker).

[2] Almost all the coal for the Cornish peninsula came from South Wales. It is not possible to determine how much reached this area at the end of the sixteenth century. In the year ending at Michaelmas, 1600, more than 4,000 tons left the harbours of Swansea and Neath for English ports (see Appendix D (ii)), and, in so far as these cargoes were actually unloaded in England, they were undoubtedly unloaded in Cornwall, Devon, or Somerset. But there is good reason to suppose that some of them were deceitfully entered for coastwise shipment, in order to escape the export duty which became effective at Christmas, 1599 (see below, vol. ii, p. 237), and that such cargoes were shipped to the Continent (*Exch. Deps. by Com.*, 12 Charles I, Mich. 15).

[3] The tax of 2*s.* per London chaldron collected at the ports between Weymouth and Bridgwater yielded £1,194 17*s.* 2*d.* during the six-month period (see Appendix G). It was collected, therefore, on nearly 11,950 chaldrons, or approximately 15,930 tons. The imports for the whole year may have approached 30,000 tons, for the coal trade of the west coast was not seasonal to anything like the same extent as that of the east coast (see below, p. 396). It is probable, moreover, that some coal unloaded in Cornwall, Devon, and Somerset escaped the collectors of this new tax.

[4] About 60,000 tons, or more, came from South Wales (cf. the discussion of Table VI in Appendix B). The remainder came from Newcastle, Whitehaven, and Workington. Some idea of the proportions furnished by the different mining

market appears even more striking when we examine the statistics of imports at particular towns. In the year 1571-2, the port book for Barnstable, the chief centre on the peninsula for distributing coal, is innocent of a coal entry ; the book for 1572-3 contains entries totalling only 48 tons ; but in 1651 the record of the tax collected at Barnstaple proves that the town imported substantially more than 5,000 tons. Plymouth had no coal imports in 1575, but received more than 1,000 tons in 1651, and 5,142 tons in 1678.[1] These figures show that, although the market for coal in the south-west developed somewhat later than in the south-east, once the development had begun, it was scarcely less rapid than that in counties along the North Sea coast.

(d) The Irish Market

The medieval shipmasters, who sailed on the North Sea, occasionally picked up, on the Tyne or the Northumberland coast, a casual lading of coal, with which to ballast their ships for the journey to some English, Flemish, French, or German port.[2] In the same way the shipmasters of the Irish Sea probably took a little coal to Ireland from Lancashire, Cumberland, or South Wales.[3] But the amount of trade along the western shores of England, before the discovery of America, was small compared with that along the eastern shores. The quantity of coal carried on the Irish Sea was probably unimportant even in comparison with the small medieval coal trade from the Tyne valley. At the beginning of Elizabeth's reign, when we first get returns in the port books, the annual exports to Ireland do not appear to have exceeded two or three thousand tons.[4] The English expeditionary force under Essex, which landed in Dublin in 1599 to subdue the national revolt, brought coal to burn in barracks. Essex himself wrote of it as an essential fuel in Ireland, where it was chiefly " spent for . . . her Majesties souldiers as also for the releefe of the poore within divers towns ".[5] Yet the quantity imported in 1600, to judge from the Lord Deputy's protest " against the imposycon sett uppon Sea Coles ", did not much exceed 5,000 tons.[6]

From the beginning of the seventeenth century the consumption increased rapidly at towns all along the Irish coast, and at Douglas

---

districts may be gained from the port book of imports at Plymouth in 1678. Out of 5,142 tons, about 300 are entered as Newcastle coal, and a few score as Cumberland coal ; all the rest come from Wales (Exch. K.R., Port Bks., 1042/23).

[1] See Appendix D (v).

[2] See above, pp. 9-10.

[3] See above, pp. 53, 61, 70. It is uncertain whether the coal, over which the Holy Trinity Gild at Dublin was granted a monopoly in 1438 (Charles Gross, The Gild Merchant, 1890, vol. i, pp. 135-8, vol. ii, pp. 65-6), was mineral fuel, or, as seems more probable, merely charcoal. (Cf. above, p. 4.)

[4] For a fuller discussion of the consumption of coal in Ireland, see Appendix B (vii).

[5] Cal. S.P. Irish, 1599-1600, pp. 102-3 (Earl of Essex to the Privy Council, 22nd July, 1599). See also Acts of the Privy Council, N.S., vol. xxxi, pp. 378-9.

[6] Lansdowne MSS., 156, no. 102.

and Ramsey on the Isle of Man.[1] Overwhelmingly the greatest demand came from Dublin, but there was soon a regular trade to Drogheda, Cork, Dundalk, Wexford, and Waterford as well, and occasional shipments were made to Londonderry, Belfast, Strangford, and Carrickfergus in the north, to Limerick and Galway in the west, and to Malahide, Carlingford, Youghal, Kinsale, Ross, and Wicklow in the south and east.[2] In 1636 the new fuel was said to be " almost the only material . . . for firing along this coast all the winter from Knockfergus [Carrickfergus] to Youghal ".[3] In the nineties of the seventeenth century, an English writer, anxious to impress upon his countrymen the importance of this overseas market, informed them that " Ireland takes off above 100,000 tuns of our coals yearly ".[4] This was, no doubt, an exaggeration. The imports, on the eve of the Irish Rebellion of the nineties, probably did not exceed 60,000 or 65,000 tons per annum. Of this Dublin appears to have consumed about two-thirds.[5]

Dublin, in fact, bore much the same relation to the Irish market as London bore to south-eastern England. Each city was the one great consuming centre in a large area. Coal, Sir William Petty remarked, had become a staple commodity of the whole population of Dublin, and the amount of coal burned might be taken as a measure of the general prosperity of the citizens.[6] Some years later Dean Swift regarded a cheap and plentiful supply as the essential basis of their welfare. " As this city, the head of this weak, poor, feeble nation, is grown so monstrous great ! (a head too big for the body) so in proportion it will require and take an exceeding great quantity of coals to serve it with firing." [7]

(e) *The Scottish Coastwise Trade*

In Scotland the market for sea-borne coal during the sixteenth and seventeenth centuries was practically confined to a few towns and villages along the east coast, for the seaboard shipments along the west coast from Glasgow and Ayrshire were of no consequence until the eighteenth century. Indeed, the market may almost be said to

[1] *P.C.R.*, vol. xlvi, p. 121 ; *Exch. K.R. Port Books*, 1448/6. The annual consumption of coal on the Isle of Man at the end of the seventeenth century, however, did not perhaps exceed one or two thousand tons.

[2] Numerous records of shipments to these ports are contained in the port books for Swansea, Neath, Llanelly, Tenby, Milford, Chester, Liverpool, and Whitehaven ; and in the Scottish *Customs and Excise Accounts* for the ports of Ayr and Glasgow.

[3] *Cal. S.P.D.*, 1633–47, p. 130.

[4] Walter Harris, *Remarks on the Affairs and Trade of England and Ireland*, 1691, p. 19.

[5] The consumption of coal in Dublin apparently increased rapidly between 1677, when Petty estimated it at 26,000 tons (*The Petty-Southwell Correspondence*, ed. Marquis of Lansdowne, 1928, pp. 23, 26), and 1684, when it amounted, he tells us, to 42,727 tons (Petty, *Economic Writings*, ed. C. H. Hull, 1899, p. 589).

[6] *Ibid.*, pp. 589, 596.

[7] Letter to the *Dublin Weekly Journal*, upon the use of Irish coal, 16th August, 1729 (in Swift's *Works*, ed. Sir W. Scott, 1824, vol. vii, p. 226).

have been limited to the broken coastline of the Firth of Forth and the Firth of Tay,[1] the district in which the political and economic life of Scotland had centred for centuries, and in which much of the population was concentrated. Here the coal trade had a very early history. At least as far back as the beginning of the fourteenth century, ships partly filled with the " black stones " which had so astonished Æneas Sylvius discharged cargoes at Perth and Dundee.[2] But, while the traffic along the two firths in sea-borne coal, carried as ballast, may have been as great, in proportion to the population living near the sea-shore, as along any coast during the Middle Ages, this traffic increased less rapidly during the sixteenth and seventeenth centuries than in other parts of Great Britain. It may be doubted whether more than 30,000 or 40,000 tons were carried coastwise to Scottish ports at the end of the seventeenth century.[3]

One reason for the relatively slow development of the coastwise trade in Scotland was the growth in the demand for Scotch coal abroad. Native buyers, who came with their boats to any of the dozen or more coal shipping ports on the coasts of East Lothian, Linlithgowshire, Stirlingshire, Clackmannan, or Fife, competed for ladings with foreign buyers, especially with Dutch merchants or their agents. The opportunity, which the coal sellers enjoyed, to play off one buyer against another arose out of special conditions brought about in 1599, when Elizabeth's ministers laid a heavy tax on the export of English coal. This led many foreign shipmasters, who had formerly purchased their fuel on the Tyne, to direct their vessels farther north to the Firth of Forth, where the impost on exported coal was unimportant, and where even this small impost could be more easily evaded than the tax at Newcastle.[4] After 1600 competition between native Scottish and foreign buyers raged furiously and continuously. It became one of the principal problems with which the Scottish Privy Council had to grapple. Without effective government intervention, such as the Council had neither the means to enforce nor the mind boldly to undertake,[5] all the winning cards remained in the hand of the foreign buyer. He was willing to pay the mine owners more for their coal than the domestic consumer could offer. Owing to the general poverty prevailing in Scotland at the beginning of the seventeenth century, most of the coastal towns, particularly those outside the district bordering the Firth of Forth, had little to exchange for the fuel exploited by the powerful colliery owners of Fife and the Lothians. As late as 1656, the traders of Aberdeen had nothing better to offer

---

[1] In this chapter coal carried on the firths is regarded as sea-borne fuel, and is considered under that head. But the coal traffic on the river Clyde is treated under the head, " river markets " (see below, p. 100).

[2] See above, p. 9 n.

[3] Account books for collieries in Fife, preserved in the Register House, Edinburgh, indicate that greater quantities of coal were exported, during the latter half of the seventeenth century, than were shipped coastwise. For the amount probably exported, see above, Table VIII.

[4] See below, vol. ii, pp. 223 sqq.

[5] The Council was composed to a considerable extent of coal-owning landlords or their relatives (see below, vol. ii, pp. 157, 224).

than a cargo of salmon.[1] And every species of fish swam under the
very wharves whence coal was shipped ! In contrast with the poor
Scots the Hollanders were the richest people of the age ; the pockets
of their merchants bulged with paper representing claims on the rich
jewels of the east, on the choicest French wines and furniture, on the
gold and silver of the Americas, and on the products of the famous
manufacturers of the Low Countries. The high tax on exports of
English coal enabled the Scottish mine owners to exact a high price from
their continental customers. Coal on the wharves along the Firth
of Forth sometimes sold for as much as six shillings a ton, while coal
on the wharves along the Tyne was seldom more than four shillings.[2]

If the " native " who brought his ship to port on the Firth of
Forth was to be served with coal as readily as the foreigner, he had to
pay a higher price than was ordinarily paid by the English shipmasters
in England. Even if the Scottish shipmaster from Aberdeen, Inverness,
Banff, Montrose, Arbroath, Perth, Dundee, North Berwick, Dunbar,
Eyemouth, or Edinburgh offered money the face value of which was
equivalent to that offered by Dutchmen, the colliery owners were still
likely to slight him ; for the Scottish currency constantly depreciated
and had no certain value, while the Dutch *riksdaler* (or " dollar "
as the Scots called it) contained after 1606 a fixed quantity of fine
silver,[3] and could be circulated in Scotland at a rate above native
money.[4] In 1631 pressure was brought upon the Privy Council to
remove this incentive to favour foreigners, and a proclamation was
published prohibiting the acceptance of foreign tokens in payment for
coal and salt. But this proved ineffective.[5] Particular " masters
of coal and salt " succeeded in getting exemption from the proclamation ; and all of them took exemption, whether it was granted or not.[6]
Patriotic appeals to favour the native buyer failed to move them.
They only obeyed the impulse which has normally guided the merchant,
and his descendant, the modern capitalist, when, " preferring thair
immoderat gayne to the Cristeane bandis of cheritie and love
quhairwith thay aucht to be unite with thair nighbouris, thay prefer
strangearis unto thame, insofar as, quhen his Majesteis subjectis to
thair grite chargeis have frauchted veshellis and send thame up the
River of Forthe, for coillis, and thair servandis haveing instantly
required thair dispatche for reddy payment thay ar refuised, post-
ponnit, and deferrit till the whole strangearis lyand thair . . . be
first outred and laidnit, swa that the cuntrie veshellis . . . being

---

[1] Tucker, *Report upon the Settlement of the Revenues of Excise and
Customs in Scotland*, 1656, pp. 6–8.

[2] See Appendix E. The ruling price on the Tyne during the seventeenth
century was 10*s.* and 11*s.* per Newcastle chaldron, of from 2 to 2·65 tons (cf.
below, vol. ii, pp. 91–2). This price included a shilling or more for " keelage ",
an expense which the Scottish seller escaped, because he could load sea-going
vessels directly at his wharves.

[3] C. F. Dunbar, *Chapters on the Theory and History of Banking*, 1891,
pp. 83–4.

[4] *Privy Council Register of Scotland*, 2nd Ser., vol. iv, pp. xxv–vii.

[5] *Ibid.*, vol. iv, pp. 287, 298, 302.

[6] *Ibid.*, vol. iv, pp. 555–6 ; vol. v, pp. xix, 341–2 ; vol. vi, pp. 263–4.

past all hoip to be ansuerit so lang as thair is ane foreyner to be dis-patcheit, ar in end returnit emptie." [1]

From the frequent protests of the Scottish burghs against the preference which the mine owners showed to foreigners, it is clear that this picture of the natives returning northward empty-handed, while the Dutch sailed out of the Firth of Forth with their ships full of coal, is not altogether an imaginary one.[2] These conditions account to some extent for the slow expansion of the Scottish coastwise market. The burghers of towns along the coasts had the will, but not the means, to buy. According to their spokesman who addressed the Privy Council, the low tax on the export of coal from Scotland benefited only the mine owners and their foreign customers. At the beginning of the eighteenth century, after this tax had been raised to a figure which equalled that levied on English coal exported from Durham and Northumberland, most of the advantage which foreigners had in buying on the Firth of Forth disappeared, and the proportion of all water-borne Scotch coal consumed along the native coasts rapidly increased.[3]

The main factor which prevented any great expansion of the Scottish coastwise trade in the seventeenth century was probably not the advantage possessed by the foreign buyer, but the presence of coal seams near the coasts in those districts where the population was concentrated. Owing to the early exploitation of the excellent seams which lie beneath a substantial part of the area bordering the Forth,[4] most of the chief towns and villages—such as Haddington, Linlithgow, Falkirk, Stirling, Dunfermline, and Kirkcaldy—received their supplies of fuel from local pits sunk within a mile or so of their dwelling houses and of the shops of their artisans. Among the large burghs of the Firth of Forth district, only Edinburgh, with its port of Leith, found an advantage, in that age of expensive land carriage, in importing some of its coal. This is an important exception. Seventeenth-century Edinburgh, though far behind London or even Dublin, in point of population, was much the largest and richest town in Scotland, and the principal consuming centre for coal. But there is reason to believe that the proportion of all fuel burned in Edinburgh which came by water diminished greatly during the seventeenth century. Water-borne coal was brought in ships along the coast from collieries at Seton, Elphinston, Fawside, Whitehill, and Preston in East Lothian, and across the Forth from Alloa, Culross, Dysart, and Wemyss. The price at these collieries was forced up by the competition of the foreign buyers. At the same time, new " land-sale " coal pits were sunk in Midlothian, between the Esk valley and the town itself. For every vessel laden with coal which sailed into the harbour of Leith, there must have been hundreds of pack-horses and carts driven along the

---

[1] *Ibid.*, 1st Ser., vol. xii, p. 606 (Act concerning the sale of coal, November 29th, 1621).

[2] This controversy is discussed in detail below, vol. ii, pp. 224 sqq.

[3] Out of 56,000 tons of coal shipped from Alloa in 1790, only 6,000 tons appears to have reached foreign countries (Sinclair, *Statistical Account of Scotland*, vol. viii, pp. 637–8).

[4] See above, p. 43.

winding, muddy paths converging from the east and south-east upon the capital. There remained, along the coasts of Scotland, only the district bordering the Firth of Tay where a considerable market for water-borne fuel could develop unhampered by competition from land-sale collieries.

(f) The River Valley Markets

We have estimated that between 200,000 and 300,000 tons, or something approaching twenty per cent of all coal carried by water at the end of the seventeenth century, never reached the sea at all, but was unloaded from keels or lighters or sailing barges at various points along the few navigable rivers with collieries close beside their banks—the Severn, the Wye and the Trent, the Tyne, the Wear and the Clyde.[1] Nature had made these rivers navigable, at least to flat-bottomed lighters, as far up as the coal mines. The amount of water traffic in coal depended upon the breadth, depth, and current of the river, and upon the size and industrial activity of the towns and villages situated on or near the banks, at a sufficient distance from the collieries to make it cheaper to load coal into boats for transport up or down stream than simply to haul it overland in carts or wagons. Newcastle got its fuel by land carriage, except some of that burned in the glass furnaces and lime kilns which lined the bank of the Tyne,[2] for many of the houses stood some distance up the hill that rises sharply from the river, and more labour was required to unload keels at the wharves and pull coal up the steep streets, than to bring it down the paths from the pits above the town. Sunderland, on the other hand, being situated at the mouth of the river Wear, at least six miles by land carriage from the nearest important colliery, could get the fuel for the increasing number of its domestic hearths, and for its growing salt manufacture, more cheaply by water carriage in keels floated down-stream. Nearly all the fuel used in the growing salt works of South Shields[3] came by river. The output of most Tyneside collieries contained a considerable proportion of an inferior grade, which the mine owners could not sell for shipment coastwise unless they mingled it surreptitiously with the better grades.[4] It proved perfectly suitable for the manufacture of salt ; [5] indeed the local inhabitants called it " pan-coal ", after the salt pans under which it was burned. More and more of this inferior grade was loaded into keels to be carried down-stream to the pans at the mouth of the Tyne, where the sea-water was boiled into salt.[6] In the last quarter of the seventeenth century the traffic in pan-coal on the rivers Tyne and Wear probably amounted to at least 70,000 tons per annum, while perhaps another 20,000 tons was carried for use in other industries and as household fuel.

[1] See above, Table VI, p. 79.
[2] See below, p. 230.
[3] See below, pp. 176–7, 208.
[4] See below, vol. ii, pp. 241 sqq.
[5] See below, p. 112.
[6] As early as 1605 thirty keels were constantly employed in carrying coal to the salt pans on the Tyne (J. Brand, *History of Newcastle*, 1789, vol. ii, p. 22). Cf. Defoe, *Tour*, 1769 ed., vol. iii, p. 237.

While the coal carried on the Tyne and Wear for local consumption may be considered as a sort of by-product turned out by the great " sea-sale " collieries, that carried on the Severn and Trent provided the principal source of revenue for the collieries of Coalbrookdale, in Shropshire, and for those of the district on the southern outskirts of Sherwood Forest just west of the town of Nottingham. During the Middle Ages the Severn valley from Shrewsbury to Gloucester had become one of the richest districts in the whole of England. Professor Gras has shown that corn prices were lower here than elsewhere.[1] Medieval crafts of every sort were pursued with a busy vigour in the chief towns,[2] for the broad, deep, gentle-flowing river Severn, together with its tributary, the Avon, naturally navigable as far as the town of Stratford,[3] united the entire district into a single market. By means of water carriage, the products of one town enjoyed a ready sale in the fairs of all the others. When, during the second half of the sixteenth century, coal began to come into general use, ever greater quantities were shipped on the river until it became, in the seventeenth century, " the most common freight." [4] In Coalbrookdale the seams run close to both banks of the Severn for several miles, and so near are some of these seams to the surface of the valley, as it slopes up from the river bed, that to this day local cottagers frequently cut through them in hewing out their cellars.[5] Consequently coal could be mined and brought to wharves on the river very cheaply.[6] It could be freighted thence up-stream to Shrewsbury, and down-stream to Bridgnorth, Bewdley, Worcester, Tewkesbury, and Gloucester, for consumption in those towns and in the neighbouring villages and country houses. This could be done more cheaply than on the other, less commodious rivers of Great Britain ; for it was possible to navigate the Severn with sailing boats instead of lighters propelled by oars,[7] and therefore with a substantial economy in labour. At Tewkesbury, fifty miles from the mines, coal sold in 1678 for less than six shillings a ton,[8] or about a third of the price in London or Dublin, and less than the price in Bristol or Nottingham, towns standing within sight of the mounds of dirt and black stones which indicated the presence of working

---

[1] N. S. B. Gras, *Evolution of the English Corn Market*, 1915, pp. 46-7.

[2] Cf. *V.C.H. Worcestershire*, vol. ii, pp. 271, 294-300 ; *V.C.H. Gloucestershire*, vol. ii, p. 202.

[3] A. Yarranton, *England's Improvement by Sea and Land*, 1677, p. 64. Navigation on the Avon was improved by the installing of locks, authorized by an Act of Parliament in 1664, but the works were swept away by a flood a few years later, and were not reconstructed until 1771 (J. Priestley, *Hist. Account of the Navigable Rivers, Canals, and Railways throughout Great Britain*, 1831, p. 42 ; and cf. below, pp. 258-9).

[4] *V.C.H. Shropshire*, p. 454. This statement was made in 1739, but the fact was true much earlier (cf. above, p. 65).

[5] *V.C.H. Shropshire*, pp. 454-5.

[6] For further reference to the advantages which the mines of Coalbrookdale had over the mines in other districts, see below, pp. 350-1 n., 365.

[7] See below, p. 393.

[8] 7s. 6d. per chaldron (*Hist. MSS. Com., Report on MSS. of the Duke of Portland*, vol. ii, p. 300). I assume that this was the London chaldron of about 1⅓ tons (see Appendix C (i)).

collieries. In addition to being an area of low corn prices, the Severn basin became an area of low coal prices. At the beginning of the seventeenth century the market for coal appears to have grown almost as rapidly as the market in the Thames valley.[1] By the time of the Civil War the towns of Shrewsbury, Bridgnorth, Bewdley, Worcester, Tewkesbury, and Gloucester had all become absolutely dependent upon the collieries of Coalbrookdale for their necessary fuel, as is shown by the plight of their citizens when the parliamentary armies held the mines and refused to allow coal to pass down-stream.[2] Probably there was more traffic on the Severn at the end of the seventeenth century than on any other river in Europe, with the exception of the Meuse, which flowed through the important mining area between Namur and Liége down into the web of dykes, canals, and populous villages and towns of the Low Countries.[3]

The Severn river traffic was not swelled to any appreciable extent by coal from the Forest of Dean. This may have been due partly to the difficulty of competing with the cheap product of the collieries in Coalbrookdale (for at Monmouth and Lydbrook on the river Wye, within a few miles by water of the collieries of the Forest, coal cost at least as much as Shropshire coal at Tewkesbury) ; [4] but it was also due to the essential conservatism of the mine owners in the Forest. Most of the mines remained the property of working miners, owing to the force of custom entrenched behind special legal rights, such as prevailed in no other colliery district.[5] As working miners, the owners' ambition, unlike that of the great landlords and merchants who came into possession of most of the mines in other districts, was simply to gain a comfortable livelihood, rather than to increase their profits by reaching out for new markets. The Miners' Court in the Forest of Dean, an institution unique in the annals of English coal mining, actually strove to curtail the sale of coal outside the bounds of the Forest, by charging " foreigners " a higher price than natives of the district.[6] And, in so far as shipments from points along the Severn

---

[1] *Star. Chamb. Proc.*, *James I*, 86/18, 294/25, 310/16.

[2] See below, vol. ii, p. 288.

[3] Out of eighteen officials who were to have been appointed, under an Act of 1695, to collect a duty on all coal shipped upon the inland rivers of England, seven were to have established themselves on the Severn, a larger number than were to have been sent to any other river (*Treasury Board Papers*, 34/51 ; and cf. below, vol. ii, p. 311). If the number of officials can be taken as an indication of the traffic, we may assume that nearly 40 per cent of the coal shipped upon inland rivers went up or down the Severn. The shipments are said to have exceeded 100,000 tons in 1758 (*Gents. Magazine*, vol. xxviii, 1758, p. 277), and, as there are many indications that Coalbrookdale was one of the most advanced colliery districts before 1700 (cf. above, p. 65), it is probable that the shipments did not fall much short of this figure at the end of the seventeenth century. For the coal traffic on the Meuse and Scheldt, see below, pp. 128–9.

[4] In 1719 the best Forest of Dean coal sold for 9s. a ton in Monmouth and 8s. a ton in Lydbrook (*V.C.H. Gloucestershire*, vol. ii, pp. 227, 229), as compared with the 6s. paid for a ton of Shropshire coal sold in Tewkesbury in 1678.

[5] The special conditions under which coal was mined in the Forest of Dean are discussed at some length below, pp. 277–81, 290.

[6] *V.C.H. Gloucestershire*, vol. ii, p. 228. In other districts there were numerous, and sometimes successful, attempts to restrict the shipment of coal, but nowhere, except in the Forest of Dean, were these attempts supported by

were concerned, this policy appears to have been successful. We have no evidence of loadings either at Lydney or Newnham,[1] on the northern bank, just outside the bounds of the Forest. If no shipments were made here, that may have been due to the high cost of bringing fuel from the pits to the water. On the west side, where the Forest of Dean is bounded by the Wye, which Defoe called " a very noble River ",[2] there was no such difficulty. Coal could be loaded into small barges at various wharves, and it was regularly carried up-stream, during the reign of Charles II, to Redbrook, Monmouth, Lydbrook, Welsh Bicknor, and still farther north to villages along the river Lugg, which empties into the Wye below Hereford.[3] It might be supposed that this up-river traffic was accompanied by shipments of coal down-stream, beyond the point where the Wye joins the Severn, three miles below Chepstow ; for Chepstow, as Defoe explains, was " the Sea-port for all the Towns seated on [the Wye] and the Lug, and where their Commerce seems to centre." [4] But Forest of Dean coal was rarely shipped farther south than Chepstow. In fact, the shipments down-stream do not appear to have supplied that busy place with all the fuel it required. Its port books reveal occasional imports of Welsh coal from Glamorganshire, and of Kingswood coal from Bristol.[5] The market for water-borne coal from the Forest of Dean was, it seems, purely a local, up-river market, confined to the valleys of the Wye and Lugg.

Although the Trent was somewhat less cheap to navigate than the Severn, and although the collieries west of Nottingham were not quite as conveniently placed as those of Shropshire for loading coal into boats beside the river bank, still the traffic in fuel on the Trent at the end of the seventeenth century did not perhaps fall far short of that on the Severn. A traveller between Nottingham and Hull, where Midland coal came into competition with that from the north of England, would have met with nearly as many vessels bearing this fuel, and would have seen nearly as many heaps piled up on the wharves of the principal towns, as a traveller between Shrewsbury and Gloucester. "The Trent," wrote Defoe in 1726, "is navigable by Ships of good Burden," probably sailing vessels similar to those used on the Severn, " as high as Gainsborough, which is near forty miles from the Humber by the River." He adds that "Barges, without the Help of Locks or Stops, go as high as Nottingham", but points out in another place that " the Town has been at a great Expence in making the Trent navigable here for Vessels or Barges of Burden", suitable for

the colliery owners, or by a body representing them (cf. below, vol. ii, pp. 211 sqq.).

[1] This does not prove that no shipments of coal by river were made on the eastern and southern sides of the Forest. The mines within Sir John Winter's manor of Lydney were worked for profit in the reign of Charles I (see below, vol. ii, pp. 11-2), and some of the produce may have found its way on to the Severn, but, if the traffic in Gloucestershire coal on that river had been extensive, we should hear more about it.

[2] Defoe, *Tour*, 1748 ed., vol. ii, p. 354.

[3] *V.C.H. Gloucestershire*, vol. ii, p. 227 ; *Treasury Board Papers*, 34/51.

[4] Defoe, *op. cit.*, vol. ii, p. 356.

[5] Three chaldrons of " seacoles " were shipped from Chepstow in 1620, and again in 1633 (*Exch. K.R. Port Books*, 1273/13, 1275/11) ; but imports of coal were more frequent than exports, though we find very few of either.

carrying " their heavy and bulky Goods ".[1] At the beginning of the seventeenth century, when there is evidence of an important increase in the coal traffic upon the river, " the barges, boates, or keeles," loaded with coal, often went aground on the sandbars above and below the town, to the despair of the merchants and colliery operators who owned them ; [2] and the expense to which Defoe refers was probably undertaken to relieve the distress of these owners. The town of Nottingham had an important interest in the success of the coal traffic, without which the citizens were likely to be left with insufficient supplies of grain. In 1620 the justices of the peace had " no fear of future scarcity, since other counties who send up the Trent for coals bring in corn whenever it is needed." [3]

Most of the fuel carried by water from the Nottinghamshire collieries appears to have been consumed in that populous, fertile part of the Trent valley which lay between the Humber and Newark, where the river changes its course from an easterly to a northerly direction.[4] One Thomas Stringer, of Ivychurch, Wiltshire, who has left in manuscript " A book concerning the time ", written about 1655, estimated the quantity of coal that could be profitably absorbed by " Newarke and the townes adiacent " at 16,000 tons per annum.[5] We have no means of estimating the quantity which reached Gainsborough, for consumption in that town and the neighbourhood as far east as Lincoln, or the quantity which was unloaded from the barges at Gainsborough to be reloaded into vessels of greater burden for carriage to points still farther down the Trent valley, and even to points in the valley of the Humber, which is joined by the Trent below Goole. But there is every reason to believe that the coal traffic below Newark was of considerable and growing importance throughout the seventeenth century. If coal enjoyed a less ready sale at Gainsborough than at Worcester or Tewkesbury, this was due almost entirely to the higher price on the Trent ; for, owing chiefly to the heavy labour of propelling lighters with oars for so long a distance, coal at the wharves in Gainsborough cost in 1605 at least 9s. 4d. a ton, or half as much again as in Tewkesbury seventy years later, when prices generally were somewhat higher.[6] Where, as in the Trent valley, population was

---

[1] Defoe, op. cit., vol. iii, pp. 58, 63.

[2] Hist. MSS. Com., Report on MSS. of Lord Middleton, pp. 172–3, 175–6.

[3] S.P.D., James I, vol. cxiii, no. 22.

[4] The town of Nottingham probably received its coal supplies overland, rather than by water carriage, from the neighbouring collieries of Wollaton and Bilborough (see above, p. 60). Notwithstanding the proximity of these collieries, the price of coal was high enough to cause concern to the Corporation of Nottingham, and to lead that body to encourage projects for finding coal within the town lands (see below, vol. ii, p. 19).

[5] Additional MSS., 33509. See f. 10b : " An undeniable proportion of Coles to bee led from the pitts neere Nottingham to the Trent side, and soe by boats to Newarke and the Townes adiacent."

[6] Hist. MSS. Com., Report on MSS. of Lord Middleton, pp. 171–2. This paper, which sets forth a project for bringing coal from Nottinghamshire to London (see below, p. 110), describes the manner in which coal was brought down the Trent, and has served as a basis for the account of this traffic given in the text. The cost of Nottinghamshire coal shipped to Hull was estimated at 13s. 2d. per ton.

scattered in a number of small towns or villages, a price of 9s. a ton was likely to restrict the amount of coal consumed per head to a smaller quantity than a price of 18s. a ton in a great city like London.[1] And the quality of the coal dug near Nottingham was such that it burned more rapidly than Newcastle coal, and was therefore more expensive to use for cooking or for warming the house.[2]

When we have mentioned the Trent, the Wye, the Severn, the Tyne, the Wear, and the Clyde (down which, although it was only navigable to ships of less than ten tons displacement, some coal from collieries at Cambuslang, Rutherglen, and Old Monkland was brought in shallow barges to Glasgow and Dumbarton [3]), we have exhausted the list of rivers extensively used for the inland water carriage of coal in the seventeenth century.[4] None of these rivers, except the Severn, offered facilities for transport by boat such as could not have been improved by proper dredging. Yet in 1695 the central government contemplated the appointment of eighteen officials (each at a salary of £50 per annum and with the equipment and personnel essential to his office) solely for the purpose of collecting a duty of 3s. 6d. per ton, to be levied upon coal shipped on inland rivers.[5] It is remarkable that, in a country naturally so poor in navigable streams, the river traffic in coal should have assumed such proportions. The setting up of so expensive an office could hardly have been considered, unless the Midland traffic in water-borne coal had been already important. And, when we add to that traffic the tens of thousands of tons that were carried on the Tyne to supply the salt-pans at Shields, it seems probable that, at the end of the seventeenth century, more coal was carried on inland rivers than was shipped overseas, to Ireland and foreign countries.

(ii) *Markets Reached by Land Carriage*

What was being done to tap those " inexhaustible stores " [6] of inland coal which had no access to water transport ? From one colliery

---

[1] See above, pp. 88–9.

[2] See above, p. 12 n., and below, p. 119.

[3] Tucker, *Report upon the . . . Revenues of Excise and Customs in Scotland,* p. 38. A wharf for storing fuel mined at the collieries owned by the Hamilton family was built on the Clyde at Broomilaw before the end of the seventeenth century (see above, p. 50). The rebuilding of the wharf in 1716 marked the beginning of a more important traffic in coal upon the river Clyde. See also *Hamilton MSS.*, charter of November 16th, 1661 ; Sinclair, *Statistical Account of Scotland,* vol. v, p. 257, and vol. ix, p. 8. Small quantities of coal were shipped from Glasgow to Ireland during the seventeenth century (*Exchequer Rolls of Scotland,* vols. xviii, xx, *passim* ; *Customs and Excise, Misc. Accts. and Papers,* 1620–1, and *Customs and Foreign Excise Collectors' Accts.*, 1665–72 (General Register House, Edinburgh) ; and see above, p. 49).

[4] Some coal, brought overland from the West Riding, was freighted up and down the Ouse in Yorkshire, where an official was to have been stationed to collect the proposed duty on inland river coal.

[5] *Treasury Board Papers,* 34/51. Cf. below, pt. v, ch. iv (iii).

[6] Defoe, *Tour,* 1769 ed., vol. iii, p. 128.

district, at least, the traffic in coal borne away overland in two-wheeled carts, or on the backs of pack-horses, assumed such proportions during the seventeenth century as almost to rival the trade down the great river thoroughfares, the Severn and the Trent.[1] Watling Street, the ancient highway connecting London with the Irish Sea at Chester, together with the equally ancient Fosse Way, which it crosses just east of Nuneaton, provided an outlet to the south and east for fuel from the Warwickshire coalfield, which occupies an area of about sixty square miles to the north of Coventry, and which is the Midland field nearest to London. Between Tamworth and Nuneaton seams ran close to the surface all along Watling Street ; and, at the end of the sixteenth century, substantial sums of capital were invested for the first time in these mines.[2] In 1632 pits were actually dug under " the Rode through which the great droves of Cattell doe dayly passe upp to London from Lancashire, Shropshire [and] Cheshire." Underground excavations threatened to damage the highway and interrupt traffic.[3] Tradesmen and other travellers thus became familiar with Warwickshire coal, and encouraged its carriage along the great Roman roads, in tracing which, Defoe wrote, " we necessarily come to the principal Towns, that either are or have been ".[4] Coal was carried in some quantities fifteen miles north-east to Leicester, and southwards into Northamptonshire, southern Warwickshire, and even Oxfordshire.[5] The sale for consumption at a distance from the field itself was facilitated, not only by the passage of through travellers along Watling Street and the Fosse Way, but by the resort of traders to Coventry, a market-town of capital importance in the sixteenth and seventeenth centuries, and a great centre for cloth, brought thither from all the adjoining counties to be dyed and dressed.[6] Visitors on business, even from distant parts, were tempted to provide themselves with a few sacks of coal from the neighbouring pits, just as early traders to Newcastle ballasted their ships with coal for the return voyage. Other " strangers " came primarily to purchase coal. According to the testimony of several townsmen in 1682, that trade had added largely to the number of persons visiting Coventry.[7]

In no other part of Great Britain do we find a considerable market for the " land-sale " of coal, as it was called, at such long distances from the pits as apparently existed during the seventeenth century in the country towns and villages to the south and east of Coventry. The situation of the Warwickshire collieries at the extreme southern end of the Midland coalfields, with the nearest navigable water, the

---

[1] For a more complete description of the manner of carriage overland, see below, pp. 381–4.

[2] See above, p. 67.

[3] *P.C.R.*, vol. xlii, pp. 65–6 (letter to the justices of the peace of Warwick County next adjoining to Griff Lane, June 1st, 1632).

[4] Defoe, *op. cit.*, vol. ii, p. 415. On the importance of Watling Street and the Fosse Way see *ibid.*, vol. ii, *passim*.

[5] H. Stocks, *Records of the Borough of Leicester*, vol. iv, 1923, p. 241 ; *P.C.R.*, vol. xxxviii, p. 424 ; *Exch. Deps. by Com.*, 36 Charles II, Mich. 43. For a purchase of Warwickshire coal at Wormleighton, see Thorold Rogers, *Hist. of Agriculture and Prices*, vol. vi, p. 369.

[6] Cf. *V.C.H. Warwickshire*, vol. ii, pp. 251 sqq.

[7] *Exch. Deps. by Com.*, 36 Charles II, Mich. 43.

Cambridgeshire Ouse, fifty miles to the east, and the Thames, fifty miles to the south,[1] prevented other fields from competing with the product of these collieries, either by water or land sale ;[2] the abundantly travelled roads facilitated the carriage of commodities, even of so bulky an article as coal ; and the greater concentration of population in Warwickshire, Northamptonshire, and Oxfordshire than in most districts farther to the north stimulated the demand. All three factors combined to make profitable what was, for those days, a very long haul of fuel overground.

Generally the cost of transport prohibited any considerable sale at distances of more than ten or, at the most, fifteen miles from the collieries. Lord Harley, in 1725, commented upon the streams of pack-horses that he encountered all along his route from Darlington to Bishop Auckland.[3] A regular trade in coal from south Durham into the Tees valley had existed at least as early as the fifteenth century. Yet, in spite of the antiquity of this trade, Newcastle coal, carried fifty miles by water, and loaded several times, competed, during the Stuart period, at Stockton with south Durham coal brought directly by pack-horses from the mines.[4] To reach the town of York, Newcastle coal had to be carried at least two hundred miles by water, and loaded five separate times—at the pithead into carts, at the wharves from carts into keels, from the keels into sea-going colliers anchored in the Tyne, from the colliers into river lighters on the Humber, and from the lighters into carts at York for carriage to the house of the consumer or merchant. Yet, until the deepening of the Aire and Calder at the beginning of the eighteenth century, York received some of its coal from the north of England, in spite of the fact that supplies could be had by land carriage from the mines of the West Riding only twenty miles away. It is well known that few roads were fit for heavy traffic in England in the seventeenth and early eighteenth centuries.[5] If

---

[1] The Avon to the south was navigable to within 15 miles of Coventry (see above, p. 96), but it may be doubted whether Shropshire coal, brought down the Severn and up the Avon, was any cheaper at Stratford than Warwickshire coal brought overland.

[2] Coal from south Staffordshire may have competed with Warwickshire coal in Oxfordshire. At the end of the seventeenth century, the accounts for Magdalen College begin to contain purchases of Midland, as well as of Newcastle, coal. This is stated to be " Wedgebury " coal (Rogers, *op. cit.*, vol. v, p. 388). The name strongly suggests Wednesbury, although the reference may conceivably be to Hawkesbury, north of Coventry, where there were at the time important coal pits.

[3] Lord Harley, *Journeys in England* (printed *Hist. MSS. Com., Report on the MSS. of the Duke of Portland*, vol. vi, pp. 100 sqq.).

[4] Stockton regularly imported small quantities of Newcastle coal during the seventeenth century (see Appendix D (v)).

[5] See Sidney and Beatrice Webb, *Story of the King's Highway*, esp. ch. v ; also Arnold Toynbee, *Lectures on the Industrial Revolution*, 3rd ed., 1908, pp. 28–9. In a recent article Professor Gay has suggested that Arthur Young's writings (which have served as the main source for most modern research on the condition of the roads in the late eighteenth century) can be interpreted as showing that substantial improvements in road travel had been made during his life-time (" Arthur Young on English Roads," in *Quart. Journ. of Econ.*, vol. xli, 1927, pp. 545–51). But there is general agreement among all modern authorities that most roads during the seventeenth and the early eighteenth centuries were in a deplorable condition.

the lanes of London were so full of ruts as to lead Montesquieu to exclaim, " il faut faire son testament avant d'aller en fiacre,"[1] if Watling Street itself was passable by carts only at certain seasons of the year, the state of smaller routes, safe enough for sure-footed pack-horses, with their small loads, must have been such as seldom to permit the passage of any cumbersome vehicle.[2] The price of coal seems almost to have doubled with every two miles it was carried from the mines. At Darlington, coal brought ten miles from Etherley colliery cost four times the pithead price.[3] As Sir Robert Southwell pointed out in a paper that he read before the Royal Society in 1675, the expense of carrying coal three hundred miles by water was ordinarily no greater than that of carrying it fifteen or twenty miles by land, and in the case of the Newcastle-London trade no greater than that of carrying it three or four miles by land.[4] The coal of Somerset, which was mined in manors along the famous Fosse Way, to the east and north of the Mendip Hills, found so ready a sale along this road and its bypaths, in the towns of Wells, Glastonbury, Frome, Warminster, and Bath (where the growing leisured class of England already tended to congregate), that " to goe to Mendip for coles " became a common expression locally.[5] But it must be remembered that all these towns stood within a ten-mile radius of the mining district, so that coal sold in all of them for a much lower price than in London.[6] Beyond the ten-mile radius, there was probably no substantial market for Mendip fuel.

If the expense of land carriage proved so great as generally to prohibit the haulage of coal for any considerable distance, if land carriage was so vastly more costly than water carriage, how is it possible to reach the conclusion that the land-sale of coal in Great Britain at the end of the seventeenth century exceeded the water sale ?[7] The explanation is to be found in the manner in which Nature had spread coal seams, many of them rising to within a few feet of the surface, through a considerable part of Great Britain. By drawing a line on the map from Leeds to Preston, extending it south to Liverpool, thence east to Manchester, south through Newcastle-under-Lyme and Stafford to Wolverhampton and Stourbridge, thence east and north through Birmingham, Coventry, Leicester, and Nottingham, and so, through Rotherham and Pontefract, back to Leeds, we encircle a solid block of territory in the very heart of England, about six thousand square miles in area, in the whole of which there was, by the

[1] *Notes sur l'Angleterre*, 1729.
[2] The subject is discussed in detail below, pp. 381–2.
[3] Lord Harley, *Journeys in England*, as cited above.
[4] Birch, *History of the Royal Society*, vol. iii, pp. 207–10.
[5] *V.C.H. Somerset*, p. 381 ; *Hist. MSS. Com., Report on the MSS. in Various Collections*, vol. i, p. 114.
[6] The actual cost of freight by collier from Newcastle to London was less than the cost of hauling coal ten miles by land (Birch, *loc. cit.*), but in addition to freight, the buyer in the capital paid for the carriage by wagon and keel from the pit to the ship at Newcastle, for the carriage by lighter and cart from the ship to the house in London, and for taxes collected at both ports (cf. below, pp. 347–8 ; vol. ii, 102 n., 126 n., 308 sqq.). Consequently the price in London was higher than at places ten miles from the pits.
[7] See above, Tables I and VI.

end of the seventeenth century, scarcely a spot not within a morning's ride of a working coal pit.[1]   Between two lines, the one drawn from St. Andrews through Dumbarton to Largs on the Firth of Clyde, the other from Girvan by way of Abington and Peebles to Dunbar on the North Sea, there is another district covering almost as great an area, of which the same may be said.   And it may be said of two other large blocks of territory: the first, a strip along the North Sea from Berwick to Stockton, including nearly three thousand square miles, or the larger part of the counties of Northumberland and Durham ;   the second, a strip along the Bristol Channel, from St. David's Harbour on the Irish Sea to Gloucester on the river Severn.   When we include the districts around other small mining fields, such as those of Somerset, Kingswood Chase, Shropshire, North Wales, and Cumberland, outside the main coal-producing areas, it appears that the population of at least a fifth of all the territory of Great Britain was in a position to receive coal by land carriage at a lower—generally at a very much lower—price than that paid for water-borne coal in London.[2]   No other country in the world had such great natural advantages with respect to the land-sale of coal.

The coal-producing territory did not include what were still, at the end of the seventeenth century, the most thickly settled parts of England.   Probably the number of persons in England, Scotland, and Wales combined, who lived within fifteen miles of the pits, did not equal the number living in the east and south-east.   But it is also probable that the consumption per head in the neighbourhood of the pits was at least twice as great as in the latter region.   While the price fifteen miles away was almost prohibitive, except for rich country gentlemen or the inhabitants of populous towns, where all substitute fuels were dearer, this price diminished with astonishing rapidity as one approached the mines.   In Monmouthshire and the Forest of Dean, Yarranton tells us in 1677, manorial tenants no longer found it worth while to cut down their own timber trees for fuel, " because in all these places there are Pit coals very cheap." [3]   " Tis common," wrote Defoe fifty years later, " in the meanest Cot to see a good Fire." [4] Until the reign of Charles I, the price in the neighbourhood of Newcastle was less than two shillings per ton, as against ten or fifteen shillings in London.[5]   Coal sold for two pence a barrel (or about fifteen pence

---

[1] The coal seams east of a line drawn from Nottingham through Rotherham to Pontefract were, of course, too deep under the surface to be accessible to the seventeenth-century miners.

[2] The total area of the island of Great Britain is above 90,000 square miles. The coal-producing area, as I have defined it, included about 6,000 square miles in the Midlands, 5,000 in Scotland, 3,000 in Durham and Northumberland, 2,500 in South Wales and Monmouthshire, and perhaps 2,000 more in various counties containing small coalfields ;   or, in all, more than 18,000 square miles of territory.

[3] *England's Improvement by Sea and Land*, vol. i, p. 57.

[4] *Tour*, 1769 ed., vol. ii, p. 357.   Coal sold at the pits in the Forest of Dean in Defoe's time for 2*d*. a horseload, less than 2*s*. per ton.

[5] Sir Wm. Brereton, " Travels in Holland . . . England, Scotland, and Ireland, 1634–5," in *Publications of Chetham Soc.*, vol. i, 1844, pp. 85–6 ; and below, Appendix E.

a ton) at Neath in the early seventeenth century.[1] It was common
for the lords of manors, or the lessees of their coal, to make an especially
advantageous price for their tenants. Within the manor of Etal,
just south of the Tweed, the tenants were allowed coal in 1584 at a
penny a horseload of about two hundredweight, a very low price
even in terms of modern currency.[2] Nearly all the miners received
" firecoal " free, as a part of their wages.[3]

Under these conditions, the market for coal within a radius of two
or three miles from a colliery was bound to be considerable, even though
the local population had not assembled in towns. While a poor
man living in the country would have collected his own brushwood for
a fire before consenting to pay half the price charged to members of
the growing London populace, he was likely to leave the brushwood
lying on the ground if he could have coal delivered at his door for a
sixth of the London price, or, as was often the case, could eliminate
the cost of carriage by taking his own sack to fill at his neighbour's
pit. In Pembrokeshire, at the beginning of the seventeenth century,
" for the most part those that dwell neere the coale or that may have
it carried by water with ease, use most coale fiers in their kitchens
and some in their halls." [4] In the course of a letter written in 1584
to Walsingham from Staffordshire, Sir Ralph Sadler comments upon
the fact that coal is " much used in this countrey, and compted their
best fewell." [5] One south Durham small tenant-farmer, named Curle,
who could afford to be lavish in burning coal, because he received a
free supply in return for granting a colliery owner wayleave through
his holding, used twenty tons a year, or nearly as much as was purchased
for the mansion of a well-to-do lord or merchant in London.[6] In 1580,
a Scottish family, living near the coal pits of Pencaitland in East
Lothian, was provided in a similar way with eight horseloads (of
about two cwt. each) per week, so that its annual supply may have
reached thirty tons.[7] During the seventeenth century coal became
the common fuel of most of the country farmers in the Lothians, Fife,
Stirlingshire, Lanarkshire, and Ayrshire, but in the Highlands of
Scotland, even in the eighteenth century, the farmers were forced to
get along almost entirely without it, owing to the excessive price,
though they had the greatest difficulty in finding any other fuel with
which to warm their huts, and had often to perform their husbandry
without any fertilizer because they had no peat or brushwood to spare
for burning lime.[8]

In the coal-producing districts colliery owners did not depend
exclusively, or even predominantly, upon the country inhabitants
to buy the sacks or cartloads of coal offered for sale locally. Sprinkled

---

[1] G. G. Francis, *Smelting of Copper in the Swansea District*, 1867, p. 83.

[2] *Augm. Partics. for Leases*, 112, ff. 12, 15, 27, 30. For an estimate of the
probable weight of a horseload of coal see Appendix C (iii).

[3] See below, vol. ii, pp. 187–8.

[4] Owen, *Description of Pembrokeshire*, pt. i, p. 86.

[5] Cited Galloway, *Annals of Coal Mining*, p. 116.

[6] *Exch. Spec. Com.*, no. 5276. Cf. with the provision made in 1637 for the
Earl of Rutland's town mansion (see below, p. 198).

[7] *Hist. MSS. Com.*, *Report on the MSS. of Col. Home*, p. 187.

[8] Sinclair, *Statistical Account of Scotland, passim*.

all through these districts were growing towns, providing an ever-increasing market for the output of the mines. A place of five thousand inhabitants would hardly be considered a town to-day ; but we must remember that before the fifteenth century there were probably not a dozen centres in the whole island of Great Britain with as large a population as this, and that (with the exception of London, which numbered perhaps forty thousand souls in the late Middle Ages) there was probably not a single centre with as many as 12,000 inhabitants.[1] Some day historical research undoubtedly will show that during the sixteenth and seventeenth centuries there occurred a very important movement from the country to the towns, and that the proportion of urban to rural population was much greater in 1700 than in 1500.[2] Historical research may also prove that there was, at

---

[1] George Chalmers, *An Estimate of the Comparative Strength of Great Britain*, 1804 ed., pp. 16 sqq. The well-known estimate of 1377, based on the poll-tax returns for that year, gives London a population of 34,971, York 10,872, Bristol 9,517, Plymouth 7,255, and Coventry 7,225. There are only seven towns with a population of more than 5,000 persons, three other towns with a population between 4,000 and 5,000, and eight more with a population between 3,000 and 4,000.

[2] Gregory King, writing in 1696, estimated the population of London at 530,000, that of all the other " cities and market-towns " of England and Wales at 870,000 ; so that—if his estimate is near the mark—approximately one-fourth of the population lived in what were then called towns. So far as I am aware, no estimate exists of the relative proportion of urban to rural inhabitants at the end of the fifteenth century, and, if we are to compare conditions at the end of the seventeenth century with those in 1377, then we ought to know what was the population of Gregory King's smallest market-town. Let us assume that it contained at least a thousand inhabitants. In 1377 not a tenth of the population of England and Wales seems to have lived in places as large as this ; out of a population of about two millions and a half, all but 200,000 lived in tiny villages containing fewer than a thousand souls. It would appear, therefore, that the progress of opulence before 1700, as measured by Adam Smith (in the third book of *The Wealth of Nations*) in relation to the growth of towns, was marked ; and, from our general knowledge of the history of agriculture and manufactures (see below, pp. 165–91, 226–33, 237), it seems more than probable that most of this progress occurred after 1500, and particularly after the accession of Elizabeth.
 According to the generally accepted view, no English town, except London, increased very rapidly in size until the eighteenth century. Doubtless progress before 1700 was slow in comparison with the extraordinarily rapid growth of provincial towns between 1750 and 1850. But has there not been a tendency since the time of Macaulay to minimize such progress as actually took place before 1700 ? Macaulay based some of his figures of the population in 1688 upon contemporary estimates ; but, in his selection and interpretation of sources, he was guided too much by his desire to stress the tremendous growth of wealth and population since the period about which he chose to write his history. (Cf., for instance, his interpretation of Pepys' description of Bristol with the actual entry in the *Diary*.) He gives Birmingham 4,000 inhabitants in 1685, but, according to another estimate, the population of this town grew from about 5,000 in 1650 to about 15,000 in 1700 (Hamilton, *The English Brass and Copper Industries to 1800*, 1926, p. 125). While most of Macaulay's other estimates are probably nearer the mark, even a small under-estimate for the end of the seventeenth century may warp one's view considerably in comparing the growth of towns between 1700 and 1750 with the growth between 1550 and 1700.
 The contrast is striking between the picture painted by Defoe, in his *Complete English Tradesman* and in his *Tour through Great Britain*, of a country teeming with busy market-towns, and the picture painted by Macaulay of a few

the same time, a significant growth in the population of many villages, in which the inhabitants were employed an increasing proportion of their time in the rising rural industries rather than in husbandry, so that the new centres, while not towns, were assuredly not, in the medieval sense of the word, villages. It will probably be found that these changes were more marked in the colliery districts than in any other part of Great Britain except the Thames valley.[1] It is almost certain that many of the busiest places in these districts—Manchester, Liverpool, Leeds, Wakefield, Halifax, Sheffield, Wigan, and Sunderland (to name only a few)—had grown from the smallest of villages to flourishing centres of trade and population. Other places, like Newcastle, Edinburgh, Glasgow, and Bristol, already towns of some importance at the beginning of the sixteenth century, probably tripled or quadrupled in population before the beginning of the eighteenth. It is the number of growing towns in the mining districts, rather than the size of any particular place, that is impressive. Those already named, and many more, stood within ten miles of important collieries. In some cases pits were dug within the municipal lands, and, only a few steps from the doors of the citizens, coal could be had for a low enough price to insure a prodigious consumption.

The markets for coal provided by towns in and near the colliery districts were almost as completely creations of the sixteenth and seventeenth centuries as the markets for water-borne coal. While there is plenty of evidence that small quantities of coal had been burned during the thirteenth, fourteenth, and fifteenth centuries in towns and villages within a few miles of the outcrops, close investigation shows that the common grades of coal never came into general use until the supply of other fuels grew scarce. And in most towns a genuine scarcity set in in the second half of the sixteenth century.[2] According to Chambers, the historian of Scottish manners, peat remained the chief fuel in Edinburgh in 1563 ;[3] but in 1595 the coal traffic

trading centres, which his swinging rhetoric reduces to insignificance. This contrast can hardly be due altogether to changes that occurred during the few years that separate the England depicted by Macaulay from the England of Defoe. And, even after we have made allowance for Defoe's well-known habit of exaggerating (he puts the population of London at a million and a half), abundant evidence remains to show that the increase in the urban population between the end of the fifteenth and the beginning of the eighteenth centuries was only less remarkable than the increase after 1700 which Macaulay and Toynbee and Mantoux have stressed.

[1] Between 1700 and 1750 the five counties where population apparently grew most rapidly—Warwickshire, Staffordshire, Durham, Lancashire, and the West Riding of Yorkshire (Toynbee, *Industrial Revolution*, 1908 ed., p. 10)—were among the most important coal-producing areas.

[2] See below, pp. 158–61.

[3] R. Chambers, *Domestic Annals of Scotland*, 1858, vol. i, p. 24. There is reason to believe that, owing to the particularly agreeable fire which could be made with some grades of Scotch coal (see above, p. 12, and below, pp. 117–9), and to the lack of suitable firewood, such as was found in most parts of England, coal was burned, as something of a luxury, by the Court and nobility of the northern kingdom, while the common people in the Lothians kept to peat and brushwood until their price rose above that of coal. Elsewhere, as we shall see (below, pp. 196–8), the order of adoption was generally the reverse ; it was the common people who first burned the new fuel, which was shunned by the rich (cf. Galloway, *op. cit.*, pp. 26–7).

into the Scottish capital was large enough to cause the people of several mining villages in Midlothian to petition the Scottish Privy Council to repair the bridge called " Lady Brigend " ; [1] and by 1620 regular " caryaris of coillis " made it their business to buy fuel at the pits (or " on the hill " as it was called), and transport it in carts, or on the backs of pack-horses, to Edinburgh and all the other towns of East- and Mid-Lothian, where it had become almost the universal fuel.[2] At Bristol, in 1566, " all manner of fewell is good cheap ", and the existence of " a myne of sea cole " near at hand is mentioned as an afterthought ; but by 1620 the poor of the town have all taken to burning this mineral as the sole means by which they can maintain their fires, and it was not long before the same fumes were issuing from the hearths of the rich.[3] A visitor to Bristol in the reign of Charles II could hardly have walked along the narrow lanes, so strikingly depicted by Macaulay, without flattening himself against the walls of houses a number of times to get out of the way of two-wheeled carts filled with sacks of coal to be deposited at the doors of the townspeople.[4] The cheapness of this fuel—which, according to Defoe, was delivered for 7s., 8s., and 9s. per chaldron [5] (or from 5s. to 7s. per ton), about a third of the price in London—must have encouraged them to burn it in great quantities. In Leicester, a place situated farther away from collieries than Bristol or Edinburgh, there is no mention of coal in the very complete records of the Corporation until 1569 ; but, before the end of Elizabeth's reign, it has become indispensable to the maintenance of the poor during the winter.[6] Although " seacoal " was known in Coventry in the fifteenth century, and probably long before,[7] the demand for it became pressing only during the last two decades of Elizabeth's reign.[8] In the seventeenth century, the colliery owners of Warwickshire depended for their market mainly upon Coventry,[9] notwithstanding the special opportunity they had to transport fuel farther to the south and east. Coal was regarded as an essential commodity by the citizens of Manchester at least as early as 1615, and by the citizens of Wigan at least as early as 1602.[10]

Enough has now been said to show that the increase in the demand for coal was not confined to particular parts of the British Isles, any more than the increase in output was confined to particular mining districts. Before the eighteenth century, the new fuel had come to supply the great majority of the population with its necessary firing.[11]

---

[1] *Privy Council Register of Scotland*, 1st Ser., vol. v, p. 227.
[2] *Ibid.*, vol. xii, pp. xx–xxiv, 387–8, 418–19, 433–5, 466–7, 474.
[3] Galloway, *op. cit.*, pp. 118, 212–14 ; *Egerton MSS.*, 2044, ff. 12, 18.
[4] Macaulay, *Hist. of England*, ch. iii. Cf. *A New Present State of England*, 1727 (?), vol. i, p. 205.
[5] Defoe, *Tour*, 1769 ed., vol. ii, p. 313.
[6] Mary Bateson, *Records of the Borough of Leicester*, vol. iii, p. 128, and *passim*. See also the subsequent volume of records edited by Helen Stocks.
[7] M. D. Harris, *The Coventry Leet Book*, p. 339.
[8] *V.C.H. Warwickshire*, pp. 265–6.
[9] *Exch. Deps. by Com.*, 36 Charles II, Mich. 43.
[10] *Duchy of Lancs. Pleadings*, 195/S/10 ; *Star Chamb. Proc., James I*, 106/7.
[11] Cf. below, p. 222.

The growth in the consumption per head, while most rapid in the large towns and in the coalfields, was everywhere remarkable. The general expansion in output during the period from 1550–1700 can only be attributed to a change in the habits with respect to fuel of the entire British people.

(iii) *Inter-Regional Competition : The Different Grades of Coal*

A survey of coal markets in the sixteenth and seventeenth centuries would be incomplete without a study of the conditions upon which depended a monopoly over the sale of coal in a particular market, such as that possessed by the colliery owners of Durham and Northumberland over the south-east. Such a study leads inevitably to the question how far the market controlled by a given producing region was contracted or expanded ; or, in other words, to what extent the various regions competed for markets ?

Two collieries, or two colliery districts, could ordinarily compete with one another only in markets accessible to each on fairly even terms with respect to the expense of carriage. Such conditions prevailed in the Low Countries, particularly in that amphibious district at the mouth of the Rhine and the Scheldt, accessible by water transport to the coal of several regions. There was constant competition for this market between the colliery owners of Scotland, of the Tyne valley, and of Liége.[1]

There was also competition between the colliery owners of various districts for the Irish market. Costs of carriage across the Irish Sea were roughly equal from Ayrshire, from Cumberland, from Flintshire, from Pembrokeshire, and from Lancashire ; and in all these counties coal seams could be worked within six miles of the water. Owing to the situation of Ayrshire and South Wales, that part of the Irish coastline along which the colliery owners of these two districts could compete with the coal from the other districts was limited : to southern Ireland in the case of South Wales, to Ulster in the case of Ayrshire. Dublin, where nearly two-thirds of all the coal was burned, remained effectively beyond the reach of both. The struggle between the mine owners of Flintshire, of Cumberland, and of southern Lancashire for control of the Dublin market apparently proceeded on fairly even terms during the first half of the seventeenth century. Although coal from the pits at St. Helens, Prescot, Whiston, Huyton, and Sutton had to be hauled in carts five miles and more to the wharves in the tiny port of Liverpool, it competed successfully with coal loaded into vessels a few score yards from the mines at Mostyn and Bagillt on the Dee estuary, and at Workington in Cumberland. In 1630 the shipments of coal from Liverpool to Ireland amounted to about 4,000 tons, or almost half as much as the shipments from Flintshire.[2] By this time the shipments from Cumberland (which had been more important than those from Liverpool, though less

---

[1] Cf. below, vol. ii, pp. 228–9.   [2] See Appendix D (iii).

important than those from Flintshire, during the reign of James I) probably exceeded those from Liverpool and from Flintshire combined. By 1675, as is shown in a letter addressed by the English Privy Council to the Lord-Lieutenant of Ireland, Sir John Lowther had driven practically all his rivals out of the Dublin market.[1] Administrative skill and sufficient resources, combined with the natural advantages which the region offered for mining and bringing coal cheaply to the waterside, enabled Lowther to sell to the shipmasters at a lower price than the colliery owners and merchants of Flintshire and Lancashire ; and, for at least a century, the Lowther family was able to maintain something closely approaching a monopoly over the sale of coal in Dublin. But, in northern Ireland, the Lowthers had to compete to an increasing extent with coal from Ayrshire, and, in the south, coal from Pembrokeshire and Glamorganshire had a great advantage over that from mining fields farther to the north.

While we can speak of competition between the colliery owners of several districts for the Irish and the foreign markets, it is hardly possible to find such competition in connection with the markets within the island of Great Britain itself. Along the fringes between two markets, as in the lower valley of the Humber, where Durham and Northumberland coal brought by sea encountered Nottingham-shire coal brought down the Trent, or as on the English Channel at Southampton and in the islands of Guernsey and Jersey, where north-country encountered Welsh coal, there was competition. But this competition did not extend to the heart of these markets, as it did in Holland, and, during at least a part of the seventeenth century, in Ireland. Neither the supremacy of Welsh coal in the south-west, nor the supremacy of Shropshire coal in the Severn valley, was seriously challenged. On no occasion, except during the Civil War, when parliamentary warships blockaded the Tyne, did London receive a considerable proportion of its coal supplies from other fields than Durham and Northumberland.[2] A group of London merchants combined with certain Midland gentlemen in 1605 to finance the carriage of Nottinghamshire coal down the Trent and Humber to Hull, and thence by sea to the Thames ; but, though they arranged to provide the necessary river barges and ships, nothing came of the project. Quarrels among the adventurers were perhaps the immediate cause for its failure, but what was bound to prevent the success of any project of the sort was the high cost of carriage. In spite of all possible economies, Midland coal could not be sold (as the calculations of these projectors show) under 22s. 2d. a ton in the Pool in London, against a prevailing price of less than 15s. a ton for New-castle coal during the reign of James I.[3] Welsh colliery owners were under a similar disadvantage when they tried to compete with Durham and Northumberland colliery owners in the south-east. In the eighties of the sixteenth century a load of Pembrokeshire " culm ", or anthracite, was sent to London for Burghley's inspection, and there

---

[1] *P.C.R.*, vol. lxiv, p. 480.
[2] Cf. below, vol. ii, pp. 286–8.
[3] *Hist. MSS. Com., Report on the MSS. of Lord Middleton*, pp. xi, 171–2 ; and see below, vol. ii, p. 79.

is no doubt that, as Owen explained, " great use " would have been made of it in the Thames valley, if " the passinge " round the Cornish peninsula and through the English Channel " were not soe tediouse ".[1] According to Sir Robert Southwell's estimate, the freight on coal from London to Land's End in 1675 was three times as much as the freight from Newcastle to London, owing to the delays involved in the former voyage.[2]

Nevertheless some " culm " was burned along the east coast. Records in the port books show that London, King's Lynn, and Yarmouth all received small, though regular, annual shipments from Milford Haven during the greater part of the seventeenth century. How can we account for this persistent movement of culm northward along the east coast, against the prevailing movement of coal, which was southward ?

Cost of transport, while the paramount factor, was not the only one in determining where the output of one producing region could find a market. The quality of the coal was already of some importance. As all persons who have had any connection with the trade are aware, there are numerous grades of coal, ranging, according to their chemical constituents and properties, from anthracite to lignite ; and not all grades will serve equally well for all purposes.[3] By the seventeenth century, the uses for which British coal was burned were sufficiently various to make it a matter of concern to traders what grades they could obtain, and to mine owners what grades their mines yielded.[4] In every important coalfield men were employed to sort, or " riddle ", the output from each colliery ; [5] and the problem of disposing of all grades at a profit, even when there was no competition from other coalfields, had become acute. At Newcastle, as the Lord Mayor and Aldermen of London reported in 1616, after an investigation undertaken for the Privy Council, are " several sorts of coals—sometimes arising from one and the same mine ".[6] Natural science had not yet developed sufficiently for men to make a very exact chemical analysis of the fuel ; generally they distinguished between different grades by the size of the block, by the brightness of the flame, by the odour of the fumes, or by the damage done to the materials with which the fumes came in contact. The only certain method was to try the coals in the fire. " Theare are two sorts of coales which are called sea coales," wrote in 1589 a petitioner for a patent to restrain the export from the Tyne of the better sort, " th'one much better then th' other, as by experience and use of them from tyme to tyme is founde." [7]

[1] Owen, *Description of Pembrokeshire*, pt. i, p. 88.

[2] Birch, *History of the Royal Society*, vol. iii, pp. 207–10.

[3] For an excellent discussion of the subject, intelligible to the lay reader, see H. S. Jevons, *British Coal Trade*, 1915, ch. iii.   See also J. Percy, *Metallurgy*, 1861, vol. i, pp. 267, 302–10.

[4] Cf. below, vol. ii, pp. 240 sqq.

[5] *S.P.D., James I*, vol. lxxxvii, no. 66, I ; Owen, *op. cit.*, p. 90 ; Dud Dudley, *Mettallum Martis*, 1665 ; and see below, p. 348, and vol. ii, pp. 242–3.

[6] F. W. Dendy, *Extracts from the Records of the Company of Hostmen of Newcastle-upon-Tyne*, 1901, p. 63, note. See also a statement made in 1627 by the London coal dealers to the same effect (*S.P.D., Charles I*, vol. l xviii, no. 48).

[7] *Lansdowne MSS.*, 59, no. 71. Cf. below, pt. v, ch. ii (i).

In using the word " better ", he was speaking from the point of view of the domestic consumer. It would appear that the " better " grade of Newcastle coal, consisting of the larger and less sulphurous lumps, was sought chiefly by the householder for use in the kitchen and the fireplace, by the beer brewer, the soap boiler, the glass maker, and other manufacturers who required a clear bright flame rather than an intense, smouldering heat.[1]   The smaller lumps, which were more noxious in the burning, were regarded as equally satisfactory for heating brick kilns and dyers' vats, and were actually preferred for most of the rough iron work performed by the smith, and for the making of lime.[2] The growth of the salt industry during the seventeenth century permitted the economy of using a still poorer grade, which had hitherto gone to waste in the underground passages of the mine, or had been cast aside at the surface by " riddlers " or " weallers ", except when it was mingled with other grades to deceive the buyer.[3] Contemporary authorities seem to agree that this " pan " coal (as it was called to distinguish it from the two other grades, called " ship " coal) was seldom " fitt .for any other thinge " than salt making.[4] They describe it as " a sort of Crusty, drossy mouldering Coale . . . taken from the upper part of the Mine ".[5]

The colliery owners of nearly all districts in the seventeenth century came to distinguish between at least three grades of mineral fuel, though the distinction did not always follow the same lines as in the Tyne valley.[6] In Pembrokeshire and the Forest of Dean, " riddlers " separated the output from the pits into " fire " coal, consisting of the largest blocks, " smith " coal, and " lime " coal. For each of these grades the local dealers asked a different price. At Liége and Mons, the dealers recognized only two sizes, the large and the small ; but they subdivided the larger blocks, which they called " houille ", into two grades according to their burning properties.[7] On both sides of the Firth of Forth, the fuel for shipment by sea was ordinarily divided into " great " coal, particularly suited for domestic use and for beer-brewers, soap boilers, glass makers, etc., and " small " coal, which was

---

[1] Cf. below, pp. 119-20.

[2] Cf. Dendy, *op. cit.*, p. 63, note ; Boislisle, *Correspondance des Contrôleurs Généraux*, vol. iii, no. 496 (Letter of M. Doujat, *intendant* in Hainaut, concerning the uses of coal in the Belgian provinces, July 9th, 1709).

[3] See below, pt. v, ch. iii (i).

[4] *S.P.D. Charles I*, vol. clxxx, no. 58 ; vol. cclxxxix, no. 109. The common opinion in the Newcastle district seems to have been that this grade was useful only for salt making (*Lansdowne MSS.*, 213, no. 30, and 253, no. 217 ; the latter is printed in M. A. Richardson, *Reprints of Rare Tracts*, 1847-9, vol. iii) ; although it is referred to on at least one occasion (*Lansdowne MSS.*, 198, no. 276) as also providing a satisfactory fuel for the burning of lime. We also hear of a " useless part of the coal ", serviceable only for salt making, in connection with the mines of Ayrshire and Staffordshire (Sinclair, *Statistical Account of Scotland*, vol. vii, pp. 10-11 ; John Collins, *Salt and Fishery*, 1682, pp. 10-11) ; so it would appear that the grade known as " pan coal " in the Newcastle region had its equivalent in other fields.

[5] *Lansdowne MSS.*, 253, no. 217.

[6] The line of division was arbitrarily drawn, of course, being based mainly, as appears in the text, upon the size of the blocks.

[7] Boislisle, *op. cit.*, vol. iii, no. 496. The word " houille " originated in the Liége coal field.

regarded as more appropriate for the operations of smiths. Indeed, it was usually called " smiddy " coal. Besides these grades the Scottish mine owners also distinguished between at least two further types for use in the home market—" stocken coall," chiefly burned in lime kilns, and " panwood ", used exclusively in the making of salt.[1]

No prudent adventurer invested his money in mines during the seventeenth century, without examining the nature of the demand that existed in the market where he planned to sell his output ; nor did he invest his money without grading the coal he could hope to obtain from his pits, usually according to its size. " The smallest [lumps,]" reasoned Sir Ralph Delaval, the admiral, in planning for the disposal of the fuel from one of his mines at Hartley, near the coast of Northumberland, " will serve for lime burning and the rounder will please the cook because they make a quick fire and a constant heat." [2]

The prices obtained for different grades in each producing region depended upon the nature of the market, and the supply of coals of various sizes. Since the major portion of all coal shipped coastwise from the Tyne valley during the last years of the sixteenth century, and throughout the seventeenth, was consumed as domestic fuel in the south-eastern counties,[3] the larger blocks generally sold at the highest rate, and the chief problem for the mine owner was the disposal of his small coals at a profitable price.[4] This second grade of " ship " coal could be sold to the exporter, for along the coast of France and Holland the demand came chiefly not from the domestic, but from the industrial, consumer ; and, among the industrial consumers, dyers and smiths, who actually preferred the smaller blocks, were very numerous. The smith, in fact, would not touch the larger pieces. This is shown by the observation of a French *intendant* in the St. Etienne district, where, as late as the beginning of the eighteenth century, almost the only market for coal was provided by the local forges. He reported that there was scarcely any sale except for what he called the " sand " (*sable*) of coal.[5] It is not surprising, therefore, that along the northern coast of the Continent, as one dealer in Newcastle coal put it, the smaller lumps " sell as well as those which we call the best ".[6] If the market for Newcastle coal on the Continent had approached in importance that in south-eastern England, the colliery owners could have disposed of both grades with ease. But the continental market was constantly being shut off. During the first Dutch War (1652–3), Englishmen

---

[1] *Hamilton MSS.*, no. 592 (2), bdl. 5 ; *Accounts of Kincardine and Tulliallan Coal and Salt Works*, 1679–80 (General Register House, Edinburgh).

[2] *Additional MSS.*, 21,948, f. 64 ; as quoted *A History of Northumberland*, vol. viii, p. 23.

[3] Cf. below, pt. ii, ch. iii. A considerable portion of the coal burned for industrial purposes in and about London was used in beer brewing, soap boiling, and glass making, all of which processes required a grade similar to that preferred by the domestic consumer.

[4] They were rejected on many occasions both by the domestic consumer and the brewer (*Lansdowne MSS.*, 65, no. 11 ; *Additional MSS.*, 12,496, no. 87 ; and see below, vol. ii, pp. 245, 250).

[5] Boislisle, *loc. cit.* (Letter of M. Trudaine, *intendant* at Lyons, concerning the coal mines of Forez, July 16th, 1709).

[6] *Lansdowne MSS.*, 67, no. 85.

had no way of trading with the United Provinces, and the " inferior " grade of " ship " coal, " too small for the English market," could not be sold at all.[1] Even in peace time, the foreign vent had become altogether insufficient to absorb the output of small coals. For, while the share of Tyne valley coal consumed abroad shrank steadily throughout the seventeenth century, there was no way of persuading Nature to effect a corresponding shrinkage in the proportion of small coal produced.

To some, but not to a sufficient, extent, Nature did assist the north-country mine owners who were concerned over the disposal of their small coal. With the increase in the depth at which seams were worked, the proportion of larger and " better " coal yielded by the mine apparently also increased. Peter Osborne maintained that the " two sorte[s] of coale "[2] might be distinguished according to the side of the Tyne river on which they were mined. " The Southsyde coales ", he wrote in 1591, " bee the best and the fewer and lesse sulpherous, and the Northsyde coales bee the worse and the more in nomber and farre more sulpherous and full of salt peter ".[3] This can probably be explained by the greater depths to which pits had been sunk on the south bank.[4] Most natural scientists in the seventeenth century would have agreed with the generalization of Robert Plot, whose knowledge of the coal industry was gained mainly in the mining districts of Staffordshire, " that the deeper . . . [coal] is fetch't the harder and better ".[5] Few pits anywhere in Great Britain had been sunk in Plot's time to depths greater than forty fathoms. It was not until the use of " engines for raising water by fire " in the eighteenth century helped to make deeper mining fairly common, at least in Durham and Northumberland, that observers began to realize that Plot's principle had only a limited application.[6] In 1591, even on the south bank of the river Tyne, no pits had been sunk to greater depths than thirty or forty fathoms. Most of the coal came from much shallower mines, which produced greater quantities of the inferior grades. During the first part of the seventeenth century, the percentage of all coal in the lower Tyne valley mined at depths of thirty or forty fathoms undoubtedly increased. In spite of the protests of various London consumers,[7] it is reasonable to assume that the quality tended to improve.

In so far as the coal of one region had an absolute monopoly over a given market, differences in quality might give competitive

---

[1] R. Welford, *Records of the Committees for Compounding in Northumberland and Durham* (Surtees Soc., vol. cxi, 1905), p. 287.

[2] He was writing in 1591, before a distinction had been made between *three* grades. For Osborne, see below, vol. ii, pp. 213–14.

[3] *Lansdowne MSS.*, 67, no. 21.

[4] See above, pp. 25–6.

[5] *Natural History of Staffordshire*, p. 129. The small, sulphurous coal, suitable for salt making and lime burning, was taken from the upper part of the mine. Cf. G. Jars, *Voyages métallurgiques*, vol. i, p. 238 ; *Sloane MSS.*, 4812, no. 7 (John Powell, " A Rule for to know how for to find out any Stone Coal Mynes," a paper read in 1707, but apparently written some decades earlier).

[6] Jars, *Voyages métallurgiques*, vol. i, pp. 180 sqq. On the depth of the pits in the seventeenth century, see below, pp. 350–2.

[7] For the history of this controversy, see below, pt. v, ch iii (i).

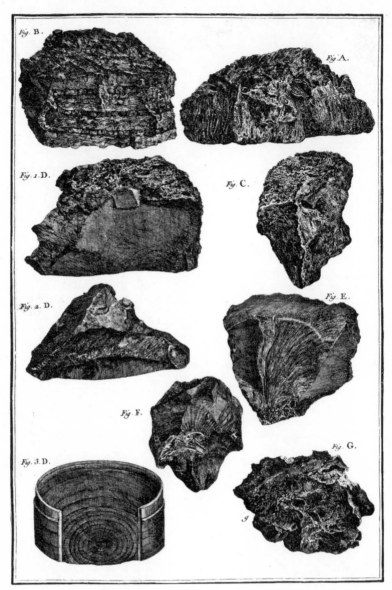

BLOCKS OF VARIOUS COALS

(Morand, *L'art d'exploiter les mines de charbon de terre*, 1768-77, Plate I)

The blocks are apparently not drawn to scale. The object of the engraving was to show the different arrangement of matter which had been observed in the coals. The following explanation is based on *Ibid.*, vol. ii, p. 1,552: A. Ordinary English coal or seacoal. Used in London in cooking and in metallurgical work. B. Scotch coal used in heating the rooms of persons of quality. C. Culm. Used especially in Cornwall in smelting. D1. Cannel. An exceedingly pure coal used for writing in place of a black crayon, and as the material for making small articles, such as inkwells, salt-cellars and snuff-boxes. D2. A piece broken off a block of cannel, to show two surfaces, one concave, the other convex. D3. A snuff-box made of cannel. E. The product obtained by distilling black amber or jet, very similar to cannel. F. A piece of bitumen of Judea, presented to show its resemblance to D1, D2, and E. G. Coal from Vienna.

advantages to particular mines within the region. They could not occasion inter-regional competition. But, when the advantage of one region over another in transport charges was not great, colliery owners in the second region might be able partially to overcome their disadvantage in the matter of transport, provided they could offer a grade not obtainable in the first. Hitherto we have spoken as if the various grades of coal in one district had their equivalents in every other district. But this is by no means true. More important than the differences in the appearance and burning properties of various coals produced from the same mine or district were the differences between the grades from different·districts. It is this fact which accounts for the imports of Welsh coal at English east-coast towns during the seventeenth century. Pembrokeshire culm was sold in London for a much higher price than Newcastle coal, but, because it was superior in quality to any fuel obtainable in Durham or Northumberland, it was preferred for certain purposes. In the seventeenth century the instructed already looked upon Scottish " great coal ", Newcastle " sea-coal ", and Pembrokeshire " culm " almost as three distinct commodities. The constitution and properties of " peacock coal " (named for its bright colours), and " cannel " (named for its candle-like flame), which were found only in certain small areas in the Midlands and Scotland, engaged the attention of natural scientists, who devised experiments to test the properties and constituents of these rare varieties, in order to compare them with the more common grades.

The cannel coal of Wigan, because of its bright, clean flame, was generally esteemed " the choicest coal in England ".[1] It was carried for distances of twelve or fifteen miles overland to places which could obtain ordinary coal much cheaper from pits near at hand, and Roger North held the opinion that, if Wigan had access to water transport, " cannel " could be freely sold in London.[2]

Anthracite, which, of all mineral fuels contains the highest percentage of carbon, is harder than other coals. The inhabitants of Pembrokeshire, where most of the British anthracite is obtained, appropriately called it " stone coal ", though this was not the sole term used to describe it, nor was this term applied to it alone.[3]

---

[1] G[uy] M[eige], *The New State of England*, 1691, pt. i, p. 128 ; *A New Present State of England*, 1727 (?), vol. i, p. 139 ; and R. Blome, *Britannia*, as cited by Crofton, *Lancashire and Cheshire Coal Mining Records*, p. 63.

[2] Defoe, *Tour*, 1769 ed., vol. iii, p. 282. The remark of North, which is repeated by Defoe, is quoted by Galloway, *Annals*, p. 188. The river Douglas, which enters the Ribble below Preston, was first made navigable in 1727, giving cannel its outlet to the sea (Mantoux, *The Industrial Revolution*, 1928, p. 126), but, contrary to the expectations of North and Defoe, it did not invade the London market.

[3] The mineral fuel found in Pembrokeshire was also called " culm ", " sea coal ", etc. The word " stone coal " was applied to coal in other parts of the British Isles where no anthracite has ever been found. In Germany *Steinkohle* became a generic term for all mineral fuel except lignite. While in some districts the name was probably coined to describe a particular grade distinguished from the others by its hardness, we cannot assume that it was always used, like the words " cannel " and " peacock ", as the title for one special grade. Like the terms " sea coal ", " pit coal ", and " earth coal ", the word

The seams of South Wales are hardest in the extreme south-west, and what may properly be called anthracite mines probably could not be found east of Llanelly, in Carmarthenshire.[1] Owen, the chief authority on Pembrokeshire in the time of Elizabeth, notes in 1603 that " the further Est the vaynes runne the softer groweth both the coal and lymestone, and the easier to be digged ".[2] Hard coal can be burned with far less smoke and dirt than the ordinary bituminous grades, and it has always been esteemed for domestic uses. It " is so pure ", observed Owen, " that fine camricke or lawne is usuallie dried by it without any staine or bleamishe ".[3] In spite of these advantages, anthracite does not appear to have been much used during the sixteenth and seventeenth centuries for heating halls and chambers, except in sparsely settled Pembrokeshire and western Carmarthenshire, where little soft coal was found. Along the coasts of Cornwall and Devon, whither the relatively low cost of carriage should have permitted anthracite to be sent on something like even terms with the bituminous coal from Glamorgan, most of the supplies came from Swansea or Neath, rather than from Tenby or Milford Haven.[4] The smallness of the blocks of anthracite (which is often little more than a powder) made it unsuitable to burn in the open fireplaces of English houses, notwithstanding the acknowledged advantages which it possessed over soft coal. Even the biggest pieces, for which the name " stone coal " was generally reserved, could not be easily kindled or kept burning in the hearths without supplies of dried branches and soft coal ; and we find the inhabitants of Pembrokeshire occasionally importing by sea small quantities of bituminous coal, mined in the district behind Swansea Bay, to aid them in making their own fires.[5] If the English had adopted stoves instead of fireplaces for heating their houses,[6] the output of Pembrokeshire coal in the seventeenth century would have been much greater than it was, for anthracite was found especially suitable for burning in stoves.

The small pieces of stone coal, wrote Defoe, " are stiled Culm, which is very useful in drying Malt, and is the cheapest and

---

" stone coal " was sometimes used to indicate a particular grade, sometimes as a title for coal generally. When the mines themselves prove so inconsistent in the nature of the fuel they yield, consistency in naming the various grades is hardly to be expected at so early a stage in the history of the coal industry, and in the development of scientific knowledge. (Cf. Appendix P.) For a description of French *charbon de pierre*, see Savary des Bruslons, *Dictionnaire de commerce*, 1741 ed., vol. ii, p. 220.

[1] It is not possible to draw a sharp line on the map separating the anthracite from the bituminous area, since one grade shades almost imperceptibly into another as we move from west to east. At a given point on the map, the deeper seams tend to be the harder (cf. H. S. Jevons, *British Coal Trade*, pp. 97, 663) ; and, as only the uppermost seams were worked in the seventeenth century, it is likely that bituminous mines were found farther to the west, nearer to Pembrokeshire, than is the case to-day.

[2] Owen, *op. cit.*, pt. i, p. 89.

[3] *Ibid.*, p. 88.

[4] *Exch. K.R. Port Bks.* for South Wales. During the latter half of the seventeenth century, however, the proportion of anthracite reaching the south-west increased.

[5] *Ibid.*, 1310/10 ; *S.P.D., Interregnum*, vol. xvii, no. 93.

[6] See below, pp. 199–200.

best Firing in the World for Hot-houses and Garden stoves, burning long with a bright red colour, and very little Flame or Smoak ; affording at the same time a strong and equal Heat." [1] Pembroke-shire was " peculiarly noted " for this culm.[2] Although some large pieces were shipped by sea under the title of " stone coal", "sea coal ", " pit coal ", or simply coal, the chief shipments were of " colne, beinge the small ashes of that cole ", according to the description made of it in 1567 by John Barrett, "master workeman at the . . . pitts of Jeffreston."[3] The special advantages of obtaining culm for use in hot-houses may well explain, both why the output of the Pembrokeshire mines was carried to all parts of the western world (even to parts where the fuel from another region had a virtual monopoly), and why it was wanted only in small quantities. Cargoes left Milford Haven destined for Flanders, Holland, Scotland, and Norway, and for ports along the east coast of England, as well as for French, Spanish, Portuguese, and West Indian ports, in which Welsh coal, by virtue of the situation of the South Wales field, might be expected to undersell the product from other British fields.[4] But the annual shipments of culm to any one port seldom exceeded a hundred tons. The export of 448 tons in 1681 from Milford Haven to Rotterdam would appear to be almost a record.[5]

Scotch " great " coal, like Welsh culm, found a small but certain market during the seventeenth century along the English east and south coasts, although the cost of carriage from the Firth of Forth to King's Lynn, London, or Plymouth probably exceeded the cost of carriage from the Tyne or Wear to those ports. Records of imports for the year 1605 show that London received 129 " chalders ", or approximately 200 tons, of Scotch coal.[6] Eleven years later the imports had increased to 3,678 tons at London and 952 tons at other English ports.[7] But from that time until after the middle of the eighteenth century the quantity brought into England did not materially increase, except during the blockade of Newcastle in the Civil War.[8] Shipments to England remained stationary during a century and a half in which the market for other coals in London and the south-eastern counties increased at least four-fold. Yet a few tons of Scotch great coal were sometimes unloaded in places like Plymouth, which could receive other grades much more cheaply from South Wales, Flintshire, Cumberland, or even Durham. These two

---

[1] *Tour*, 1769 ed., vol. ii, p. 366.
[2] *New Present State of England*, vol. i, p. 313.
[3] *Exch. Spec. Com.*, no. 3493.
[4] *Exch. K.R. Port Bks.*, 1273/10, 11 ; 1283/1 ; 1293/2 ; 1310/4, 9.
[5] *Ibid.*, 1310/4.
[6] *Additional MSS.*, 34318, f. 29. The year in question ended at Michaelmas, 1605.
[7] *Exch. Decrees*, Ser. 4, vol. ii, f. 114*d*. The year in question ended at Michaelmas, 1616.
[8] Custom house returns for London from 1745 to 1765 show annual imports of from 3,000 to 6,000 tons of Scotch coal (H. S. Jevons, *British Coal Trade*, p. 153). At the end of the seventeenth century, the imports at London were apparently somewhat less than in the late years of James I's reign. According to John Houghton (*Husbandry and Trade*, vol. ii, p. 150 ; vol. iv, p. 161), 1,473 tons reached London during the year 1694.

facts suggest that great coal did not compete directly with the ordinary grades of " sea coal " from Newcastle or Sunderland, or of Welsh coal from Swansea or Neath, but that, like culm, it served some special purpose, for which its burning properties made it particularly suitable.

Contemporary descriptions indicate that great coal was much less objectionable to burn in the house than other varieties of mineral fuel. Indeed there is some reason to think that, if all collieries had yielded only this grade, Adam Smith might never have made his famous comparison between wood and coal to the disadvantage of the latter. Great coal, known abroad simply as " charbon d'écosse ", or Scotch coal, is said to have burned with a bright flame approaching in cheerfulness that obtained from cannel. It apparently gave off, like Welsh anthracite, relatively few noxious smells and little smoke. Unlike anthracite, which because of its powdery nature was ill-suited to the English fire-place, great coal, as the name indicates, consisted of unusually large blocks. When, during James I's reign, imports of fuel from Scotland first took on some importance, the officials of the port of London found that their age-old method of measuring incoming shipments, adopted when Newcastle seacoal was the only mineral fuel that reached the city, could not be satisfactorily applied to great coal, because of its size. Coalmeters were accustomed to board the incoming colliers, where they measured the coal by shovelling it into vats, made to contain nine bushels, and counting the vats emptied over the ship's side.[1] But a vat would take a much larger quantity of Newcastle than of Scotch coal, because the first variety was made up for the most part of much smaller blocks. It was found necessary in 1614 to set up a special measurage for Scotch coal, and to enter it in the city books by its weight in metric tons.[2] The large blocks of Scotch coal were soft enough to kindle readily, and in every way they were preferred by the domestic consumer. This explains why the nobility in Scotland, and the rich burghers of the Aachen district, where a similar grade was found, began to burn coal in their houses at a period when the inhabitants of other districts united to oppose its use even by limeburners.[3] At a later period, when the merchants proposed to introduce coal as domestic fuel among a population hitherto entirely dependent for their house fires upon other combustibles, it was found prudent to break them in on Scotch coal. " Un charbon d'Ecosse flambant de prix modique," to be tried in the hearths of the Parisians, was brought up the Seine in 1714 to Paris, which had hitherto received by canal, from various French provinces, small quantities of coal for some of its artisans, but which had avoided altogether the domestic use of coal, and had prided itself upon the absence of smoke such as choked the streets of London.[4]

If Scotch coal made so much pleasanter a domestic fire than Newcastle coal, why could it not compete more effectively in London,

---

[1] Cf. below, p. 430, and vol. ii, p. 253.

[2] *Exch. Spec. Com.*, no. 4127 ; *City of London Repertories*, vol. xxxi, pt. i, f. 301. Cf. Appendix C (i).

[3] See above, p. 12 ; below, pp. 157-8.

[4] Boislisle, *Correspondance des Contrôleurs Généraux*, vol. iii, no. 1743 (Letter from M. Bignon, *prévot des marchands de Paris*, November 28th, 1714).

where the principal market during the seventeenth century was provided by the domestic consumer ? The answer is given by a Frenchman who travelled in England at the end of the seventeenth century. He tells us that Scotch burns more rapidly than Newcastle coal, and is more expensive.[1] " Le charbon d'Ecosse, qui est plus bitumineux que pyriteux," wrote in 1786 Aléon Dulac, a French authority on mineral fuel, " dure fort peu ; celui de Newcastle se consume si lentement, qu'on dit communément qu'il fait trois feux." [2] Only the richest citizens of London cared to pay the price of warming their rooms with the former. Beginning with the first Stuart reign, the sovereign imported for his palace at Westminster a regular supply from Scotland. Wealthy citizens like Lionel Cranfield followed suit, and stocked their cellars mainly with wood, charcoal, and Scotch coal, avoiding the despised " sea coal " except for heating the servants' quarters.[3] Throughout the seventeenth century Scotch coal retained its small niche in the market of the south-east as an article of luxury.[4]

In parts of Holland, Flanders, and northern France, Scotch coal could apparently undersell Belgian, and, after the high export tax was imposed in 1599,[5] English coal. We know that many foreign manufacturers took to burning it and apparently found it well suited to their purposes. William Brown, a merchant of Edinburgh, who travelled through the Low Countries during 1682 and 1683 in the interest of the Scottish colliery owners, reported that in Flanders, " the great coalls sell well with the brewers, salters, soape boylers, and most all trades except the smiths, especially in Ghent, Bridges [Bruges], Ostende, Newport, Dixmude, Ipre, etc." [6] North of the Scheldt brewers and soapmakers had long made free use of Scotch coal, but the smiths insisted on having Newcastle coal, " as it is peculiarly fit for making anchors and all kinds of Great Ironwork." [7]

---

[1] Misson, *Mémoires . . . par un voyageur en Angleterre*, 1698, p. 50.

[2] *Observations générales sur le charbon minéral*, 1786, f. 41 ; preserved in the Bibliothèque Nationale, Paris, *Manuscrits français*, no. 11857. This book, which remains in manuscript, contains much interesting information concerning coal mining both in England and in France. According to Savary des Bruslons, Newcastle coal was " plus léger " (*Dictionnaire de commerce*, 1741 ed., art. " Charbon ").

[3] *Denmilne MSS.*, vol. ix, no. 23 ; *Sackville MSS.* (Earl of Middlesex's kitchen ledger, 1628). Cf. below, p. 198.

[4] See Vanbrugh, *The Relapse*, 1697, act iii, sc. iii, where Sir Tunbelly Clumsey, desiring to impress Young Fashion, orders his servants to " run in a-doors quickly : Get a Scotch Coal Fire in the great Parlour ; set all the Turkey-work Chairs in their places ; get the great Brass Candlesticks out, and be sure stick the Sockets full of laurel ".

[5] See below, vol. ii, pp. 219–20.

[6] *Hist. MSS. Com., Report on the MSS. of the Earl of Mar and Kellie*, pp. 212–13. It is difficult to understand why the Dutch salters should have wanted Scotch great coal, when salters generally burned an inferior grade of small coal (see above, p. 112). Possibly the inclusion of the word " salters " is a mistake made in printing the manuscript.

[7] *S.P.D., Interregnum*, vol. clxxx, no. 121. Cf. Jars, *Voyages métallurgiques*, vol. i, pp. 181–2, for the same opinion. Scotch coal, he wrote, " est moins bon que celui de Newcastle pour la forge, parcequ'il ne colle pas autant, ce qui prove qu'il est moins bitumineux ; mais il est bien préférable pour brûler dans les appartements."

When we remember that it was the small sulphurous grade of Newcastle coal that these ironworkers wanted, and that smiths in the St. Etienne district used only the *sable* of coal, we can readily understand why the ironworkers along the northern coast of the Continent did not look with favour upon Scotch great coal, which was " plus bitumineux que pyriteux ", and which was so bulky that the London meters found it impracticable to measure it in their vats. While the smiths did not find this grade suitable for working up smelted ores into finished metal wares, the absence of sulphur made it preferable to common coal for smelting, and in 1622 Gerard de Malynes was under the impression that " in the lower parts of Germanie, about Acon [Aachen] and Collogne . . . they melt great quantitie of yron stone " with a coal, which resembles Scotch coal, in that, among other things, " it doth not cake as our [English] coales [do]." [1] Whether the soapboilers and brewers, who wanted a grade as free as possible from sulphur, actually purchased Scotch in preference to Newcastle coal, when there was nothing to choose between them in price, or whether, as with the domestic consumer, the undoubted advantages of the former were counter-balanced by the greater economy of the latter, it is not possible to say.[2] All we have is the general statement, made by a customs inspector who visited Newcastle and Sunderland in 1696, that there is " as good a Coal for the Hollanders use in Scotland as at these Ports ".[3] It is certain that, except for a brief period during James I's reign, when Scotch coal was regarded as the only suitable coal for use in glass furnaces [4] (and was imported at London for that purpose as well as for heating the chambers of gentlemen of quality),[5] there was no industrial use for which it was indispensable.

There was only one other coal that appears to have been marketed in nearly as many parts of the world as Welsh culm, or Scotch great coal, and that was the coal of Kingswood Chase, exported through the port of Bristol in small intermittent quantities during the late sixteenth and the seventeenth centuries. Although the shipments of this fuel from Bristol never amounted to more than 2,000 tons a year, it was carried in vessels bound not only for local ports in the Bristol Channel, but for Irish, French, Portuguese, West Indian, and American ports, and even for German ports in the North Sea.[6] Small boats loaded with Bristol coal sometimes made their way up the Severn as far as Gloucester, a place served with culm from Milford Haven or Tenby, with bituminous coal from Swansea or Neath, and with

---

[1] Malynes, *Lex Mercatoria*, 1622, p. 271. On the question of the adoption of coal in the smelting process, see below, pp. 245-6, 249-51.

[2] According to Savary des Bruslons (*Dictionnaire de commerce*, art. " charbon "), many artisans in the north of France found a mixture of Scotch with Newcastle coal best suited to their work.

[3] *Treasury Board Papers*, 39/31.

[4] See below, p. 219.

[5] *Denmilne MSS.*, vol. ix, no. 13 (Letter of the Scottish Privy Council to King James I, July 22nd, 1619) : " Our coale quhilk formarlie wer forbidden is daylie now transported to England for making of glass and for others [sic] uses without the quhilk no glass can be maid there."

[6] *Exch. K.R. Port Bks.*, 1136/3, 1137/1, 1160/3 ; *Exch. K.R. Customs Accts.*, 171/26 ; Taylor, *Archaeology of the Coal Trade*, p. 220 ; *P.C.R.*, vol. xxxvi, p. 185 ; Savary des Bruslons, *op. cit.*, art. " Commerce, St. Malo ".

Shropshire coal from Broseley and Benthall.[1] On the other hand, Gloucester occasionally shipped coal to Bristol.[2]

Coal from Kingswood Chase had to be hauled several miles overland to the waterside, and must have cost much more at the loading wharves than either Shropshire or Welsh coal. Its carriage in competition with both, and the occasional import of other coals at Bristol, suggest that, like culm, it must have possessed special properties; and this hypothesis is strengthened by the title " stone coal " sometimes given to it in the port books of exports from Bristol.[3] The grades of coal found in Kingswood Chase and Somerset to-day vary more widely in quality than those of most other fields, ranging from a hard steam coal to a soft bituminous variety ;[4] and a reference in 1611 to " mynes of seacole and stone cole " near Bristol suggests that several grades were mined at that time.[5] The exports may have been mainly of a variety approaching Pembrokeshire anthracite in hardness, and serving for special purposes to which it was better suited than the softer grades produced at mines in the Severn valley.

The export of coal from Bristol might be explained in a different way. Until after the Restoration Bristol remained the chief trading port of the British Isles next to London. Ships from Bordeaux laden with French wines, ships from the Mediterranean laden with the spices and jewellery of the East, ships from across the Atlantic laden with sugar, tobacco, and other raw produce of the new plantations put in here to discharge their cargoes and to reload with the cloth and other manufactured articles which were being turned out in increasing quantities all through the valley of the Severn. If a shipmaster, loading for the return voyage, wanted to fill out his cargo with coal, which could serve as ballast until he found a market for it, he was likely to buy it in Bristol, rather than to incur the delay and expense of a special trip to some Welsh port, in order to take advantage of the lower wharfside price he could there obtain. The port book records show that nearly all the coal exported from Bristol went in small lots of from five to ten tons, in vessels laden with other commodities. As the town had trading relations with ports all over the world, the advantage of getting a supplementary cargo of fuel, without a special trip to another harbour, is enough to explain why the foreign market for Bristol coal was as wide as its capacity was small. Defoe pointed out long ago that trade begets more trade, and that what brought merchants in ever growing numbers to Bristol, and more particularly to London, were the facilities these places afforded for the ready

---

[1] *Exch. K.R. Port Bks.*, 1136/7, 11, 24 ; 1249/9 ; *Star Chamb. Proc., James I*, 160/39.

[2] *Exch. K.R. Port Bks.*, 1249/9, 10.

[3] *Ibid.*, 1160/1.

[4] H. S. Jevons, *British Coal Trade*, p. 79 ; *V.C.H. Somerset*, vol. ii, p. 353.

[5] *Exchequer Decrees*, Series 4, vol. i, p. 223. Another document of the time of James I refers, however, to " stone coale *alias* sea coale " (*Harleian MSS.*, 368, no. 7 ; printed Galloway, *Annals of Coal Mining*, pp. 212–13), suggesting that these were interchangeable terms in the Bristol field, and that the use of both cannot be considered as proof that anthracite as well as bituminous coal was mined in this district in the seventeenth century.

marketing and the ready purchase of every kind of commodity.[1] For this reason, although the price of north-country coal on the Thames was at least double the price on the Tyne,[2] some shipmasters who traded with foreign countries preferred to take their ladings at London rather than at Newcastle, a town affording a much smaller market and having far fewer goods to sell.

Special market conditions, such as existed at Bristol, caused an even smaller exception to the general rule that dealers and consumers purchased their fuel where it could be had cheapest, than did the fact that the grades of coal produced in some mining fields were especially well-suited for certain purposes. Costs of transport remained the factor of greatest importance in determining what markets the coal of each region should command. No sufficient revolution in methods of transport was effected, during the period with which we are specially concerned, to alter extensively the markets controlled by these regions. The areas served by most of them were much the same in 1700 as in 1550. What had changed was the amount of coal consumed in each of these areas.

Our survey of production and of consumption has brought out the fact that there was a remarkable increase in the amount of coal produced, and that this increase resulted almost entirely from the demands of the home market. We now proceed to compare the expansion of the British coal industry in the century and a half following the accession of Elizabeth with the progress which has been made since that time, and to determine whether during our period any corresponding growth in mining can be observed on the Continent.

---

[1] *The Complete English Tradesman*, 1726, esp. chap. xxvi.
[2] See Appendix E.

CHAPTER III

THE LEAD GAINED BY THE BRITISH COAL INDUSTRY

THE first three quarters of the nineteenth century have usually been regarded as incomparably the period of most rapid expansion in British coal mining. But, measured by the rate of increase in the use of coal, the period from 1550 to 1700 may, without qualification, be compared to it. If the figures given in the preceding chapters are approximately correct, it appears that less than one hundred-weight per head was consumed annually before the accession of Elizabeth. At the end of the seventeenth century the consumption per head was about nine hundredweight. The rate of increase was clearly as rapid in the earlier period as during the nineteenth century, for to-day not more than four or five tons per head is burned annually within the country.

It is difficult, therefore, to agree with Jevons' conclusion " that *the rate of multiplication* [*in coal produced*] *is in recent years many times as great as during preceding centuries* ".[1] His statement is based upon two tables which he compiled of the " Vend of coal from Newcastle " and of " Coal imported into London ". But, if these tables enable him to establish his point, it is only because he was unable to obtain statistics of the " vend " prior to 1609, or of the London imports prior to 1650. If we add to his table the figure for the " vend " for 1564, compiled from the record of shipments contained in the account books of the town of Newcastle, we get the following result [2] :—

TABLE IX

Year.	Vend from the Newcastle Coalfield. Tons.	Increase in fifty years or as for fifty years. Tons.	Rate of Increase per cent as for fifty years.
1564	32,951 [3]	—	—
1609	251,764	243,126	738
1660	537,000	279,643	110
1700	650,000	141,250	27
1750	1,193,467	543,467	84
1800	2,520,075	1,326,608	111
1864	18,349,867	12,367,025	372

[1] W. S. Jevons, *The Coal Question*, ed. 1906, p. 263.
[2] Under the heading " Vend from the Newcastle coalfield " Jevons includes coal shipped from other ports besides Newcastle. My figure for 1564 (see Appendix D) does not include any other port besides Newcastle, but the additional shipments from other ports at the beginning of Elizabeth's reign do not appear to have reached 1,000 tons all told (cf. above, pp. 29, 36). Jevons' figure for 1864 includes 7,562,963 tons carried out of the northern counties by the railroad.
[3] The italics indicate additions to Jevons' table.

The figures show that the rate of increase was actually greater during the last half of the sixteenth century than during the first half of the nineteenth.

If we carry the figures of imports at London back into the sixteenth century, and add them to Jevons' table, we find again that his statement concerning the rate of multiplication during the nineteenth century does not hold good :—

TABLE X [1]

Year.	Total quantity of coal imported into London. Tons.	Increase in fifty years or as for fifty years. Tons.	Rate per cent increase per cent as for fifty years.
1586	*23,867*	—	—
1606	*73,984*	*125,297*	*525*
1650	216,000	*161,378*	*119*
1700	428,100 [2]	212,100	98
1750	688,700	260,600	61
1800	1,099,000	410,300	60
1850	3,638,883	2,539,883	231
1865	5,909,940	7,570,190	404

If, as appears to be the case, the expansion of the British coal industry in the century following the accession of Elizabeth is scarcely less remarkable than the expansion in the nineteenth century, it is evidently important, if we are concerned with the influence which the coming of a coal economy has had upon the course of economic and social history, to examine this early development. The importance of such a study becomes still more apparent when we compare the growth of the coal industry in England and Scotland with that on the Continent.

The best way, perhaps, of estimating the influence of coal upon British history is to contrast the development of economic and social life in Great Britain with that in countries which made no comparable use of coal. For this purpose the period from 1550 to 1700 offers an exceptional opportunity. In no other country can there be found a parallel expansion in coal mining.

In France, although Colbert and other mercantilist statesmen made the encouragement of native coal mining a part of their general policy of attempting to relieve the French nation from its dependence on the foreigner for raw materials,[3] very little progress had been achieved in colliery enterprise by the beginning of the eighteenth century. A general survey of the state of the coal mines was submitted

---

[1] The figures in italics, which indicate additions to Jevons' table, have been reached by adding up the entries in the London port books (see Appendix D (iv)).

[2] It will be noted that this figure and that for 1750 do not agree with the figures for London imports given below in Appendix D (iv). Jevons' figures were taken, of course, from a different source. If mine are substituted, it will be found that the rate of increase in the first half of the eighteenth century is even slower than Jevons' table indicates.

[3] *Discours sur les manufactures du Royaume*, 1663, in P. Clément, *Lettres, instructions et mémoires de Colbert*, 1861, vol. ii, pp. cclvii–xii ; and cf. below, pp. 234–5, 273.

in 1709 by the *intendants*, in response to a request by the *Contrôleur Général* for detailed information on the point. M. Trudaine, the *intendant* at Lyons, supplied him with almost all the details which either he or the future historian could wish for concerning the colliery district of Forez, generally regarded as forming, with Lyonnais, the principal coal-producing region in seventeenth-century France. The coal dug there could be carried by water to the populous valleys of the Rhone and the Loire. But M. Trudaine found the mining industry seriously depressed. Many of the twenty-six mines near St. Etienne had temporarily ceased producing altogether, and none of them was working up to capacity. The total output did not exceed 10,000 tons a year, although it might have reached 25,000 tons if a market could have been found to absorb the product of all the mines working full time.[1] It is probable that the collieries of Lyonnais, at Rive-de-Gier, which are not mentioned in M. Trudaine's report, but which had furnished fuel to the Rhone valley at least a century earlier,[2] produced more coal than those of Forez. They were situated nearer to Lyons—the principal market. Between them, the mines of Lyonnais and Forez perhaps produced from 25,000 to 50,000 tons annually at the end of Louis XIV's reign.

The output of all the remaining collieries in France was probably not greater than that of this single district. In the small Anjou field, about Chalonnes and Saumur, near the river Loire, few of the pits that had been dug were producing much coal, according to M. Turgot, *intendant* at Tours.[3] Some mining had been undertaken in Auvergne prior to 1709, and an effort had been made to deepen the river Allier from Brioude to Pont-du-Chateau, near Clermont, in order to bring Auvergne coal to Paris,[4] but the results had been no more satisfactory than in Anjou,[5] and no report on Auvergne is to be found among the letters sent in response to the request of the *Contrôleur Général*. Neither do we find mention, in any of these letters, of coal pits in Brittany near St. Brieuc, nor in Champagne, nor near Moulins in Bourbonnais, nor in Nivernais, nor near Auxerre in Bourgogne, nor at Carmaux and other places in the diocese of Albi, nor in the Alais

---

[1] Boislisle, *Correspondance des Contrôleurs Généraux*, vol. iii, no. 496 (Letter from M. Trudaine, July 16th, 1709). According to M. Trudaine, 35 hewers, working 230 days a year, raised 100,000 " charges ". According to M. Rouff (see Appendix C), the " charge " in Forez was ordinarily a human load of about 70 lb., but the term " charge " was sometimes applied to the load a horse could carry on its back (about 2 cwt.), and that is probably the meaning in this case. If the " charge " was only 70 lb., then the average output per man employed in Forez would be less than 100 tons per annum, which is a smaller output than we should expect at this stage of mining (cf. below, vol. ii, pp. 136 n., 138).

[2] Rouff, *Les mines de charbon en France*, p. xiii.

[3] Boislisle, *loc. cit.* The annual output of the coal mines of Anjou has never exceeded 20,000 tons, a figure reached there in the middle of the nineteenth century (see O. Couffon, *Les mines de charbon en Anjou*, 1911, p. 14).

[4] *Archives Nationales A.D.*, xi, 8 (" Depar du Provost des Marchands et Eschevins de la Ville de Paris ", April 18th, 1681), Clément, *op. cit.*, vol. iv. p. 437 ; vol. vii, p. 292.

[5] Boislisle, *op. cit.*, vol. ii, p. 482 (*Mémoire du sieur des Casaux du Hallay, député de Nantes, sur l'état du commerce en général*, March 4th, 1701).

district. Yet we have evidence that attempts had been made to work coal seams in all these parts before 1700.[1] Great hopes were built on the discovery, made in the nineties of the seventeenth century, of two coal mines in the parishes of Hardinghen and Réty near Boulogne,[2] but their combined annual output, according to the letter written in 1709 by M. de Bernage, *intendant* at Amiens, was less than 1,000 tons.[3] According to an English writer, the produce of these mines was " a dark coloured dirty stuff . . . [which] does not deserve the name of coal ".[4] Not until 1714 was a seam found in the important northern field, and in 1709 M. de Bernières, *intendant* for Flanders, could inform the *Contrôleur Général* " Il n'y a aucune mine de ce charbon de terre dans mon département, et celui qui s'y consomme vient des environs de Mons et de Charleroi ".[5] From these reports it seems probable that the output of coal in France at the end of the seventeenth century did not amount to more than 50,000 or 75,000 tons per annum,[6] a smaller quantity than was mined in the single manor of Whickham, near the Tyne, on the eve of the Civil War.[7] The greater part of such coal as was burned in the French kingdom came by sea or river from abroad—from England, Scotland, Belgian Hainaut, and in small quantities overland from the neighbourhood of Chambéry in Savoy.[8]

German political organization was not sufficiently centralized in the early eighteenth century to provide us with so complete a survey of the early collieries as the French have done. But German scholarship has been busy making up for the former backwardness of German political organization, and we are not without a considerable supply of useful statistics concerning the output of coal in various regions. In mining more costly minerals, such as copper and silver, the Germans were in advance of all other countries in the Middle Ages, and no doubt more mineral fuel had been raised than in France ;[9] but it is certain that, before 1700, the German coal industry was of very small importance in comparison with the English or even the Belgian. In the Saar basin, a district rich in wood, coal diggings are

---

[1] Boislisle, *op. cit.*, vol. i, nos. 694, 811, 1003, 1515, 1573 ; vol. ii, nos. 410, 557, 797 ; Clément, *op. cit.*, vol. iv, pp. 595–6. I have also found a reference to the " charbon de terre de Rouerque " (*ibid.*, vol. iii, pp. 380–1).

[2] Boislisle, *op. cit.*, vol. i, no. 1762 (M. Bignon, *intendant* at Amiens, to *Contrôleur Général*, October 5th, 1698).

[3] 4,000 to 5,300 *barils* (Boislisle, *op. cit.*, vol. iii, no. 496). For the *baril* see Appendix C.

[4] *A State of the Coal-Trade to Foreign Parts, by Way of Memorial to a supposed very Great Assembly*, 1744–5.

[5] Boislisle, *op. cit.*, vol. iii, no. 496 ; Decamps, *Mémoire historique sur l'industrie houillère de Mons*, 1879, p. 55.

[6] On the eve of the Revolution, according to Leseure, the annual production of coal in France amounted to 382,000 tons. There is abundant evidence that the industry had expanded enormously during the eighteenth century, and Rouff has shown that this figure falls far short of the truth (*Les mines de charbon en France*, p. 431 ; cf. pp. 422–31).

[7] See below, p. 361 n. ; and Appendix I. Several other collieries, besides Whickham, listed in the Appendix, were in the manor of Whickham.

[8] Boislisle, *op. cit.*, vol. iii, no. 1818 (M. d'Angervilliers, *intendant* in Dauphiné, to the *Contrôleur Général*, April 18th, 1715).

[9] Cf. C. Dieudonné, *Statistique du département du Nord*, 1804, vol. i, pp. 155–6.

heard of round Sulzbach-Duttweiler at an early date, and an effort was made in the sixteenth century to provide an outlet for the coal of the Saar by making a canal to the Rhine. But industry in this district suffered a serious set-back as a result of the Thirty Years' War ; and the combined annual output of all the mines there was only 15,000 tons as late as 1767, thirteen years after Kurfürst Wilhelm Heinrich had begun to interest himself in their exploitation.[1] Lower Silesia did not produce more than 10,000 tons of coal in 1740, Ibbenbüren produced about 9,000 tons in 1747, and Minden about 22,000 tons in 1770.[2] Doubtless the output in each of these districts was considerably smaller in 1700.

The chief centres of the German coal industry before 1700 were Saxony, the Ruhr district, and the Wurmrevier, the district round Aachen. According to one estimate, the output in Saxony in 1695 exceeded 30,000 tons, and doubtless several thousand tons more were produced in the Duchy of Magdeburg, where we hear of a number of coal pits at Wettin, operated chiefly to supply fuel for the salt manufacture at Halle.[3] Although it has been generally believed that the Ruhr mining industry was little developed before 1784, when Oberbergrat von Stein threw open the great Dutch market to Westphalian coal, by making the river Ruhr navigable, we have a record of an annual production of more than 100,000 tons as early as 1737.[4] The indications are that the Ruhr collieries were growing very rapidly at this time, and it may be doubted whether the output at the beginning of the eighteenth century was half as large.[5] We have found no record of early production in the Wurmrevier, and, as the seams of this district are geologically a continuation of those in the adjoining Belgian provinces, we propose to consider it as a part of the great coal basin extending westward along the valleys of the Meuse and Sambre. It seems unlikely that the output of all the German fields, other than the Wurmrevier, exceeded 100,000 tons, or 150,000 tons at the most, at the end of the seventeenth century, and probably 70 or 80 per cent of this coal was produced in Saxony and the Ruhr district. Germany, according to the author of the most

---

[1] 9527 " fuder ", i.e. 1500 kilograms (A. Hasslacher, *Geschichtliche Entwickelung des Steinkohlenbergbaues im Saargebiete*, 1884, pp. 401, 410, 424–50). For the " fuder " see Appendix C.

[2] These figures are taken from Hue, *Die Bergarbeiter*, vol. i, p. 354. The best survey of the early German coal industry with which I am familiar is that contained in this work, pp. 342–55.

[3] The estimate for Saxony is that given in Hue, *loc. cit.* On the colliery at Wettin, cf. Jars, *Voyages métallurgiques*, vol. i, pp. 314–19 ; J. P. Bünting, *Sylva Subterranea*, 1693, Introduction, and chaps. xiii and xxvi ; K. T. von Inama-Sternegg, *Deutsche Wirtschaftsgeschichte*, 1879–1901, vol. iii, pt. ii, p. 145. According to Jars, the output of coal at Wettin in 1765 was about 2,400 " wispel ", or approximately 10,000 tons (cf. Appendix C). On the coal industry of Saxony, see Bünting, *op. cit.*, chap. xiii.

[4] Hue, *op. cit.*, vol. i, p. 354 ; Beck, *Geschichte des Eisen*, vol. ii, p. 105, vol. iii, pp. 921, 958, 960. Some Ruhr coal was shipped on the Rhine long before the deepening of the river Ruhr.

[5] The output in the *Grafschaft* Mark increased, according to the figures cited by Hue (*loc. cit.*), from 116,968 tons in 1737 to 685,000 tons in 1790. Elsewhere, however (p. 347), he gives much smaller figures than these.

complete account of the German miner, was by no means a coal-burning country even as late as the middle of the eighteenth century.[1]

It is only in the Belgian provinces that the seventeenth-century coal industry can bear comparison with that of the large British fields. The chief centres of mining were at Liége and Mons, although the Wurm field was of some importance also, and collieries could be found about Charleroi and along the valley of the Sambre towards Namur. From Liége, coal was carried down the Meuse to Holland, but large quantities were used in the growing manufactures about the town itself. In 1812 the annual production from 140 collieries in the immediate vicinity was rather more than 500,000 tons, of which nearly two-thirds were consumed on the spot and the remainder exported, chiefly to Holland.[2] Already in the period from 1600 to 1648, according to Professor Pirenne, the extraction of coal at Liége " prend une importance croissante ".[3] At the beginning of the eighteenth century, the annual shipments of Liége coal to the Netherlands are said to have been worth about 100,000 *ducats*, or roughly £20,000, and it is therefore probable that nearly 50,000 tons were shipped each year, and perhaps 100,000 tons more consumed in the neighbourhood.[4] Some fuel was carried up the wide, slow-flowing Meuse as far even as Sedan.[5]

But for the most part this up-river market was served not from Liége but from mines near Charleroi. Here were collieries in the town itself, at Mons-sur-Marchienne, Darmet, Roux, Jumet, Gilly, Montigny-sur-Sambre, and at other places as far down as Namur.[6] Notwithstanding that the rivers were impassable nearly half the year,[7] in 1709 the Charleroi region was said to furnish about 2,000 tons annually to the French province of Hainaut ; a larger quantity being consumed in the country belonging to Spain, between the Sambre and the Meuse, and in the forges of the " Comté de Namur ", whither coal could be brought by way of the Sambre and the Meuse.

Mons coal, like Liége coal, had access to an important river market, for the Haine was made navigable, in the sixteenth century, to its junction with the Scheldt at Condé.[8] In addition to the increasing coal traffic in flat-bottomed river barges, loaded by women bearers who carried the fuel from more than a dozen collieries down to the left bank of the Haine, a considerable quantity of the large blocks

[1] Hue, *op. cit.*, p. 351.

[2] 5,050,350 " quintaux métrique " (see Appendix C). L. F. Thomassin, *Mémoire statistique du département de l'Ourthe*, 1879, pp. 416 sqq.

[3] *Histoire de Belgique*, vol. iv, p. 427.

[4] P. L. Berkenmeyer, *Recueil géographique et historique des choses les plus remarquables* . . . 1729, p. 182. The price of coal in Holland undoubtedly varied according to the distance from Liége. If the prices along the Meuse were much the same as those along the Severn and the Trent, where the conditions of transport were similar, they probably amounted to from 6s. to 15s. per ton (cf. *H.M.C.*, *Report on MSS. of Lord Middleton*, pp. 171–2 ; and above, pp. 96, 99).

[5] Boislisle, *op. cit.*, vol. i, no. 1490 ; *Archives Nationales A.D.*, xi, 8 (*Arrêt du Conseil d'Etat*, October 31st, 1672). Apparently in 1709 no Liége coal came as far as the French frontier (Boislisle, vol. iii, no. 496 : M. Doujat, *intendant* in Hainaut, to the *Contrôleur Général*, July 9th, 1709).

[6] Boislisle, *loc. cit.*

[7] E. Bidaut, *Mines de houille de l'arrondissement de Charleroi*, 1845, pp. 156–7.

[8] G. Arnould, *Mémoire sur le bassin houiller du couchant de Mons*, 1877, p. 87.

of coal was carried overland " par chariots " to Condé.[1] Thence it was carried in boats down the Scheldt to Tournai, to be distributed overland to Lille, Roubaix, Tourcoing, and Douai, or transported still farther down stream to Ghent, and various towns in southern Holland. The quantity annually brought to Condé at the end of the seventeenth century appears to have approached 50,000 tons, and it may perhaps be assumed that the output of the Mons collieries had reached approximately 100,000 tons a year.[2]

It is apparent that the coal industry already played an important part in the economy of what is now southern Belgium. But the annual output of Belgian coal at the end of the seventeenth century could hardly have been more than a third of that of Durham and Northumberland. The rate of increase in production in the Belgian provinces between 1550 and 1700 must have been much slower than in Great Britain. At the beginning of the sixteenth century, mining in the Liége, and perhaps in the Mons, district had been pushed more intensively than in any other part of the world.[3] At the end of the seventeenth century this was no longer the case ; the Belgian coalfields had surrendered first place to the British. And outside the Belgian provinces, there were few colliery districts of importance. It may be doubted whether the entire annual production of the Continent amounted to more than a sixth of the annual production of Great Britain.[4]

In view of the evidence that the expansion of British coal mining in the sixteenth and seventeenth centuries was on a scale without any parallel in the continental coal fields, it is natural to ask to what

---

[1] The principal collieries along the Haine were to be found in the seventeenth century, as to-day, at Frameries, Wasmes, Warquignies, Boussu, Dour, Quiévrain, Elouges, and Carignon (Boislisle, *op. cit.*, vol. iii, no. 496).

[2] About 600 " boatloads " of coal were shipped to Condé each year (Vauban, *Mémoire sur le charbon*, 1699, in Boislisle, *op. cit.*, vol. i, no. 1874). Each boatload was said to contain 1,200 *wagues* of coal, and if this is the *vague* of 144 lb. referred to by M. Rouff (*Les mines de charbon en France*, p. lx), then the total shipments to Condé amounted to about 50,000 tons per annum. In addition large quantities of coal were carried overland and consumed locally in the Mons district. It should be pointed out, however, that 80 tons of coal seems a large cargo for a river boat at this period.

[3] See above, pp. 13–4.

[4] We assume that the annual production in the Belgian provinces, including the Wurmrevier, was between 300,000 and 500,000 tons, that that of Germany was between 100,000 and 150,000 tons, and that of France between 50,000 and 75,000 tons. In other parts of the world coal mining can hardly be said to have begun. In Sweden, the coal of Helsingborg was first seriously exploited in 1741, while the rich coal measures of Bohemia were apparently neglected until an even later time (Decamps, *op. cit.*, p. 57). A Russian coal mine was first opened in the Donetz basin at the very end of the eighteenth century (Beck, *op. cit.*, vol. iii, p. 1148). In North America, coal had been discovered in several places before 1700—in Nova Scotia, Greenland, Illinois, and probably elsewhere as well (Pepys, *Diary*, ed. H. B. Wheatley, 1904, vol. vii, p. 103 ; Clément, *Lettres . . . de Colbert*, vol. iii, pt. ii, pp. 402–3, 539–40 ; Birch, *History of the Royal Society*, 1756–7, vol. i, pp. 199–201). But, although imperialist statesmen were already interested in gaining control of the mineral resources of the new world (see below, vol. ii, p. 202), it is unlikely that much coal had been dug across the Atlantic. Wood was too abundant for that (cf. below, pp. 156–7).

extent the peculiar development of economic and social life in England and Scotland during the seventeenth century can be explained in terms of coal. Such a question leads to an examination of the influence of coal upon industrial development, upon the ownership of natural resources, upon the industrial and financial organization for the carrying on of economic life, and finally upon public policy. But, before examining these subjects, we must ask why it was that the coal industry should have assumed during the seventeenth century so much greater an importance in Great Britain than on the Continent.

PART II

COAL AND INDUSTRIALISM

PART II

COAL AND INDUSTRIALISM

CHAPTER I

CAUSES FOR THE EXPANSION OF THE INDUSTRY

THE first and most obvious explanation of the rapid expansion in coal mining, which appears to have begun in the reign of Elizabeth, is to be found in the natural advantages which the British enjoyed for moving coal to all the chief centres of population. Nowhere else were outcropping seams spread through so many sections of the country. Further, the facilities for the carriage of mineral fuel in Great Britain were superior to those of any other country having large coal deposits. Transport by water, it was remarked by Adam Smith, is the principal nourishment to commerce and consequently to the division of labour. For coal, of all goods the bulkiest in proportion to its value, water transport is a necessity. In Great Britain coal outcrops were found in South Wales, in Flintshire, in Cumberland, in Northumberland, in Ayrshire, and along the Firth of Forth, on the very edge of the sea, and, in addition, along five navigable rivers, the Trent, Wye, Severn, Tyne, and Wear. Coal could be carried by water to almost any part of the country from these seams. Sir Robert Southwell, in 1675, estimated that the average distance of a plot of land from the sea was only twenty-four miles in England as compared with eighty-six miles in France, and, doubtless, even more in Germany.[1] Holland, of course, possessed facilities for water transport unrivalled even by England,[2] but Holland had no coal of her own. On the Continent natural conditions were far less favourable to early coal mining. Outcrops were scarce, and outcrops accessible to water even scarcer. It was the lack of access to navigable rivers which held back the development of the Ruhr and Saar coalfields, while the situation of Liége and Mons permitted the Belgian fields to be more intensively exploited.

The natural facilities for transporting British coal help to explain why the industry developed earlier in England and Scotland than in continental countries. It still remains to consider why a sensational expansion should have begun in the second half of the sixteenth century.

(i) Changes in the Ownership of Mineral Property

If capital was to be found to meet the increased expenses of deeper mining, without which no great expansion in the coal industry

---

[1] Birch, *History of the Royal Society*, vol. iii, pp. 207–10.
[2] Southwell estimated that in the case of Holland the average distance of any plot of land from navigable water was only one quarter of a mile.

could take place, attractive mineral concessions had to be offered. The leases had to be long enough to permit a return on a heavier original investment than was needed in connection with the small medieval diggings. To balance the hazard of a flooded pit, an explosion, or a failure to reach thick beds of coal, the terms had to be sufficiently easy, and the royalties exacted sufficiently low, to offer a prospect of large rewards. The rights of surface owners to demand compensation for damage to their land from subsidence, or from the débris cast up in sinking shafts, or for over- and underground way-leave, had to be defined and limited. It was all the better for the entrepreneur if he could settle his grant once and for all with a single lord of the soil, and not be troubled with a host of individual claims on the part of small tenants. The first deep coal mines were no less of an adventure than the first efforts to span the American continent with railways, and the acquisition on easy terms of the right to make use of the soil was as great an inducement to the early English capitalist as to the American promoter in the nineteenth century.

Before the reign of Henry VIII such terms were not to be obtained. A considerable portion of all the lands in which coal had already been mined, or in which it was destined to be mined in the future, belonged to the Church. This was especially true in the case of the most important of the early coal fields, that of Durham and Northumberland. The Bishop of Durham, unquestionably the greatest of early coal owners, maintained a title to all minerals, not in the lands of freeholders, within the County Palatine. That title gave him control of many manors richly stored with mineral wealth, including several of those which lined the south bank of the Tyne. Whickham and Gateshead, the principal source of the fuel which warmed the southern part of the kingdom during the reigns of Elizabeth and the first two Stuarts, were in his hands.[1] The Monastery of Tynemouth owned the manors of Benwell, Elswick, and Denton, the sites of the first three important " sea-sale " collieries on the north bank. Much of the coal shipped from the Northumberland coast at the end of the seventeenth century came from lands which had once belonged to the Monasteries of Tynemouth and Newminster and to Brinkburne Priory. These and other ecclesiastical bodies, such as Finchale Priory and the Nunnery of St. Bartholomew at Newcastle, possessed further valuable mineral-bearing property in the two northern counties. Elsewhere in England the Church was also seized of lands containing important coal seams. Wenlock Priory, for example, owned the manors of Broseley, Madeley, and Shirlett, on the banks of the Severn in Coalbrookdale ; in Cumberland the Monastery of St. Bees owned the manor of St. Bees, which was to furnish, in the second half of the seventeenth century, much of the coal mined by the Lowthers to supply Dublin ; in Glamorgan portions of at least two of the manors, Gower and Margam, which were to

---

[1] Although the coal in certain parts of these manors belonged to various ecclesiastical foundations, who apparently held the land and minerals as freeholders of the Bishop (see Appendix M).

supply coal for the growing trade with Cornwall and Devon, belonged to the Church.[1]

The rôle of the Church in all phases of medieval economic life was important, and the ecclesiastical authorities possessed a large part of all the land of England, but their holdings of coal were perhaps proportionately more extensive than their holdings of any other kind of property. They controlled what were, from the standpoint of the future development of the coal trade, the key positions. In Glamorgan, Cumberland, and Shropshire they held just those manors from which were one day to come the chief coal supplies for Ireland, for the coastal towns of Somerset, Devon, and Cornwall, and for the populous lower basin of the Severn. Above all, the ecclesiastics owned in Durham and Northumberland most of those seams which were destined to provide nearly half the entire output of Great Britain in the seventeenth century. When we remember that even to-day, after the confiscation, at the Reformation, of the larger part of its property, the Church is still the largest single royalty owner in Great Britain, enjoying (in the person of the Ecclesiastical Commissioners) nearly £400,000 annually, or at least seven per cent of all coal royalties,[2] we get some idea of the proportion of such revenue as might have accrued to the Church, had its original possessions remained intact.

If the ecclesiastics had continued unmolested in the enjoyment of their original holdings, they would perhaps have hindered the growth of coal mining under Elizabeth and the Stuarts, by retarding the investment of the large capital sums necessary to this growth. The arguments in favour of such a view are three. Religious bodies were not themselves prepared to invest heavily in their own mines. The conditions which they offered lessees were not so favourable as to induce others to invest heavily in them. The Church was not entirely in harmony with the municipal trading class, which controlled the sale of coal, and this opposition sometimes led to a conflict over privileges, which made trade difficult.

In medieval Belgium practically all the coal in the basin west of Mons was owned by powerful ecclesiastics,[3] and rich monastic houses are said to have taken part in the construction of some of the adits for draining pits in the Liége district ;[4] but investigators have not yet provided sufficient data to make it possible to determine the probable influence of ecclesiastical ownership upon continental mining. We must fall back upon British experience. There are those who believe that in England the monks were not only the improvers of the arts

[1] See Appendix M (i). Further investigation would probably reveal a great many other examples of important coal-bearing lands in the hands of religious foundations, besides those given in the Appendix. See also J. B. Simpson, *Coal Mining by the Monks*, 1910

[2] *Report of Coal Industry Commission*, 1919, vol. ii, p. 584 ; vol. iii, appx. 1. £370,000 was the figure given by the Ecclesiastical Commissioners as their income from coal royalties in the year 1917.

[3] Decamps, *Mémoire historique sur l'industrie houillère de Mons*, pp. 58 sqq.

[4] Malherbe, *Historique de la houille dans le Pays de Liége*, pp. 298-9.

but also the great cultivators of land.[1] It has been said that, in Yorkshire, they were as ready as the lay landholders to develop their mineral resources during the thirteenth and fourteenth centuries.[2] As we shall see later, neither of these opinions is necessarily in opposition to the view that, had religious organizations continued to enjoy the same power over property that was theirs in the Middle Ages, they would have retarded the expansion of the coal industry. This view does not imply either criticism or approval of the Church. One may argue that we should be no worse off had we done without coal altogether. Or one may believe that the alleged evils associated with the growth of the coal industry are the evils of civilization, and would have developed, though possibly in different forms, even without mineral fuel. Speculation on such questions is not, perhaps, as far outside the province of the historian as has sometimes been suggested. But here we are concerned simply with a preliminary question—whether the power of the Church over mineral property in the Middle Ages actually did tend to retard the development of the coal industry ?

There is no doubt that ecclesiastics dug, burned, and sold coal in many parts of Great Britain during the Middle Ages. Yet it is plain that the medieval coal industry was of small importance. There is evidence that ecclesiastical landowners were much more conservative than temporal landowners in working their mines. Among the considerable number of accounts of ecclesiastical coal pits that are extant there is no record to show that the Church ever spent, before the reign of Elizabeth, more than £60 on a single adit for drainage.[3] If the deeper seams, which alone could supply coal in large quantities, were to be reached, investments of many hundreds and even thousands of pounds were required.[4] Churchmen were not disposed to invest large sums of their own in mining, and they did not encourage their lessees to do so. The common monastic coal pits probably resembled the mines in France at the beginning of the eighteenth century, when the resources of small landholders who worked the coal permitted them to buy only the poorest equipment and to make only the shallowest diggings.[5] Before the Reformation there was a noticeable tendency towards a diminution in the revenue obtained by the Finchale and Durham monks from coal works in the Wear valley [6]—a tendency which suggests that the brethren were not prepared to seek for deeper seams at higher costs. The mine at Culross in Fifeshire, which became one of the most important in Scotland after it was taken over by Sir George Bruce in 1575, had been worked by the monks of Culross Abbey for hundreds of years, but, before its acquisition by Bruce, it was reported to have lain " long in disuetude, in so much that we have neither large nor small coal for our own house fire ".[7] When the

---

[1] Cosmo Innes, *Lectures on Scotch Legal Antiquities*, 1872, p. 240.
[2] *V.C.H. Yorkshire*, vol. ii, p. 340.
[3] Galloway, *Annals of Coal Mining*, pp. 70–1. This amount was spent by the Finchale monks in 1428.
[4] See below, pt. iv, ch. i (i).
[5] Cf. below, pp. 284, 318.
[6] J. B. Simpson, *Coal Mining by the Monks*, pp. 13 sqq.
[7] A. S. Cunningham, *Mining in the Kingdom of Fife*, p. 5.

lands of Tranent, once owned by the Church, came into the hands of the Scottish Crown, the revenue from the coal mines increased.[1]   After 1500, at any rate, it appears that the ecclesiastics were generally far less ready than lay landowners to push coal mining in their own holdings.

Religious institutions did not always work their own coal ; sometimes they leased it.   That was the common practice in the Bishopric of Durham from the beginning of the fourteenth century, except for a short period in the sixties of the fifteenth century, when a group of mines in the south, at Raby, were operated directly by the Bishop's officials, probably owing to some failure by the lessee to conform to the terms of his lease.[2]   These terms were usually so unfavourable to the " farmer ", or tenant, as to discourage enterprise. To offer mineral property for a short term, at a high rent, and with rigid restrictions upon output, was not the way to attract such capital as might be available to finance a considerable colliery.   Let us examine more closely a lease granted by the Bishop, in 1478, of the " mine of sea coal " in " Rabey, Tofts, Caldhirst, Hertkeld, and Hetherclogh " manors in south Durham, to Sir William Eure for eleven years, at an annual rent of £150.[3]   All expenses are to be borne by Eure, who may, however, take timber and granite from the Bishop's land for the support of the pits.   His output is limited to 31 tons per day at Raby, 27 tons at Tofts, 5 tons at Hertkeld, and not more than 2 additional tons at any other mines he may operate.[4]   He is to work only " on days or nights which are usually ordinary and working days in the Bishopric of Durham ".   If noxious gases, called the " styff ", prevent his getting his full quota in a working day, he may make up this shortage at the same pit, but only on the following working day.   He must not touch the seam underground within four rods of the boundary of the Bishop's manor.[5]   Out of all coal raised, the Bishop is entitled to " so much . . . as it may please [him] " to burn in his lodging at Auckland, and elsewhere in the Bishopric, at 4d. per cartload, a sum which no more than covered the cost of mining.   In addition, the Dean of Auckland is entitled, by virtue of his tithes on these mines, to " so much coal as it may please him to expend in his lodging at Auckland ", at 2d. per cartload.[6]

---

[1] *Exchequer Rolls of Scotland, passim.*

[2] *V.C.H. Durham,* vol. ii, p. 324.

[3] *Durham Curs. Recs.,* 54, *Roll of Bishop William Dudley,* m. 7, no. 20.

[4] The lease does not limit the output by tons, but by "corfes" or "scopes". I have arrived at the figures given in the text by assuming that a corf of coal weighed about 1 cwt. (see Appendix C (iii)).

[5] The lease requires him to leave sufficient "forbarres", four rods in breadth.

[6] The pithead price at Raby in 1580 was 1s. 4d. per wainload (of about 17½ cwt.) to inhabitants of the Bishopric, and 1s. 8d. to outsiders (*Exch. Deps. by Com.,* 22 and 23 Eliz., Mich. 5) ; and, as the wainload was probably not then a great deal larger than the earlier cartload (*ibid.,* 23 Eliz., East. 6, and Appendix C), it is clear that the Bishop and the Dean got their coal at cost price, or less, under the terms of this fifteenth-century lease.   The Dean of Auckland was accustomed to take 70 wainloads yearly for tithe coal.   Provisions of this sort were common in medieval leases.   The lessee of the Earl of Westmorland's mine in Brancepeth (Durham) was obliged to furnish the bailiff of the manor

The terms of the agreement make clear the functions of the Bishop's supervisor and officials, " to be deputed when and as often as it may please us ". They are to be the sole judges of any damage done to property, or waste of coal, and may order Eure, at any time they see fit, to " correct and amend " at his own expense such damage or waste, even if it require more than two months to carry out their orders. The supervisor has to pass upon the method of sinking pits and leaving " forbarres ", or pillars. No timber or granite can be taken without his permission. He checks the output at each pit, and can, at his entire discretion, measure the " corfes " or " scopes " by which this output is determined, to make sure that none contain more than two and a half bushels of coal.

This lease was clearly a one-sided contract, in which the supervisors, who represented one of the parties—the Bishop—had powers which are usually delegated to an impartial tribunal. It is even doubtful whether there was any appeal from their rulings. The output of the lessee was much more narrowly restricted than would appear from a superficial perusal of this grant. Working days, according to the custom of the Bishopric, were interspersed with numerous holidays. No mining was done on the Sabbath, nor for a period of about forty days in midsummer. There was no provision for making up the short output of one mine from the produce of another. While new shafts were being dug, or while repair work ordered by the supervisor was being carried out, the pits must have yielded little coal, for the Bishop inserted in the lease a clause permitting himself or the Dean to get their fuel at Hertkeld, " if coal cannot be won and gained at . . . Rabey and Tofts ", the two mines which were presumed to furnish from 85 to 90 per cent of the output. Bishops did not hesitate to make the most of their unlimited right to fuel; when Wolsey occupied the episcopal chair, he took advantage of a similar clause in another lease to insist upon the delivery of one ton every working day from each pit at Whickham and Gateshead.[1] The market was very uncertain, and, if the lessee could not sell his output, he was still bound to pay full rent. Although, in the case of the mines in south Durham, Eure was granted a practical monopoly of the coal trade into Yorkshire,[2] consumers were likely to burn wood or turf if the price of coal rose too much. And coal could be carried overland only with the greatest difficulty after " Hallowtide ", because of the " evelness of the wethir in the winter ".[3] If we assume that there were 250 actual working days in the year, the maximum output permitted was 16,000 tons. In 1461, when the Bishop worked the mines himself, his output for a half year did not quite reach 3,000

with sixty " fothers " (probably about 50 tons) of coal a year, to burn in his house (*Augm. Partics. for Leases*, 39/78). In 1539, the parson of Gateshead, in a dispute over tithes, was granted the right to tithe coals from the lessees of the Bishop. He might work each pit for three working days a year, one before Christmas, one before Candlemas, and one before Michaelmas, paying only the wages of the workmen, and providing tools (Galloway, *Annals*, pp. 82–3).

[1] M. Dunn, *View of the Coal Trade*, p. 12.

[2] The Bishop agreed to allow no coal mining or carrying away of coal, within two miles of these pits, by anyone except Eure or his agents.

[3] *Exch. Deps. by Com.*, 22 and 23 Eliz., Mich. 5 ; and cf. below, p. 396.

tons [1], and it seems probable that when the mines were leased the production was not much greater.

An annual production of 6,000 tons meant in effect a royalty of 6*d.* per ton, which then amounted to at least a third of the pit-head price. When we consider that to-day the average royalty amounts to about one-thirtieth of the pit-head price,[2] we see how hard a bargain the Bishop had driven. His lessees had little more opportunity to profit than had his paid officials, and we find, indeed, the same person in the rôle of lessee at one time and of paid official at another. A certain William Thomlinson of Gateshead was made Clerk of all the Bishop's mines of coal, lead, and iron in 1529, and was still acting in a similar capacity in 1542, though in 1534 he had become one of the " farmers ", or lessees, of the Bishop's coal mines in Gateshead, mines which he continued to farm as late as the reign of Philip and Mary.[3] The terms offered in the Bishopric were perhaps too hard to attract outside capitalists.

There is no reason to suppose that the lease just considered was a peculiarly harsh one. The terms granted by other ecclesiastics in Durham and Northumberland, while probably not so sternly enforced, were based upon the same principles : a short term, a limited output, and a high royalty rent. In the case of a one-year grant of coal at Trillesden and Spennymoor, made by the Prior of Durham in 1447, the lessee was limited in two ways. Only one pit, already dug, might be worked ; only three " pikkes " (hewers) might labour in this pit on a working day, and each hewer might get no more than five tons (sixty " scopes "). At the same time, the Prior reserved the right to appoint viewers, " as oft as hym likes ", to see that the digging was done without prejudice to himself.[4] In an earlier seven-year lease of a pit near Aldengrange,[5] the lessee was obliged to supply all the coal for the Priory of Durham, and for the tenants of the Prior's manor of Beaurepair. Leases made by the monks of Tynemouth just previous to the Dissolution were for seven years, and the rents were fixed in accordance with the output per working day permitted by the lease—£25 a year where the output was limited to twenty tons a day.[6] Such a rent represented a royalty per ton of perhaps 3*d.*, or about one-sixth of the pit-head price.[7] A lease granted by the Monastery of Byland in 1537, of the " mynes and myneralls of leade and cooles " in the manor of Nidderdale in Yorkshire, reserved to the monks, in payment of rent, three loads out of every twenty

---

[1] *V.C.H. Durham,* vol. ii, p. 324.

[2] Cf. below, p. 327.

[3] Galloway, *op. cit.,* p. 85 ; Welford, *History of Newcastle,* vol. ii, pp. 147, 194 ; *Court of Augm. Proc.,* 3/45.

[4] J. Raine, *Historiae Dunelmis Scriptores Tres (Surtees Soc.,* vol. ix, 1839), appx., pp. cccxii–iii, cited by Galloway (*op. cit.,* pp. 69–70) and by J. B. Simpson (*Coal Mining by the Monks,* p. 25), who is wrong in stating that this is the earliest example we possess of a colliery lease (see succeeding footnote).

[5] Made in 1398 (*Greenwell MSS.,* D. 84).

[6] *Court of Augm. Proc.,* 3/89.

[7] The rent could not have been less than 1¼*d.* even if the full output was achieved each day.

extracted.[1]   In Nottinghamshire the monks of Beauvale limited the output of their lessees, by restricting the number of miners who might work in the pit.[2]

Not only did the Church restrict at every turn the freedom of lessees, it resisted attempts made by town merchants to control (and, through control, to expand) the trade in coal.   This resistance appears in the early struggle between ecclesiastical and municipal interests, over the right to ship coal and other commodities from the Tyne.

As has been pointed out by Galloway,[3] the merchants of Newcastle had a special interest in controlling shipping on the river.   When, in 1213, the town was first incorporated as a royal borough, King John's charter confirmed a title, which was later said to have existed from time immemorial, to " the water of Tyne, and the soil of the same by the water covered ", from a place called Spar Hawk in the sea to a place called Hedwin Streams, above which the river was not navigable, " saving the rents, prises, and assessments of the king in the port of the said town ".[4]   The precise meaning of this clause was a source of continual controversy.   Newcastle burgesses interpreted it in the sense that no ships entering or leaving the river could load or unload elsewhere than at the town wharves, unless the master had been granted an exempting licence by the Corporation.   But the Church refused to recognize this interpretation.   The Bishop of Durham claimed an immemorial right " to load and unload fish, flesh, wood, coals, and all other vendibles in any part of the water of Tyne ",[5] and claimed also the actual ownership of the southern half of the river from Stanley Burn to the sea.[6]   The Prior of Tynemouth asserted his privilege to carry on, without interference from the burgesses,[7] his own shipping at North Shields on the Northumberland side.

Each of the three contending parties had a special interest in the coal trade.   Bishop and Prior desired to market themselves, or through their lessees, the product of their extensive mineral holdings. The burgesses, although at first they had little direct connection with the mining of coal, were not willing to allow a product of growing commercial importance to escape their control.   In the fourteenth century they claimed a right of " hosting " merchant strangers ; that is to say they insisted that no commodity might be sold to, or purchased from, a visiting shipper, unless one party to the transaction was a " freeman " of Newcastle.   If a " foreigner " sailed into the

---

[1] I. S. Leadam, *Select Cases in the Court of Requests* (*Selden Soc.*, vol. xii, 1898), pp. 201–4.

[2] *V.C.H. Nottinghamshire*, vol. ii, p. 325.

[3] *Op. cit.*, pp. 22–3.

[4] Inquisition taken at Newcastle, January 4th, 1447, printed in Welford, *History of Newcastle and Gateshead*, vol. i, pp. 316–18.

[5] *Ibid.*, p. 33.   [6] *Ibid.*, p. 56.

[7] I. S. Leadam, *Select Cases in the Star Chamber* (*Selden Society*, vol. xxv, 1910), vol. ii, pp. xciii–v.   The Prior of Durham was less active in prosecuting his interest in the bank of the river at South Shields, an interest which, however, did not at this time have to do with the coal trade (Welford, *op. cit.*, vol. i, pp. 141, 262).

river wishing to trade corn for coal, he must sell his corn and buy his coal from a burgess.[1]   Naturally, if this alleged right was enforced, it meant that neither the Bishop, nor the Prior, nor their lessees, could sell coal for export to anyone except a citizen of Newcastle. The citizens acquired, in 1351, some hold over mineral property, by a royal grant of the ownership of coal under the Castle Field and the Frith [2] (common lands outside the town walls), and by leases of certain coal mines in the neighbourhood.[3] These holdings served to whet their appetite for a monopoly over the trade.

During the fourteenth and fifteenth centuries the struggle between the Church and the town was waged with fairly even odds, and the net result was that neither side gained a permanent ascendency. The burgesses were blocked from obtaining large interests in the mining of coal, and their alleged exclusive right to sell it never went long unchallenged.   Unquestionably the most pronounced result of this struggle was that it held back, by mutual jealousy and even sabotage, the development of the trade.   Between 1314 and 1393, the policy of the Crown was, on the whole, favourable to the Bishop, and in 1393 a royal charter confirmed his right to half the river, with power to load and unload ships with coal or other merchandise.[4]   The townsmen, however, were not daunted.   They seized ships loading on the south side of the Tyne, and forced the masters to unload at Newcastle the coal with which they were about to set to sea.   The very frequency with which the King had to intervene, to secure the rights of the lessees of the Bishop's coal mines in Whickham and Gateshead, is proof of the energy of the burgesses.   By 1447 the town was again asserting its right to trade in all " seacoal " sold for export,[5] and during the latter part of the fifteenth century the Bishop had little success in reaching the royal ear, although the charter of 1393 still had effect in law.

The course of the town's struggle with Tynemouth Monastery was somewhat different.   As early as 1267 the jealousy of the burgesses was aroused by the Prior's gains from the trade, and they raided his wharf at North Shields, beat his servants, set fire to his houses, and carried off a vessel laden with coal.[6] They were able successfully to plead in defence of their conduct that, if an independent trade were permitted the Tynemouth monks, it would prejudice the King's tolls

---

[1] Welford, *op. cit.*, vol. i, pp. 119–20 ; see also F. W. Dendy's introduction to *Extracts from the Records of the Company of Hostmen of Newcastle-upon-Tyne* (*Surtees Society*, vol. cv, 1901), where the whole subject of " hosting " at Newcastle is discussed.

[2] Galloway, *op. cit.*, pp. 40–2 and map facing p. 41 ; Welford, *op. cit.*, vol. i, pp. 135–6.

[3] Lease in 1334 to Richard Scot of Newcastle by Sir Robert Delaval, of the sea coal of " le Chester in Benwell " (Craster, *Index to Waterford MSS.*, vol. iii, p. 64).   In 1444 John Burke of Newcastle is spoken of as having been one of the " farmers " of coal mines at Whickham (*Deputy Keeper's Report of the Public Records*, no. 44, p. 318).

[4] Welford, *op. cit.*, vol. i, pp. 33, 96–7, 124–5, 201, 216, 246–7 ; Galloway, *op. cit.*, pp. 46–7, 50 ; J. Brand, *History of Newcastle*, vol. ii, pp. 12, 256.

[5] Welford, *op. cit.*, pp. 316–18.

[6] *A History of Northumberland*, vol. viii, pp. 286–7.

at Newcastle, tolls which were raised by means of a tax on exports and imports. In July 1292, judgment was given in Parliament against the Prior, and he was ordered to tear down his wharf.[1] The monks were no more disposed to accept this decision as final than the burgesses had been to acknowledge the power of the Bishop over the southern half of the river.[2] A hundred years later the Prior had again built wharves at North Shields, and had begun to carry on commerce which prejudiced the trade of Newcastle—so the burgesses alleged in 1447—to the extent of £340 a year.[3] Perhaps as a result of this protest, a royal grant, made to the Prior in 1436, permitting him to take tolls and customs from ships at North Shields, outside the jurisdiction of Newcastle, was withdrawn by an Act of resumption in 1450. But Edward IV, in 1462, granted the Prior the right to load and unload cargoes of coal and salt from Tynemouth manor without interruption by the burgesses, a right which was upheld in 1512 in the well-known Star Chamber case, in which the Prior and Brethren of Tynemouth opposed the Mayor and Citizens of Newcastle.[4]

It appears, on the one hand, that the Church was not overzealous in developing, or encouraging the development of, its own mineral resources ; and, on the other hand, that the desire of the Bishop and the Prior to trade independently of Newcastle resulted in a conflict which could hardly have been favourable to the expansion of the coal industry in the Tyne valley.

During the sixteenth century a series of great social and economic changes placed the prospective investor in coal mines in a position to obtain concessions on far better terms, and put an end to the conflict between the Newcastle traders and their clerical rivals. Property on an enormous scale was taken from its owners and transferred to new hands, reverting in the process temporarily to the Crown. The dissolution of the monasteries brought into the hands of the State a very considerable part of the mineral property of the country. The confiscation of the property of the religious gilds under the Act of 1547 added further coal-bearing territory to the possessions of the Crown, which also acquired, by attainder, a number of great lay estates : notably the lordships of Brancepeth and Winlaton in Durham, previously the property of the Earl of Westmorland, the manors of the Earl of Northumberland in Cumberland, Durham, and Northumberland, the manors of the Duke of Clarence in Yorkshire, and the larger part of Carmarthenshire and Pembrokeshire, which had once belonged to Rice Griffith.

The first table in Appendix M, which should be read in connection with what follows, gives a list of numerous properties, large and

---

[1] Leadam, *Select Cases in the Star Chamber*, vol. ii, p. xciv.
[2] In 1338 the Tynemouth monks rented a " staith " (wharf) at Newcastle from the Corporation for £40 a year, from which they proposed to ship coal won in their manor of Elswick (J. B. Simpson, *Coal Mining by the Monks*, p. 7 ; Galloway, *Annals*, pp. 38–9).
[3] Welford, *op. cit.*, vol. i, pp. 317–18.
[4] Leadam, *op. cit.*, vol. ii, pp. xciii–v, 68 sqq.

small, which came into the hands of the Crown during the sixteenth century. Under all these lands were easily workable coal seams. This list includes, of course, only a very small portion of the coal-bearing manors which changed hands during the sixteenth century, when confiscation followed upon confiscation.

What was done with this property ? Clearly it was of the utmost consequence to future generations into whose hands it should fall. A State has seldom had such power over its economic destiny. Yet one looks in vain for any consciously pursued policy on the part of Tudor governments. As Professor Pollard has written, " it was not the Crown which profited from the change ; its feeble control over wealthy landlords left the proceeds of the revolution at their mercy ".[1] To attain their ends, these landlords employed every weapon of self-interest, from advice to definite corruption. Sir William Herbert, one of the guardians of the thirteen-year-old Edward VI, secured from the boy-king a grant of enormous possessions in South Wales, including much of the territory lying between Chepstow and Swansea, which has subsequently proved so rich in coal resources.[2]

To the extent that lands already secularized were handed over to another subject, as was the case with much of the property in the Herbert grant, the changes generally had little effect upon the exploitation of minerals. It was different when a layman was substituted for an ecclesiastical landlord. Less conservative in his attitude towards commercial development, the layman was more ready to encourage the development of his coal resources, by granting leases on favourable terms. Moreover, property owned by him was more likely to come on the market than property belonging to a corporation. Lay holdings could nearly always be purchased for a price, but ecclesiastical holdings were rarely for sale. The lay holder might be driven by circumstances to sell at a sacrifice, and to see his land snapped up by a new proprietor, conscious of the potential value of the mineral resources, and intent upon developing them. A case in point is that of the manor of St. Bees, granted by the Crown to Sir Thomas Chaloner in 1553.[3] The Chaloners took no interest in developing the mineral wealth which lay beneath the surface within a mile or two of the Cumberland coast, and Sir Thomas' son sold the property for £2,000 to Thomas Wilbergh and Gerard Lowther. Lowther and his brothers, aware of the opportunity offered by the growing demand for coal in Dublin, bought out Wilbergh, and, by the liberal investment of money, developed the mines about White-haven so successfully that, in the next generation, Sir John Lowther

---

[1] A. F. Pollard, *Political History of England, 1547–1607*, p. 20.
[2] *Pat. Roll*, 4 Edw. VI, pt. 9, printed in *Report of Coal Commission*, 1919, vol. iii, app. 81, pp. 233–5.
[3] The story as told in the text is taken from the testimony given to Exchequer Commissioners in the case of Gerard Lowther *v.* Thos. Wilbergh (*Exch. Deps. by Com.*, 14 Jas. I, Mich. 26). Fletcher (*Archæology of the West Cumberland Coal Trade*, pp. 270–2) has it that Sir John Lowther made the purchase, while it is related in *V.C.H. Cumberland* (vol. ii, p. 359) that Wilbergh originally made the purchase from Chaloner, but that the Lowthers took a mortgage from him which they subsequently foreclosed.

succeeded in capturing most of the Irish trade from his competitors round Chester and Liverpool.[1] Had the lands behind Whitehaven remained in the possession of the Monastery of St. Bees, the Lowthers would not have been able to make their purchases, and the coal resources might have remained undeveloped until much later.

The Baron of Culross in Fife, where Sir George Bruce opened his "matchless" colliery in the reign of James VI, the manor of Broseley in Shropshire, from which the Clifford family drew the first large supplies of coal for transport on the Severn, the manor of Bedworth, which remained for at least a century after the death of Elizabeth the seat of the most important colliery in Warwickshire, the manor of Lumley in Durham, which was the seat of one of the three chief coal mines of the Wear Valley on the eve of the Civil War, were all taken from the Church at the time of the Reformation. Throughout the country we find that it is in manors lost by the Church that the principal mining enterprises of the late sixteenth and early seventeenth centuries are started.

Not all these manors had actually come into private hands before they became important seats of the growing coal industry. Much of the progress was made in the lands which the Crown retained until after Elizabeth's reign. In such manors the development of mining was encouraged by the easy terms upon which it was possible to obtain leases of coal.[2]

Whether concerned with a mine in Northumberland, Durham, Yorkshire, Warwickshire, or Glamorgan, the Crown lease had certain well-defined characteristics. Usually drawn for a term of twenty-one years, and never apparently for a shorter period, these leases offered a sufficient length of tenure to invite deeper mining, and the investment of capital in new drainage devices.[3] Seldom was a limit put on the output. In return for a fine and a fixed annual rent, the tenant commonly acquired an unlimited right to take coal within the territory leased him. If no coal was found, it was easy for him to get exempted from his obligation to pay rent. During the reign of Elizabeth, commissioners for the Queen were generally so impressed with the hazards of coal mining, and so anxious to encourage colliery enterprise as a means of relieving the country of its shortage of wood, that they fixed the rents low enough to encourage even the most careful investor to apply for a lease, which he could hold as a speculation if he was unwilling to risk his money in the sinking of shafts and the purchase of mining equipment.[4] Under a Crown lease the adventurer was free from that onerous supervision to which medieval lessees had been subjected by ecclesiastical officials.

There were, it is true, clauses binding the tenant to dig " according

---

[1] See above, p. 110.

[2] See the examples in Appendix M (ii).

[3] When the Crown took over the possessions of Tynemouth Monastery, there was still in being a seven-year lease of a coal pit in Bebside and Cowpen manors, granted by the Prior to Nicholas Mitford and John Preston. The Crown immediately regranted the coal pit to the same individuals on a twenty-one year tenure (*A History of Northumberland*, vol. ix, pp. 223–4).

[4] Cf. below, p. 323, and vol. ii, pp. 35–8.

to the order of minery ", " to leave good and sufficient pillars ", and not to work to the " disinheritance of the Queen "—this last a stipulation intended to prevent wastage of mineral wealth through inefficient mining methods, and to hinder the tenant from taking coal out of all proportion to his rent. Actions were occasionally brought against a lessee on the ground that he was " disinheriting " the Crown, but it was difficult to prosecute under such vague stipulations, and, once a lease had been granted, inspections by commissioners for the Crown were infrequent and superficial. In Durham and Northumberland the Crown did employ certain permanent officials to supervise the mining of coal in the royal domains. We hear in 1557 of William Dixson, " deputie viewer of the King's cole mynes in Benwell ",[1] and in 1581 of Edward Bulmer, " viewer of the Queen's mines in and near Newcastle ".[2] But there is no evidence, among the abundant prosecutions undertaken on behalf of the Crown in the Court of Augmentations, that such officials were overzealous in asserting the right of the Crown to share in the spoils of coal mining. Bulmer was a yeoman of Newcastle ; Dixson also was a local man ; and it is not unlikely that both stood in greater fear of their influential fellow townsmen, who worked the Crown mines, than of the royal power. After all, London was far from the coal fields—seventy-three hours were required to bring a letter post haste to Newcastle—and the effectiveness of Crown administration perhaps diminished in proportion to the distance from the capital.[3]

So weak was this administration, or so eager to make investments in mining attractive, that when monastic lands were transferred to the Crown with leases of coal mines still in being, tenants were able to wriggle out of certain stipulations of the original leases before they expired.[4] James Lawson, a brother of the last prioress of the Nunnery of St. Bartholomew, was bound, under the terms of the lease granted to him by the Nunnery, to pay an annual rent of £16 for certain mines in Gateshead, and to confine himself to an output of twenty score " corvis of coles " (or 15 to 20 tons) per working day.[5] He was denounced by Thomas Thomlinson, and other lessees of the adjoining mines belonging to the Bishop of Durham, who alleged that, since 1544, when the property of the prioress had been confiscated by the Crown, Lawson had extracted from the pits " great store of coal ", sometimes worth more than two hundred marks (£133 6s. 8d.) a week,

---

[1] *Court of Augm. Proc.*, 38/2.

[2] *Augm. Partics. for Leases*, 112/46.

[3] There was no regular mail service between Newcastle and the capital until the London–Edinburgh post was established in 1635 (Welford, *History of Newcastle and Gateshead*, vol. iii, p. 328).

[4] With the decline in ecclesiastical power, the right to exact tithes on coal mines was also challenged. Thus the " farmers " of the dissolved Deanery of Auckland continued to assert the Dean's title to tithe coals from Raby pits, but the right was denied by the Bishop's lessees as early as 1568 (*Exch. Deps. by Com.*, 22 and 23 Eliz., Mich. 5 ; 23 Eliz., East. 6). Cf. below, vol. ii, p. 278.

[5] For the cases in which Lawson was engaged, first against William Inskip and William Thomlinson, then against Thomas Thomlinson (William's son), and, finally, against William Inskip and his son Vincent, see *Court of Augm. Proc.*, 3/45, 7/11, 14/67, 14/76, and *Augm. Misc. Bks.*, vol. cxi, ff. 16–19.

ten to fifteen times the quantity permitted under the terms of his lease ; and, further, that he had concealed his obligation to pay rent. From his answer to these charges, it is plain that Lawson subjected the lease to a most liberal interpretation. He claimed the right to make up short output " by any policy, labour and means ", including the digging of as many new pits as he might choose.[1] He believed himself freed from his obligation to pay rent, because, he alleged, he had been prevented by Thomlinson from getting his full quota every day.

As a result of the readiness of the Crown to lease mines, most of its tenants were firmly entrenched for periods of at least twenty-one years at a fixed annual rent, when there came, in the decade 1580 to 1590, a great increase in the demand for coal. Many of these tenants found themselves able greatly to increase their output, and consequently their profits, without any increase in the rent due to the Crown. Instead of the landowner taking for himself a third or more of the pithead price of coal, as he had done in the Bishopric of Durham during the Middle Ages, he now received on the average, to judge from a survey made in 1611 of the King's coal mines in Durham and Northumberland, less than one-seventh as much as the profit made by his tenants.[2]

TABLE XI

Value [3] of Crown Mines in Durham and Northumberland in 1611

Manor.	Description of colliery.	Annual rent paid.			Annual value.		
		£	s.	d.	£	s.	d.
Northumberland—							
Benwell	2 seacoal pits [4]	20	0	0	120	0	0
Wylam	1 landcoal pit		18	6	13	6	8
Amble	1 landcoal pit	2	1	0	10	0	0
Tynemouth	1 seacoal pit	3	0	0	50	0	0
Cowpen	2 seacoal pits	5	0	0	40	0	0
Elswick	2 seacoal pits	21	0	0	120	0	0
Durham—							
Tudhoe	1 landcoal pit	3	0	0	30	0	0
Cockfield	1 landcoal pit	4	0	0	40	0	0
Thornley	1 landcoal pit	3	0	0	10	0	0
Chopwell	1 seacoal pit	4	0	0	20	0	0
Eldon	1 landcoal pit	100	0	0	400	0	0
Crown Freehold [5] in							
Gateshead	3 seacoal pits	17	0	0	133	7	4

[1] Lawson's actual output for a period of one and a half years after Christmas, 1544, is said to have been 3,057 chaldrons (*Augm. Misc. Bks.*, vol. cxi, ff. 16–19).

[2] *S.P.D., Jas. I*, vol. lviii, no. 19. Other mines besides those in the table are mentioned, but none of them was being worked.

[3] The term " value " was used in the sense of profit accruing to the lessees who owned the colliery (see Appendix B (viii)).

[4] The terms " seacoal " and " landcoal " indicate coal carried by sea and coal raised for local sale.

[5] Although the Bishop of Durham was the overlord of the manors of Whickham and Gateshead, within these manors certain freeholds once held of the Bishop by small monastic houses now belonged to the Crown, which had taken over the title of these houses after the dissolution of the monasteries. As freeholder, therefore, the Crown owned the minerals as well as the surface in certain parts of Gateshead and Whickham.

Manor.	Description of colliery.	Annual rent paid. £ s. d.	Annual value. £ s. d.
Brinkburne Freehold in Whickham	1 seacoal pit	1 6 8	20 0 0
Greenlaw Freehold in Whickham	coalmines	5 0 0	400 0 0[1]

There are even more striking examples of the discrepancy between rent and " value " in the case of other coal mines belonging to the Crown. A colliery at Denton, farmed from the Crown at a rent of £10 a year, was judged by one of the barons of the Exchequer in 1610 to be worth £1,200 annually to the operators.[2] The mines under Brinkburne freehold in Whickham produced about 25,000 tons of coal in the four years from 1619 to 1623, but paid only £1 6s. 8d. annually in rent to the King.[3] In 1621 Robert Brandling of Felling,[4] surveyor of coal mines, filed a petition against Sir Peter Riddell, mayor of Newcastle, and others, alleging that they rented the King's mines at a thirtieth of their " value ". For a lease of one mine, worth £1,200 a year, they paid only £16 in rent, while for another, worth £1,000, they paid £20 in rent, so that, for these and other mines worth in all from £3,000 to £4,000 they paid less than £100 in rent.[5] In Warwickshire the lessees of " coale mynes within the manors of Bedworth and Griff " paid the Crown only 20s. per annum for their lease, taken about the year 1597, " being very well worth £300 per annum ".[6] Coal in Great Britain has never been subject to such low royalties, as when for a brief spell the State owned a very considerable proportion of all the mine als.[7]

In spite of efforts by the Crown to stiffen the rents, some of the colliery operators succeeded in keeping them low until at least the Civil War. Shrewd entrepreneurs, in asking for concessions at a small annual rent, offered to pay fairly large sums down in fines ; and the Crown, pressed for immediate cash, was trapped into accepting what proved often enough a bad bargain. By paying a fine of £100, John Lyons, a Yarmouth merchant heavily interested in the coal and salt trade, succeeded in getting his lease of Greenlaw colliery, on the south bank of the Tyne, at the small rent of £5 a year. When Crown commissioners threatened in 1611 to bring an action against him for defrauding the King of his just share in the spoils of mining, Lyons (together with his partners, Richardson and Chapman) agreed to pay the increased rent of £40, together with a new fine of one thousand marks, " to avoyde law ".[8] If their annual profits amounted

---

[1] This figure for Greenlaw is confirmed in *Lansdowne MSS.*, 156, no. 109 (" Touching certeine landes and Colemines in Grenelawe in the Bishoprick of Durham, June 1st, 1611 ").
[2] *S.P.D., Jas. I*, vol. lviii, nos. 17–18.
[3] *Palat. of Durham, Bills and Answers*, bdl. 24 (Alex. Stevenson *v.* Thos. Liddell and others).
[4] Cf. below, vol. ii, pp. 128–9.
[5] *S.P.D., Jas. I*, vol. cxvi, no. 74.
[6] *Chanc. Proc., James I*, F. 4/53.
[7] For a detailed discussion of the proportion between the royalty and the selling price of coal, see below, pp. 319 sqq.
[8] *Lansdowne MSS.*, 156, no. 109.

to £400 or more,[1] they could well afford to make this bargain. It was for a fine of £100 in 1623 that Alexander Stephenson, a page of the bed chamber, secured a lease for three lives of Brinkburne colliery, at a rent of only £1 6s. 8d. ; and this lease was still in being fifty-five years later.[2]

When two or more prospective colliery owners bid against each other for a concession, neither was likely to get it at so low a rent.[3] But, as long as the demand for mineral fuel was increasing rapidly, any terms which fixed the rent without regard to output were likely to prove favourable to the lessee.

Although the Crown acquired a vast amount of mineral property as a result of the dissolution of the monasteries, the Church retained many holdings in which were valuable coal deposits. In Durham the Bishop still owned a great number of manors. But his powers were weakened until, at the end of the sixteenth century, he conserved only a vestige of the wealth which once went with his office, and was able neither to exact high rents from the lessees of his coal mines, nor to withstand the will of the Newcastle merchants to control all trade on the river Tyne.

In the reign of Henry VIII a brief attempt to revive the Bishop's power, already on the wane, accompanied the appointment as Bishop in 1523 of no less a personage than Wolsey himself. The temporal chancellor of the Bishopric, William Franklyn, mindful of the interest of his illustrious master, informed Wolsey that, after examining the Bishop's titles, preserved in the Exchequer at Durham and the Tower of London, he had convinced himself that " we may shippe the said colis on the bishopriche syde " of the Tyne. Franklyn proposed to contract directly with some London merchants to carry away the Bishop's coal from the wharves opposite Newcastle.[4] Urgent state business, however, prevented Wolsey, during the six years in which he occupied the see, from paying so much as a visit to Auckland or Durham, and Franklyn apparently could not succeed in interesting his chief in these commercial projects. In any case, the fall of Wolsey in 1529 put an end to what was the final attempt to assert a right on behalf of the Bishop to trade independently of Newcastle. The right

---

[1] See above, Table XI.

[2] *Palat. of Durham, Bills and Answers*, bdl. 24 ; *Parl. Surveys, Durham*, no. 8 ; *Exch. Deps. by Com.*, 29 Chas. II, East. 20.

[3] See, for instance, *S.P.D., Jas. I*, vol. lviii, nos. 17, 18.

[4] William Hutchinson, *History and Antiquities of the County Palatine of Durham*, 1785, vol. i, p. 404, note. Franklyn goes on : " The marchaunts of New Castell wil be lothe t'aplie thereunto howbeit your privilegis and graunts be clere inow and it is no reason that they shuld enforce your grace to sell your colis only unto theym at their own prices and they to utter the same agyn at their own libertie bothe to Englishmen and straungers at prices onresonable as they have doon heretofore." In 1522 a ship was loaded with coal at the Bishop's staith in Gateshead (Welford, *History of Newcastle and Gateshead*, vol. ii, p. 66), and in 1523 Franklyn, Sir William Bulmer, and Sir William Eure were appointed King's commissioners to seize a prize ship driven into the Tyne on behalf of the Bishop, " as of right belonging to him, allowed by ancient grants and confirmations from the crown " (*ibid.*, vol. ii, p. 75).

of the town to a monopoly of all shipping on the Tyne, hitherto based only on the early town charters, was now confirmed by an Act of Parliament.[1] A monopoly is not, of course, generally regarded as favourable to the expansion of industry, but, in this case, the triumph of Newcastle, by ending a conflict which was in itself a hindrance to trade, contributed to the growth of the coal industry. The decline of the Bishop's power had another and more important influence upon this growth. Like the dissolution of the monasteries, it helped the local commercial interests to obtain control over minerals formerly worked or leased by the Church.

Crown policy after 1529 aimed at diminishing the economic strength of the Bishops by substituting, wherever possible, fixed annual stipends, payable by the Crown, for episcopal estates administered by the Bishops themselves, and likely, with the growth of population and national wealth, to increase in value.[2] The Crown wished, in short, to make the ecclesiastics mere pensioners, dependent for their livelihood on royal goodwill, instead of allowing them to continue as great landholders whose wealth might enable them to influence State policy. This programme, it is true, was never completely carried out. During the minority of Edward VI, the Duke of Northumberland attempted to dismember the see of Durham, and secured the passage of " an acte for the uniting and annexing of the Towne of Gateside to the Towne of Newcastell uppon Tyne ", but Mary succeeded in repealing this statute two years later.[3] By peacefully, yet methodically, curtailing the Bishop's powers, Elizabeth accomplished far more to reduce the strength of his position than Northumberland had accomplished by legislation. The aged Tunstall was deposed a few months before his death, and his successor, James Pilkington, formerly master of St. John's College, Cambridge, was the first protestant prelate to occupy the see. A number of manors, including Gateshead, were detained by the Crown until 1566, when, as a result of continued protests from the new Bishop, they were restored to him.[4] Elizabeth, however, confiscated each year £1,000 of his revenue,[5] and refused to recognize his claim to the Earl of Westmorland's vast Durham possessions, with their valuable coal deposits, which the Crown had seized in 1570 after the Rebellion of the Earls.[6] Pilkington was in no position to push his claims. He wrote to Cecil in 1566,

---

[1] 21 Henry VIII, cap. 18 ; see also Welford, *op. cit.*, vol. ii, p. 117. Franklyn had suggested to Wolsey that he should obtain the passage of an Act of Parliament, confirming the Bishop's right to ship his own goods without interference by the citizens of Newcastle (Hutchinson, *loc. cit.*).

[2] A. F. Pollard, *History of England, 1547–1603*, p. 17.

[3] 7 Edward VI, cap. 10 ; 1 Mary, session 3, cap. 3 ; Welford, *op. cit.*, vol. ii, pp. 305–9.

[4] Brand, *History of Newcastle*, vol. i, p. 480, note ; *Lansdowne MSS.*, 8, nos. 84, 85, 87.

[5] *S.P.D.*, *Eliz.*, vol. xxxix, no. 81. According to a commonly accepted version of the affair, Elizabeth excused her action on the ground that Pilkington, having been rich enough to provide his daughter with a marriage dowry of £10,000, equal to that of a princess, could well afford to part with some of his revenue (Pollard, *op. cit.*, p. 360, note).

[6] Hutchinson, *op. cit.*, vol. i, p. 452 ; G. T. Lapsley, *County Palatine of Durham*, 1900, pp. 48–9.

" I am now at the phissicions for maindemant [mendment] . . . [and] can not so well attend as I wold ".[1] He seems, in fact, to have been a weak man, incapable of managing his financial affairs, for he left the palaces and public edifices of the see in such a ruinous state, that his successor began proceedings against his executors for dilapidations.[2]

Nor did the Bishop retain his power to exact high rents from his lessees. As late as 1552 his pits at Grewborn and Raby in south Durham had been leased for £150 a year, but in 1568 it was possible to find an undertaker only by reducing the rent at Raby to £10.[3] At the same time the Bishop was obliged to go to law against the Gascoignes of Ravensworth, who were encroaching on his mining rights in the great waste of Chester-le-Street.[4] The most serious curtailment of the ecclesiastical power came during the tenancy of Bishop Richard Barnes (1577–87), when the Church lost all control over coal mining in the manor of Whickham, which was destined to produce 100,000 tons per annum and upwards during the seventeenth century, a larger output than that of any area of equal size in the world.[5] At the same time, the Bishop lost his control over the manor of Gateshead, which was also well stored with minerals, and which provided the best site for shipping coal of any manor in the Tyne valley. As long as the Bishop held the reversion of short-term mineral leases in Whickham, there remained a chance that he might attempt to resurrect his old power. And, although the Newcastle burgesses enjoyed by Act of Parliament the right to force all vessels to load at the town " and no where ells ", as long as Gateshead remained under the control of the Bishop they could not load directly into sea-going colliers from staithes within the manor.[6] It is no wonder that the Newcastle citizens should have made repeated attempts to bring Gateshead and Whickham within their own jurisdiction. These were met with opposition from the citizens of Gateshead, who sent a petition to the Speaker of the House of Commons in 1576, begging him to intercede against the passage of a bill for uniting Gateshead and Whickham with Newcastle. " If these towns shall be annexed ", they complained, the men of Newcastle " may put all their cattle to eat with Gateshead . . . and they may have the coal of Gateshead Moor, which will be worth, if they may win the same, ten thousand pounds, which were to the disinheritance of the see of Durham ".[7]

There is reason to believe that Thomas Sutton took advantage of this conflict to make himself one of the richest men in England.[8]

[1] *Lansdowne MSS.*, 8, no. 87.

[2] Hutchinson, *op. cit.*, vol. i, p. 458.

[3] Galloway, *Annals of Coal Mining*, pp. 105–6.

[4] *Lansdowne MSS.*, 8, no. 87 ; *Palat. of Durham, Bills and Answers*, bdl. 1 (Attorney General of the Bishopric *v.* Gascoigne and others).

[5] See below, pp. 361–3.

[6] Until the end of the sixteenth century many of the ships were of such light burden that they could be brought alongside the wharves. (Cf. below, pp. 416–7.)

[7] Welford, *op. cit.*, vol. ii, p. 479 ; see also pp. 293, 305, 476, 504.

[8] On Sutton, see *Dict. Nat. Biog.* ; Galloway, *op. cit.*, pp. 93–4 ; John Wilford, *Memorials and Characters of divers Eminent and Worthy Persons*, 1741, pp. 617–23.

Through his appointment to the office of Master of the Ordnance at Berwick he had become familiar with the desire of the citizens of Newcastle to control all mineral property along the Tyne. He saw clearly the great future which awaited the local coal industry. He possessed immense influence at Court through the patronage of Leicester and Warwick, the Queen's favourites, and was doubtless aware of the desire of the Crown to reduce the revenues of the Bishop.

Under Bishop Pilkington four coal mines in Whickham had been leased separately to three Newcastle merchants—Bertram Anderson, Edward Lewen, and Andrew Crofton. They paid between them £100 in rent, while another £10 was received from William Blakiston, of Gibside.[1] Bishop Barnes, however, by a new lease granted by him to the Queen, " did take Mr. Sutton bounde by obligacon for the true answering of the said several rents above said during the time of years then remaining unexpireed, over and besides the rent of £117 15s. 8d. reserved in the new Lease ".[2] In these words, one of the Bishop's agents explained Sutton's connection with what was about to become the greatest colliery in England.

The first lease secured by Queen Elizabeth was for a term of 79 years from February 1st, 1577, but this was superseded on April 26th, 1582, by another, for 99 years.[3] From the passage quoted above it is clear, however, that it was Sutton, and not the Queen, who bound himself to pay the Bishop rent for the manors and the coal mines. While retaining nominal ownership, the Bishop had in effect renounced for a century every right he possessed in these manors, in return for a fixed yearly income which bore no relation to the true value of the property. Sutton, who took over temporarily all the rights belonging to the Crown as lessee, seems to have permitted the Newcastle merchants to pool their interest in the minerals, and to operate most of the mines within the Bishop's holdings as a single colliery.[4] There was every reason why the Crown, already convinced of the need to develop the coal fields, should have welcomed such a scheme.[5] Although Sutton was in a strong position as lessee, he was willing, it would seem, to sell his position at a price, for he negotiated a deal whereby on November 12th, 1583, the Queen assigned her lease to Henry Anderson and William Selby, the two most influential merchants of Newcastle.[6]

---

[1] Bertram Anderson's lease was later held by Henry Anderson and Crofton's lease by Edward White (*Lansdowne MSS.*, 66, no. 87).

[2] " A Valuation of the Lordships of Whickham and Gateshead part of the possessions belonging to Bishop Barnes ", drawn up May 25th, 1591, by John Bathe, " Auditor of the aforesaid premises " (*ibid.*, 66, no. 84).

[3] *Ibid.*, 66, no. 87. Mention is made of another lease, dated June 20th, 1581 (Brand, *op. cit.*, vol. i, p. 80) ; but this apparently did nothing more than confirm the first.

[4] Cf. below, p. 361, and vol. ii, pp. 63–4, 120 sqq.

[5] Much wider monopolies were granted by Elizabeth in the other extractive industries.

[6] Welford, *op. cit.*, vol. iii, p. 18. See also *Palat. of Durham, Bills and Answers*, bdl. 2 (Selby and others *v.* Hedworth and others, August 1st, 1605), wherein mention is made of this conveyance. The conveyance runs : " Know ye that we, for the good and faithful service done unto us by our trusty and well beloved servant, Thomas Sutton . . . and at his humble suit and petition

Sutton is said to have received £12,000 for this transfer,[1] and, whatever doubt may be cast on that assertion, a sum of at least £5,500 was paid by the citizens of Newcastle,[2] many of whom contributed towards the purchase money, and thereby obtained an interest as hypothetical shareholders in what was well named the " Grand Lease ".[3] As the conveyance of 1583 to Anderson and Selby was made by the Crown " without any rent, service, or account whatsoever therefore to be had ", and at the " humble suit and petition " of Thomas Sutton, " for the good and faithful service done unto us by [him] ", one is perhaps justified in thinking that it was Sutton whom the citizens paid. He may have received a considerable profit from the mines before he sold out. A few years later it was stated by the Lord Mayor of London that the price of coal at Newcastle rose from 4 to 6 shillings a chaldron while Sutton held the Grand Lease.[4] Legend has it that he departed from the north with two horseloads of money, and during the decade 1590 to 1600 he was famous in London for his great wealth.[5] His biographers have been unable to discover any source other than the coal trade from which he could have received sums large enough to lay the foundations of his immense fortune.

Sutton's activities served to conceal the very transfer which had been dreaded by all partisans of the Bishop. A direct grant of Gateshead to the Mayor and Corporation of Newcastle could not have been pushed through the House of Commons without a struggle, for there

---

to us made in that behalf . . . have accepted and received . . . the said two several indentures ", i.e. the 79 year and the 99 year leases. " And further, know ye that we, for the consideration aforesaid . . . have given . . . unto our well beloved subjects Henry Anderson and William Selby, of our town of Newcastle-upon-Tyne, aldermen, not only the said indentures . . . but also all our estates, rights, titles, interests, uses, possessions, terms of years, claims and demands whatsoever, of, in, and to the said manors or lordships ". Although Sutton is said to have returned to London in 1580 (Galloway, *Annals*, p. 94), it is possible that he took an active part in the administration of the manors as late as 1583.

[1] Galloway, *op. cit.*, p. 93.

[2] *H.M.C., Report on the MSS. of the Marquis of Salisbury*, vol. iv, pp. 208–10 (" An abstract of Abuses committed at Newcastle ", 1592). Abuse no. 2 reads, " The town disbursed at least £5,500 for the ground lease, besides great sums in procuring a new charter to make the town capable thereof." It is probable that the co-partners advanced further money on their own account. The sum of £5,500 is also mentioned in a letter from certain citizens of Newcastle to the Privy Council, May 31st, 1597 (*S.P.D., Eliz.*, vol. cclxiii, no. 72, I).

[3] There are records of several Newcastle wills which show holdings in this lease ; see, for instance, those of Robert Lambe, Ralph Cole (who was owed by " the company of the town of Newcastle for money towards the purchase of Gateshead ", £2 8s.), Wm. Jennison, Robt. Barker, Roger Nicholson, John Gibson, and Wm. Greenwell, in the years 1585 to 1597 (Welford, *op. cit.*, vol. iii, pp. 32, 37, 43, 60, 67, 96, 110).

[4] *Lansdowne MSS.*, 65, no. 11.

[5] Galloway, *Annals*, p. 94. £50,000 was stated to have been the size of his fortune. This would make Sutton almost a millionaire in terms of present-day currency. Dr. Leslie Hotson tells me that, from an examination of the *Recognizances for Debt* in the Public Record Office, he is convinced that in the last decade of the sixteenth century there could hardly have been a richer Englishman than Sutton. The pages of these *Recognizances* are littered with his credits. Unfortunately the *Recognizances* for the years during which he held the Grand Lease have not been preserved.

was much opposition to it in London as well as in the Bishopric.[1]
But the same result had been accomplished without an Act of Parliament, by causing the Bishop to alienate almost everything except his
ultimate title to the ownership of the manors.[2] The interest of the town
of Newcastle was partly concealed by making over the lease to two
individuals,[3] one of whom, Anderson, was the least objectionable
of the leading citizens from the point of view of the London opposition.[4]
In fact, however, the lease was largely bought with the common money
of the town, and it was regarded by the burgesses as belonging to
them, as is proved by the struggle which ensued between the faction
which controlled the coal mines and another group, representing the
remaining citizens who had no direct share in the profits.[5] The inhabitants of the Bishopric, discovering that they had been tricked, and
realizing their true position, set forth their grievances in the following
appeal.

" The lease is verie preiuditiall and hurtfull, To the Bishop,
The poore Tenants and To all the Gentlemen of this Countrie.

" a. To the Bishop because he is like to lose not onelie the Rent
. . . reserved by certaine Leases of Coale and way leave . . .
but also all Fynes, Eschetes and Royalties of great value ; and yet
he must repair Tyne brigg left of purpose forth of that Lease.

" b. To the poore Tenants, who instead of one have gotten
many Lordes hard to please, and are greeved to have their Recordes
kept in another County.

" c. To the Gentlemen of this Contrie, Because they can not
have that Libertie upon the water which they had under the Bishop ".[6]

While the new lessees are said to have reaped almost immediately
a yearly profit of from £1,500 to £1,600 [7] from their coal pits,

---

[1] Welford, *op. cit.*, vol. ii, pp. 293, 305, 476, 504, and esp. p. 481.

[2] He retained the right to incorporate the mysteries of Gateshead (*ibid.*,
vol. iii, pp. 101–2).

[3] In 1595 Anderson and Selby, as the Bishop's lessees, were still collecting
the rent from copyholders and tenants of the manor of Whickham (Anderson
and Selby *v.* Nich. Watson, in *Palat. of Durham, Bills and Answers*, bdl. 1).

[4] Sir William Fleetwood wrote to Burghley in 1576 : " The town of Newcastle are all papists, save Anderson, and yet he is so knit in such sort with the
papists, that *aiunt, ait ; negant, negat* " (Welford, *op. cit.*, vol. ii, p. 481 ; and
cf. below, vol. ii, pp. 122–3).

[5] See below, vol. ii, pp. 121–5. The town paid for bookkeeping in connection with the lease as early as 1593 (Welford, *op. cit.*, vol. iii, p. 85).

[6] " Causes why the Lease of the Mannors of Whickham and Gateshead
and the Coalmynes there deserveth no favor " (*Lansdowne MSS.*, 66, no. 86).

[7] *Ibid.* Sixty years later Ralph Gardiner (*England's Grievance discovered
in relation to the Coal-Trade*, 1655), claimed that, during the last years of the
sixteenth century, the lease had been worth £50,000 a year to the Newcastle
burgesses. This opinion was generally accepted in the eighteenth century
(*Allan MSS.* (Library of the Dean and Chapter of Durham), vol. vii, p. 41), but
the figure is absurdly high. The cash received on all the coal annually shipped
from the Tyne in 1590 could not have amounted to more than half the sum.
In 1591, nine years after the lease came into the hands of the Newcastle merchants,
one of the colliery owners referred to the large purchase price, " which the . . .
buyers ", so he pretended, " have not yet recovered nor are like to doe theis many
yeres " (*Lansdowne MSS.*, 67, no. 22). This statement involves an underestimate
of the profits of the " grand lessees " no less flagrant, perhaps, than Ralph
Gardiner's overestimate.

a profit which doubtless grew larger during the reign of James I, they actually paid no royalty for their mineral rights, apart from the purchase money. According to the original grant from the Bishop to the Queen, she or her nominee was bound to pay £110 annually for the mines, in addition to the £117 15s. 8d. collected in rents from the copyhold and freehold tenants, who now paid their dues to the Bishop's lessees instead of to the Bishop. The former sum was regularly paid until 1587, when Bishop Barnes died, and the episcopal see was left vacant for two years.[1] In spite of continual protests from the succeeding Bishop, Matthew Hutton (1589–94),[2] the rent of £110 was apparently never recovered after that date.[3] Not only were the Bishops left without any pretension to a direct interest in the Tyne trade ;[4] they were destined to have no share in the financial harvest which accompanied the boom in coal mining that began during the seventies and eighties of the sixteenth century. A few collieries continued to pay revenues into the see, but the total receipts from all the Bishop's coal mines in south as well as in north Durham in 1635, when the annual output must have been about fifteen times as great as in the reign of Henry VIII, were only £278 7s. 5d., or less than had been paid for a lease of mines in Whickham and Gateshead in the fourteenth century.[5] A seventeenth-century Bishop, going through

---

[1] *Lansdowne MSS.*, 66, no. 84.

[2] *Ibid.*, 86, no. 85. In 1588–9 the auditor of the premises claimed this rent for the Bishop on behalf of the Queen, the see then being vacant (*Ministers' Accounts, Eliz.*, bdls. 661–2).

[3] In 1635 we find that only £117 15s. 8d., the rent due from the landholders, was being paid (*Eccles Comrs., Ministers' Accts.*, 190357).

[4] All the important collieries near the mouth of the river Wear were within manors outside the Bishop's jurisdiction, although he still exercised rights over the river. It was the Bishop, not the King, who conferred on Sunderland a charter of incorporation in 1634 (M. Dunn, *View of the Coal Trade*, p. 15) ; it was the Bishop who issued commissions for measuring the Wear coal keels in 1667–9 (*Cal. of Treas. Bks.*, vol. v, pp. 441, 1212). Similar commissions for measuring keels on the Tyne were issued by the Crown.

[5] *Eccles. Comrs., Ministers' Accts.*, 190357. In 1635 the episcopal revenue from mines was obtained as follows :—

Colliery.	Lessee.	Rent.			
		£	s.	d.	
Carterthorne . .	Lord Eure . .	70	0	0	
Grewborn . . .	Mayor of Durham City	20	0	0	
Hargill . . .	Mayor of Durham City	18	0	0	
Tallawe . . .	Mr. Hugh Wright .		13	4	
Knitsley and Clewborne	Sir Tim Wittingham .	1	6	9	
Finlay . . .	Mr. William Killinghall	1	0	0	and 20 wainloads of coal.
Urpeth . . .	Mr. Edward Liveley .	2	13	4	
Ryton . . .	Sir Thos. Tempest and Partners.	12	0	0	(£5 6s. 8d. per annum and £1 6s. 8d. for each of 5 pits.)
Charlowe . . .	Sir Tim Wittingham .		13	4	
Hummock Moor .	Mr. Francis James and Partners.	6	0	0	and 40 cartloads of coal to spend at Auckland, at 4d. the load.
Morton . . .	Sir William Bellasis .		13	4	

the episcopal records at Auckland, must have felt some exasperation on reading Franklyn's statement, written in 1523, that the Bishopric might be improved " by 1,000 marks a yere only in cole and led ".[1]

We have now examined the easy terms granted to lessees of mineral property under lands in Durham and Northumberland, owned either by the Bishop or by the Queen. There is no doubt that the two, between them, possessed nearly all the important mining sites in the Tyne basin at the end of the sixteenth century. In a document dated 1591, and written apparently by a mine owner, we read :—

" The moste parte of the coale Mynes doe belonge unto her Majestie and are lett unto particular farmers who both paie unto her Majestie rent and fine . . .

" The residue of the Coale Mynes for the moste parte did belonge to the late Buishopp of Durham who leased them unto her Majestie, from whome certeine perticuler men did buy the same for a verie great some of money ".[2]

It is clear that the transfer of ownership and authority from the Church to the Crown had enabled the Newcastle traders to obtain that monopoly over the collieries of the Tyne valley, and the trade in coal, which they had been seeking. Their attitude towards the expansion of the coal industry was as progressive as that of the Church appears to have been conservative.

Outside the Tyne basin, it has not been possible to follow in such detail the influence of the transfer of mineral property upon the terms of leases, and upon the expansion of the coal industry. What is known suggests that the decline of the ecclesiastical power everywhere

Colliery.	Lessee.	Rent.		
		£	s.	d.
Blackburn . .	Mr. Thos. Liddell .	20	0	0
Whickham and Gateshead.	Town of Newcastle .	117	15	8
Wayleave for coal carriage from Ravensworth to Gateshead.	(?)		5	0
Chester . .	Mr. Edward Liveley .	5	0	0
Lanchester . .	Sir Tim Wittingham .	1	6	8
Plawsworth . .	Widow Hutchinson .	1	0	0
		£278	7	5

It should be noted that the rent from Whickham and Gateshead was not paid for the coal mines, but was merely the sum of the surface rents owed by the Bishop's tenants at the time the lease was granted.

[1] Hutchinson, *History of Durham*, vol. i, p. 405 note.

[2] *Lansdowne MSS.*, 67, no. 22. Other Church lands, not confiscated by the Crown, belonged to the Dean and Chapter of Durham. They continued to operate pits at Rainton and Spennymore, and to grant coal on lease at Jarrow and near Raby, but such power as they had over the northern trade was exercised through the land they owned at South Shields, where an important salt industry grew up. (See below, vol. ii, p. 127; *S.P.D., Eliz.*, vol. ccxxxviii, no. 18; *Palat. of Durham, Decrees and Orders*, vol. i, ff. 96, 380 sqq. (Hutchinson *v.* Metcalfe, April 3rd, 1637); *ibid., Bills and Answers*, bdl. 4 (Morecroft *v.* Hutchinson, June 5th, 1624); Welford, *History of Newcastle and Gateshead*, vol. iii, p. 294). After the dissolution of the monasteries, leases granted by the Bishop of Durham had to be confirmed by the Dean and Chapter (*Hunter MSS.*, vol. xi, no. 16).

favoured the development of mining.[1] In those lands over which the Church retained its power, economic enterprise was discouraged. Even as late as the eighteenth century, Defoe explained the backwardness of building near Wolverhampton on the ground that " the land, for the chief Part, being the Property of the Church, . . . the Tenure [is] not such as to encourage People to lay out their money upon it ".[2]

The Reformation in England was accompanied by extensive transfers of mineral property, for which there is no parallel on the Continent. These transfers facilitated the exploitation of the coal mines ; and they help to explain, therefore, the remarkable expansion of the industry which occurred at the end of the sixteenth century. At the same time the enclosure movement, which began in England before the sixteenth century, but which first became a great social issue in the period of the Reformation, was gradually reducing the number of small yeomen landholders. One result was to encourage large-scale colliery enterprise, by making it possible to obtain extensive mineral concessions without having to deal with a host of small tenants. The advantage which lords of manors derived from using their land for pasture instead of for arable farming has been generally, and doubtless rightly, regarded as the most important motive for the early enclosures. Until after the middle of the sixteenth century, the enclosure of commons and wastes, and the eviction of copyholders and other customary tenants, were almost never undertaken for the purpose of consolidating mineral property, and restricting the mineral rights of small landholders. But, after the accession of Elizabeth, these motives begin to be important. The enclosure movement, as it develops in the period from 1550 to 1700, should be regarded increasingly not only as a cause for the expansion of the coal industry, but as an effect of it.[3] We have chosen to emphasize, in Part III, the latter aspect. Nevertheless we should not lose sight of the fact that the enclosures which preceded, and accompanied, and were doubtless facilitated by, the confiscation of Church lands, were partly responsible for those changes in the ownership of mineral property which were so much more extensive in England than in any continental country.

(ii) *The Timber Crisis*

The transfer of mineral property at the time of the Reformation removed one obstacle to the investment of capital in mining, and natural advantages favoured the development of this industry at an earlier period in Great Britain than on the Continent. But these two considerations, taken together, do not explain at all completely why there should have been a revolutionary growth of the coal industry in the late sixteenth and early seventeenth centuries. A more important explanation is to be found in the concurrent diminution in the supply of wood, and the consequent rise in the price of billets, faggots, and charcoal. Adam Smith recorded an observation, the truth

---

[1] See above, pp. 134–7, 139–40, 142–4.　　[2] *Tour*, 1769 ed., vol. ii, p. 411.
[3] See below, esp. pp. 342–3.

of which time scarcely modifies, when he wrote : " Coals are a less agreeable fuel than wood : they are said to be less wholesome. The expense of coals, therefore, at the place where they are consumed, must generally be somewhat less than that of wood ".[1] So strong had been the prejudice against coal that its use in London had been forbidden to limeburners in Southwark by a proclamation of 1307.[2] " An intolerable smell diffuses itself throughout the neighbouring places ", we read in the preamble, " and the air is greatly infected, to the annoyance of the magistrates, citizens and others there dwelling and to the injury of their bodily health ".

This prejudice was deep-seated as well as of long standing. Shakespeare made Master Seacole a grubby, dirty fellow ; in 1578 Queen Elizabeth is said to have stayed away from London because of the " noysomme smells " of coal smoke.[3] In that year a brewer and a dyer were committed to prison for burning coal in Westminster,[4] and, as late as 1641, brewers who dwelt near the palace might be sentenced if they made too free use of coal during the residence of the royal family at Whitehall.[5] Evelyn favoured the removal of the shops of all such artisans to a point beyond Greenwich, in order to rid London of the " Columns and Clouds of Smoake, which are belched forth from the sooty Throates of [those shops], . . . rendring [the city] in a few Moments like the Picture of Troy sacked by the Greeks, or the approches of Mount-Hecla ".[6] " The physicians in

---

[1] *Wealth of Nations*, bk. i, ch. xi.
[2] *Cal. of Close Rolls*, 1302–7, p. 537.
[3] *S.P.D., Eliz.*, vol. cxxvii, no. 68.
[4] *City of London Repertories*, vol. xix, f. 412b.
[5] In 1623 a bill had been introduced prohibiting the burning of seacoal in brewhouses within a mile of any building in which the King's court, or the court of the Prince of Wales, should be held, or in any street west of London Bridge. Although the bill never became law, several brewers were prosecuted, fined, and forced to abandon their houses, for burning seacoal in the period between 1635 and 1641. One of them, at least, was committed to the Fleet ; but he was pardoned soon after. (*Lords' Journals*, vol. iii, p. 269 ; *Cal. S.P.D.*, 1635–6, p. 161 ; *H.M.C.*, *4th Report*, appx., pp. 54, 67.)
[6] Evelyn, *Fumifugium ; or the Inconveniency of the Aer and Smoke of London dissipated*, 1661, pt. ii. After 1641 the Government apparently abandoned all direct effort to prevent the burning of coal in the city. By this time the smoke nuisance had already become serious in London, and, as is common in our own day, private persons brought forward projects for purifying the air. The best known of these projects was advanced by Evelyn, in the tract just cited. Tim Nourse, his follower in this crusade (*Campania Felix ; or a Discourse on the Fuel of London*, 1700), wrote : " That of all the Cities perhaps in Europe there is not a more nasty and a more unpleasant place " than London, primarily because of the " burning of seacoal ".
It is obvious that the prejudice against the new fuel survived its adoption by the Londoners (see John Graunt, *Natural and Political Observations upon the Bills of Mortality*, 1662, p. 70) ; obscured, however, by those persons who always contrive to give for a bad cause a worse reason, and who insisted that, far from polluting the air, coal smoke served to clear and purge it (Defoe, *Journal of the Plague Year*, ed. J. C. Dent, 1927, p. 216).
On the Continent the prejudice against coal fires was even more pronounced. In France an edict was issued in the sixteenth century that no one should burn mineral fuel in Paris upon any pretence whatsoever (E. Meunier, *Etude sur la houille du bassin Franco-Belge*, 1896, p. 55). In Germany its adoption was retarded by the belief of doctors that the smoke caused apoplexy (Kircher,

Dublin ", wrote Dean Swift some decades later, " make it their constant practice to remove their patients to some purer air, near the suburbs, out of the smoke of the city, which in winter is so thick, and cloudy enough to stifle men and beasts, so great an influence that it affects even the blossom and bloom of the flowers in the spring ".[1]   Sheer necessity alone could have forced the population to adopt a fuel regarded by many as not only disagreeable, but actually noxious.

That necessity became pressing at the beginning of Elizabeth's reign. It has been calculated (though too much stress must not be laid upon the calculation) that between 1500 and 1550 the price of firewood actually fell slightly. Whether that is so, or not, it is certain that after 1550 the price rose far more rapidly than the average price of all commodities. In his compilation of index numbers for the sixteenth and seventeenth centuries, Wiebe gives the following figures for the cost of firewood, as compared with the average cost of all commodities for which he could find price statistics.[2]

	1451–1500	1531–40	1551–60	1583–92
General Prices	100	105	132	198
Price of Firewood	100	94	163	277

	1603–12	1613–22	1623–32	1633–42
General Prices	251	257	282	291
Price of Firewood	366	457	677	780

While the scant materials available for a compilation of this kind make it unsafe to vouch for the accuracy of these figures, such information as can be obtained from other sources goes to confirm them. Between 1550 and 1605 the price paid by the dyers of London for their dye wood rose more than five-fold.[3] In 1536 a load of charcoal on the Earl of Rutland's estate in Rutlandshire cost 3s. 4d. ; in 1586 the same sized load cost 25s.[4] So rapid an increase in the cost of any commodity in common use must have been almost without precedent in the history of western civilization.

Apart from the evidence afforded by this rise in the price of timber, there are sufficient reasons for believing that the English and Scottish countryside was being stripped of its woods. During the reigns of Elizabeth and James I, dozens of commissions were sent out by the Court of Exchequer to investigate the spoil taking place in every county, and from every county came the same lament of deforestation. By 1585, the famous weald which stretched through Sussex, Surrey, and Kent had " been greatly decayed and spoiled ", and, concluded the

---

Mundus Subterraneus, Amsterdam, 1665, vol. ii, p. 60 ; Fouillon, *Historia Leodiennis*, vol. i, p. 405 ; both cited by Decamps, *Mémoire historique sur l'industrie houillère de Mons*, pp. 56–7). As late as 1762 a councillor of the ducal chamber of Brunswick and Lüneburg ascribed the high death-rate from consumption in London to the burning of coal (Hue, *Die Bergarbeiter*, vol. i, p. 351).

   [1] *Dublin Weekly Journal*, August 9th, 1729, in Swift, *Works*, ed. Sir Walter Scott, vol. vii, 1824, p. 222.

   [2] Wiebe, *Zur Geschichte der Preisrevolution des 16 und 17 Jahrhunderts*, p. 375.

   [3] From 2s. to 10s. and 12s. (*Star Chamb. Proc., James I*, 8/15).

   [4] *Lansdowne MSS.*, 83, f. 227.

framers of a statute designed to check the evil, "will in short time be utterly consumed and wasted if some convenient remedy therin be not timely provided ".[1] Seven years before, the London Company of Brewers had petitioned Walsingham " that wood cannot be gotten to sarve all your . . . orators withoute the rewin and greate decaie of the whole common Weale of the Citie ".[2] An agent who collected the revenues which the Bishop of London derived from his estates in Middlesex complained in 1598 that, while earlier bishops had obtained yearly profits of from £400 to £2,500 by selling their wood to London traders, " the nowe Bishop hathe been driven to bestowe at the least £220 in tymber for the necessary repayring of his houses, and is nowe driven to burne seacoles ".[3] Along the coasts, wrote the anonymous author of a petition urging Queen Elizabeth to prohibit the shipment of mineral fuel abroad, " all the country villages round about the land within twentie myles of the Sea are for the most part dryven to burne of theis coales, for that most part of the woods are consumed ".[4] In Devonshire in 1610 " tymber for buyldinge and other necessaries for husbandrie are alreaddy growen soe extreame deare" that the inhabitants had to obtain supplies from Ireland, Wales, and Flanders.[5] Of Pembrokeshire, Owen, who knew by heart the condition of his native county, wrote in 1603 : " this Countrie groneth with the generalle complainte of other countries of the decreasinge of wood ".[6]

In the neighbourhood of Burnley, in Lancashire, the shortage had been felt even before the reign of Elizabeth. James Roberts, an aged husbandman, deposed in 1526 that neither he nor any other inhabitant had " any nede in tyme past to get Colis for there fuell by Reason they hadde plenty of woode from the forests and turves at theyre liberty which now be decayed and Restrayned from theym ".[7] Farther north, at Cliviger, in 1587, no wood was to be had " for divers miles . . . thereabouts ".[8] Even the bordermen in northern Northumberland lacked supplies. In the manor of Bamburgh, according to a Crown survey of 1575, " great woods hath beene, but now utterly decayed and no wood at all remaineth thereon ".[9] The whole country round Etal and Kyloe was described in 1609 as " barren both of wood and of coal ".[10] To such desperate straits were the inhabitants of some parts of Scotland driven by 1621, " that nomberis of thame bothe to burgh and land hes bene constrayned not onlie to cutt doun and distroy thair policie and planting, bot thair movable tymmer

---

[1] 27 Eliz., cap. 19.
[2] *S.P.D., Eliz.*, vol. cxxvii, no. 68.
[3] *Ibid.*, vol. cclxvi, no. 119.
[4] *H.M.C., Report on the MSS. of the Marquis of Salisbury*, vol. xiv, pp. 330–1. Another copy of this document, wrongly dated, is *S.P.D., Charles I, Addenda*, vol. xli, no. 85 (see below, vol. ii, p. 219 n.).
[5] *Harleian MSS.*, 6838, no. 32.
[6] Owen, *Description of Pembrokeshire*, pt. i, pp. 86, 145–6.
[7] *Duchy of Lancs. Deps.*, 19/T/3.
[8] *Duchy of Lancs. Spec. Com.*, no. 418.
[9] *Augm. Partics. for Leases*, file 112, ff. 12, 15, 27, 30.
[10] *Star Chamb. Proc., James I*, 154/12.

worke, to mak fyre of it . . . and in mony placeis the trade of brewing and baiking for want of fyre is neglectit and cassen up ".[1]

Thirteen years earlier, the Scottish Privy Council had written to James I that all the wood in Scotland could not supply a hundredth part of the uses of this nation ; " whereby your maiesteis subiectis has bene constrayned . . . to mak their provision . . . of tymmer . . . frome foreyne pairtis to supplee this defect ".[2] By this time, too, substantial imports of foreign timber had begun to reach most of the English sea towns, those along the east coast receiving supplies chiefly from Norway.[3] Tyne colliery owners came to depend, at least in part, upon foreign boards for pit props, as is proved by their successful petition of 1622, asking the Privy Council to lift the restraint just imposed on trade in foreign ships, in so far as it concerned the discharge at Newcastle of Baltic timber, " the want whereof about their coal works is so great . . . that without it coales can not be wrought ".[4] Supplies from abroad were already carried as far inland as Halifax, about which town towards the end of James I's reign forests were so " decayed and scant ", that " from Newcastle firewood and other tymber " (received, doubtless, from Norway), " readie formed and fitt for buildinge may be had cheaper then the western men of Yorkshire can work their own tymber trees ".[5] London is said to have been rebuilt, after the fire of 1666, largely with wood from abroad,[6] and in eighteenth-century Edinburgh, to quote Adam Smith, " there [was] not perhaps a single stick of Scotch timber ".[7]

Before the end of Elizabeth's reign, the shortage had become so acute through the greater part of England that the Crown attempted to check it, by prohibiting the taking of timber from the royal forests without special warrant. The concern which the Government felt over the timber crisis can be seen from a letter which Burghley himself addressed to the Countess of Rutland in 1594. The countess, who possessed a warrant entitling her to be supplied with timber, had complained that the woodward within Sherwood Forest, refused to deliver it to her. In reply Burghley, who had lately been informed of " the great decaye and Spoile of wood within that Forest ", pointed out that the countess had already received warrants for four hundred and twenty trees for the repair of the castle and mills at Newark, " which is a verie great number towards the repaire only of the Castle and Mylls, especially wher as it hath bene informed unto me and Complained of that the greatest nomber of the said trees have not been imployed to the use they were allowed for but sold (by such

[1] *Privy Council Register of Scotland*, 1st Series, vol. xii, p. 605.

[2] P. Hume Brown, *Scotland before 1700*, pp. 273–4.

[3] *Acts of the Privy Council*, 1615–16, pp. 262–3 ; *Exch. K.R. Port Bks.*, *passim*. Ports in South Wales received imports of timber from Ireland and France. Wood from Norway reached Glasgow as early as 1656 (Tucker, *Report on Excise and Customs*, p. 38).

[4] *P.C.R.*, vol. xxxi, p. 500.

[5] *Denmilne MSS.*, vol. xxxlii, no. 15.

[6] Bosse, *Norwegens Volkswirtschaft*, vol. i, p. 270, cited Sombart, *Der Moderne Kapitalismus*, 1916, etc., vol. ii, p. 1139.

[7] *Wealth of Nations*, bk. 1, chap. 11.

as your Ladyship putt in trust) for money and converted the same to the owners private use, which . . . is a verie foule disceit and abuse toward mee and wrong to hir Majestie, which shall make mee more careful both in granting my warrants hereafter and in seeing them imployed to the use they are granted for. And for the better preservation of the wood there I mynd to make a generall staye without uppon verie especiall occasion. And therefore I pray Your Ladyship to pardon me that I write no more unto them therin then I have already donne but leave it to their owne good discretion, considering the state of the Forrest and spoile already made . . ." [1]

All the evidence suggests that between the accession of Elizabeth and the Civil War, England, Wales, and Scotland faced an acute shortage of wood, which was common to most parts of the island rather than limited to special areas, and which we may describe as a national crisis without laying ourselves open to a charge of exaggeration.[2] There is good reason to suppose that this general exhaustion of the forests in Britain preceded a similar general exhaustion on the Continent of Europe. Such statistical evidence as is available suggests that wood prices rose somewhat less rapidly abroad than in England during the century from 1550 to 1650.[3] One meets, it is true, with early complaints of deforestation in special areas on the Continent. In the fourteenth century Venetian lime and brick makers were rationed by town ordinance in their supplies of firewood. A scarcity of charcoal for iron smelting, and of timber for pit props in copper and silver mines, was felt in Bavaria as early as 1463, and in Bohemia about 1550 ; while the number of blast furnaces in Siegerland was reduced by half between 1563 and 1616, owing to the high price of wood.[4] The Norwegian government took measures to protect the timber resources of the country even before the end of the sixteenth century.[5] Petitions concerning " déboisement ", submitted by the citizens of Nevers, caused the French King in 1560 to order the demolition of all forges within three leagues of the town.[6] But in France the decline in the timber supplies does not appear to be mentioned as a national problem until 1638, when Charles Lamberville, in putting forward proposals for constructing canals to distribute coal throughout the country,

---

[1] *Lansdowne MSS.*, 103, no. 80. For a similar scandal concerning the extravagant burning of wood after the Restoration, by Lady Clayton, wife of the Warden of Merton College, Oxford, see W. T. Layton, *The Discovery of Gas Lighting*, 1926, p. 10, a reference for which I am indebted to Professor Gay.

[2] Cf. R. G. Albion, *Forests and Sea Power*, 1926, ch. iii. I find myself unable to agree that the timber problem became acute only " during the Restoration " (p. 97). The figures for the price of oak, given by Professor Albion (p. 91), suggest that timber prices rose much more rapidly in the century and a half preceding 1660 than in the century and a half following. There are reasons for thinking that the problem of deforestation was less acute at the end than at the beginning of the seventeenth century (see below, pp. 221–2).

[3] See Wiebe, *op. cit.*, *passim*. D'Avenel's work does not indicate that there was in France a great discrepancy between the rise in general prices and the rise in the price of firewood, such as occurred in England (*Histoire économique*, vol. iii, 1913, pp. 366 sqq., vol. v, 1919, pp. 481–5, and *passim*).

[4] W. Sombart, *op. cit.*, vol. ii, pp. 1145–8.

[5] *Ibid.*, p. 1148.

[6] *Ibid.*, p. 1145.

called attention to the need for conserving the forests for purposes other than firewood.[1]

By the end of the seventeenth century, when a German, Johann Philip Bünting, published the first treatise entirely devoted to coal,[2] the timber crisis had become general in all the countries of western Europe. William Brown, an Edinburgh merchant who travelled abroad in 1680 in quest of markets for the produce of the Scottish collieries, described Flanders, Brabant, and Artois as then " destitute of wood ".[3] " All the forests and woods which were near the river banks and which could be easily transported are almost exhausted ", wrote in 1701 the Sieur des Casaux du Hallay, *député* for Nantes, " scarcely any timber remains except at long distances inland, and the carriage of that will cost too much for one to be able to make use of it ; so that the wood for land and sea construction work, as well as for heating, is extremely scarce and dear, and will be still more so in the future. It is to be feared that these conditions may become so critical that we shall be obliged to import Baltic timber ".[4] Letters from the *intendants* in all *départements* tell a similar story.[5] What had previously been a few isolated voices in separate districts becomes a full chorus of complaint. The same national investigations into the spoil of wood, and the same search for substitutes, which were undertaken in England and Scotland at the end of Elizabeth's reign, and during that of James I, are first met with in France at the end of the seventeenth and during the early eighteenth centuries.[6] Holland, alone among continental countries, appears to have suffered from a general timber shortage at the beginning of the seventeenth century ; and Holland was probably the chief coal-consuming country on the Continent.[7] Elsewhere, native forests probably continued to provide almost all the fuel required for nearly a hundred years after the shortage of wood had become a serious problem in most parts of Great Britain.

It was not an accident that the inquest held by the French *Contrôleur Général des Finances* in 1701 into the destruction of the forests was followed, in 1709, by a demand for information concerning the condition of the coal mines.[8] In England and Scotland the rapid substitution of coal for firewood began in Elizabeth's reign ; and, the higher the price of timber rose, the greater became the distance from the pits at which the new fuel could be marketed, and, consequently, the larger the output of the collieries. Turf is said to have provided a temporary solution of the fuel problem in Holland in the seventeenth century ;[9] but, although peat and other substitutes

---

[1] *Alphabet des Terres à brusler et à Charbon de Forge.* My attention was called to this interesting tract by Professor Hauser.
[2] *Sylva Subterranea, oder Vortreffliche Nutzbarkeit des Unterirdischen Waldes der Stein-Kohlen*, 1693.
[3] *H.M.C., Report on the MSS. of the Earl of Mar and Kellie*, pp. 212–13.
[4] Boislisle, *Correspondance des Contrôleurs Généraux*, vol. ii, appx., p. 498.
[5] *Ibid.*, vol. ii, no. 355.
[6] Rouff, *Les mines de charbon en France*, pp. 121–36 ; Dulac, *Observations sur le charbon minéral*, ff. 1–2.
[7] See above, p. 87.   [8] See above, pp. 124–6.   [9] J. P. Bünting, *op. cit.*

were tried in various parts of England and Scotland,[1] none of them was found in sufficient abundance to meet the requirements of a growing population. When intensively used, they probably increased in price nearly as rapidly as firewood. Just as to-day the high price of coal leads to the substitution, wherever possible, of oil or water power, so in Elizabeth's reign the high price of other fuels led to the substitution of coal, and also to a timber conservation policy, analogous to the coal conservation policy sometimes suggested to-day.

No one is likely to deny that, if the timber crisis in England and Scotland preceded the timber crisis on the Continent, this may be esteemed a principal cause of the priority obtained for the British coal industry. But those who wish to probe deeper may well ask why the problem of deforestation should have become serious in Britain sooner than on the Continent, and why it was in Elizabeth's reign that this problem assumed so grave an importance.

The early exhaustion of native timber may have been due in some measure to the natural limitation of the supply. It is probable that, in comparison with continental countries, England was poor in forests, and Scotland, at any rate, is known to have been so.[2] Knowledge of the amount of timber available per head is even less obtainable than is a comparative statement of the timber supplies of different countries. It is probable, however, that population increased somewhat more rapidly in Great Britain from 1550 to 1700 than in any continental country ; [3] and that would also help to explain why the timber crisis should have occurred when it did in England and Scotland. Contemporaries often speak of the increase of population in England during the late sixteenth and the seventeenth centuries, and most historians would agree that a good case can be made in favour of the view that a considerable increase took place. In the reign of Henry VIII, there are said to have been in England and Wales about three million souls.[4] Before 1700 their number is estimated to have reached between five and five and a half millions ; [5] and, according to Gregory King, the nation then had one inhabitant to every seven acres of land ; a density of population exceeded only in Holland and China.[6] The economic centre of gravity had shifted definitely from the south of Europe to the north.

Between 1500 and 1700 the population of London probably multiplied about ten fold.[7] Petty pointed out that, if the city

---

[1] See, for instance, Thorold Rogers, *History of Agriculture and Prices* 1866, etc., vol. iv, pp. 373–4.

[2] P. Hume Brown, *Scotland before 1700*, pp. 273–4.

[3] See below, vol. ii, p. 327.

[4] N. S. B. Gras, *Evolution of the English Corn Market*, p. 75.

[5] See above, p. 81.

[6] Gregory King, *Natural and Political Observations*, etc., 1696, in G. Chalmers, *Estimate of the Comparative Strength of Great Britain*, 1802 ed., p. 414.

[7] According to the most trustworthy estimates (see Gras, *op. cit.*, p. 75), the population of London grew as follows :—

1200–1500	40,000–50,000
1534	60,000
1605	224,275
1634	339,824
1661	460,000
1696	530,000

continued for another hundred and sixty years to grow in numbers so much more rapidly than the country generally, it would absorb all the inhabitants.[1] This concentration of population was not limited to London ; in many parts of Great Britain we find country villages developing into busy towns.[2] As a consequence, the cost of timber rose more than would have been the case had the additional inhabitants been spread evenly throughout the country, since the purveyors of wood, in order to meet the needs of the larger centres, had to draw their supplies from greater distances. It is precisely in the London area, where the growth in population was most marked, that we find the most stringent wood shortage, and the greatest expansion in the market for coal. It is easy to understand why the consumption of coal per head tended to be greater in the towns than in the country districts.

The rate of increase in the population of England during the reigns of Elizabeth and James I was hardly fast enough, however, to explain of itself either the magnitude of the timber crisis or the speed with which it developed. Nor can the additional fact that the inhabitants tended to concentrate in towns complete the explanation ; for the wood shortage was felt, not only in populous areas, but throughout the land ; in the sparsely settled valley of the Tweed, or in the lake district of Cumberland, as unmistakably (if not as insistently) as in the counties along the Thames. A final link is needed to complete our chain of causes.

This link is to be found, it may be suggested, in the increased demand for wood, both as fuel and as building material, in industry and manufacture. In order to prove this, it will be necessary, first of all, to show that a sharp expansion of native industrial enterprise did in fact occur under Elizabeth and James I, and then to give grounds for believing that this expansion made extensive inroads upon the forests. The more this subject is investigated, the more it becomes apparent that, if the adoption of coal as fuel is partly the result of attempts to develop English and Scottish manufactures, it is equally true—and much more important for an understanding of the development of modern economic life—that the success of these attempts ultimately depended on the adoption of coal as fuel, and that the necessity of substituting coal for wood at an earlier period than was found necessary in other countries had much to do with the powerful position which England had obtained among the nations of the world at the beginning of the eighteenth century. Her strength appears to have rested in some measure upon coal, a hundred years before the Coal Age is supposed to have begun.

---

[1] *Another Essay in Political Arithmetic Concerning the Growth of the City of London*, 1682, in *Economic Writings*, ed. C. H. Hull, 1899, vol. ii, p. 464.
[2] See above, pp. 106–7.

CHAPTER II

AN EARLY INDUSTRIAL REVOLUTION

Is any parallel for the rapid expansion of coal mining to be found in the history of other British industries during the sixteenth and seventeenth centuries, or was this expansion merely a result of the shortage of wood and of the growth of population? It is obvious, of course, that the increase in population must have occasioned an increased demand for manufactured goods; but was there a native development of industry and manufacture which was more than proportionate to the increase in population? Without a more detailed investigation than can be attempted in these pages, it is only possible to give a tentative answer to these questions But such information as has been found suggests that, while the expansion in coal mining was undoubtedly more rapid than that in other industries, this expansion is not an isolated phenomenon in early British economic history, but part of a general industrial development, the importance of which has not yet been fully appreciated. This industrial development not only involved a remarkable growth in the output of many commodities; there were also technical improvements and changes in organization, which, together with the evidence of a rapid growth, lead us to suggest very tentatively that the late sixteenth and seventeenth centuries may have been marked by an industrial revolution only less important than that which began towards the end of the eighteenth century. An attempt is made in succeeding chapters to explain some of the technical developments of the period, and to show how the increasing importance of coal mining involved changes in industrial organization.[1] In the present chapter we are concerned, however, only with the growth of industry.

The change which overtook English economic life between the early sixteenth and the late seventeenth centuries needs no emphasis. A diarist in the reign of Henry VIII would hardly have thanked God with the same assurance as Pepys for the monthly evidence of his advancing fortune, measured in terms of hard cash. As the new age advances, the difference of atmosphere becomes more marked. To steep oneself in English literature of the early eighteenth century, or in the accounts of foreign observers like Montesquieu or Voltaire, is to be aware of the extent to which the commercial spirit has permeated the lives, not only of town merchants, but of country squires and courtiers; of all, in short, who aspire to worldly honours. The merchant has become, in Johnson's oft-quoted phrase, " a new species of English gentleman ".

While historians have long recognized the growth during the seventeenth century, amid the débris of an outworn gild system,

---

[1] See below, pt. II, chaps. iii, iv; pt. IV generally; and Conclusion.

of an increasingly capitalistic society, in which the joint-stock company and the financier first become important, it is sometimes implied, nevertheless, that openings for capital were found primarily in trade or in colonial enterprise, and that manufactures, apart from textiles, did not develop to any marked degree before the era of the great inventions.   In the absence of any comparative statistics of the home production of salt, or soap, or beer, in the sixteenth and seventeenth centuries, this view of the course of manufacture has been encouraged by the magnitude of the progress made during and since the Industrial Revolution, which has tended to obscure the signs of an earlier development.   Confirmation of the view is found in the history of the two industries of tin mining and iron smelting.   The annual production of the Stannaries seems to have shown a slight decrease during the century from 1550 to 1650, and to have advanced very slowly during the succeeding century ; [1] while the output of iron seems to have remained stationary after the Restoration.[2]   It is desirable, therefore, to begin our discussion with the metallurgical industries, and to consider first the extractive and then the manufacturing stages.

It is possible to suggest reasons why the output of tin cannot be taken as a satisfactory measure of industrial activity during the sixteenth and seventeenth centuries.   The chief markets for it had been in the export trade and in the native manufacture of pewter, though supplies had also been required for coinage and for the making of bronze, in which tin is compounded with copper.   Now, the growth of the foreign market, which had provided a lucrative field both for the tinners and the London pewterers throughout the Middle Ages, was checked by the increasing production of tin abroad, and by the improvement in the technique of pewter-making in France and Flanders, where it had previously been inferior to that in England. As a consequence, the London pewterers (who monopolized the greater part of the native production) not only had trouble in selling their wares abroad, but met with foreign competition at home.[3]   If the export market did not actually shrink, it cannot have expanded to any considerable degree.   At home, the tin industry suffered from additional handicaps.   For domestic utensils and plate, pewter, lacking durability, was tending to be superseded by china ware and pottery, which was being turned out in north Staffordshire in increasing quantities.   Except as an ingredient in making bronze, tin was apparently little needed in the English industries of the seventeenth century ;   the process of tin plating was introduced commercially only just before 1700.[4]   Bronze goods gave way, in large measure, before brass, chiefly as a result of the discovery of " calamine ", or zinc, in 1566, by William Humfrey, founder of the Company of Mineral and Battery Works, who had set men to search for it in the region of the Mendip Hills.[5]   Hitherto no zinc had been found in

---

[1] G. R. Lewis, *The Stannaries*, Appendix J, and p. 41.
[2] See below, pp. 167–8.
[3] Lewis, *op. cit.*, pp. 47–8.
[4] Houghton, *Husbandry and Trade*, vol. ii, pp. 178 sqq.
[5] Scott, *Joint Stock Companies*, vol. ii, p. 414 ; *V.C.H. Somerset*, vol. ii, p. 389.

England, and, as brass was made by combining copper with prepared calamine, brass ware had been imported, for the most part from Nuremberg. Humfrey's discovery made possible a new native industry, and led to the substitution of brass for bronze in the casting of church-bells, cannons (which were also made from iron), cannon balls, and ordnance of all kinds. It encouraged the native manufacture of pins and wool cards (for the carding of wool), both of which could be prepared only with a supply of brass or " latten ", a species of brass, wire.[1]

While the development of the brass industry tended to depress the output of tin, it must have encouraged the output of copper. The activities of the Society of Mines Royal in Cumberland and Wales during Elizabeth's reign, and more particularly the founding of the Company of Mineral and Battery Works, indicate an effort to supply copper for the first time in large quantities.[2] Neither of these societies concerned themselves in any way with tin mining. Apart from their search for the precious metals and for copper, they were chiefly interested in lead, of which increasing supplies were now wanted for roofing houses, sheathing the hulls of men-of-war, making bullets and shot, grates, hearths, and furnaces. In the reign of Charles I the annual output of lead in England is said to have reached about 12,000 tons,[3] as compared with a recorded tin production of about 500 tons.[4] Lead, like copper, production had probably been increasing for some time. In the Mendip district the output began to increase about 1550 and reached a peak about 1670 ; in the Derbyshire district the industry is said to have been in a flourishing condition throughout the seventeenth century.[5]

If it is possible to account for the stationary condition of tin production during the sixteenth and seventeenth centuries by referring to a shrinkage in the market both for English tin and for alloys containing English tin,[6] no similar explanation could be made to account for a decline in iron production. But how far is the history of iron output really parallel to that of tin output ? The assumption that there was a definite decline in the output of iron is based on a comparison of the 300 furnaces mentioned by Dud Dudley in 1665, with

---

[1] Scott, *op. cit.*, vol. i, pp. 31, 39, 30 ; vol. ii, pp. 413–29 ; Unwin, *Industrial Organization*, pp. 164 sqq., 236 sqq.

[2] The literature concerning the early history of these two companies is already considerable. We may mention here H. Hamilton, *The English Brass and Copper Industries to 1800*, 1926 ; Scott, *op. cit.*, vol. ii, pp. 383–429 ; W. G. Collingwood, *Elizabethan Keswick*, 1912 ; C. T. Carr, *Select Charters of Trading Companies* (Selden Soc., vol. xxviii, 1913), pp. 48 sqq. ; *S.P.D., Eliz.*, vol. clxx, no. 37 ; *Lansdowne MSS.*, 5, no. 47, and 19, no. 98. A further list is given by Lewis, *op. cit.*, pp. 41–2 (note 6).

[3] 12,600 " foder " (*S.P.D., Charles I*, vol. cccxli, no. 130, (?) 1636).

[4] For the five years from 1638 to 1642 the annual production of tin in Devon and Cornwall did not exceed 1,100,000 lb. (Lewis, *op. cit.*, appx. J).

[5] J. W. Gough, *The Mines of the Mendip*, 1930, p. 112 ; *V.C.H. Derbyshire*, vol. ii, p. 331—references for which I am indebted to Mr. F. J. Fisher.

[6] Lewis (*op. cit.*, pp. 41 sqq.) mentions technical causes for the slump in the development of tin mining, e.g. the greater cost of mining materials, and the fact that depths were now reached at which drainage became a much more serious problem and expense. But these factors applied equally to coal mining (see below, pt. IV, ch. i (i)) ; yet they did not check the expansion of that industry.

the 59 furnaces which are said to have been operating in 1720.[1] But Dudley's use of round numbers suggests that he was guessing, and recent investigations have shown that most of his guesses should be swallowed with a grain of salt.[2] His figure is really of little use for purposes of comparison. On the other hand, it has recently been shown that many large furnaces were omitted from the list of 1720, and that the figure for the national iron production derived from it is, for other reasons as well, an under-estimate.[3] It is impossible, moreover, to compare the output even on the basis of the number of furnaces at work at different periods. The producing capacity of a single furnace undoubtedly increased ; we know that the average annual output of one furnace in 1720 amounted to nearly 300 tons, whereas, in the reign of Elizabeth the usual output was, perhaps, in the neighbourhood of 75 to 100 tons.[4]

When every allowance is made for Dudley's inaccuracies, for the under-estimate of early eighteenth-century production, and for the increasing output from a single furnace, the impression remains nevertheless, that there could have been no substantial increase in English iron production between 1660 and 1720 ; and there is an abundance of further evidence to support this conclusion. But what happened to the iron industry during the preceding century ? There is some reason to think that it grew rapidly in importance under Elizabeth and James I, reaching a high point about 1610, after which date the output of iron remained stationary, if it did not decline.[5] A list, drawn up in 1574,[6] enumerates 52 furnaces and 51 forges in England, and, even if this is an under-estimate,[7] they certainly multiplied in the years immediately following. Norden, in 1607, counted 140 " hammers and furnaces " in Sussex alone,[8] and an anonymous inventor in 1611 estimated for England and Wales 800 " iron mills ", of which half were in Sussex and Surrey.[9]

---

[1] Dudley, *Mettallum Martis*, 1665, p. 50 ; H. Scrivenor, *History of the Iron Trade*, 1841, p. 57. Mr. Ashton has found that the list which gives the figure 59 furnaces is dated 1720, not 1740 as Scrivenor has it (Ashton, *Iron and Steel in the Industrial Revolution*, 1924, p. 235).

[2] Ashton, *op. cit.*, pp. 10–12. See also Galloway, *Annals*, p. 256, and below, note 9.

[3] Ashton, *op. cit.*, pp. 235–6.

[4] The output contemplated for an iron mill in Cannock Chase in 1588 was 100 tons (*Lansdowne MSS.*, 56, no. 36). By an agreement made in 1589, 3,000 cords of wood were to be supplied every year for an iron works at Coyty English in Glamorganshire (*H.M.C., Report on the MSS. of Lord de L'Isle and Dudley*, vol. i, p. 29) ; and since it required about 40 cords to produce a ton of iron (see below, pp. 193), the undertakers must have planned for an annual production of about 75 tons.

[5] Production is known to have fallen off enormously in Surrey and Sussex, and also in Derbyshire (E. Straker, *Wealden Iron*, ch. viii ; *V.C.H. Derbyshire*, vol. ii, pp. 356 sqq.).

[6] *S.P.D., Eliz.*, vol. xcv, no. 15 ; vol. xcvi, no. 199.

[7] Ashton, *op. cit.*, p. 6.

[8] John Norden, *The Surveyor's Dialogue*, 1607, pp. 214–15. This statement seems to be wrongly cited by L. F. Salzman (*English Industries of the Middle Ages*, 1923, p. 40) as " 140 forges ".

[9] *Harleian MSS.*, 7009. f. 10. It is probable that this includes both forges and furnaces. Dudley's estimate, referred to above, does not include forges. According to him there were 500 of these (*op. cit.*, p. 50). Malynes (*Lex*

While the rapid increase in the smelting of iron did not continue after the first decade, or at any rate after the first quarter of the seventeenth century,[1] it is by no means likely that there occurred a similar slump in the manufacture of finished metal goods. The smelting of copper and lead, which perhaps increased no less rapidly during the reigns of Elizabeth and James I than did the smelting of iron, probably suffered a somewhat less serious depression during the remainder of the seventeenth century, for copper and particularly lead could be produced with a smaller expenditure in wood fuel than could iron. The supplies of iron and copper turned out in the native furnaces and forges were supplemented by increasing imports from abroad—especially from Sweden. At the beginning of the eighteenth century, when conditions were favourable for foreign trade, the imports of bar iron were probably about equal to the native production.[2] Even if that did not increase after the first quarter of the seventeenth century, a considerable increase may have occurred, therefore, in the amount of metal worked up by British artisans into pins, nails, razors, scissors, pocket knives, ships' anchors, wagon wheels, horses' bits, fire hearths, locks and keys, wool cards and combs, and a thousand other articles, which the ingenuity of man (or, at least, of that growing minority of mankind which went under the name of Smith) busied itself in making.

It is not possible to advance quantitative evidence in support of this view : to show, for example, that the yearly output of nails, or of horses' bits, was much greater in 1700 than in 1550. Yet a general survey of other industries which required metal tools and accessories—such as shipbuilding, coal-mining, salt making, and glass manufacture [3]—suggests that there developed in precisely this period a great new demand for metal goods ; while the writings of contemporary Englishmen and foreigners show that serious efforts were contemplated to relieve the country from its dependence upon Holland for finished iron and copper products,[4] and to supply these products by native manufacture ; efforts which—to judge from the number of artisans apparently engaged and the fame of some of their produce —must have met with substantial success. " The Skill and Neatness of our Workmen is such in Locks, Keys, Hindges, and other Curiosities of this kind ", wrote Davenant in 1699, " that our Exportations of these Commodities may in time grow very considerable ".[5]

The importance of such manufacture before the Elizabethan Age

---

Mercatoria, 1622, pp. 269–70) also refers to 800 English iron works ; which suggests that this estimate in round numbers may have been current in the early seventeenth century. Being current, it was probably highly inaccurate, and it is not impossible that Dudley was simply repeating it.

[1] Cf. below, pp. 224–5.

[2] Mr. Lipson puts the total imports of bar iron, in the second decade of the eighteenth century, at between 14,000 and 22,000 tons (*The Economic History of England*, vol. ii, 1931, p. 161) ; while, according to Mr. Ashton, the output of bar iron, at this time, was probably nearer 18,000 or 19,000 than 12,060 tons, the figure given in the list of 1720 (*op. cit.*, pp. 235–6.)

[3] See below, pp. 172–84.

[4] Houghton, *op. cit.*, vol. ii, pp. 212–13 ; Yarranton, *op. cit.*, pp. 61–4.

[5] *An Essay upon the Probable Methods of Making People Gainers in the Ballance of Trade*, 1699, p. 155.

must not be under-estimated. Smiths and metal workers of all kinds had been numerous in towns, and even in very small villages, during the Middle Ages, as Mr. Salzman's survey of medieval industries convincingly shows.[1] One has only to consider the powerful position then occupied by the goldsmiths or the pewterers among London Companies, to realize how profitable trade in products of gold and tin had already become. Both goldsmiths and pewterers, however, dealt primarily in articles of luxury, such as plate and jewellery, which only the richer citizens could afford ; and, consequently, dealers depended for gain rather on the high price of their wares than on the quantity of their output. The new demand, which developed in Elizabethan times and after, was primarily for products of the coarser metals, which were wanted for use rather than display, and which derived their value from their abundance and cheapness. The expansion in native shipbuilding [2] called for supplies of lead, brass, copper, and, above all, iron wares, for even a boat of wood could not be constructed and launched without nails to hold the planks together, wire for the masts, chains and anchors to allow her to ride at rest in harbours. In an account of 1618, showing the expenses for building ten men-of-war, five per cent of the total payments went for anchors.[3] When it came to equipping battleships, far larger quantities of metal work had to be provided than in the case of merchantmen, not only for the cannons and cannon balls, but for sheathing the hulls, either with iron or lead.[4] Between June 24th, 1660, and March 27th, 1661, a period of nine months when the nation was at peace, and when the navy probably still drew its principal stores of iron from Sussex and the Forest of Dean, one man, Richard Foley, who made his fortune as a capitalist from heavy industry in his native Staffordshire, supplied Sir George Carteret, Pepys' chief at the Navy Office, with £5,000 worth of iron wares.[5] In addition to the manufacturing regions, such as south Staffordshire, Sussex, and the Forest of Dean, London, and no doubt all the chief shipbuilding ports, had their special gun-smiths and anchor smiths, who worked primarily for the navy.[6] And, if the growth of the English navy stimulated the demand for metal wares, the expanding use of artillery and explosives in the warfare which raged, especially on the Continent, during the sixteenth and seventeenth centuries, stimulated this demand still further. Cannons of bronze, but more often of cast iron or brass, muskets and small arms made from gun metal (an alloy of zinc or tin and copper), bullets of lead and shot of iron and lead,[7] all the engines, in short, which men have invented for their own destruction, were

[1] Salzman, *op. cit.*, esp. chap. vii.   [2] See below, pp. 172–4.
[3] Cited by W. Sombart, *Der Moderne Kapitalismus*, vol. i, pp. 767–8.
[4] Apparently iron was first used for ship's sheathing, but later lead was substituted, probably because it could be supplied more cheaply (*Cal. S.P.D.*, 1660–1, p. 549).
[5] *Ibid.*
[6] *Cal. S.P. Colonial*, 1661–8, p. 427 ; *P.C.R.*, vol. lviii, pp. 116, 124.
[7] *Cal. S.P.D.*, 1653–4, p. 142 ; Houghton, *Husbandry and Trade*, vol. ii, pp. 204–6. The method of manufacturing cast iron cannons is said to have been introduced into England in 1543 by a Sussex iron master, who employed a Flemish gunsmith to produce explosive shells (Samuel Smiles, *Industrial Biography*, 1863, p. 33).

wanted in ever-increasing quantities. Saltpetre and gunpowder manufacture, which was probably introduced in England in our period,[1] involved the use of copper kettles.[2] The growth in population and in wealth, together with the remarkable rise in the cost of timber, stimulated a demand for metal in more peaceful industries. Coal mining required substantial supplies of metal for the pumping machinery which was being set up, not only at the larger collieries but also at small " land-sale " enterprises throughout the coal fields.[3] Saltmaking, alum and copperas manufacture, soap boiling, sugar refining, and the dying of cloth—all industries which were progressing rapidly [4]—required metal pans, boilers, and vats.[5] In building operations, iron bolts, locks and keys, and gates were wanted in greater quantities for protection against thieves, while lead was increasingly used instead of thatch or shingle for roofing.[6] Iron hearths, rare before the adoption of coal as fuel, were installed not only in the houses of the middle class in the principal towns, but very generally in the dwellings of humbler subjects, even in remote hamlets.[7]

Before the middle of the eighteenth century, the importance of the metal industries in England was, of course, a commonplace. " Persons, especially well informed concerning English commerce, assure me that the trade in iron, and iron and copper products, employs as much labour and yields as great a return as does wool ", wrote in 1738 a certain Monsieur Ticquet, who had come to England to study the development of heavy industry.[8] English writers from Yarranton to Defoe, in surveying the economic condition of the country, refer to the army of metal workers in terms which could hardly have been employed before the reign of Elizabeth ; and, while it may be felt that Ticquet's statement goes somewhat too far, all observers are in essential agreement as to the growing importance of this manufacture.[9] An anonymous English authority, who wrote some years earlier than Ticquet, tells us that Staffordshire iron utensils " are made . . . in the utmost beauty and perfection ".[10] In the district round Birmingham were perhaps half a dozen towns (Sedgley, Dudley, Wednesbury, Walsall, and Wolverhampton), each with a thousand or more metal workers,[11] and these artisans were scarcely less numerous about Sheffield or Wigan.[12] Even by the middle of the seventeenth

---

[1] See below, p. 185.   [2] See below, p. 210.
[3] See below, pp. 354–7, 371–2.   [4] See below, pp. 174–9, 184–6, 188.
[5] See below, pp. 206–14 *passim*.   [6] *Exch. K.R. Misc. Bks.*, vol. xxxvii, f. 317.
[7] See below, pp. 199–200.
[8] *Copie d'une lettre écrite par un françois étant en Angleterre le 1er Septembre* 1738 (*Archives Nationales*, O¹1293).
[9] Yarranton, *England's Improvement by Sea and Land*, 1677, pp. 58–9 ; Defoe, *Tour*, 1769 ed., vol. iii, p. 425. G. L. Beer (*The Commercial Policy of England Towards the American Colonies*, 1893, p. 83), ranks " iron production " third in importance among British industries at the beginning of the eighteenth century.
[10] *A New Present State of England*, c. 1727, vol. i, pp. 213–14.
[11] Yarranton, *op. cit.*, pp. 58–9 ; Houghton, *op. cit.*, vol. ii, pp. 220–1 ; *A New Present State of England, loc. cit.* As early as 1590, bar iron was brought to Walsall from as far away as Nottingham (*H.M.C., Report on the MSS. of Lord Middleton*, p. 495). Probably much Shropshire and Forest of Dean iron was also made into finished goods in Staffordshire. Cf. below, pp. 231–2.
[12] Defoe, *Tour*, 1769 ed., vol. iii, pp. 100–1, 108, 281.

century the casual village smith, who supplied the neighbouring population with nearly all its metal ware, was becoming relatively less important. Only a diminishing proportion of the goods now turned out was consumed locally, and the artisans themselves were increasingly employed by large merchants as " putting out " workmen. The new industries, which had no place in the gild organization of the old towns, naturally lent themselves with particular ease to large-scale production.[1] Not only did the artisans tend to specialize in the kind of article they turned out, but in the production of a single article, a pin of brass-wire, nine distinct operations might be performed, each by a different worker, as John Houghton showed in one of his weekly discourses concerning husbandry and trade, an example which Adam Smith selected much later for his celebrated illustration of the division of labour.[2] Such specialization indicates a highly developed stage of manufacture.

Satisfactory statistics concerning the output of finished metal goods in Great Britain will be much more difficult to compile than statistics concerning the output of coal ; and, up to the present, no effort has been made to compile them. The case in favour of a remarkable growth in the metal manufactures between 1550 and 1700 must rest, for the moment,[3] upon incomplete evidence concerning the increase in the demand for them, the increase in the number of artisans employed in the metal trades, the increase in the domestic production of iron, lead, and copper during the reigns of Elizabeth and James I, and the increase in the imports of such metals.

While the evidence with reference to the output of such goods is of an indirect nature, enough quantitative data are already available to indicate that the production of ships, of salt, and of glass in Great Britain increased far more rapidly during the period from 1550 to 1700 than did the population.

The coal trade itself created a great demand for ships to carry away the produce from the mines. Before Elizabeth's reign. the total quantity of English and Scotch coal shipped by sea probably did not exceed 40,000 to 50,000 tons a year.[4] Foreign vessels took all consignments for foreign ports, and even a part of the consignments for native ports.[5] It is said that in 1550 the traffic to London employed only two native ships,[6] and it is questionable whether the kingdom then possessed twenty coal hoys altogether. But, at the end of the seventeenth century, more than 1,600 vessels were employed in the carriage of coal by sea.[7] All the vessels engaged in the coastwise trade, and the great majority of those engaged in the foreign trade, were now British owned ; for by means of preferential duties the Government had largely succeeded in driving out aliens.[8] After the Restoration, about eight tons

---

[1] See, for instance, Henry Hamilton, *The English Brass and Copper Industries to 1800*, 1926, pp. ix, 69 sqq.

[2] Houghton, *op. cit.*, vol. ii, pp. 192–4 ; *Wealth of Nations*, bk. 1, ch. 1.

[3] But see below, pp. 201–5, 231–2.

[4] See above, Table VI.   [5] See below, vol. ii, pp. 23–4.

[6] Hylton Dale, *Coal and the London Coal Trade*, 1912, p. 5.

[7] More than 1,000 in the English east-coast trade (see below, vol. ii, p. 95), and several hundreds in other branches of the coastwise and in the export trade (cf. above, Table VI, p. 79, and below, pp. 391–2, and vol. ii, pp. 141–2).

[8] See below, vol. ii, p. 25.

in every ten exported from the Tyne and the Wear to foreign countries were carried in an English vessel.[1] River boats carrying coal were even more numerous than sea-going colliers. On the Tyne, at the beginning of the eighteenth century, there were at least four hundred keels (as compared with forty before Elizabeth's reign), and at Sunderland perhaps two hundred more, as well as a prodigious number of lighters at London to unload coal for the use of the metropolis, or to carry it farther up the Thames ; barges at ports like Bridgwater or King's Lynn for bringing Welsh or north country coal inland by river, and river colliers on the Severn and Trent to transport downstream the produce of mines round Broseley and Nottingham.[2] The total number of craft engaged, in 1700, primarily in moving coal may be estimated, without fear of exaggeration, at from three to four thousand.

According to Sir Francis Brewster, the Dublin magistrate who published a treatise on trade and navigation in 1695, nearly thirty per cent of all English vessels were at that time engaged in the coal trade.[3] Although the demand for ships in other branches of commerce had not expanded at so impressive a rate, the shipping business had received substantial encouragement from many other trades : from the fishing trade, the Baltic timber trade, the French wine trade, and from the growing traffic with India and America. Such statistics as have been brought to light suggest that there was a notable development in all branches of the native shipping industry, not only after the passage of the Navigation Acts, or the establishment of the preferential duties on particular goods which preceded those Acts, but during the reigns of Elizabeth and James I.[4] Between 1550 and 1660 the total tonnage of the Royal Navy apparently increased six fold.[5]

Native shipping increased ; but to what extent the ships were built in native yards is another question. After the imposition of

---

[1] According to information obtained from *Exch. K.R. Port Bks.*, 193/1, 195/13, 196/2, and *Sackville MSS.* (" An account for . . . Ld. Buckhurst, upon Coales exported ", June 3rd to December 25th, 1663).

[2] See below, pp. 388–9, 393 n. Four hundred coal ships are said to have plied on the Severn in 1758 (*V.C.H. Salop*, pp. 435–6).

[3] *Essays on Trade and Navigation*, p. 75. Brewster estimated that there were in all 11,000 English merchant ships, of which number 4,400 were fishing smacks, 1,900 were employed in the trade with the Straits and Portugal, 800 in the French trade, 400 in the Sound, 500 in the trade with Muscovy, and 3,000 in the coal trade. Doubtless all his figures were exaggerated.

[4] W. Sombart, *Der Moderne Kapitalismus*, 1916, vol. i, p. 762, and vol. ii, p. 300 ; *Krieg und Kapitalismus*, vol. ii, pp. 175 sqq. ; R. G. Marsden, " English Ships in the Reign of James I ", in *Trans. Roy. Hist. Soc.*, N.S., vol. xix, 1905, pp. 309–42. Merchant ships of 100 tons and more in burden are said to have increased in number as follows :—

1545	35
1577	135
1582	177
1588	183
1629	350

And the ships in this class increased not only in numbers but in size (cf. below, pp. 390–2).

[5] Sombart, *Der Moderne Kapitalismus*, vol. i, pp. 762, 766.

preferential taxes on foreign-owned vessels, a controlling interest in these vessels was frequently purchased by English adventurers, who thereby brought them under the term " native ", within the meaning of the Act. Again, during the Dutch wars, captured enemy boats were made over to English owners.[1] Consequently a large number of ships flying the English flag had, in reality, been fitted out by foreigners, and, apart from necessary repairs, had brought business to foreign rather than to native shipbuilders.[2] Nevertheless, after every allowance for this factor has been made, it is undoubtedly true that native shipbuilding was stimulated by the increase in trade, particularly by the increase in the coal trade, and that a large proportion of all sea-going vessels, and the vast majority of all harbour and river boats, were turned out in English yards, especially at Ipswich, Yarmouth, Newcastle, Bristol, Whitby, and London. Only at the very end of the seventeenth century did there set in at these centres a serious and prolonged depression in shipbuilding,[3] which was due in part to the over-stocking of the merchant marine with foreign prizes.[4]

The view that England and Scotland underwent a period of rapid industrial expansion before the eighteenth century is strengthened by the history of the salt manufacture, the second of the industries for which statistical evidence is available. Salt is made in one of three ways : by boiling down the liquid from inland brine pits or springs, by evaporating sea water collected in trenches or pans along the seashore, or by dissolving into brine a mineral known as rock salt. Before Elizabeth's reign only the first two of these methods had attained importance in the British Isles.[5]

There is no means by which to determine the early output from the brine springs at Droitwich in Worcestershire and at Northwich in Cheshire, which were described by Leland in the course of his travels during the thirties of the sixteenth century.[6] At the time of Harrison's visit, towards the middle of Elizabeth's reign, there were at Droitwich 360 furnaces, producing 1,400 or 1,500 weighs (of 40 bushels) of salt per annum.[7] For Northwich and the other seats of brine manufacture in Cheshire—Middlewich and Nantwich—we

---

[1] R. B. Westerfield, *Middlemen in English Business*, 1915, p. 229.

[2] By 14 Car. II, c. 11, all foreign-built vessels not purchased before October 1, 1662, were to be considered " aliens ", even if owned and manned by English subjects ; but it is hardly probable that the Act was rigorously enforced, and, in any case, an exception was made for " such ships as shall be taken at sea by Letters of Mart or Reprisal and Condemnation made in the Court of Admiralty as lawfull Prize ".

[3] See below, vol. ii, p. 95.

[4] Cf. below, vol. ii, pp. 95–9, where the causes of the depression are discussed at greater length.

[5] For details concerning the first two methods see Agricola, *De Re Metallica*, trans. by H. C. and L. H. Hoover, 1912, pp. 545–7.

[6] *Itinerary*, 1538, vol. v, p. 92.

[7] Each furnace yielded 4 loads of salt per annum, according to Harrison (*Description of England*, 1577, chap. xiii). Therefore 360 furnaces must have yielded 1,440 loads. The load, according to Harrison, was of 5 to 6 quarters, i.e. 40 to 48 bushels ; and, since the " weigh " (the most common measure for salt in the sixteenth and seventeenth centuries) is said to have been of 40 bushels, we may suppose that the load mentioned by Harrison (cf. also Appendix C) and the weigh were roughly equivalent.

find no estimate of production until the reign of Charles II,[1] when John Collins, the first great authority on the salt trade, puts the combined output of these three works at rather more than 500 weighs a week,[2] or let us say from 20,000 to 25,000 weighs a year. This was a considerably larger output than that of the entire country in the reign of Henry VIII.[3] It is reasonable, therefore, to suppose that the salt industry in Cheshire had been expanding rapidly. Northwich accounted for sixty per cent of the output of Cheshire when Collins wrote. As the brine manufacture at Droitwich was more famous than that at Nantwich, it is not improbable that the output of salt was as great. If that is so, it is clear that the industry in Worcestershire had also expanded greatly since Harrison's time.

The method of obtaining salt by evaporating sea water, like the method of boiling down brine, went back to an early epoch in English history ; [4] but, until the sixteenth century, it had been a casual occupation, carried on by a few poor dwellers near the coast, upon their own initiative, and with an inadequate equipment. As late as 1526, Tynemouth Monastery, which owned most of the sites in Northumberland where this manufacture was destined to develop, received only £61 in annual revenue from its sales of salt.[5] It was between 1563 and 1571, through the initiative of Cecil, that the earliest attempt was made to establish, with the help of German artisans, a heavily capitalized industry on a large scale, first at Dover, and later at Blyth on the Northumberland coast. Backed by a group of influential court officials, who obtained an exclusive patent in 1566,[6] John Mount, one of Cecil's secretaries, undertook to buy out the small men, or " Haglayers " (who had hitherto been the sole makers), and to install a number of expensive iron pans, all to be operated as a single enterprise.[7] He failed for want of capital.[8] But, during the 'nineties, groups of London, King's Lynn, and north country merchants succeeded where the politicians had given up. Within the framework

---

[1] There is extant, however, a survey (made in 1651) of the brine of 24 hours' boiling at Nantwich in Cheshire (*Augm. Parl. Surveys*, Cheshire, no. 20). *Exch. Deps. by Com.*, 6 James I, Hilary 13, deals with an interesting suit concerning the salt works at Droitwich. See also *V.C.H. Worcestershire*, pp. 256–61.

[2] John Collins, *Salt and Fishery*, 1682, p. 3. The weekly output in bushels is given by Collins as follows : Northwich, 12,214 ; Middlewich, 4,300 ; Nantwich, 4,200.

[3] See below, p. 178.

[4] Both methods antedate human records (see the Hoovers' note in their edition of Agricola's *De Re Metallica*, p. 546).

[5] *A History of Northumberland*, vol. viii, p. 114.

[6] 8 Eliz., cap. 22.

[7] The change from wood to coal fuel at the brine springs in Cheshire (see below, p. 208) involved the substitution of iron for lead pans (*Roy. Soc., Philos. Trans.*, vol. iv, 1669, pp. 1061–9), and the same substitution may have taken place in the sea-salt manufacture. The expansion of the salt industry, which was everywhere accompanied by a great increase in the size of the pans, must have stimulated considerably the demand for iron (cf. above, pp. 170–1).

[8] The full story of this attempt is well told by Edward Hughes, " The English Monopoly of Salt in the Years 1563–71 ", in *Eng. Hist. Rev.*, vol. xl, 1925, pp. 334–50. For information concerning the salt industry at Blyth from 1571 until 1605, see *Exch. Spec. Com.*, no. 4,347 ; *A History of Northumberland*, vol. ix, pp. 225–7.

of the well-known Wilkes patent,[1] which superseded that of 1566, the most up-to-date equipment obtainable was installed at Offerton on the Wear by Robert Bowes, one of an old Durham land-holding family, and John Smith, a merchant of King's Lynn, backed by the third Earl of Huntingdon.[2] Soon after, another work was set up at Seaton Delaval by members of the Delaval family, who contracted to sell the produce of their pans to John Lyons and his partners.[3]

While the salt manufacture thus became an established industry at Sunderland on the Wear, and while it flourished for a short period towards the end of the seventeenth century at Blyth, Hartley, Cullercoates, and Seaton Delaval on the Northumberland coast, the great centre throughout the period from about 1580 until after 1700 was Shields, at the mouth of the Tyne.[4] " Here . . . are the vastest salt works I have seen ", wrote in 1636 Sir William Brereton, who had set out on his travels fresh from experience of this manufacture in his native Cheshire. And, he added, " here is such a cloud of smoke as amongst these works you cannot see to walk ".[5] Many years later, Defoe told how he saw the smoke from these same fires " ascend in huge Clouds over the Hills, four miles before we came to Durham, which is at least sixteen Miles from [Shields] " ; and how, as he approached Berwick and looked back over his shoulder, he could still see it 40 miles to the south.[6] Recorded shipments of salt from the Tyne to English ports did not apparently exceed 300 or 400 weighs in the year 1561 ;[7] but, during the Commonwealth period, they averaged more than 6,000 weighs per annum,[8] although many of the pans at Shields are said to have been then given up, owing to Scottish competition, which was stimulated by Cromwell's repeal of the preferential duties against the import of Scotch salt into England.[9] By 1605, the yearly output of the entire industry in Durham and Northumberland had reached 7,600 weighs.[10] After the incorporation in 1635 of the

---

[1] The following are a few of the documents which deal with this patent : *Lansdowne MSS.*, 59, nos. 66, 67, 68, 69, 70 ; 73, nos. 48, 49 ; 74, no. 4 ; *Exch. Spec. Com.*, no. 2651.

[2] *H.M.C., Report on the MSS. of the Marquis of Salisbury*, vol. v, p. 526.

[3] *Waterford MSS.*, cf. below, vol. ii, p. 17.

[4] An attempt was made at the end of the sixteenth century to manufacture salt from sea water at Hull, Yarmouth, King's Lynn, and Southampton ; but works set up near these ports were never very successful (cf. below, pp. 227–8). The pans on the Hampshire coast produced some salt during the reigns of James I and Charles I (*Star Chamb. Proc., James I*, 171/21 ; *Exch. Deps. by Com.*, 7 Charles I, Trin. 1).

[5] Sir W. Brereton, " Travels in Holland, England, Scotland and Ireland, 1634–5 ", in *Publications of Chetham Society*, 1844, vol. i, p. 88.

[6] Defoe, *Tour*, 1769 ed., vol. iii, pp. 237, 299.

[7] Shipments of salt for six months from February 2nd to July 31st, 1561, amounted to 183 weighs (*Exch. K.R. Customs Accounts*, 111/4).

[8] 6,051 weighs in 1652, 6,101 weighs in 1655 (*Exch. K.R. Port Bks.*, bdl. 192).

[9] *Discourse concerning the Salt workers . . . of Durham and Northumberland*, temp. Charles II (*Lansdowne MSS.*, 253, no. 17). This document is printed in M. A. Richardson, *Reprints of Rare Tracts*, 1847–9, vol. iii.

[10] Brand, *History of Newcastle*, vol. ii, p. 22 (cited Welford, *Hist. of Newc. and Gateshead*, vol. iii, p. 171). Brand cites a contemporary letter of the Earl of Northumberland (now among the MSS. of the Duke) concerning the salt trade. According to this letter there were 153 salt pans in Durham and Northumberland, each of which produced about 50 weighs per annum.

" Society of Saltmakers ", and before the Civil War, the works at Shields alone turned out between 14,000 and 15,000 weighs annually, and those at Sunderland an additional 1,200 weighs.[1] Towards the end of the seventeenth century the total production from all the pans in both Durham and Northumberland probably reached 25,000 weighs.[2]

A record of progress no less remarkable is found in the history of the manufacture of salt from sea water on the Firth of Forth. In the decade 1550 to 1560 only £17 17s. 6d. was collected in customs on exported salt ; but in the decade 1570 to 1580 the revenue had risen to £1,195, notwithstanding an additional premium put on smuggling by legislation to discourage shipments abroad.[3] Salt making, like coal mining, rapidly became one of the chief occupations of the inhabitants ; and all along both banks of the Firth of Forth for seventy miles, from beyond Musselburgh round to Pittenweem, one could see clusters of small iron pans with their clouds of dirty, brownish smoke, so typical of the budding industrialism of the north. In no one spot were as many pans concentrated as at Shields ; but, on the other hand, there were many more centres here than in Durham and Northumberland, and Brereton, who had commented on the manufacture at Shields before he had visited Scotland, observed that the output " cannot be

---

[1] Sir Lionel Maddison to Sir Henry Vane, from Newcastle, November 6th, 1644 (*Cal. S.P.D.*, 1644–5, pp. 98–9). It is said that the output of salt fell off during the monopoly of the Society of Salt Makers (Davies, *An Answer to . . . the late Patentees of Salt*, 1641, p. 10, as cited by Scott, *Joint-Stock Companies*, vol. ii, p. 470) ; but, as Davies was an ardent opponent of the Society, it is possible that his zeal in attacking it may have led him to exaggerate the depression. He agrees with Maddison that at one time before the Civil War an annual output of 15,000 or more weighs was obtained at Shields. On the Society of Salt Makers of South and North Shields, see Gardiner, *History of England*, vol. viii, pp. 284–5 ; Scott, *op. cit.*, vol. i, pp. 209–10, 216–17, 219, 221–2, and vol. ii, pp. 468–70 ; C. T. Carr, *Select Charters of Trading Companies* (*Selden Soc.*, vol. xxviii, 1913), pp. 142–8, 167–72.

[2] The output at Shields still amounted to about 15,000 weighs per annum, when Lord Harley visited the pans in 1725 (*H.M.C., Report on MSS. of the Duke of Portland*, vol. vi, p. 105). He says that there were at South Shields about 200 pans, and others on the north side of the Tyne, each of which " makes two tuns and a half of salt " in a week. He adds, " the whole annual produce of these pans at Shields, etc., is about one hundred and fifty thousand pounds per annum ". But this is obviously a mistake ; it is quite impossible that, if the weekly output of one pan was two and a half tons, the annual output of more than 200 pans was only 67 tons. It is possible that Lord Harley was referring, not to the weight, but to the annual value of the product. Since the industry at Shields is said to have reached a peak of prosperity about 1686 (R. Surtees, *History of Durham*, 1816–40, vol. ii, p. 95), and to have declined steadily thereafter (*A History of Northumberland*, vol. viii, p. 21 ; and see below), we are perhaps justified in assuming that the output may have been somewhat larger at the end of the seventeenth century than in 1725. At Cullercoates in 1708 the annual production exceeded 2,000 weighs (*ibid.*, vol. viii, pp. 281–3) ; and the production at Sunderland, Blyth, Hartley, or Seaton Delaval may have been as large as at Cullercoates. Therefore, 25,000 weighs is probably not a high estimate for the total output of Durham and Northumberland at the end of the seventeenth century.

[3] Edward Hughes, *op. cit.*, in *Eng. Hist. Rev.*, vol. xl, p. 349, note. On the state of the salt manufacture in Scotland by 1578 see Bishop Leslie, " The South Countreyis of the Realme ", in P. Hume Brown, *Scotland before 1700* ; and also *Privy Council Register of Scotland*, 1st Series, vol. ii, pp. 442–3.

estimated and guessed, because the works are not easily to [be] numbered ". There were, he declared, " all along the shore at least thirty English miles from beyond Musselborough almost to Sterling, an infinite innumerable number of them ".[1] Had he visited Fife, he would have found salt works in similar abundance along the north shore of the Firth of Forth as far east as St. Andrews. The output may well have exceeded that in the north of England, because for most uses Scotch could regularly undersell English salt in the English market, unless handicapped there by a high duty. In spite of such a duty, it would appear that more than 8,000 weighs of Scotch salt per annum reached England in the years immediately preceding the Civil War,[2] at a time when, according to Brereton, " the greatest part of [Scotch] salt . . . is transported into Holland ".[3]

Besides the works in Durham and Northumberland and on the Firth of Forth, there were before 1700 other less important centres for manufacturing salt from sea water at various places on the Scottish and English coasts : notably at Brora in Sutherland, at Saltcoats in Ayrshire, at Whitehaven in Cumberland, and along the Dee Estuary in Flint and Denbighshire.[4]

In 1670 a searcher for seacoal hit upon a rock of salt in Cheshire.[5] His discovery caused an immediate rush to exploit the new mineral, which was soon after found in several other counties, and before 1700 a rock salt manufacture had developed, considerable enough to offer serious competition to the brine and sea salt makers,[6] a competition destined within another half century effectively to ruin them both.

This brief historical survey of the industry suggests that, during the hundred and fifty years which followed the accession of Elizabeth, the position of Great Britain with reference to salt underwent a revolutionary change. In the reign of Henry VIII, when the annual consumption of England and Wales amounted to some 40,000 weighs, or about 40,000 tons, from two-thirds to three-fourths of the supply had to be purchased abroad.[7] But on the eve of the Civil War, it is probable that about 40,000 tons per annum were produced on the coasts of Durham and Northumberland and the shores of the Firth

---

[1] Brereton, *op. cit.*, pp. 98, 112. According to Maddison there were 222 pans at the mouth of the Tyne before 1640, of which from 180 to 190 were usually in operation. On the Firth of Forth the greatest concentration was at Culross, where there are said to have been 50 pans working in 1663 (Sir John Sinclair, *Statistical Account of Scotland*, vol. x, p. 144). There may have been as great a difference between grades of salt as between grades of coal (cf. above, pp. 111 sqq.), but time has not permitted me to investigate the question.

[2] Carr, *op. cit.*, p. 170.

[3] Brereton, *op. cit.*, p. 112.

[4] Robert Chambers, *Domestic Annals of Scotland*, 1858, vol. i, p. 302 ; *Acts of the Parliaments of Scotland*, vol. viii, p. 609 ; *V.C.H. Cumberland*, vol. ii, p. 359.

[5] John Collins, *Salt and Fishery*, 1682, p. 4.

[6] *The Case of the Rock Salt as it now stands Burthen'd with a Higher Duty than the other sort of English made Salt ; The Case of the Refiners of Rock Salt in Lancashire, Cheshire, and Flintshire ; The Case of the Brine-Pits Truly Stated ; The Case of Rock Salt, considered in relation to a Clause now depending, for prohibiting the Refining thereof at any places, but near the Rock-Pits · Objections against the Rock Salt answered ; A reply on behalf of Rock Salt and Refineries* (all written about 1699) ; see also Scott, *op. cit.*, vol. ii, p. 470.

[7] Edward Hughes *op. cit.*, p. 334.

of Forth alone, besides a very greatly increased quantity at the midland wiches. The reigns of Elizabeth and James I appear to have been the period of most rapid expansion in the brine and sea salt industries. After the Restoration, progress was maintained. By the nineties of the seventeenth century, when Houghton estimated the annual consumption of England and Wales at 130,000 tons,[1] a considerable portion of the supply was still, it is true, imported,[2] but the imports must have been very nearly balanced by Scottish exports to Holland, and English west-coast exports to Ireland.   Taking England and Scotland together, one may estimate very tentatively the annual yield of brine salt at from 30,000 to 40,000 weighs,[3] and the yield of sea salt at from 60,000 to 80,000 weighs.[4] In addition, some quantity of rock salt was being produced by this time. The total output of all kinds of salt in Great Britain could hardly have been less than 100,000 tons, and it may have been considerably more than this. It had been a principal aim of Tudor and Stuart governments to make the country independent of foreign supplies of a commodity so essential in packing meat, preserving fish, butter, cheese, and eggs, in an era when the slowness of transport made such preservation most important.[5] That aim had been achieved, though probably less as a consequence of the protective policy pursued by successive governments,[6] than because of the natural advantages afforded for salt making by cheap supplies of coal.[7]

The development of the English and Scottish glass industry, which was also facilitated by the increasing output of coal, is less easily measured by statistics, but it was, in all probability, no less remarkable.   Records of the glazier's work are scarce during the Middle Ages, and they refer mainly to ovens in the Chiddingfold

---

[1] Houghton, *Husbandry and Trade*, vol. ii, pp. 75–6. " I rather think we use more [than 130,000 tons] ", writes Houghton. In another place, he speaks of 53,248 weighs as " all the salt that was known to be made in, and imported into England and Wales " from March 25th, 1695, to March 25th, 1696 (*ibid.*, pp. 104–5). The discrepancy is perhaps to be explained by the fact that it was not possible to obtain figures concerning the output of most of the English manufactures. But it should be pointed out that there are also objections to the larger estimate. This is based on the assumption that each family used four bushels per annum, which is of course a guess ; and Houghton's figure for the total population was higher than the one which is now usually accepted.

[2] Perhaps 25,000 weighs.   In the year from Mich., 1685, to Mich., 1686, the recorded imports of French salt at English ports amounted to nearly 17,000 weighs (Charles King, *The British Merchant*, 1743 ed., vol. i, pp. 257, 262).

[3] 20,000 to 25,000 weighs in Cheshire ;   10,000 to 20,000 weighs in Worcestershire.

[4] 25,000 weighs in Northumberland and Durham ;   25,000 to 35,000 weighs in the Firth of Forth district ;   10,000 to 20,000 weighs at all other places on the Scottish and English coasts.

[5] The fishermen were apparently the chief customers of the salt makers at Shields (*S.P.D., Charles I*, vol. cclxxxix, no. 109).

[6] Cunningham, *Growth of English Industry and Commerce, Modern Times*, 1921 ed., pt. i, pp. 229, 309 ;   *Reasons for laying a further Duty upon French, Spanish and other Foreign Salt*, (?) 1699.

[7] Davies wrote that foreign salt was cheap, because it was cheaper to evaporate salt water by the sun than by coal fires (*An Answer to the Patentees*, p. 11). This was doubtless true in southern France, where conditions were favourable, but not in Great Britain (see below, p. 206).

district of Surrey and Sussex.[1] But this, in itself, is no proof of the absence of glass makers elsewhere. Research frequently shows that industrial processes, like ideas, have an older history than that which tradition has assigned them, and that the word of a commentator is no more likely to reveal the true origin of an industry, than the stamp of a government patent that of an invention. It has been categorically asserted that no glass was made in the Tyne valley before 1615 ; [2] yet the Newcastle Corporation received a small revenue from the export of glass as early as 1574.[3] Again, it has been assumed that, until the very end of the sixteenth century, all glass vessels were imported from abroad ; [4] but Mr. Salzman refers to a contract made in 1380 for producing such articles in Surrey, and the operators of a glass furnace at Falkland in Scotland sold in 1506 and 1507—among other things—a number of flasks.[5]

In spite of this evidence of an earlier activity, it is unlikely that any great quantity of glass was made in England or Scotland before the reign of Elizabeth. Until the sixteenth century, this commodity had been in little demand for the common people, who had neither glass mirrors, nor glass cups, nor glass windows. To a far larger extent than metal, glass remained an article of luxury and artistic display. Apart from its place in the castles of the nobility, it had been used chiefly to adorn the stately windows of medieval cathedrals. Much of the labour of glass making then concerned the planning of patterns, and the forging of deep blues and reds or fiery whites. That perhaps explains the high rank given to English painted glass by Vasari, who considered it the best produced in his time.[6] Not by accident did glass making come within the purview of the illustrious author of the *Lives of the Painters* rather than within the speculative dealings of a Florentine financier ; for, like goldsmiths' work, it was still more an art than a manufacture, and the spirit in which glass had been produced was that of the creator rather than of the merchant.

During the sixteenth century, glass first became an object of common use in England. The spread of civilization brought the apothecary's bottle and the test tube of the natural scientist. Spectacle makers appeared among the London companies.[7] Even modest householders clamoured for glass windows. Glass mirrors, reputed to have been devised by Venetians in the fourteenth century, were introduced in London residences and in the houses of the country gentry.

---

[1] Salzman, *English Industries of the Middle Ages*, pp. 183 sqq.
[2] A. Hartshorne, *Old English Glasses*, 1897, pp. 177–8.
[3] *Newcastle Corporation MS. Account Bks.*, 1574–80. The port books also show that glass was exported from the Tyne before 1615 (*Exch. K.R. Port Bks.*, 187/5). Bourne (*History of Newcastle*, 1736, p. 155) was probably justified, therefore, in asserting that some of the Lorraine glass makers set up a furnace at Newcastle during Elizabeth's reign. Sidney Grazebrook (*Collections from the Genealogy of the Noble Families of Henzey, Tyttery and Tyzak*) disputes Bourne's assertion (see Welford, *History of Newcastle and Gateshead*, vol. iii, p. 208).
[4] Hartshorne, *op. cit.*, p. 147.
[5] Salzman, *op. cit.*, pp. 185–6; *Accounts of the Lord High Treasurer of Scotland*, vol. iii, pp. 161, 194.
[6] Hartshorne, *op. cit.*, p. 160.
[7] Unwin, *Gilds and Companies of London*, p. 293.

Glass drinking goblets, and vessels for all purposes, replaced the old stone cups, not only in aristocratic mansions but in cheap inns and ale houses.[1] " The poorest", wrote Harrison in 1586, "also will have glasse, if they may, but sith the Venetian is somewhat too dear for them, they content themselves with such as are made at home ".[2]

To meet these new demands the native glass industry necessarily became commercial. A preliminary attempt to establish in Surrey and Sussex a large manufacture of a more practical nature, with the aid of the so-called *gentilshommes verriers* from Lorraine, met with no more immediate success than had Cecil's scheme for importing German skilled labour for the early saltworks on the east coast;[3] and, indeed, one may observe many parallels between the development of these two industries during the sixteenth and seventeenth centuries. " As for glass makers", says an anonymous writer in 1577, " they be scant in this land ".[4] Henceforth, however, they increased rapidly in numbers, and achieved an ever more skilful mechanical technique. Before 1620 furnaces for turning out glass on a large commercial scale had been successfully established, not only in the original home of this industry in the Chiddingfold region, but in London and Newcastle-on-Tyne, at Stourbridge in Worcestershire, at Wollaton and Awsworth in Nottinghamshire, at Wemyss in Fife, at Leith in Midlothian, and even in commercially backward Ireland—to mention only a few of the most important.[5] Under a well-known patent of 1615, the Crown gave a monopoly of this manufacture in England to a group of Court favourites, among whom the most influential was Sir Robert Mansell, the notorious admiral who soon bought out all his partners.[6] In the face of strong and vocal opposition from independent glass makers,

---

[1] Hartshorne, *op. cit.*, pp. 157, 191–2.

[2] *Description of England*, 1586 ed., p. 167.

[3] The project referred to in the text did not, stictly speaking, bring about the first migration of foreigners to teach the English glass making. Some Venetian glaziers spent the years from 1549 to 1551 in England and imparted a certain amount of their technique to the natives (Hartshorne, *op. cit.*, pp. 147 sqq.). The story of the Lorraine glass makers has been fully told by Hartshorne (pp. 152 sqq.) and by Hyde Price (*English Patents of. Monopoly*, 1906, pp. 67–70). Three of the documents on which both accounts are based (*Lansdowne MSS.*, 59, nos. 72, 75, 76) have been printed by Hartshorne in appendices.

[4] Cited Galloway, *History of Coal Mining*, 1882, p. 37.

[5] Hartshorne, *op. cit.*, pp. 174–5, 177–8 ; *S.P.D.*, *James I*, vol. clxii, no. 63 (printed Hartshorne, pp. 426–31). Galloway, *op. cit.*, p. 38 ; *V.C.H. Worcestershire*, vol. ii, pp. 278–9 ; *H.M.C.*, *Report on the MSS. of Lord Middleton*, pp. 181–2, 499–501 ; *Privy Counc. Reg. of Scotland*, 1st Series, vol. xii, pp. xv–xvi, 374, 428, 439, 440, 451–2, 481, 760, 772 ; *Denmilne MSS.*, vol. ix, no. 13 ; *Lansdowne MSS.*, 59, no. 72 ; R. R. Steele, *Bibliog. of Tudor and Stuart Proclamations*, vol. ii, no. 265 ; and *Exch. Deps. by Com.*, 16 James I, East. 12. The last two documents prove that glass works were operated in Ireland in the reign of James I.

[6] Several writers have investigated the Mansell patent. For the account given in the text I have made use of the following : Gardiner, *History of England*, vol. x, see index ; Hartshorne, *op. cit.*, pp. 187 sqq., 426–33 ; Price, *op. cit.*, pp. 71–3 ; *V.C.H. Surrey*, vol. ii, pp. 301 sqq. ; Galloway, *History of Coal Mining*, pp. 37–8 ; *Additional MSS.*, 12496 (" Reasons for defence of the glass patent ", December 21st, 1621).

Mansell maintained his monopoly (at least in law) until the Civil War.[1] His principal houses were in London and Newcastle, but he apparently made agreements with rival proprietors whereby furnaces continued to operate in Nottinghamshire and at Stourbridge, and probably elsewhere as well.[2] At the same time, he employed every method he could find to crush less accommodating competitors,[3] and in 1639 he succeeded in procuring a proclamation to prohibit all further manufacture in Ireland.[4] But he was never able to suppress his far more formidable rivals at Leith, nor to have legislation passed against the import of Scotch glass into England.[5] Most of the glass produced by Mansell at Newcastle was shipped in cases to other ports ; and a considerable amount must have gone to independent glaziers, spectacle makers, hour-glass makers, and others, to be worked into finished articles, although Mansell said that he also sold window glass, beer and wine tumblers, looking-glasses and spectacles direct to consumers.[6] The quality of his output was much criticized,[7] and modern authorities generally hold—with a substantial show of reason—that his patent proved a hindrance, rather than an aid, to the development of a native manufacture.[8] During the Commonwealth period the recorded shipments of glass from the houses which he had been forced to abandon

---

[1] The patent was renewed in 1635 for a period of 21 years (Steele, *op. cit.*, no. 330).

[2] *H.M.C.*, *Report on the MSS. of Lord Middleton*, p. 500 (Sir Percival Willoughby leases to Sir Robert Mansell (December 8th, 1615) a barn, to be used for glass-works) ; *V.C.H. Worcestershire*, vol. ii, pp. 278–9. An order of the Privy Council, of April 3rd, 1616, authorized Isaack Bungard and Edward Henzey to continue their manufacture of broad glass in Surrey, in spite of Mansell's patent. They were, however, to abandon the enterprise as soon as they had used up the materials which they then had on hand (*Acts of the Privy Council*, 1615–16, pp. 469–70).

[3] *Ibid.*, pp. 471–2, April 3rd, 1616: Warrant to John Brunte to stay glass works erected in Devonshire, Staffordshire, and North Wales contrary to the patent. Price (*op. cit.*, p. 73, note) gives references to a great number of similar cases (investigated by the Privy Council between 1618 and 1625) concerning infringements of the Mansell patent.

[4] Steele, *op. cit.*, vol. ii, no. 330.

[5] In 1619 it was ruled that Scotch glass came within the meaning of the term " foreign " in the act of 1615 (*ibid.*, no. 1164), which forbade the importation of all foreign glass into England. This ruling led the Scottish Privy Council to address to the King a vigorous protest (*Denmilne MSS.*, vol. ix, no. 13), and to threaten to retaliate by prohibiting the export of Scotch coal, then used in the furnaces of the London glass makers (*ibid.*, vol. ix, no. 23 ; cf. above, p. 120). Immediately after this threat, a new proclamation was passed (February 25th, 1619–20), exempting Scotch glass from the embargo on imports (Steele, *op. cit.*, no. 1273).

[6] *The True State of the businesse of Glasse of all kindes*, 1641.

[7] See " Reasons against Mansell's Patent ", 1624 (*S.P.D., James I*, vol. clxii, no. 64, printed Hartshorne, *op. cit.*, pp. 423–6). See also *Cal. S.P.D.*, 1619–23, pp. 323, 330 (attacks on the patent in the House of Commons by Mansell's rivals, John Worrall and Isaac Bungard, etc.), and pp. 129–30 (petition of Ralph Colbourne, hour-glass maker, to be relieved from oppression at the hands of Mansell). The London Glaziers Company, on the other hand, defended the quality of Mansell's glass (*ibid.*, pp. 243, 247).

[8] This is Price's view (*op. cit.*, pp. 74–5). But Hartshorne (*op. cit.*, p. 187) thinks that the granting of this monopoly was a " wise and statesmanlike step " which saved glass manufacture from decay in the hands of " mere lawless artisans ".

at Newcastle [1] did not exceed 3,000 cases—300 tons—per annum,[2] a quantity which he had claimed to exceed as early as 1624.[3] But trade from Newcastle in the early 'fifties had not recovered from the depression caused by the Civil War.[4]

Whatever may have been the course of production during the years when Mansell held the glass patent, it is certain that the manufacture of glass in England increased very rapidly between 1580 and 1615, and that it increased again after the Civil War. Under Mansell only three furnaces had been operated at Newcastle ; in 1696 there were eleven.[5] According to an estimate made in 1589 by George Long, who petitioned for the right to start a manufacture in Ireland, there were fifteen glass furnaces in the whole of England,[6] several of which appear to have been started within a decade of the time when Long wrote. In 1696 Houghton put the number of furnaces at 90, the majority of which were probably capable of turning out a far greater amount of glass than had been possible with an Elizabethan furnace.[7] In addition, the manufacture at Leith on the Firth of Forth rivalled that of the great English glass making centres, London, Stourbridge, Newcastle, and Bristol, both in the quality of its green bottles, chemists' and apothecaries' glasses, and in the quantity produced. As a furnace at Wollaton, in Nottinghamshire, produced about 800 cases a year in 1615, and a furnace at Newcastle during the Commonwealth about 1,000 cases, it seems probable that the annual output of glass in England and Scotland at the end of the seventeenth century may have exceeded 100,000 cases, or 10,000 tons.[8] Probably not a fifteenth

[1] An entry of August 12th, 1653, in the *MS. Journal of the Common Council of Newcastle*, vol. for 1650–60, f. 141, refers to a lease to a Mr. Moyer of one of the glass houses formerly rented by Mansell. In 1658 all three of Mansell's houses were taken over by a partnership consisting of Robert Tainton, Wm. Pollicott, and others (*ibid.*, f. 480). How many years before 1653 Mansell gave up these works is uncertain, though we know from his own statement to Secretary Windebank (*Cal. S.P.D.*, 1640–1, p. 65 ; Hartshorne, *op. cit.*, p. 200), that the Scottish invasion of 1640 forced his workmen to flee, leaving 1,200 cases of glass ready for shipment. Doubtless, they returned later. Mansell, according to the *Dictionary of National Biography*, died in 1656 at the age of 83, but the entry of 1653 in the *Journal of the Common Council* refers to him as already dead. The glass patent was not revived after the Restoration.

[2] 2,706 cases were shipped in the year 1655 (*Exch. K.R. Port Bks.*, 192/9). Ten cases made a ton at Wollaton in 1615 (*H.M.C., Report on MSS. of Lord Middleton*, pp. 500–1). This glass then sold in London at 16s. per case ; in 1640 Mansell valued his window glass at £1 5s. per case (*Cal. S.P.D.*, 1640–1, p. 65).

[3] 3,000 to 4,000 cases annually (*S.P.D., James I*, vol. clxii, no. 63, printed Hartshorne, *op. cit.*, pp. 426–31 : A defence of Mansell's patent).

[4] See Appendix D (i).

[5] *Cal. S.P.D.*, 1640–1, p. 65 ; Houghton, *Husbandry and Trade*, vol. ii, p. 48.

[6] *Lansdowne MSS.*, 59, no. 72 (Petition to Burghley). Long had an interest in exaggerating the number of furnaces in England, for, in order to further his project to establish an Irish manufacture, he wished to impress Burghley with the havoc wrought by English glass houses on the native wood supply (cf. below, pp. 193, 195).

[7] Houghton, *op. cit.*, vol. ii, pp. 48–9. Not all the furnaces were at work.

[8] The output at Wollaton in 1615 is given in *H.M.C., Report on the MSS. of Lord Middleton*, pp. 500–1. The output at Newcastle is an estimate, based on the knowledge that there were three furnaces in the town and that the exports of glass amounted to approximately 3,000 cases a year. It seems probable that before the end of the seventeenth century, some of the furnaces may have produced much more than 1,000 cases annually.

as much glass had been produced before Elizabeth's reign. In a little more than a century, Great Britain had changed from a country almost altogether dependent upon foreign glass for commercial uses, to a country producing more commercial glass than could find a market at home.[1] It was no longer a question, as in Harrison's time, of being "content" with the native product ; for most English glass now equalled in quality any made abroad, and a finer type of crystal glass was turned out in England than has ever been produced elsewhere.[2] That such proficiency in making vessels to ornament the tables or sideboards of rich gentlemen should be obtained at the expense of losing the art which is written in the windows of Fairford Church scarcely concerned the merchants who financed these new furnaces, and whose object was to sell their produce in as large quantities as possible. As in the case of the salt manufacture, the command which the native manufacturers obtained over the domestic market was undoubtedly due to the advantage they possessed in getting cheap supplies of fuel, rather than to the protective policy of the Stuart government.[3]

Everyone must be struck by the parallel development in the three industries just reviewed. In the case of shipbuilding, salt making, and glass manufacture the reign of Elizabeth marks the beginning of a period of far more rapid progress than any which had gone before ; and such quantitative evidence as it has been possible to obtain suggests that the increase in output in all three industries, during the period from 1560 to 1700, was only less rapid than the increase in the output of coal. Time has not permitted us to attempt to make similar quantitative estimates concerning the development of other industries. But the evidence at our disposal suggests that the period was one of rapid expansion in the output of most manufactured commodities.

Much has been written by historians about the establishment of "new" processes of manufacture in Elizabethan England, and, while it may be doubted whether the alum and copperas industries, which provided dyers with ingredients indispensable for their trade,[4] the manufacture of saltpetre and gunpowder, or soap making can properly be described as perfectly novel, it seems clear that the end of the sixteenth century, and the beginning of the seventeenth, saw the first systematic attempt to develop them. Bristol dyers made use of an "Alym de Wyght" as early as 1346, so that it is inaccurate to say that alum stone was discovered in England in Elizabeth's reign.[5] But there can be no doubt that the dyers had been hitherto

[1] In the year ending Michælmas, 1686, 15,750 glass bottles were exported to France (Charles King, *The British Merchant*, 1743 ed., vol. i, p. 276).

[2] Hartshorne, *op. cit.*, p. 184 ; and cf. below, Appendix H.

[3] See below, pp. 218–20, 228.

[4] The chief use of alum was to fix and render dyes more beautiful. It was also used in making gold ware and paper. Copperas was useful in dyeing black, and also as an ingredient in the preparation of writing ink.

[5] Salzman, *English Industries of the Middle Ages*, p. 208. For the reference to the discovery in Elizabeth's reign, see *S.P.D., Eliz.*, vol. xxxvi, no. 72.

largely dependent upon " Romish " alum. The development, during the reign of James I, of important alum works, near the supply of newly found alum stones on the Yorkshire coast, and at Hartlepool in Durham, was a long step towards relieving England from its dependence upon foreign supplies. Much Italian alum was still smuggled into the country, notwithstanding proclamations against imports, and this illicit traffic was likely to continue as long as the manufacturers persisted in their policy of subsidizing the export of alum, by charging a higher price for their product at home than abroad. But the failure of the early patentees was not followed by a failure of the industry. There is abundant evidence in the port books of Whitby, Hartlepool, and Sunderland that the alum works along the north-east coast produced alum in increasing quantities during the seventeenth century.[1] And, before the Civil War, we hear also of a large plant at Queenborough in Kent for the manufacture of copperas.[2] The output of alum and copperas in 1655 was already worth about £50,000 a year.[3]

Charcoal, brimstone, and saltpetre are required for making gunpowder, and, of the three, saltpetre is by far the most important ingredient. It is generally assumed, though we do not know whether the assumption is justifiable, that no gunpowder had been manufactured in England before the establishment of a mill in Surrey in 1554 or 1555. The strategic importance of adequate native supplies of both gunpowder and saltpetre was soon realized by advisers of the Crown, who did what they could, by attempting to shut out foreign supplies, and by declaring a royal monopoly over the collection and preparation of saltpetre, to encourage the native industry. The most famous mills for making gunpowder were those of the Evelyn family at Long Ditton, founded by George Evelyn in Elizabeth's reign; but before the reign of Charles I there were others in England, especially in the neighbourhood of London, and in Scotland. At first these mills probably depended on supplies of Indian saltpetre. But the so-called commissioners for saltpetre had planned a considerable manufacture at least as early as 1588.[4] Armed with a royal warrant, their agents ruthlessly entered private holdings in all parts of the country, to search the incrustations of old buildings for the material requisite for their trade. By 1591, London was receiving supplies from

---

[1] *Exch. K.R. Port Bks.*, 148/3, 4, 16 ; 200. For the history of the early English alum manufacture see J. Beckmann, *History of Inventions*, vol. i, 1797, pp. 184–96 ; Hyde Price, *The English Patents of Monopoly*, pp. 82–101 ; Sombart, *Der Moderne Kapitalismus*, 1916, vol. ii, p. 878 ; W. Cunningham, *History of Industry and Commerce, Modern Times*, 1921 ed., pt. i, p. 293 ; W. R. Scott, *Joint-Stock Companies*, vol. ii, pp. 475–6 ; *V.C.H. Yorkshire*, vol. ii, p. 384. Among the very extensive source material I have met with, mention may be made here of *Lansdowne MSS.*, 152, which is practically devoted to the alum industry ; *Exch. Deps. by Com.*, 9 Charles I, Mich. 11, and 11 and 12 Charles I, Hilary 17 ; *Exch. Spec. Com.*, nos. 2684, 4465, 5789, 6548 ; *Star Chamb. Proc.*, *James I*, 30/18 ; *S.P.D.*, *James I*, vol. lxxiv, nos. 19–21.

[2] Sir William Brereton, *op. cit.*, p. 2. Copperas was certainly manufactured in England as early as 1593 (*S.P.D.*, *Eliz.*, vol. ccxlvi, no. 65).

[3] *S.P.D.*, *Interregnum*, vol. xciv, no. 106.

[4] *Lansdowne MSS.*, 58, no. 63 ; *Harleian MSS.*, 1926, no. 118.

Hull ; and henceforth the industry developed rapidly in many parts of England, and, somewhat later, in Scotland.[1]

Soap making had not been an important native industry before the reign of Elizabeth ; and there is no doubt that the growth of textile manufactures had been handicapped by the lack of a commodity so essential in scouring the wool before it could be made into cloth. In the thirties of the seventeenth century between 5,000 and 10,000 tons of soap was being produced annually for the English market.[2] Before the end of the century it was estimated that about 5,000 tons was consumed in London alone.[3]

The expansion of the trade in alum and copperas, saltpetre and gunpowder, and in soap, was accompanied by the introduction of tobacco pipe making as an important industry at Broseley, on the Severn, and in London, and by attempts to refine the raw sugar which was brought into the country in increasing quantities from the Colonies.[4] Beer brewing, the distillation of spirits, and the preparation of the ingredients used in both assumed a greater importance in the national economy. Hops, once largely imported from Artois, are said to have been first planted in England in 1524. By 1700, the preparation both of hops and malt [5] had reached a higher degree of perfection at home than anywhere on the Continent.[6] The country was now independent of foreign trade for nearly every kind of drink except wine, and certain native beverages, such as Dorset beer or Bristol " milk ", were thought to rival the most delicious draughts to be obtained abroad.

While the introduction of a number of new, or quasi-new processes, resulted in the development of manufactures, some of which had scarcely existed in Great Britain before the sixteenth century, conditions within the country were bringing about a great increase in the amount of manufacturing done in connection with two of the most ancient industries, building and textiles. If the rising price of timber encouraged the substitution of lead for wooden roofs, it offered an even greater encouragement to the erection of houses built partly

---

[1] *Exch. K.R. Port Bks.*, 9/3. On the saltpetre and gunpowder industry see Beckmann, *op. cit.*, vol. ii, pp. 487 sqq., 509 sqq. ; Cunningham, *op. cit.*, pp. 60–1, 291 ; *V.C.H. Surrey*, vol. ii, p. 246 ; John Evelyn, *Misc. Works*, 1825, p. 689 ; *Dict. Nat. Biog.*, under John Evelyn ; Scott, *op. cit.*, vol. ii, pp. 471–2 ; Agricola, *De Re Metallica*, ed. Hoover, pp. 561–4 ; *Privy Counc. Reg. of Scotland*, 1st Series, vol. x, p. 82, and vol. xi, pp. 275, 306, 322 ; 2nd Series, vol. i, p. 377, and vol. ii, pp. xxxii–iii, 333–4 ; *Acts of the Privy Council*, N.S., vol. xi, pp. 246, 249 ; *S.P.D., Charles I*, vol. clxxx, nos. 8–15 ; vol. cccxli, no. 68 ; vol. cccliv, no. 168 ; vol. ccclxxvi, nos. 147–154 ; vol. Dxxix, nos. 58, 88 ; vol. Dxxx, no. 45 ; vol. Dxxxi, nos. 43–5 ; vol. Dxxxviii, no. 51 ; *Cal. S.P.D.*, 1645–7, p. 411 ; *ibid.*, 1651–2, pp. 253, 256, 274, 285.
[2] Gardiner, *History of England*, vol. viii, pp. 71–3 ; *S.P.D., Charles I*, vol. cclxxix, no. 72.
[3] Houghton, *Husbandry and Trade*, vol. i, pp. 349–50.
[4] On pipe-making, see *V.C.H. Shropshire*, pp. 440–1, and below, p. 218. In spite of the effort made to develop sugar refining in England and Scotland, at least a third of all sugar imported from the Colonies, at the end of the seventeenth century, was re-exported raw (G. L. Beer, *Commercial Policy of England towards the American Colonies*, 1893, p. 56).
[5] For the improvement in the quality of malt, see below, pp. 215–6.
[6] Beckmann, *op. cit.*, vol. ii, pp. 384–6.

of brick and stone and mortar, instead of wood. Harrison commented on the change in building materials that had taken place during his lifetime.[1] As a result of the change, two manufactures, those of brick and lime, assumed an importance out of all proportion to that which they had possessed when stone construction work had been largely reserved for the great cathedrals and the castles of rich landowners. Brickmaking became the constant occupation of considerable numbers of workmen, and the limekilns were not only set up in towns, but were scattered in increasing numbers over the countryside. For lime was also wanted in ever greater quantities to nourish arable ground, now that the advantages of artificial fertilization were beginning to be understood.[2] Before 1700 nearly all husbandmen, even in remote districts where a primitive economic life still prevailed, had to have a good supply of lime " for manuring [their] . . . lands and [for] the reparacons of their several tenements ".[3] Every large farm had its kiln. We have only to read contemporary treatises on husbandry, and accounts of travellers or discourses of natural scientists, to be convinced that lime was becoming a product of rapidly increasing importance throughout the seventeenth century.[4]

Materials are not available with which to measure even roughly the rate of growth in the textile industries. To-day the generally accepted view is that they have increased in magnitude from the thirteenth or fourteenth centuries down to the present day, and that progress was steady but relatively slow until 1660, when the jog trot becomes a gallop, to turn into a race in the period of the Industrial Revolution. There can be little doubt that the progress of woollen cloth manufacture differed from that of most of the industries discussed in this chapter. In the Middle Ages its relative importance was much greater. Records of cloth exportation suggest that the woollen industry developed rapidly between the middle of the fourteenth and the beginning of the sixteenth centuries.[5] But it would be a mistake to assume, because the growth in exports of cloth during the late sixteenth and seventeenth centuries was less striking,[6] that the progress in clothmaking had slackened. It is not in the export trade that we are likely to find a full reflection of the growth of industrial enterprise during the seventeenth century. Production was carried on above all for the domestic market, which appears to have exhibited new wants, many of which were fulfilled. The last half of the sixteenth

---

[1] Harrison, *Description of Britain*, 1577, bk. ii, chap. x. Cf. Owen, *Description of Pembrokeshire*, p. 76 ; Sombart, *op. cit.*, vol. i, p. 775 ; F. J. Furnivall, *Harrison's Description of England* (New Shakspere Soc., Series vi, no. 5), p. xlii.

[2] Fitzherbert, *Surveyinge*, 1539, chap. xxxiv.

[3] *Star Chamb. Proc., Henry VIII*, vol. v, f. 21.

[4] *Chanc. Proc., James I.*, C/24/5 ; *H.M.C., 4th Report*, appx., p. 475, *b* ; Thos. Pennant, *Tour in Scotland*, 1771, p. 3 ; Sinclair, *Statistical Account of Scotland*, vol. i, p. 373 ; vol. v, p. 346 ; vol. xii, pp. 102–3, and *passim* ; MS. *Journal of the Common Council of Newcastle*, vol. for 1650–60, ff. 469, 484, and *passim*. Some English lime was exported to the Continent as early as Elizabeth's reign, notwithstanding the heavy cost of transport (*Lansdowne MSS.*, 79, no. 90).

[5] A. Friis, *Alderman Cockayne's Project and the Cloth Trade*, 1927, pp. 10–12.

[6] H. Heaton, *The Yorkshire Woollen and Worsted Industries*, 1920, p. 258 ; cf. chap. vi.

and the early years of the seventeenth centuries were evidently a period of great importance for textile manufacture as well as for other industries. When Cunningham made his generalization that England had been successively a great corn-growing, a great wool-raising, and a great coal-mining country, it was the period of the Tudors and Stuarts with which he specially associated the ascendancy of wool, and the earliest historian of the worsted manufacture tells us that the accession of Elizabeth marks " a fresh starting point ".[1] A recent authority, speaking for Lancashire, thinks that the closing years of the sixteenth century and the first years of the seventeenth were marked by changes in the textile manufactures second in importance "only to those of the last third of the eighteenth century".[2] While the actual growth in the output of cloth was probably slower than that of most other commodities discussed in this chapter, it is not unlikely that certain branches of the textile industries such as dying, which must have been stimulated by the progress of alum manufacture, progressed almost as rapidly as glass melting and salt boiling.[3]

Whatever may have been the actual course of textile production, it seems certain that the period from 1560 to 1700 was one of unprecedented growth in many industries besides coal mining. By 1700 Britain had already become a country of " manufactures ". If one were to take at random a certain number of tracts on economic subjects published between 1550 and 1575, and an equal number published between 1675 and 1700, it is probable that the word " manufacture " would appear ten times in the latter to once in the former. By 1700 some writers had a consciousness of Britain's industrial destiny which is too commonly supposed to have originated with Adam Smith. When the mercantilist doctrine enjoyed the fullest measure of prestige, we find an enlightened—if perhaps ambiguous—thinker urging the importance of " taking off all Restraints, and giving due liberty to Manufacturers and alluring them Home ; in encouraging and improving those advantages which are in a manner peculiar to us ; in discouraging and clogging those Trades which draw away our Treasure ; in keeping a good Correspondence with those Kingdoms and Countries whence we derive Materials for our Manufactures, and those which take off our Natural Products, Manufactures and Artificial Commodities ".[4]

What is new and significant in this quotation is the proposition that England shall live, to some extent, on her surplus manufactures. That would have been a purely hypothetical proposition a hundred and fifty years earlier ; for then most British industries could only partly supply the home market for finished goods, and, however

---

[1] John James, *History of the Worsted Manufacture in England*, 1857, pp. 104–5.

[2] Wadsworth and Mann, *The Cotton Trade and Industrial Lancashire*, 1931, p. 11.

[3] On the use of soap in textile manufacture, see Misson, *Mémoires . . . par . . . un voyageur en Angleterre*, 1698, p. 383.

[4] Walter Harris, *Remarks on the Affairs and Trade of England and Ireland*, 1691, p. 56.

valiantly statesmen and their advisers might strive to prevent all export of wool, lead, or tin in a raw or half-finished state, such exports were inevitable, if the country was to pay for finished goods imported. The industrial backwardness of Scotland before the seventeenth century is notorious. More than ninety per cent of all articles exported as late as 1614 were either raw or unfinished.[1] " For the country, I must confess it is too good for those that inhabit it and too bad for others to be at the charge of conquering it ", wrote a certain English visitor, Sir Anthony Welldon, who is said to have suffered banishment from Court for his blasphemous reflections on the King's native land.[2] And although England, closer to the Continent, stood a step higher in the industrial scale, it would probably have seemed at the beginning of the sixteenth century as primitive, judged by the sophisticated standards of a trader from Florence or Antwerp, as Scotland a century later seemed to Welldon. By 1700, however, England and, to a less degree, Scotland had taken rank with the leading countries of the world, both in the quantity and the quality of their manufactured goods ; and this development is reflected in the writings of contemporary English authors who touch on economic affairs.

What has this early growth of British manufactures to do with coal ?   The answer is that without coal it could never have taken place.   Without coal an expansion would have begun, but it could hardly have continued.   Already by the end of Elizabeth's reign the development of industry had begun to make great inroads upon the supply of timber, and it is evident that what was causing a critical shortage was the increase of industrial enterprise.

---

[1] *H.M.C., Report on the MSS. of the Earl of Mar and Kellie*, pp. 70–4, has a " table of Scottish produce exported yearly ", as submitted to James I in 1614.   The total value of all articles exported is given as £736,986 Scots (£1 sterling equals roughly £12 Scots).   " Wairris and commodaties that the land yieldis " account for £375,085 ; " Commodaties that the sea randeris yeirlie ", £153,354 ; " Foirrane wairis re-exported ", £39,047.   " Commodaties that ar maid and wrocht in the countries quhairby the peopill ar sett to labour " amount to £169,097 ; but only £50,000 to £60,000 worth of these commodities (notably salt and gloves) were exported as finished goods.

[2] *Ibid., Report on the MSS. of Lord Middleton*, pp. 184–5.

CHAPTER III

THE SUBSTITUTION OF COAL FOR WOOD

THE expansion of industry, and particularly the expansion of the woollen industry, diminished the space available for planting new trees. Ever since the time of Thomas More, who was among the first to comment on the increase in sheep farming, much attention had been paid to the conversion of arable fields to pasture, especially in the sixteenth century. But, if the wool growers encroached upon the corn fields, they also encroached upon the woodlands. They did so not only directly, but also indirectly, since, for acres of arable turned to grass, acres of waste including some woodland were turned to arable.[1] Notwithstanding persistent complaints of the decay of husbandry, it seems likely that not less land was devoted to arable farming in 1650 than in 1500 ; for the native population still depended, except in time of famine, on English-grown grain, and, although the spread of more efficient methods of fertilization had no doubt increased the yield per acre, there was now a considerably larger population to feed.[2] It may be suggested, therefore, that, at least until 1700, it was less the supply of grain than the supply of timber which suffered through the extension of enclosure.

In all counties near the sea, writes an anonymous authority towards the end of Elizabeth's reign, " most of the woods are consumed, and the ground converted to corn and pasture ".[3] When Owen examined old Pembrokeshire records in 1603, he discovered that what then were great corn fields, once had been forests.[4] Landowners thought it far more profitable, after their trees had been cut down, to cultivate the ground or set sheep to graze upon it, than to plant new trees, which took years to mature. All through the seventeenth century they continued to allow old woodlands to be turned into corn fields or pastures, in spite of the lamentations of Evelyn and others who deplored the decline of forestry, and did so with the full approval of the school of economic thinkers represented by John Houghton, who saw in wood planting a positive waste, and who advocated the destruction of all trees and shrubberies within twelve miles of the sea or of navigable rivers, in order that the land might be converted to other uses.[5] Sheep raising, as has often been shown,

---

[1] E. F. Gay, " Inclosures in England in the Sixteenth Century ", in *Quart. Journ. of Econ.*, vol. xvii (1903), p. 589.

[2] Possibly the complaints of the decay of husbandry were due rather to the increase in the number of persons who had to have bread than to any falling off in the supply of grain.

[3] *H.M.C., Report on the MSS. of the Marquis of Salisbury*, vol. xiv, pp. 330–1.

[4] *Description of Pembrokeshire*, p. 86.

[5] Evelyn, *Sylva, or a discourse of Forest Trees, and the propagation of Timber*, 1664 ; Houghton, *Husbandry and Trade*, vol. i, pp. 99 sqq., and vol. iv, pp. 258 sqq., 399–400.

provided lords of manors with one powerful motive for enclosing land, and these enclosures were the more serious for the peasant, because they cut him off from supplies of wood and turf, which in former times he had been permitted by custom freely to take from the commons to repair his house and make his fire.[1] Thus enclosures often precipitated a timber crisis in country districts.

If the growth of the woollen industry in particular discouraged the planting of trees, the demands of industry in general quickly drained the existing forests of most of their timber. For in that age wood was the raw material of all industry to an extent which it is difficult for us now to conceive. Charcoal had to be mixed with saltpetre in preparing gunpowder. From the bark of trees workmen extracted a sap, then indispensable in making pitch and tar, with which to caulk the hulls of ships, and from wood ashes came potash, an essential constituent for the production of soap, glass, and saltpetre. The principal drain caused by the expansion of industry arose, of course, not from such relatively unimportant uses of wood as these, but from the demands for it as a building material and as a fuel.

Ours has been often called an age of coal and iron, and it is perhaps no less appropriate to call the sixteenth and seventeenth centuries an age of timber.[2] Metal still entered only to a small extent into the construction of containers of any kind, of looms or spinning wheels, of carts or coaches, of ships or small river craft. All tools were of wood, except the actual cutting edge or striking face. Muskets had wooden shafts and ramrods, cannons wooden carriages. Although housebuilders worked with brick, stone, and lead more than in the past, they continued to use much timber ;[3] and the Great Fire is an unhappy testimonial of the continued predominance of wooden construction. The expansion of every industry meant an increased demand for timber ; first in buildings, then in tools and machines, finally in containers and conveyances to carry away the product. Sometimes this demand took the form of a deep bite into the native stock of trees, such as was required to build a long bridge, or a large sea-going vessel like a man-of-war, whose hull, decks, and sides might be fashioned out of four thousand great oaks ;[4] sometimes it took the form of a mere nibble, as in making a beer barrel or a farmer's scythe. Enough nibbles, however, could make a bite, as is shown by an Act of Parliament passed in 1593, which required all exporters of beer either to return the original barrels, or to bring back from abroad an equivalent quantity of clap-board fit to make casks.[5]

Even more disastrous to forest conservation than the demand for building timber in industry, was the demand for charcoal and wood to burn. Few, if any, manufactured goods can be produced

[1] See, e.g., H. Stocks, *Records of the Borough of Leicester, 1603–60*, pp. 240–1.
[2] Cf. W. Sombart, *Der Moderne Kapitalismus*, 1916, vol. ii, pp. 1138–40; *Stow's Annals*, ed. E. Howe, 1631, p. 1024.
[3] Even in building churches and cathedrals (*Cal. of Letters and Papers of Henry VIII*, vol. xiv, pt. ii, p. 109; W. Cunningham, *Growth of Industry and Commerce, Modern Times*, 1921 ed., pt. i, p. 523, note ; Owen, *op. cit.*, p. 76).
[4] Sombart, *op. cit.*, vol. i, p. 768.
[5] 35 Eliz., cap. 11.

without fuel; generally it is an important element in the cost of production; sometimes it is the chief element. Shipbuilding in the Elizabethan age—to take one example—required heat, not only to smelt, forge, and shape all metal work, but to dry planks and bend them into proper shapes to fit the curving hulls and sides.

The production of woollen cloth, which to the layman might appear to require next to no fuel, involved the consumption of timber on a considerable scale. On the one hand, there were a number of articles, essential to the trade, which could be made only with the aid of fire: such as shears for clipping the fleece from the sheep's back, alum for fixing the dyes and rendering them more luminous, soap for scouring and scrubbing the wool, wire cards and combs for carding and combing it. On the other hand, clothiers themselves required fuel to carry out several necessary steps in clothmaking; the copper cauldron in which they cleaned the wool after sorting, the leaden vat in which they dyed it, and, finally, the stove in which, while combing, they had constantly to warm and re-warm the teeth of their combs, all had to be heated.[1] The manufacture of these cauldrons and vats also involved a large consumption of fuel. Again, when it came to working cloth into wearing apparel, fuel had to be provided for steaming felt hats, making buttons, and many other purposes. It is not surprising to find that at Wollaton in Nottinghamshire, in 1589, the woods had been sensibly diminished by use in " molting ", clothing, dyeing, and " jarnsey " [jersey].[2] Some decades later, inhabitants of the little town of Cranbrook in Kent complained to the Privy Council of a crisis in their trade of cloth-making, brought on by the high price asked of the dyers for firewood to burn beneath their vats.[3] In short, cheap fuel supplies were of considerable importance to the textile manufacturers.

It is unnecessary to dwell at length on the many thousand uses for firewood in early industry.[4] It is sufficient to point out that no change could be wrought in ore or metal without the aid of fuel; that substantial quantities of wood and charcoal were being consumed in making starch, refining sugar, baking bread, firing pottery, tiles, bricks, and tobacco pipes, drying malt and hops, and boiling soap. Attention has been called to the persistent complaints of the exhaustion of timber supplies. It is evident that the process must have been greatly accelerated by the extension of fuel-consuming industries.

There are cases in which we can follow in some detail the drain on the timber supply caused by the expanding manufactures. In the late sixteenth and early seventeenth centuries ovens, furnaces, and boilers appear to have been (as for some time after they continued to be) amazingly spendthrift, owing partly, no doubt, to a lack of experience in methods of economizing fuel.[5] Three London brewers are said to have consumed 2,000 wagon loads of firewood in the year 1578, and, if the other brewers used as much, the annual

---

[1] Heaton, *Yorkshire Woollen and Worsted Industries*, pp. 333–5.
[2] *H.M.C.*, *Report on the MSS. of Lord Middleton*, p. 499.
[3] *S.P.D.*, *Charles I*, vol. ccclxiii, nos. 55, 56.
[4] For further information on the point see below, pp. 202 sqq.
[5] Cf. below, pp. 217–8.

consumption of wood for brewing in the city must have approached 20,000 wagon loads.[1] More than one load was required to boil down two hundredweight of saltpetre, or to bake 2,000 bricks.[2] On an average about four loads were used at Droitwich in the 'seventies to reduce brine water to a ton of salt,[3] and at least as much had to be provided to burn the limestone to make a ton of lime. Since two wagon loads probably cleared away the trunk, if not the branches, of a fair-sized tree, it is evident that every increase in the quantity of bricks or saltpetre, lime or salt manufactured with wood fuel involved serious new encroachments upon native timber resources.

But the chief wastage was caused by glass makers and smelters. Glass making required just the sort of timber that best served the shipbuilders. According to Sir Robert Mansell " no wood [was] fit to make glasse, under twenty years groath ".[4] One glazier, Giacomo Verzelini, a fugitive Venetian, is reported to have used 400,000 billets of wood, amounting perhaps to 2,000 wagon loads, per annum.[5] That is a greater quantity than most iron works consumed ; but iron works were, of course, far more numerous than glass works. One of them in Glamorgan burned, in making charcoal, at least 3,000 cords of wood annually.[6] To convert ore into a ton of bar iron, in Sussex, Shropshire, Nottinghamshire, Staffordshire, Monmouthshire, or the Forest of Dean, apparently required about 12 loads of charcoal, made from 41 cords (5,125 cubic feet) of wood, the equivalent of eight young beech trees, one foot square " at the stubbe ".[7] A

---

[1] *S.P.D.*, *Eliz.*, vol. cxxvii, no. 68. Seven years later the brewers in the " Cittie, Subbarbs, and Westminster " are said the have numbered twenty-six (*Lansdowne MSS.*, 71, no. 28).

[2] It was estimated in 1589 that the wood consumed in preparing 5,000 cwt. of saltpetre amounted to 3,000 wagon loads (*Lansdowne MSS.*, 56, no. 63). On bricks see Sir Balthazar Gerbier, *Counsel and Advice to all Builders*, 1663, pp. 52–3.

[3] 6,000 wagon loads of wood were spent in making 1,600 wagon loads of salt (Harrison, *Description of Britain*, 1577, bk. iii, chap. xvii). The boiling down of seawater required even more fuel than the boiling down of brine (see below, p. 207).

[4] *S.P.D.*, *James I*, vol. clxii, no. 63 ; printed Hartshorne, *Old English Glasses*, pp. 426–31.

[5] *Ibid.*, pp. 174–5.

[6] *H.M.C.*, *Report on the MSS. of Lord de L'Isle and Dudley*, p. 29.

[7] This included the expense of wood both at furnace and forge. See " The Charge of the makinge of a tonne of Iron in Canke Wood " (i.e. Cannock Chase), November 25th, 1588 (*Lansdowne MSS.*, 56, no. 36).

" [At the Furnes] :—

Wood for 7 lode of Coles for blowinge one tonne and a half sowes at $3\frac{1}{2}$ Cordes to everie lode of Coles $24\frac{1}{2}$ Cordes at 12d. *le* corde	24/6
For cutting $24\frac{1}{2}$ cordes : at 5d. *le* Corde . . .	10/2$\frac{1}{2}$
For Colinge those 5 lodes at 2/– *le* lode . . .	10/–
For Carriege of the said 5 lodes at 2/– *le* lode . .	10/–
For Carriege the Cordes into the pitt at $\frac{1}{2}d$. *le* corde .	8d. [*sic*]
Colliers wages at 100/– per annum	1/–
Founders wages at 10/– *le* foundrey . . .	2/6
The Fyllers wages for the lyke tyme . . .	2/6
Repairinge the furnes bellowe lethers and such like .	2/6
Clarkes wages at £10 per annum	2/–
Myne readie laied at the pitt 5 lode at 9/– the lode .	45/–

ton of wrought lead, tin, or copper represented a loss in wood nearly as large. And, as the annual output of metal of all kinds may have exceeded 35,000 tons early in the seventeenth century,[1] it is safe to assume that smelters then felled at least 200,000 stout trees a year. One anonymous writer in 1611 estimated the annual cost of firewood consumed in iron works alone at £800,000.[2] " Nature ", wrote Evelyn, " has thought fit to produce this wasting ore more plentifully in woodlands than any other ground, and to enrich our forests to their own destruction ".[3]

" There is one man ", wrote an inhabitant of Weardale in 1629, " whose dwellinge place is within twenty miles of the cittye of Durhame, which hath brought to the grounde . . . above 30,000 oakes in his life tyme ; and (if hee live longe) it is to be doubted that hee will not leave us so much tymber or other woode in this whole county as will repaire one of our churches, if it should fall, his iron and lead works do so fast consume the same ".[4] Such a spoil of timber, whether pursued by smelters or by glaziers, could end

[At the] Fordge :—

Wood for 5 lode of Coles for hammeringe one tonne at 3½ Cordes to everie lode of Coles 17½ Cordes at 12d. le Corde	17/6
Cuttinge of 17½ Cordes	7/3
For Colinge 5 lodes 2/-	10/-
Carrienge of everie lode of Coles from the pitt to the hammer 2/-	10/-
Hammermans wages	16/8
Carrienge to the heape at ½ le lode	2½d.
Labourers and extraordinary charges . . .	2/6
Repairinge etc.	1/-
Keep of the Iron at 100/- per annum . .	1/-
Carrienge one tonne and a half sowes from the Fordge to the Fornes at 2/- le tonne . .	3/-
Total	£9

The estimate of the number of beech trees used in making twelve loads of charcoal is based on *Exch. Dep. by Com.*, 22 Eliz., Trinity 4 (as cited Salzman, *English Industries of the Middle Ages*, p. 40).

The assumption that the wood required for smelting a ton in Staffordshire was about the average amount generally required at this time is based on evidence concerning the consumption of wood at furnaces and forges in Sussex (as given in *V.C.H. Sussex*, vol. ii, p. 247), Wyre forest, Shropshire (*V.C.H. Shropshire*, p. 460), Nottinghamshire (*H.M.C.*, *Report on the MSS. of Lord Middleton*, pp. 494–5), Monmouthshire (*Exch. Spec. Com.*, no. 1518), and the Forest of Dean (*Harleian MSS.*, 7009, no. 16). To make a ton of ordnance is said to have required ten cords of wood less than to make a ton of bar iron.

[1] In 1636 the annual output of English lead is said to have reached 12,000 tons (see above, p. 167). The output of iron in 1611 was probably even greater (see above, pp. 168, 169 n.). Less than 500 tons of smelted tin were then produced, according to the Stannary records (Lewis, *The Stannaries*, Appendix J). I have found no estimate of copper production.

[2] *Harleian MSS.*, 7009, no. 5. He estimated the number of mills at 800, and said that "falling, billeting, and coaleing and carriage" cost each mill £1,000 per annum. His figures are no doubt highly exaggerated.

[3] Evelyn, *Sylva*, quoted John Holland, *The History and Description of Fossil Fuel*, 1835, p. 323.

[4] A. L., *A Relation of some Abuses which are committed against the Commonwealth*, 1629, p. 9 (printed in *Camden Miscellany*, vol. iii, 1855).

only in the complete exhaustion of local forests, unless, as in Warwickshire, the commons, incensed over the rising price of wood, " rose in tumults and expelled the glass makers by force ".[1]

" As the woods about [here] decay ", relates a native of Worcester-shire in the nineties of the sixteenth century, " so the glasshouses remove and follow the woods with small charge ".[2]   That was the common practice throughout England and Scotland.   Norden in 1607 observes that, during the past thirty years, such havoc has been wrought by iron and glass makers in the famous Wealds of Sussex, Surrey, and Kent, " the grand nursery of . . . Oak and Beech ", that he fears " lest fewe yeeres more, as pestilent as the former, will leave fewe good trees standing in those Welds ".[3]   In Sutherland and Ross-shire, George Hay, one of the most active promoters of early Scottish industry, is said to have " kept a colony and manufactury of Englishmen making iron and casting great guns, until the wood of it was spent ".[4]   One reason for the decline of copper smelting at Keswick towards the end of Elizabeth's reign was the exhaustion of the woods fit for charcoal.[5]

Obviously the country could not go on indefinitely exhausting a commodity then ranked among the chief national assets. " It hath bene . . . esteemed as a principall Patrimonie of this our Realme ", begins a royal proclamation of 1615, " that it hath yeelded goodly quantitie . . . of Wood and Timber . . . ; the timber is not only great and large in height and bulk, but hath also that toughness and heart, as it is not subject to rive or cleave, and thereby of excellent use for shipping, as if God Almightie, which had ordained this Nation to be mighty by Sea and navigation, had in his providence indued the same with the principall materiall conducing thereunto ".[6]   Some decades earlier, spoil made by iron works and glass furnaces had begun to alarm Elizabethan statesmen, for every year enough trees were destroyed by smelting to rebuild the entire royal navy.[7]   The consumption of wood in glass manufacture, the proclamation of 1615 proceeds, " does so import the state of this Our Kingdome as it were the lesse evil to reduce the times unto the ancient manner of drinking in Stone, and of Latice-windowes, then to suffer the losse of such a treasure ".[8]   And, indeed, had there been no other way of meeting the problem of deforestation, the burst of industrial enterprise which accompanied the later years of Elizabeth's reign, must have died in its infancy.

To substitute other material for timber in building or carpentry work was to leave the problem unsolved, so long as it required more wood to forge the iron spokes of a wheel than to make them with

---

[1] *V.C.H. Warwickshire*, vol. ii, p. 244.

[2] *Ibid.*, vol. ii, pp. 248–9.

[3] *The Surveyor's Dialogue*, p. 214

[4] D. W. Kemp, *Notes on Early Iron Smelting in Sutherland*, 1887, p. 5 ; W. I. MacAdam, " Notes on the Ancient Iron Industry of Scotland ", in *Proc. Soc. of Antiq. of Scot.*, vol. xxi, 1887, p. 89.

[5] W. G. Collingwood, *Elizabethan Keswick*, 1912, pp. 13–14, 93.

[6] Royal Proclamation of May 23rd, 1615, printed Hartshorne, *Old English Glasses*, pp. 413–14.

[7] *Ibid.*, pp. 157–9   [8] *Ibid.*, p. 413.

timber, and more wood to bake the materials needed to erect a house in brick, than to build it of timber throughout. Only by using a substitute for wood fuel could the country hope to maintain and expand its manufactures. "Of colemines", observed Harrison in 1577, " we have such plentie in the north and westerne parts of our Island as may suffice for all the realme . . . ; and so must they doo hereafter in deed, if wood be not better cherished than it is at this present ".[1] The want of timber, as an anonymous writer maintained a few decades later, " is and can only be supplied by our use and preservation of our Cole Mines ".[2] Thus it was that the replacement of wood by coal became a principal aim of state policy and inventive effort.

It was undoubtedly the rapid spread of the use of coal in warming rooms, in cooking food, and in " all other common services requisite in housekeeping," [3] that occasioned the chief demand for it, and that relieved in greatest measure the drain on the timber supplies of Britain during the seventeenth century. The relative cost of coal as compared with other fuels determined the extent to which it replaced them in the domestic hearth and the kitchen oven. As substitutes for coal were generally dearest where population was most dense, more coal was probably burned per head in London than in Cambridge, when its price was the same in both places.[4] Again, the use of coal spread more slowly among the rich than among the poor, who could less afford to pay the extra price required to obtain wood, and who had to content themselves with a less pleasant fuel.[5]

In London, where population tended more and more to concentrate, wood remained almost the only fuel burned, even by the poorest housekeepers, until the general rise in timber prices after 1540, notwithstanding that the Company of Fishmongers had agreed in 1446 to distribute every year to the needy 30 or 40 tons of coal at less than cost price.[6] During a serious fuel crisis in the winter of 1542–3 the Lord Mayor wrote : " I am daily at every wharf where wood lyeth and distribute to the poor at a reasonable price as much as will go around, to the great loss of the owners [of wood] as they affirm ".[7] He said nothing about coal.[8] Not until 1554 did the Common Council vote a fifteenth, to be levied on all citizens, " towards provision of sea coles from time to time to be provided and brought to this city to be kept in a stock for ever as well for the

[1] *Description of Britain*, 1577, bk. iii, chap. xvi.

[2] *S.P.D., James I*, vol. clxxx, no. 77.

[3] *Lansdowne MSS.*, 59, no. 71.

[4] Cf. above, pp. 88–9.

[5] For the probable exception to this rule in the case of Scotch coal, see above, pp. 12, 118–9.

[6] 132 " quarters " (H. Humpherus, *History of the Company of Watermen and Lightermen*, 1887, vol. i, p. 49).

[7] To the Privy Council (*MS. Journals of the Common Council of the City of London*, vol. xv, f. 6).

[8] But at about this time the supplies of mineral fuel provided for the poorer citizens must have been increased. Sir John Allen, a lord mayor, who died in 1544, " gave 500 markes to bee a stocke for sea coale " (Stow, *Survey of London*, ed. C. L. Kingsford, 1908, vol. i, p. 112). Cf. below, vol. ii, pp. 260–1.

succor of the poor when need shall require as for all other inhabitants ".[1]
Only when Harrison wrote, in 1577, could it be said that the " greatest
trade " in coal had begun " to grow from the forge into the kitchen
and hall ".[2]   In 1595 a complaint against the rising price of coal was
made on behalf " of the poor who do use the same for their chiefest
fuel", and a few years later Attorney-General Hobart said that
without coal the " poore laboringe persons and ordinarie handy
crafts men are not hable to maynetayne theire houses and famylies ".[3]
Henceforth, the citizens of the capital adopted the new fuel with a
rapidity of which the statistics of London imports, as given in an
earlier chapter, are sufficient evidence.[4] Coal is described at the time
of the Civil War as a commodity " absolutely necessary to the
maintenance and support of life ", and by 1676, according to Petty,
it was generally burned in chambers in London.[5]

It would be a mistake to think that its use for domestic fuel
was confined to the large towns. The growing market [6] for Welsh
coal towards the end of the seventeenth century throughout the
south-western counties of England, where it was used " as well for
burning of lime as alsoe for fireinge ",[7] indicates that country householders, even at a considerable distance from the pits, had begun
to turn to coal. Along the east coast of Ireland from Carrickfergus
to Youghal coal is said to have been by 1636 " almost the only
material used for firing ".[8] It had become indispensable in a small
port like Aldeburgh by the middle of the seventeenth century, for
during the Civil War, when the supply of Newcastle coal was shut off,
the inhabitants, being unable to get enough timber for their fires,
were driven to burn flag and heath.[9] Christopher Merret, the
surveyor of the port of Boston, has left an account, written in the
last decade of the seventeenth century, telling how the poor people
of Lincolnshire, unable to buy fuel, wander along the seashore, in

---

[1] *MS. Journals of the Common Council of the City of London*, vol. xvi,
f. 279*b*.   This action of the city government may have been brought about by the
Privy Council, which had ordered in 1551 a survey of all the London wharves,
for the purpose of estimating the total stock of " sea coal " then in the City
(*Acts of the Privy Council*, N.S., vol. iii, pp. 313–14).

[2] *Description of Britain*, 1577, chap. xvi. Coal was not yet very extensively
burned in London, however, as Harrison himself implies in another passage.

[3] *Acts of the Privy Council*, N.S., vol. xxv, pp. 31–2 ; *Star Chamb. Proc.*,
James I, 13/2.

[4] See above, Table II, and below, Appendix D (iv).   Not all coal imported
was consumed within the city (see above, p. 83 n.), nor was all coal consumed
within the city burned in domestic fires (see below, pp. 200 sqq.).

[5] Firth and Rait, *Acts of the Interregnum*, vol. i, p. 171 ; Sir Wm. Petty,
Economic Writings, ed. C. H. Hull, 1899, vol. i, p. 304.

[6] R. G. Gough, *Anecdotes of British Topography*, 1768, p. 296, note 6 ;
Elizabeth Godfrey, *Home Life under the Stuarts, 1603–49*, 1903, p. 237. There
remained, of course, inland districts remote from the pits, like the Scottish
Highlands, rural Hertfordshire, or Oxfordshire, where wood and peat remained
almost the universal fuels. See J. P. Bünting, *Sylva Subterranea*, 1693, chap.
xiii ; P. Kalm, *Account of his Visit to England in 1748*, trans. by J. Lucas,
1892, p. 355 ; Adam Smith, *Wealth of Nations*, 1920 ed., p. 167.

[7] *Exch. Deps. by Com.*, 24 Charles II, East. 24 ; and see above, chap. ii,
p. 187.

[8] *Cal. S.P. Irish*, 1633–47, p. 130.

[9] *H.M.C., Rep. on MSS. in Various Collections*, vol. iv, p. 286.

cold weather, and " sweep together a black small substance—I suppose 'tis Coales broken—wherewith they make Fires, by leaving open a Hole in their Chimneies for the Air to blow it ".[1]

Its use, at first confined to the poorer households, spread gradually among the rich. Throughout the seventeenth century, the King and some old-fashioned aristocrats clung to wood for heating their halls and private chambers,[2] though not their kitchen ovens, in spite of the praises which certain English writers began to bestow upon what had once been thought the meanest of all fuels. When, in 1623, a ship was stocked for the Prince of Wales, charcoal was the only combustible provided,[3] although Martin Frobisher had supplied himself with 14 tons of coal for his voyage in 1577, and although coal had become the fuel ordinarily used by seamen.[4] For a time rich Londoners were able to supply the needs of their entire establishments by buying wood, charcoal, and Scotch " pitcoal ",[5] but occasional payments for north of England " seacoal " creep into Lionel Cranfield's kitchen ledger in 1621.[6] By 1637 " seacoal " had become a much more important item, at least in the fuel expenses for the Earl of Rutland's town mansion. His steward's " usual provision " for one year included 30 tons of " seacoal ", 6 tons of Scotch coal, 26 loads of Kentish faggots and 12,000 " billets ".[7] Between the accession of James I and the Civil War the prejudice against common Newcastle coal was largely overcome. In 1644 men recalled a time when " some fine Nosed City Dames used to tell their husbands : O Husband ! We shall never bee well, we nor our Children, whilst wee live in the smell of this Cities Seacole smoke ; Pray, a Countrey house for our health, that we may get out of this stinking Seacole smell ". But, the blockade of Newcastle having deprived them of the fuel they had hitherto despised, " how many of these fine Nosed Dames now cry, Would to God we had Seacoale, O the want of Fire undoes us ! O the sweet Seacoale fire we used to have, how we want them now, no fire to your Seacoale ! "[8]

---

[1] *Royal Society, Philosophical Transactions*, vol. xix, 1695–7, pp. 357–62.

[2] Misson, *Mémoires . . . par un voyageur en Angleterre*, 1698, pp. 33–4, 52. According to Latimer, wood or charcoal was commonly burned in the Council House at Bristol throughout the seventeenth century, although the townsmen generally had adopted the coal fire (*Annals of Bristol in the Seventeenth Century*, 1900, p. 29).

[3] *Sackville MSS.* (Account of money spent for a voyage by the Prince of Wales, 1623).

[4] *Exch. K.R. Misc. Bks.*, Series i, vol. xxxv, for which reference I am indebted to Dr. Conyers Read. See also *Acts of the Parliaments of Scotland*, vol. ii, p. 543; *P.C.R.*, vol. xxxvi, f. 185, and vol. liv, f. 167; *Cal. S.P.D.*, 1658–9, p. 210; *Cal. of Treas. Bks.*, vol. vi, p. 259.

[5] See above, p. 119.

[6] *Sackville MSS.* (" Earl of Middlesex Kitchen Ledger ").

[7] *H.M.C., Rep. on the MSS. of the Duke of Rutland*, vol. i, pp. 499–500. For his castle at Belvoir the Duke's " yearly provision of Fuel " in 1612 had been 20 wagon loads of charcoal and 200 wagon loads of " Pitt Coles " (*ibid.*, vol. iv, pp. 480 sqq.). If the " pit coal " for Belvoir Castle came from Nottinghamshire, it was probably a grade resembling Scotch coal, and therefore pleasanter to burn than seacoal (cf. below, p. 219).

[8] *Artificiall Fire, or Coale for Riche and Poore*, 1644. Cf. below, pt. v, ch. i.

For cooking, north of England coal came to be considered, after the Restoration, " far beyond Wood, as yielding not only a more even, but more piercing Heat ".[1] It is " by far to be prefer'd before either Wood or Turf ", we learn from another Englishman interested in the culinary art, " as it affords a more even Heat than the former, and more intense than either ".[2] Later, when coal from the St. Etienne district (a grade resembling that of Newcastle) was substituted for wood in the kitchen fires of the famous Lyons epicures, many decided that coal " makes meat more succulent by rendering the juices thicker ".[3] In a well-to-do London household, like that of the Pepys, coal had become by 1660 the common fuel for all purposes, though charcoal was burned, as a luxury, in the dining room and the bedroom, when it could be obtained.[4]

It was enough to astonish the foreigner at the end of the seventeenth century that coal should be burned at all in private houses, either for heating or cooking. Except in the Belgian provinces, there were no large towns on the Continent where the inhabitants made coal fires. Even though coal had some market among manufacturers in many districts, the prejudice against its use within the household was destined to persist for decades in most European cities.[5]

With so smoky a fuel, the English had to abandon their crude early habit of building the family fire in the centre of the room and allowing the fumes to circle about before they escaped through an opening in the roof. Probably the introduction of coal led to the increased building of chimneys, which so impressed Harrison when he wrote in 1577.[6] In 1618 a petition was presented to the Lord Mayor and Aldermen on behalf of 200 city chimney sweeps,[7] so that in London chimneys must have been by this time almost universal. Henceforth even poor country houses were seldom built without them, as the sponsors of the unpopular hearth tax knew full well. Instead of constructing open fireplaces with chimneys, the English might have

---

[1] G[uy] M[eige], *The New State of England*, 1691, vol. ii, pp. 32–4.

[2] *The New Present State of England*, 1727 (?), vol. ii, p. 29.

[3] Aléon Dulac, *Observations sur le charbon minéral*, 1786. " Le cuisinier, quoyque françois," wrote in 1738 a French authority on the use of coal in England (*Archives Nationales*, 0¹1293), " de mon avis le préfére pour le rost [rôti] et pour faire quantité de ragout, au charbon de bois ".

[4] *Diary*, ed. H. B. Wheatley, 1893, vol. i, p. 32; vol. iii, pp. 13, 342–3, 345.

[5] The first serious attempt to introduce coal as a household fuel in Paris is said to have occurred in 1714 (Savary, *Dictionnaire du commerce*, art. " charbon "; see also *H.M.C.*, *Rep. on the Stuart Papers*, vol. ii, p. 271). In 1770 Jars spoke of the use of coal for domestic purposes at Lyons, St. Etienne, and St. Chamond as a recent development (*Sur le charbon*, f. 4). In 1755 coal was never burned for cooking in Westphalia, and only in the *Ämtern* of Bochum and Hattingen for heating houses. As late as 1762 in Brunswick, mineral fuel apparently never served for domestic purposes, although it had long been used by limeburners and other artisans. The prejudice against its use for house fuel persisted as late as 1796 in Munich, and in Bavaria generally (Hue, *Die Bergarbeiter*, vol. i, pp. 345, 351).

[6] See above, p. 197. For the early history of the chimney see J. Beckmann, *History of Inventions*, vol. i, pp. 295–314.

[7] *Analytical Index to Remembrancia MSS.*, 1878, p. 67.

adopted the stove, which had already come into general use in the German and Scandinavian countries, and which Montaigne found so agreeable on his visit to Switzerland. By means of these stoves one could maintain a comfortable, even temperature throughout the room and avoid any fumes.[1] But the English opposed the stove on the same ground that they oppose the modern central-heating plant : " it gives a chilling impression to go into the sharp air after being so warm in stove rooms ".[2] From a Frenchman's description of the manner in which the English made their coal fires at the end of the seventeenth century, it is evident that their system of heating houses has been hardly modified in two and a half centuries.

" They put into the Chimney certain Iron Stoves about half a foot high, with a plate of Iron behind and beneath ; before and on each side are Bars placed, and fastened like the Wiers of a Cage, all of Iron. This they fill with Coal, small or great, as they run, and in the middle they put an handful of Small-Coal, which they set Fire to with a bit of linnen or paper. As soon as this Small-Coal begins to burn, they make Use of the Bellows, and in less than two Minutes the other Coal takes Fire. You must blow a little longer after this, 'till the Fire is a little spread round about, and then you hang up the Bellows. In proportion as the Coal grows hotter it dissolves, becomes glutinous, and sticks together. To keep it up and revive it, you now and then give it a Stir with a long Piece of Iron made on purpose. As it burns out you must throw on more Coals, and thus with a little Pains you have a Fire all day long ".[3]

Macaulay expressed a common opinion of his day when he wrote that, before the eighteenth century, coal " was very little used in any species of manufacture " ; [4] and the truth of that opinion has been pretty generally accepted by subsequent writers. Yet the statement of Macaulay would not have been accepted by most of the contemporaries of Milton or even of Shakespeare, and one is left, after a study of industry in the seventeenth century, with the impression that, as a generalization, it is hardly defensible. " Yt is generally knowne ", argued Robert Bewick, one of the leading Newcastle colliery owners, in a case brought against him in 1619 for the corrupt mingling of an inferior grade with the best coals, " that Seacole is useful to be spent in other places [than halls and kitchens] and for . . . other purposes ".[5] " Within our Realme of England "

---

[1] There was, however, according to Montaigne, " une certene odeur d'air qui vous frappe en entrant ", and this odour the English found unpleasant. *Journal de voyage, 1580–1*, 1909 ed., pp. 92–3.

[2] G[uy] M[eige], *The New State of England*, p. 32.

[3] H. M. de V., *Mémoires* . . . *par un voyageur en Angleterre*, 1698, pp. 50–1. The version is that given in J. Ozell's translation, *M. Misson's Memoirs and Observations in his Travels over England*, 1719, pp. 37–8.

[4] *History of England*, Philadelphia, 1879, vol. i, p. 247.

[5] *Star Chamb. Proc., James I*, 56/10.

runs a proclamation drafted in 1590, " the use of coales is of late yeares greatlie augmented, not onlye for fuell, but also to serve divers tradesmen and Artificers ".[1]

Who then were these tradesmen and what were their trades ? One may conveniently consider them under three separate headings : (i) trades in which coal had been burned before the sixteenth century ; (ii) trades in which the substitution of coal for wood involved no serious technical problem ; and (iii) trades in which this substitution did present such a problem. Data concerning the quantities burned in the various trades are unfortunately meagre. The statistical conclusions which have been reached concerning the consumption of coal in industry are, consequently, only hypothetical.

(i) *Trades in which coal had been burned before the sixteenth century.*

It has been shown that during the Middle Ages there were only two artisans who made any considerable use of coal—the limeburner and the smith.[2] To most readers these are little more than names. As indicated earlier, the term smith was applied in medieval times to persons performing almost every kind of work in metal.[3] For our present purpose, the expression " smiths' work " is used to cover all work done in iron, lead, silver, gold, copper, tin, or zinc, after the raw ore had been converted into metal.

If there is any aspect of the medieval coal industry that compels the special attention of the historian, it is surely its relation to smiths' work, and more particularly to the making of " weapons of war ". In Scotland, most of the records of coal purchases in medieval accounts of the Crown are connected with the manufacture of arms and marine equipment from iron, or with the melting down of lead into bullets and shot.[4] In London the Clerk of the Works in the Tower was perhaps the chief early trader in coal, and his stock may well have been kept for purposes of defence.[5] Coal is associated with the medieval armament industries at Essen and St. Etienne.[6] At Liége the fabrication of arms preceded the manufacture of gunpowder in the fourteenth century,[7] and a nineteenth-century scholar has stressed the fact that in the Belgian provinces the primitive

---

[1] *Lansdowne MSS.*, 65, no. 9.
[2] See above, pp. 11–2.
[3] Cf. above, p. 169.
[4] *Exchequer Rolls of Scotland, passim ; Accounts of the Lord High Treasurer, passim.* For a description of the manner of making shot in England at the end of the seventeenth century see Houghton, *Husbandry and Trade*, vol. ii, pp. 204–6. As late as 1653, iron shot was apparently still made with the aid of a charcoal fire (see below, p. 203, n. 5).
[5] See above, pp. 10–1.
[6] Hue, *Die Bergarbeiter*, vol. i, p. 347 ; L. J. Gras, *Essai sur l'histoire de la quincaillerie et petite métallurgie à Saint-Etienne*, 1904, pp. 2–4.
[7] F. Henaux, *Fabrique d'armes de Liége*, 1858. Gunpowder was known, however, in the thirteenth century (cf. Beckmann, *History of Inventions*, vol. ii, pp. 487 sqq.).

armament industry left its traces in just those spots where the medieval miners dug coal.[1]

However much one may wish to stress this relation between coal and the medieval manufacture of weapons of war, one is bound to remember that in the Middle Ages the armament industry itself was of small importance in comparison with its importance during the sixteenth and seventeenth centuries,[2] when for the first time gunpowder came to be extensively used. We have referred to the development of new and quasi-new processes in the British metal trades, and have given reasons for thinking that the production not only of arms but of all kinds of small metal goods, increased greatly between the accession of Elizabeth and the Revolution of 1688.[3]

Harrison's remark, that the "greatest trade" in coal began "to grow from the forge into the kitchen and the hall",[4] must not be taken to mean that coal left the forge. On the contrary, its use in working up metals into finished goods increased more than in proportion to the increased output of such goods. With the extraordinary rise in timber prices, a steadily diminishing number of smiths could afford to produce their wares, as once they had done, exclusively with charcoal and wood fuel. They found mineral fuel so much more economical to burn,[5] that along the northern coast of France, as early as 1552, according to an English merchant, Thomas Barnabe, smiths who import coal from the north of England, "can lyve no more without [it] than the fysh without water—they can nother make stele worke, nor metall worke, nor wyer worke, nor goldsmythe worke, nor gunnes, nor no manner of thinge that passeth the fier".[6] While Barnabe, in his desire to call attention to the export of English coal,[7] may have stolen something of a march on the facts, his description, had it been made a century later, would not have been an exaggeration. Under Louis XIV the forges on both sides of the frontier between France and Savoy came temporarily to a standstill, because the Savoyards, who had coal but no iron, refused to bring coal across the border, and the French, who had iron but no coal, retaliated by preventing the export of iron.[8] Upon the threat of a Dutch war in 1671, Colbert told the *intendant* at Rochefort to obtain a two years' supply of coal for the marine arsenals there and at Brest, since, once

---

[1] Decamps, *Mémoire historique sur l'industrie houillère de Mons*, pp. 18–19. According to Henaux the armament workers at Liége were organized in a gild before the miners. One would think that, if the armament industry settled at Liége because of the supplies of mineral fuel, the coal miners would have been organized before the armament workers.

[2] This was as true of the Liége armament industry as of the English (Pirenne, *Histoire de Belgique*, vol. iii, p. 249).

[3] See above, pp. 169–72.

[4] See above, p. 197.

[5] Francis Bacon, *Sylva Sylvarum*, 10th ed., 1676, p. 164.

[6] Tawney and Power, *Tudor Economic Documents*, 1924, vol. ii, pp. 99–100.

[7] See below, vol. ii, p. 216.

[8] Boislisle, *Correspondance des Contrôleurs Généraux*, vol. iii, no. 1818 (Letter of M. d'Angervilliers, *intendant* in Dauphiné). A similar situation arose on the two sides of the boundary between the Spanish Netherlands and the province of Hainaut, which was temporarily French in 1710 ; a war being waged between the masters of forges in Hainaut, who had coal but no iron, and those

war commenced, it might be impossible to get through from England
the commodities essential for their maintenance.[1]

Before this time a great number of English metal workers had
become equally dependent on mineral fuel. In 1546 a Master of the
Ordnance was requested to provide ammunition for a convoy for
" vessels laden with coals for the pieces beyond seas ".[2] A supply
of " seacoal " was thought necessary for the coppersmiths at Keswick
early in Elizabeth's reign, even though the supply had to be carried
so far overland as to cost twelve times the pithead price.[3] At Wigan
in 1602 coal was a " fyering . . . without which . . . divers artyficers
and other tradesmen cannot use their trades and occupations " ;
and, as we know from another source, many of these artificers and
tradesmen were metal workers.[4] During the second Dutch War,
in 1665, the Master of the King's Ordnance in London reports " that
for want of Coales the Forges Imployed . . . with Gun-smiths and
others have been enforced to cease from working for some dayes " ; [5]
and John Timbrell, who is " employed in makeing Anchors and other
Ironworks, for the use of his Majesties Navy, att Portsmouth ", petitions
the Crown because he " wants Coales . . . for . . . carryeing on . . .
the . . . workes ".[6]

In certain districts the smiths had abandoned charcoal fires
altogether, for many of their operations. Yarranton, writing in
1677, denied that the forges in Staffordshire, Worcestershire, Shropshire,
Warwickshire, and Derbyshire prejudiced the timber supplies, because
the iron workers—so he said—burned only coal.[7] Apparently this was

---

of the Spanish Netherlands, who had iron but no coal (*ibid.*, vol. iii, no. 797).
Where coal was only to be had at a prohibitive price, the masters of forges
continued to burn wood, or at least mixed only small quantities of coal
with their wood (*ibid.*, vol. ii, no. 355, Letter of M. Foucault, *intendant* at
Caen).
    [1] P. Clément, *Lettres, instructions et mémoires de Colbert*, vol. iii, pp. 380–1.
An inventory taken in 1677 of the French naval arsenals (*ibid.*, vol. iii, pt. ii,
p. 696) shows that all had on hand stocks of coal ; Rochefort had 14 tons ;
Brest, 110 tons ; Havre, 1½ tons ; Dunkirk, 25 tons ; Toulon, 700 tons.
    [2] *Cal. of Letters and Papers of Henry VIII*, vol. xxi, pt. ii, no. 387.
    [3] W. G. Collingwood, *Elizabethan Keswick*, pp. 31, 168 sqq. ; *H.M.C.*,
Rep. on the MSS. of the Marquis of Salisbury, vol. x, p. 217. " Sea Coles "
were also purchased for the copper works near Swansea in 1586 (C. B. Wilkins,
A History of the South Wales Coal Trade, 1888, p. 18).
    [4] *Duchy of Lancs. Pleadings*, 195/S/10 ; D. Sinclair, *History of Wigan*,
1882, vol. i, p. 89.
    [5] *P.C.R.*, vol. lviii, p. 124. It is certain that charcoal was not meant, for
the Privy Council instructed the Lord Mayor to provide 150 *chaldrons*. Charcoal
was not sold by the chaldron. The London gunmakers had experienced
similar difficulties during the coal shortage of 1643 (*History of the Town of
Newcastle* (Anon.), 1801, p. 460). While coal seems to have been by the middle of
the seventeenth century indispensable for forging the metal parts of guns, it
was not always burned in the process of making shot out of iron. For that
purpose the metallurgical workers near Gloucester still employed charcoal
as late as 1653 (*Cal. S.P.D.*, 1653–4, p. 142), although they perhaps mixed some
mineral coal with it (cf. below, p. 250).
    [6] *P.C.R.*, vol. lviii, p. 116. The reference is certainly to mineral coal (see
preceding footnote). See also *Cal. S.P. Colonial*, 1661–8, p. 427.
    [7] Yarranton, *England's Improvement by Sea and Land*, 1677, pp. 58–60,
147–9.

true also in the neighbourhood of Sheffield, for Mr. Lloyd, in his interesting book on the cutlery trades, tells us that, although most of the forests there had been cut down during the sixteenth and seventeenth centuries, "the activity of the iron industry was undiminished ".[1]

By means of various technical improvements, which we discuss in the next chapter, it became possible in the seventeenth century to use coal in several metallurgical processes, which had hitherto depended exclusively upon charcoal, because contact with the fumes of seacoal damaged the product. Even in producing the finer type of finished metal work, coal, which had been considered in medieval times altogether too crude a fuel,[2] began to replace wood. From a petition addressed to the Crown in 1610 by Sir William Slingsby of Yorkshire, himself a colliery owner in Northumberland, we learn that iron wire was already drawn in a " Pit Coale " fire.[3] Slingsby proposed among other things to substitute mineral for wood fuel in drawing copper and " latten " wire and in all " battery work ".[4] Under the Commonwealth, coal, as well as wood, was supplied to the mint for making gold and silver coin.[5]

The total annual consumption of mineral fuel in Great Britain towards the end of the seventeenth century in all kinds of metal work is a matter for speculation. To-day in Paris, a small smithy, worked by the owner and one or two assistants, burns about twelve tons of coal a year. In the district north and west of Birmingham during the reign of Charles II, one could count by the thousand [6] forges which were not very different, perhaps, from those in modern Paris ; and, as these Staffordshire forges are said to have burned no other fuel than coal, one may assume that tens of thousands of tons were there consumed each year.[7] Around Sheffield metal workers were almost as numerous. Forges in small or large numbers were to be found, also, in nearly every town and village through the Midlands and along the coasts. In making a list of London artisans who are prejudiced by taxes on coal, an anonymous writer in the last decade of the seventeenth century puts first the smiths, a " numerous, laborious,

---

[1] G. I. H. Lloyd, *The Cutlery Trades*, 1913, p. 69. At the beginning of the eighteenth century there were 18 collieries in the neighbourhood (*ibid.*, p. 75).

[2] See above, pp. 11-2.

[3] *Harleian MSS.*, 7009, no. 4

[4] A term which may be taken to apply to hammered brass and copper vessels generally.

[5] *Cal. S.P.D.*, 1651, pp. 234, 488 ; *Exch. K.R. Accounts, Supplementary*, 620/62 ("Edward Lole his bill for sea Coles and Billets delivered into his Matyes Mynt ", 1640). As early as 1623, seacoal had been mixed with charcoal and wood in making " Angell Gold, Crowne Gold, and silver moneys " (*Sackville MSS.* : Accounts of the making of these moneys, July, 1623–July, 1624). The " Coles " referred to in an " estimate of the charges of refining the base money received into the Mint ", 1560–1 (*Lansdowne MSS.*, 4, no. 58), are probably charcoals.

[6] According to Dud Dudley there were " twenty thousand Smiths or Naylors at the least " (*Mettallum Martis*, 1665, p. 53). That is no doubt an exaggeration, but the shops of these workers were certainly very numerous (see above, p. 171).

[7] According to Jorden (*Discourse of Natural Bathes*, 1631, p. 23), the smiths were accustomed to throw water upon their fires of mineral coal, the water making the coal cake and bake together, whereas otherwise the blast of the forge might blow it away. Water also made the fires last longer.

and ingenious " people, who " use great store of seacoal ".[1] When we consider that the amount of metal wrought into finished products in England and Wales probably approached 60,000 or 70,000 tons,[2] and that there was also some metallurgical manufacture in Scotland, it seems hardly possible that by this time less than 200,000 tons of coal were consumed annually by all the metal workers in Great Britain.[3] The actual consumption may have been a good deal larger than this.

An estimate of the quantity burned in lime making in Great Britain is even more difficult to make, for the industry was scattered throughout the country, rather than concentrated in special districts.[4] Although Sir Balthazar Gerbier wrote in 1663 that wood was a more desirable fuel than coal for the lime makers, because it enabled greater quantities of lime to be burned at one time, we know that limeburners had made use of coal for centuries, and, before the end of the seventeenth century, it had become almost the universal fuel for the innumerable limekilns which served the husbandman in the country and the builder in the towns.[5] Want of " see coles " is said to have caused a decay in the husbandry of Cambridgeshire as early as the reign of Henry VIII.[6] In addition, towns like Newcastle and London had kilns producing considerable quantities of lime for sale in the neighbourhood, and even for export abroad.[7] At Gravesend, near the mouth of the Thames, Sir William Brereton described in 1634 a kiln in which " the fire extinguisheth not from one end of the year to the other ".[8] As early as 1640, a shortage of coal engendered a serious crisis in the building trades of the capital. Various owners of limekilns, having been refused supplies by the coal merchants, petitioned the Privy Council that they were " bound by covenant to serve . . . with lime . . . all the Bricklayers and others . . . who are now building for persons of quality ; which they cannot performe unlesse they may have coale to keepe in their fyers, soe that of necessity the said Buildings must suffer ".[9] The lime market increased rapidly with the increase in brick and stone construction work, and

---

[1] *Reasons for the taking off the Tax of Five Shillings per Chaldron on Coals*, printed Appendix H.

[2] Mr. Ashton is inclined to estimate the iron production early in the eighteenth century at rather more than 25,000 tons—25,000 tons of pig, and a small amount of bar iron which did not pass through the blast furnace at all (Ashton, *Iron and Steel in the Industrial Revolution*, pp. 235–6). In addition, from 14,000 to 22,000 tons of bar iron was imported in favourable years for trade (see above, p. 169 n.). It is probable that at least as much lead was being produced at the end of the seventeenth century as in the 'thirties when the output was estimated at about 12,000 tons. The native production of copper must have been considerable, and it was supplemented by substantial imports. The output of tin was nearly 1,500 tons at the end of the seventeenth century (Lewis, *Stannaries*, p. 256).

[3] For attempts to introduce coal in the smelting of ores during the seventeenth century, see below, pp. 245–6, 249–51.

[4] Cf. above, p. 187.

[5] Gerbier, *Counsel and Advise to all Builders*, 1663, pp. 56–7.

[6] *Star Chamb. Proc., Henry VIII*, vol. v, f. 21.

[7] MS. *Journal of the Common Council of Newcastle*, 1650–8, ff. 469, 484 ; *Lansdowne MSS.*, 79, no. 90.

[8] Brereton, " Travels in Holland, etc.", in *Publications of Chetham Soc.*, vol. i, 1844, pp. 1–2.

[9] *P.C.R.*, vol. lii, p. 730.

with the improvement in the methods of husbandry. Since something like half a ton of coal was probably required to make a chaldron of lime to serve as mortar or fertilizer,[1] the quantity burned annually in the lime industry at the end of the seventeenth century must have amounted to tens of thousands of tons.

(ii) *Trades in which the substitution of coal for wood involved no serious technical problem*

In certain parts of Great Britain a little coal undoubtedly had been burned in salt making before the sixteenth century. Seacoal, picked up on the shores of Yorkshire, is said to have served the medieval salters along that coast;[2] and, since small quantities of salt were made by the evaporation of sea water close to the early coal workings in Northumberland, Fife, and the Lothians, it is tempting to assume a connection between the two industries in those counties. Yet in the Firth of Forth district, if not in the north of England, salt making almost certainly preceded the digging of coal.[3] If the makers were in no hurry for their product, sea water gathered in troughs on the coast could be reduced to salt by the mere heat of the sun,[4] though this must have been a tedious process, especially in the north where the sun so seldom shines uninterruptedly for many hours, and it was certainly abandoned at all the important works in England long before the salters in southern France gave it up. We cannot be sure that, even where coal could be had cheaply, the first artificial fuel adopted in Great Britain was not wood.[5] Not until the sixteenth century do we get abundant references to the use of mineral fuel for heating the pans in which the sea water, or the brine from natural springs, was boiled down to salt. In Germany, where, during the seventeenth century, this manufacture probably provided the largest single market for coal, it was shipped down the river Saale, as early as 1517, to heat

---

[1] The chief expenses in lime making were the cost of kiln and utensils, coal, and limestone. For the quantity of lime required in building see below, pp. 217–8. For information concerning the coal consumed in lime burning in connection with castle building in Scotland, see *Works Accounts* (General Register House, Edinburgh), vols. vi, xi, xii. For information concerning the coal used for making lime in connection with fortress building on the French coast, see Clément, *Lettres instructions et mémoires de Colbert*, vol. v, pp. 198–9 ; Boislisle, *Correspondance des Contrôleurs-Généraux*, vol. i, nos. 372, 1890.

[2] *V.C.H. Yorkshire*, vol. ii, p. 387.

[3] There were salt works in Scotland in the reign of King David, 1124–53 (R. N. Boyd, *Coal Pits and Pitmen*, 1892, p. 18). The Domesday Book mentions salt pans along the coasts of Kent, Surrey, and Sussex (Galloway, *Annals of Coal Mining*, pp. 11, 63–4). It is well known that salt was a very important commodity in ancient civilizations.

[4] Cf. Agricola, *De Re Metallica*, ed. Hoover, pp. 547 sqq.

[5] " Panwood " was a common term for the mineral coal burned at the Scottish salt works in the seventeenth century (*Accounts of the Torry Coal and Salt Works* (General Register House, Edinburgh), 1679 ; *MSS. of the Duke of Hamilton*, Charters, no. 154, ii (1644).) It is possible to argue that the word " panwood " itself indicates that at an early period in Scotland wood, rather than coal, was burned to heat the pans. If " panwood " originally applied to mineral fuel, it is the only instance I know of in which the word wood was used to mean coal.

the pans at Wettin near Halle.[1] Before the end of the sixteenth century the owners of a salt plant at Soden, just east of Frankfort-on-Main, purchased supplies of coal, which were said to have been brought by river transport down the Meuse and up the Rhine, all the way from Liége.[2] Another plant at Allendorf on the Weser depended for fuel upon lignite dug in the Harz mountains.[3]

Coal served the salt makers in Scotland and at the mouth of the Wear and the Tyne, at least from the beginning of the sixteenth century.[4] By 1605 the annual consumption of " pan coal " (as it had come to be called by the English) in the " works " within the counties of Durham and Northumberland had reached about 50,000 tons.[5] A pan full of salt water had to be boiled dry eight successive times before the salt was ready for sale ; to accomplish this the fire had to be kept burning at white heat three days and a half.[6] About six tons of coal were consumed in making a single ton of salt.[7] Water from the brine springs was far more saliferous than sea water, so that at inland works fewer boilings, and a much smaller quantity of coal, served to produce a ton of salt. At the Cheshire " wiches " less than one ton sufficed ;[8] at Droitwich, in Worcestershire, where the brine was weaker, at least two tons must have been required, for there the pans were boiled four or five times.[9]

---

[1] O. Hue, *Die Bergarbeiter*, vol. i, p. 349. The salt manufacture at Halle took on a great importance during the seventeenth century, when the majority of shares in the enterprise were owned by the King of Prussia.

[2] *Ibid.*, p. 351. Later lignite from pits near Soden was substituted.

[3] L. Beck, *Geschichte des Eisen*, vol. ii, pp. 102–3 ; K. T. von Inama-Sternegg, *Deutsche Wirtschaftsgeschichte*, 1909, vol. iii, pt. ii, p. 115.

[4] *Acts of the Parliaments of Scotland*, vol. iii, p. 93b ; Galloway, *op. cit.*, pp. 64–5 ; *A History of Northumberland*, vol. ix, pp. 223–4. Salt making with mineral fuel is said to have been uncommon in the Tyne area before the sixteenth century (E. Hughes, " The English Monopoly of Salt in the years 1563–71", in *Eng. Hist. Rev.*, vol. xl, 1925, pp. 334–50).

[5] Brand, *History of Newcastle*, vol. ii, p. 22 (taken from a manuscript in the Duke of Northumberland's collection). According to the writer of this document there were within the two counties 153 pans, each of which consumed 16 tens (160 Newcastle chaldrons, or nearly 350 tons) per annum. The term " pancoal " came into general usage in the Tyne district only at the beginning of the seventeenth century (*Exch. Spec. Com.*, no. 5037).

[6] Lord Harley, " Journeys in England ", in *H.M.C., Rep. on MSS. of the Duke of Portland*, vol. vi, p. 105.

[7] I have arrived at this figure by comparing the output of salt and the consumption of coal as given in Brand, *op. cit.*, vol. ii, p. 22, and in *Cal. S.P.D.*, 1644–5, pp. 98–9. Lord Harley believed that the quantity of coal required was even larger (*loc. cit.*), but he was probably misinformed.

[8] John Collins, *Salt and Fishery*, p. 5. At Northwich 1,488 horseloads of coal were required to produce 12,214 bushels of salt, i.e. (allowing 40 bushels to one weigh of salt, 8 loads to one ton of coal) about 12 cwt. to produce 1 weigh. At Middlewich 632 loads of coal were required to produce 4,300 bushels of salt, i.e. slightly more than 15 cwt. to produce 1 weigh. At Nantwich 1,216 loads of coal were required to produce 4,200 bushels of salt, i.e. nearly 1½ tons of coal to produce 1 weigh. Taking the three wiches together 3,336 loads of coal were required to produce 20,714 bushels of salt, i.e. slightly more than 16 cwt. to 1 weigh.

[9] *Ibid.*, p. 3. See also Plot, *Natural History of Staffordshire*, p. 128. We know that at Droitwich, before the substitution of coal for wood-fuel, 4 wagon loads of wood were required to make a ton of salt. (See above, p. 193.) One ton of coal usually replaced about 2 loads of wood in industrial processes. (See below, p. 218.)

At the works in Worcestershire and Cheshire, however, wood was burned exclusively for more than a century after coal had become the common fuel for salt making in Scotland and the north of England, because all the brine springs were situated more than twelve miles overland from the nearest colliery. In spite of the long haul, pit coal largely replaced wood at the springs in Cheshire shortly before 1669,[1] and at the springs in Worcestershire between 1665 and 1678.[2] Before 1700 coal had become almost the only fuel used, not only in the manufacture of salt from sea water and from brine, but also in the manufacture of rock salt, the rock being carried to the coal pits, where it was dissolved in water and then evaporated like brine.[3]

Sir Lionel Maddison (whose informing letter on the mining industry of the north of England we have had occasion already to cite) estimated the annual consumption of coal in the salt works of Shields and Sunderland just before the Civil War at about 90,000 tons : 80,000 tons in the works at Shields, 10,000 tons in the works at Sunderland.[4] At the end of the seventeenth century the salt industry of the Tyne and Wear was at least as important as before the Civil War, and, in addition, a considerable manufacture had been established at Blyth, Cullercoates, and Seaton Delaval. Consequently the annual consumption of coal in salt making within the counties of Durham and Northumberland could hardly have been less than 125,000 tons between 1680 and 1690. Round the Firth of Forth pans were even more numerous and at least as spendthrift of fuel. In 1679 one colliery at Tulliallan produced 15,000 tons for the salters.[5] By that time (if one may venture to estimate it) the annual consumption in all the salt works of the Forth may have reached 150,000 tons. John Collins tells us in 1682 that the brine springs of Northwich, Middlewich, and Nantwich in Cheshire used 417 tons a week,[6] so that the consumption there perhaps approached 20,000 tons a year. It seems highly probable that, at the end of the seventeenth century, more than 300,000 tons were burned every year by the salt manufacturers of England and Scotland combined.[7]

---

[1] *Royal Soc., Philosophical Transactions*, vol. iv, 1669, pp. 1061–7 (Dr. Wm. Jackson, *Saltmaking at Nantwich*).

[2] Between the publication of Dud Dudley's *Metallum Martis*, and the article on the salt works at Droitwich by Dr. Thomas Rastell (*Phil. Trans.*, vol. xii, 1678, pp. 1059–64).

[3] *The Case of the Refiners of Rock Salt in Lancashire, Cheshire and Flintshire*, 1698 (?) *The Case of the Brine Pits truly stated*, 1698 (?).

[4] 3,900 " tens " (*Cal. S.P.D.*, 1644–5, pp. 98–9). For the coal contained, in a " ten " see Appendix C.

[5] *Register, Edinburgh : Kincardine and Tulliallan Coal and Salt Works, Accounts*, 1679–80. Professor Jevons estimates the annual consumption of coal in salt making at Culross in 1630 at 45,000 tons (H. S. Jevons, *The British Coal Trade*, 1920, p. 152).

[6] See above, p. 207, n. 8 (the figures are for one week's manufacture).

[7] From 100,000 to 150,000 tons per annum in Durham and Northumberland ; an equally large quantity on the Firth of Forth ; about 40,000 tons at the brine pits ; and perhaps 30,000 tons at the sea-salt pans of the north-west coast (Cumberland and Ayrshire), and in the new manufacture of rock salt (see below, p. 228). In reaching the figure for the brine pits, I have assumed that the annual consumption of coal at the Cheshire wiches was nearly 20,000

Unless the new taxes on coal used by them were withdrawn, protested " divers makers of alum and copperas " in 1655, the industry would be ruined.[1] Sir Paul Pindar, a farmer of the Crown alum works near the Yorkshire coast, made representations to the county justices in 1637 to the effect that his business had been brought to a standstill, since they had forbidden shipmasters to unload coal at Whitby because of the plague at Sunderland.[2] When these Yorkshire works were started during James I's reign, wood had already grown so dear that the alum manufacturers had been obliged to depend upon supplies of coal.[3] For " calcining ", or heating, the alum stone (in much the same manner that iron or copper ore were " calcined " prior to the first smelting), " wood and furzes ", as well as the cinders of coal, were used as late as 1678 ; but in the subsequent boilings a raw coal fire sufficed. Calcined alum stone, dissolved in pits of water, and mixed with kelp and urine, had to be twice boiled down (like the brine for making salt) in great metal pans, nine feet long by five broad and two and a half deep, before the alum was ready for sale.[4] So slow was the process that eight pans were required to make alum at the rate of five tons a week.[5] At least three tons of coal, and probably more, were consumed in producing one ton of alum.[6] According to Maddison, various works along the coasts of Durham and Yorkshire consumed upwards of 7,000 tons of Sunderland coal a year during the decade preceding the Civil War.[7] It is not possible to estimate the quantity

---

tons (cf. above, p. 66 n.). At Droitwich, where the output of salt was perhaps as large as at Northwich (see above, p. 175) and where at least two tons of coal were probably required to produce a ton of salt (see above, p. 207), the consumption of coal may have exceeded 20,000 tons.

[1] *S.P.D. Interregnum*, vol. xciv, no. 106. The same complaint had been made by a farmer of the alum works (Sir John Gibson) in 1637 (*Cal. S.P.D.*, 1637, p. 210).

[2] *S.P.D., Charles I*, vol. ccclvii, no. 85. Actually most of the coal was unloaded not at Whitby but on the south bank of the river Tees.

[3] *Lansdowne MSS.*, 152, no. 6, f. 57.

[4] For a full description of the process of alum manufacture see *Royal Soc., Philosophical Transactions*, vol. xii, 1678, pp. 1052–6 (*An Account of the English Alum Works*, by Daniel Colwall, Esq.). See also *Lansdowne MSS.*, 152, no. 6, f. 63.

[5] *Harleian MSS.*, 7009, f. 9 ; *Exch. Spec. Com.*, no. 5789 (see testimony of witnesses). The Commissioners found that, in the alum works at Guisborough, Skelton, and Loftus, 64 pans would be needed to produce the projected output of 1,800 tons per annum. The pans were apparently placed side by side under cover in what was called an alum house.

[6] *S.P.D., James I*, vol. lxxiv, no. 19. Two thousand Newcastle chaldrons (about 4,500 tons) were to be supplied to make 1,500 tons of alum. Though it was hoped to produce 1,800 tons of alum per annum at the Yorkshire works, the ordinary output did not exceed 1,000 tons between 1609 and 1618 (*Lansdowne MSS.*, 152, no. 6, f. 57). In one year (1613–14), however, 2,422 Newcastle chaldrons (more than 5,000 tons) of coal were imported at Whitby and Coatham (*V.C.H. Yorkshire*, vol. ii, p. 384). One suspects, therefore, that the provision of 2,000 chaldrons may have proved an insufficient quantity of fuel to produce 1,500 tons of alum.

[7] *Cal. S.P.D.*, 1644–5, pp. 98–9. For confirmation of Maddison's estimate see *Exch. Deps. by Com.*, 9 Charles I, Mich. 11 (testimony of Sir John Hedworth). The alum works at Hartlepool, as well as those in Yorkshire, got coal from Sunderland, and the 3,000 chaldrons may have included supplies for both.

shipped from Sunderland after the Restoration.[1] There was probably a considerable development of the manufacture at Hartlepool, and an expansion of the Scottish industry, which had been started somewhat later than the English. Ten thousand tons would seem to be a conservative estimate of the annual consumption of coal in British alum manufacture at the end of the seventeenth century.

Copperas stones, placed in water, had to ripen for five or six years in long wooden reservoirs, before the solution was drained into a great boiler (made to contain 12 tons of the liquid) and heated, together with a certain amount of iron dust, over a coal fire, sometimes for twenty days. Fresh liquid was pumped in as the solution came to a boil, and more iron dust added from time to time. When Sir William Brereton sailed along the south-east coast in 1634, he described a copperas works at Queenborough which consumed about 300 tons of coal a year.[2] After the Restoration the most important copperas works was at Deptford, and there greater quantities of coal must have been burned than at Queenborough.[3] Both the copperas and the alum manufacturers used the inferior grade known in the Newcastle district as " pancoal ".[4]

Saltpetre, as it came from the incrustations of old buildings, was allowed to dissolve, like alum and copperas, in water, and was prepared in tubs for boiling in copper kettles, set in a furnace of brickwork. " To save charges in fewel ", wrote John Houghton in his weekly bulletin, the solution from an extra tub is made to " dribble at a tap below as fast into the copper as the force of the fire doth waste your liquor ".[5] Great quantities of fuel must have been consumed, for the first boiling sometimes lasted sixty hours, and it was followed by two subsequent boilings, after the hot liquid had been passed through wood ashes.[6] The substitution of coal for wood had been contemplated in 1589, and in 1596 upwards of 40 tons of coal from Wollaton colliery was delivered to John Foxe of Nottingham, " saltpeterman ".[7] In 1627 Mr. Evelyn (probably the father of

---

[1] *Exch. K.R. Port Bks.*, bdl. 198. About 500 Newcastle chaldrons of coal for " Mr. Conyers alum works " at Whitby were imported in the year 1677–8 and again in the year 1678–9. Apparently the manufacture at Mulgrave, near Whitby, was seriously handicapped during the Interregnum. An agent for the alum works there, Thomas Shipton, who testified before Exchequer commissioners in 1662 (*Exch. Spec. Com.*, no. 6548), gave the following statistics for the output at Mulgrave :—

	Tons.
1652–5 .	400 to 500
1656–8 .	700
1659 . .	900
1660 . .	1,000
1661 . .	1,100

A stock of 900 chaldrons of coal was on hand at the works in 1662.

[2] See above, p. 185.

[3] *Royal Soc., Phil. Trans.*, vol. xii, pp. 1056–9 (*An Account of the way of making English Green Copperas*, by Daniel Colwall, Esq.).

[4] *S.P.D., Interregnum*, vol. xciv, no. 106.

[5] Houghton, *Husbandry and Trade*, vol. ii, p. 115.

[6] *Ibid.*, pp. 115–19. Common salt was a by-product of the first boiling, and was used for salting beef, bacon, tongues, etc.

[7] *H.M.C., Report on the MSS. of Lord Middleton*, pp. 163–4.

the diarist) protested to the House of Lords Committee for Defence of the Kingdom, that he " can get no seacoals, and cannot therefore proceed with the making of gunpowder ".[1] Henceforth coal was the principal fuel used in the manufacture, which was carried on in many counties, although as late as 1663 the saltpetre itself was burned prior to the boilings (a process corresponding to the " calcining " of alum stone) over a wood fire.[2] Gunpowder, as it came from the mill (where it had been compounded out of saltpetre, brimstone, and charcoal), was laid on " drying sieves ", and placed in a small room, about 20 feet square, called the " stove ". By the middle of the seventeenth century, if not earlier, the practice was to heat this room by a coal fire, for when war with the Dutch broke out in 1665, the " powder makers " of London were reported to " stand in great need of Sea Cole for carrying on of that Worke ", and, notwithstanding the shortage of domestic fuel felt by the citizens, the Lord Mayor was asked to supply them at once with 240 tons.[3] Though it probably required nearly five tons of coal to produce a ton of saltpetre suitable for preparing gunpowder,[4] the total amount of coal thus consumed was unimportant when compared with the consumption in salt making, for the manufacture of saltpetre and gunpowder, like the manufacture of alum and copperas, remained a small industry. An output of only 300 tons of saltpetre per annum was contemplated in 1589,[5] when commissioners for this manufacture were first furnished with a royal warrant, and there is evidence that by 1636 the output was little more than had been contemplated in 1589.[6]

Undoubtedly the amount of coal burned in the manufactures or alum, copperas, saltpetre, and gunpowder combined was only a fraction of the amount burned in boiling soap, refining sugar, making

---

[1] *Lords' Journals*, vol. iii, p. 547. In 1628 coal was brought from pits at Westerleigh in Kingswood Chase to Tetbury for the manufacture of saltpetre (*S.P.D.*, *Charles I*, vol. cxxi, no. 10). See also the demand made in 1640, by one Sikes, deputy for making saltpetre in the counties of Essex, Suffolk, and Norfolk, for a supply of coal which dealers in Maldon had refused to sell him, and without which he could not proceed with the saltpetre manufacture (*P.C.R.*, vol. liii, pp. 27–8 ; *Cal. S.P.D.*, 1640–1, p. 79).

[2] Birch, *History of the Royal Society*, vol. i, pp. 281–2.

[3] *P.C.R.*, vol. lviii, p. 105. In 1638 the gunpowder makers at Henley, Windsor, and other towns along the Thames refer to their need for coal, a need which arises only once in every six or seven years, " which is when the workes are in those parts " (*S.P.D.*, *Charles I*, vol. cccxciii, no. 13).

[4] One load of wood was required to produce 2 cwt. of saltpetre before the substitution of coal for wood. A ton of coal generally took the place of two or more loads of wood in processes of this sort. (See below, p. 218.)

[5] *Lansdowne MSS.*, 58, no. 63, wherein the probable consumption of wood is estimated at 3,000 wagon loads. Assuming that one load was spent in producing 2 cwt. of saltpetre, we have a projected output of 300 tons of saltpetre.

[6] *S.P.D.*, *Charles I*, vol. cccxli, no. 68. In 1636 the weekly output was apportioned between the " saltpetremen " in the various counties as follows : Gloucs., Worcs., Hereford, Wilts., Dorset, Somerset, each 7 cwt. ; Salop, Devon, Cornwall, and six Welsh counties, each 9 cwt. ; Ches., Lancs., Cumb., Westmor., Flint, and North Wales, each 2 cwt. ; Yorks, Northumb., Bishopric of Durham, Town of Newcastle, Town of Hull, each 3 cwt. In addition some saltpetre may have been manufactured in London.

starch and candles, and preparing preserved foods. Sugar, soap, starch, and candles (the universal means of lighting both private and public halls in London and mines in the provinces [1]) had come before 1700 to be commodities in general demand throughout the British Isles, among the poor as well as the rich.[2] Alum and copperas, on the other hand, found their principal market among the dyers, while saltpetre and gunpowder were required almost solely for armed defence, which did not yet constitute the crushing burden upon the national exchequer that it does to-day. The value of the soap consumed in London during one year in the latter part of the seventeenth century was estimated at £121,875.[3] It is doubtful whether the value of all the alum, copperas, and gunpowder produced in Great Britain was very much greater. Larger quantities of fuel were probably consumed in soap-boiling, therefore, than in making alum, even if it required more fuel to produce a ton of alum than to produce a ton of soap.

We lack information concerning the quantity of coal burned in any one of this financially more important group of trades. Two patentees obtained a licence in 1615 to make oil and pith out of vines, and soap with a "seacoal" fire.[4] Soon after, the soap-boilers were numbered among the chief "spenders" of the new fuel.[5] Good soap was made usually from olive oil and tallow, "boiled for many hours together," and then mixed with potash, "first slack'd with lime".[6] It is uncertain when coal was first used in the British sugar industry, but there can be little doubt that such use was general before 1650, for, during the second half of the seventeenth century, sugar refiners were among the chief French consumers of coal, which they imported in large measure from England and Scotland.[7] As for the chandlers, they needed fuel to heat their tallow, and one may assume that they used coal, for they are to be counted among the most conspicuous London retailers of that commodity shortly after the Civil War.[8] Coal had become essential to those in charge of supplying the navy with food and drink. In preparation for the Dutch War of 1665–7, Denis Gawden, "surveyor of the King's marine

---

[1] Houghton, op. cit., vol. i, pp. 343–4.

[2] The place occupied by sugar refining in the national economy is rather more uncertain than that of the other industries mentioned. In spite of the efforts of the Crown to prevent it, some refined sugar was imported, and a considerable part of the supply brought from the West Indies was re-exported raw (F. W. Pitman, The Development of the British West Indies, 1700–1763, 1917, p. 340 ; see above, p. 186 n.). Nevertheless, there were well over a hundred refineries in Great Britain by the middle of the eighteenth century, when a depression in the manufacture is said to have affected about 1,800 persons (Pitman, op. cit., p. 340). It may be doubted whether as large a number of workmen were employed in making alum and copperas combined.

[3] Houghton, op. cit., vol. i, pp. 349–50.   [4] Cal. S.P.D., 1611–18, p. 291.

[5] Rymer, Foedera, vol. xix, f. 561.   [6] Houghton, op. cit., vol. i, p. 349.

[7] Boislisle, Correspondance des Contrôleurs Généraux, vol. ii, no. 355, note, and appx. p. 498. Coal would hardly have been imported from Britain for sugar refining in France had it not been burned by the British sugar refiners. A new method of sugar refining was explained to the English Privy Council in 1615 (Acts of the Privy Council, 1615–16, p. 284), and this may have involved the substitution of coal for wood. For a description of sugar refining in England see Houghton, op. cit., vol. ii, pp. 314–20.

[8] See below, vol. ii, p. 103.

victualls ", arranged " to bring seacoal to the several victualling ports for his Majesty's service ", and hired a vessel to carry 80 chaldrons, or more than 100 tons, to London for " boyling pickle ".[1] At about the same time, two " brewers for the King's Navy " received 110 chaldrons from the Lord Mayor out of the stock which he held for the relief of the poor.[2]

Dyers and brewers are invariably mentioned during the sixteenth and seventeenth centuries in any list of the great consumers of coal in London.[3] Both tradesmen must have made some use of it at the beginning of the fourteenth century, for they were cited along with the limeburners in the proclamation of 1307, forbidding all but smiths to burn this black mineral in the capital.[4] It would be a mistake, no doubt, to suggest that either the dyers or the brewers felt themselves bound to keep to the letter of the proclamation. But they were hindered in burning coal by another prejudice, which did not touch the limeburners, and which proved perhaps a stronger deterrent than any law. It arose from the prevailing opinion that the smell and the dirt of a seacoal fire would be transmitted to the taste of ale and to the texture of cloth. This prejudice (like the prejudice that mineral fuel, if used in cooking, would contaminate food) probably prevented any wholesale adoption of coal in dyeing and brewing in England before Elizabethan times ; [5] though in Flanders the progressive textile manufacturers had already discovered the advantages of the new fuel. These advantages were perceived by an English trader, William Cholmeley, as early as 1553. " I saye ", he wrote, " that we have plentye of sea cole in many partes of this realme, so that we may in most partis . . . have them to serve our turne in dyinge as well as the Flemmingis have, and as good cheape, for they burne and occupye none other fuell then coles that are dygged out of the grounde, like as our smythes doe. Our dyeing therefore should not be wastful to our wodis, but rather a preserveyng, by staying the Newcastell colys at home, for then shoulde oure dyars, that do nowe wast much wode in dyinge disceytful coloures, burne no wode at all, and yet shoulde they dye as true and perfect coloures, and to them more benefytt ".[6]

---

[1] P.C.R., vol. lviii, p. 127.

[2] Ibid., vol. lviii, pp. 115, 116, 124.

[3] See for instance S.P.D., Charles II, vol. cxix, no. 24 ; P.C.R., vol. lviii, p. 124 ; Rymer, Foedera, vol. xix, f. 561 ; The Present State of the Coal Trade, 1703 ; Tim Nourse, Campania Felix, 1700. Evelyn (in Fumifugium, 1661, pt. i, p. 15) enumerates as the chief consumers of coal in London : " Brewers, Diers, Lime-burners, Salt and Sope-boylers. . . . "

[4] A. Anderson, Origin of Commerce, 1801, vol. i, p. 148.

[5] The brewers, however, are mentioned, in a paper written by one William Dyoss in 1628, as being, with the blacksmiths, the only users of coal in London in " ancient tyme " (Egerton MSS., 2533, f. 24). Doubtless his statement was based on hearsay. In the seventeenth century, the beer brewers, like the soap boilers and the sugar refiners, used a better grade of mineral fuel than the " pancoal " which served for salt and alum making (Additional MSS., 12496, f. 87, and see above, p. 210). As this better grade was not to be had in London during the Middle Ages (see above, pp. 12–3), it may be doubted whether coal was used much in brewing before the sixteenth century.

[6] The Request and Suite of a True-hearted Englishman, in Tawney and Power, Tudor Economic Documents, vol. iii, p. 144.

Whatever effective restraints on the use of coal in dyeing may have existed before the reign of Elizabeth evidently disappeared with the rise in timber prices, for in 1578 the London Company of Brewers sent a petition to Walsingham, pointing out that its members, together with " diars, hat-makers " and others (not specifically named), " have long sithens altered there furnasses and fierie places and turned the same to the use and burninge of Sea Coale ".[1]

In the textile industry the new fuel was used principally for heating the dyers' vats, the cauldrons in which clothiers scoured their wool, cotton, linen, or silk, and the kilns in which they dried flax or yarn, or steamed hats.   Clothmakers of the town of Cranbrook, in the heart of Kent, complained to the Privy Council in 1637 against John Brown, Crown Commissioner for making brass or iron ordnance and shot, on the ground that he had raised the prices they paid for their wood, by burning large quantities in his furnaces.   Brown insisted, in his defence, that while his own work could not be done without wood fuel, there was no similar necessity for burning it in the making and dyeing of cloth, which could be done at Cranbrook more cheaply with New- castle coal, brought overland from Maidstone or Newenden.   To convince the Council, Brown submitted samples of the material worked in Cranbrook to four London clothiers, who certified that " this wool of these several colours " may be dyed with seacoal as well as with wood, or better.   "We here in London", the certificate proceeds, dye wool, cloths, silks, and other stuffs with seacoals, " by which we find many advantages both for profit and commodity ".[2]

Coventry had been for centuries a leading textile centre, and, during Charles II's reign, much cloth was " woaded, boyled, and maddered " to be made into hats and caps.[3]   Richard Waterfall, a hatter of Coventry, told Commissioners appointed to investigate a dispute over a mine in Warwickshire that he " spends a greate quantitie of coles yearely ".[4]

The textile manufacture should probably be ranked among the half dozen seventeenth-century trades which absorbed the largest quantities of coal.   Although the fuel costs in the production of cloth were a far smaller portion of the total costs than the fuel costs in other industries, such as lime burning or soap boiling (where, in turn, fuel costs were smaller than in salt or alum making), the textile manufacture was so overwhelmingly the greatest British industry under the Stuarts, that the total amount of fuel consumed in it may well have been greater than that consumed either in lime burning or in soap boiling, just as the total amount consumed in soap boiling may have been greater than that consumed in alum making.

Brewing also must be ranked among the chief coal-consuming industries before the Industrial Revolution.[5]   In the latter half of the seventeenth century no London breweries could operate without

[1] S.P.D., Eliz., vol. cxxvii, no. 68.
[2] S.P.D., Charles I, vol. ccclxiii, nos. 55–7 ;   vol. ccclxxi, no. 23.
[3] V.C.H. Warwickshire, vol. ii, pp. 265–6.
[4] Exch. Deps. by Com., 36 Charles II, Mich. 43.
[5] In France as well as in Great Britain (Boislisle, Correspondance des Contrôleurs Généraux, vol. iii, no. 1175).

supplies of coal, and some of the brewers purchased as many as 500 tons a year for carrying on their business.[1] Fuel was needed not only for heating the vessels in which beer and ale were brewed, but also for distilling spirits or vinegar, and for refining cider and other beverages.[2] It was needed, moreover, in preparing the materials from which these liquors were made. But the substitution of coal for wood in the drying of malt and hops presented a technical problem of some magnitude. We shall consider this substitution, therefore, under our third heading.

(iii) *Trades in which the substitution of coal for wood involved technical problems*

We have referred to the opinion that coal might transmit its unpleasant properties to beer or to cloth as a prejudice ; and a prejudice it was, whenever, as in brewing or dyeing, the product was separated by means of a cauldron or vat from direct contact with the flames and smoke. Where, however—as in drying malt or in smelting metal— fuel had to be mixed in with the grain or the ore, the noxious, gaseous content of coal might really damage the taste or the quality of the product. While a rapid rise in timber prices led almost automatically—if gradually—to the general adoption of mineral fuel in trades like brewing or dyeing, there were a large number of industrial operations in which coal could not be substituted at all until some technical alteration had been made in the process of manufacture. It was necessary either to free the coal from its damaging properties, or to invent a device to protect the raw material from the flames and fumes.

When, in 1610, Sir William Slingsby and his partners petitioned for a patent to introduce mineral fuel in various processes, they distinguished between those manufactures in which pit coal had been successfully tried in place of wood, and those in which attempts at substitution hitherto had failed. In the first group they included all boilings—" Bear, Dies, Allom, Sea Salt ", etc. In the second group they put the baking of " Malt, Brede, Brycke, Tyles, Pottes ", and the melting of " Bell mettal, Copper, Brass, Iron, Leade, and Glass ".[3] Already a share of human ingenuity was directed towards finding the technical alterations necessary before coal could become the universal fuel.

Attempts to dry malt in kilns with raw coal gave unsatisfactory results,[4] except in Pembrokeshire, where the smokeless anthracite proved an eminently satisfactory fuel for the maltster.[5] Straw served

---

[1] *P.C.R.*, vol. lviii, p. 116.

[2] The casks in which the cider was put were scented by a mixture of brimstone, burnt alum, and water, " melted over hot coals " (Houghton, *op. cit.*, vol. i, p. 156).

[3] *Harleian MSS.*, 7009, f. 9.

[4] In 1635 one Sir Nicholas Halse received a patent for a new kiln, designed " principally for the Sweet and Speedy Drying of Mault and Hoppes by the use of Sea Coales . . . without touch of Smoake " (*Patent Specifications*, no. 85 ; see also *ibid.*, 71B).

[5] Owen, *Description of Pembrokeshire*, pt. i, p. 87.

as the chief firing until about the middle of the seventeenth century, when maltsters in Derbyshire hit on a method of ridding coal of part of its gaseous content by reducing it to what they called " Coaks " —nothing more nor less than crude coke.[1] Henceforth nearly all beer brewed in Derbyshire was made from malt dried with these " Coaks ", which, until the end of the seventeenth century, were obtained almost exclusively from a special grade of hard coal dug near the town of Derby.[2] So successful proved the new process that Derbyshire beer was actually preferred to other brews, and became famous throughout England.[3] Scottish maltsters also made use, as early as 1662, of " charcoal made of pitcoal ", but they preferred peat when it could be had.[4]

As late as 1635, public " bakers and baking cookes " are said to have burned " noe other fewell but wood, which they obteyne att deere rates ". Henry Sibthorpe was granted, therefore, a patent for a new oven which one " may heat . . . with Seacoales, or any other Coales digged out of the Earth ". Another patent for the same purpose was granted in the following year.[5] These patents do not prove that no coal had been used in baking bread before 1635. Investigation shows that, in many cases, would-be inventors, in their desire to obtain patents, might swear to the absolute novelty of a process which did not contain a single new idea.

In baking bricks and tiles the technical problem would appear to have been less serious than in malting and bread making. Artisans succeeded in producing both bricks and tiles from clay burnt in the midst of a fire of raw coal. Biringuccio, the Italian author of one of the first important books on industrial technique, writes in 1540 that mineral fuel is employed in brick making, as well as in smith's work and in limeburning.[6] A decree of 1585 recommends the burning of coal in brick work near Alfeld in the Harz Mountains.[7] One must doubt, therefore, whether it was strictly true (as Slingsby and his partners wished Privy Councillors to think) that the new fuel had never served the English brick makers before 1610. A method of baking bricks with coal was known to Rovenzon when he published

---

[1] See below, p. 249.

[2] Plot speaks of this process as if it were carried on also in Staffordshire (*Natural History of Staffordshire*, p. 128). At the end of the seventeenth century, coke of local manufacture was used for drying malt along the Lincolnshire coast (*Royal Soc., Philos. Trans.*, vol. xix, 1695-7, pp. 343 sqq.).

[3] Houghton, *op. cit.*, vol. i, pp. 108-9 ; Galloway, *History of Coal Mining*, p. 47 ; *V.C.H. Derbyshire*, vol. ii, p. 354. By the end of the seventeenth century hops were dried with any available fuel (Houghton, *op. cit.*, vol. ii, pp. 452-4).

[4] Birch, *History of the Royal Soc.*, vol. i, pp. 169-70.

[5] *Patent Specifications*, nos. 86, 94.

[6] *Pirotecnica*, Venice, 1540 ; cited Beck, *Geschichte des Eisen*, vol. ii, p. 101. But at Calais in 1540 bricks were made with firewood, although lime to be used in the same construction work was made with coal (*Cal. of Letters and Papers of Henry VIII*, vol. xvi, no. 98).

[7] O. Hue, *Die Bergarbeiter*, vol. i, pp. 350-1. The method of brick baking with coal could hardly have been common knowledge, however, in sixteenth-century Germany. In 1763 Liége brickburners were brought to Bavaria to develop there the manufacture with mineral fuel (*ibid.*, vol. i, p. 344), so that the successful accomplishment of this operation must hitherto have been limited to certain areas.

his *Treatise of Metallica* in 1613 ; and in 1618 the Venetian ambassador describes in detail the process as it was then carried on, probably in the neighbourhood of London. " At the bottom of the kilns ", he writes in a letter to the Doge and Senate, " they have a number of small furnaces for the sole purpose of lighting the fire, and when they pile the bricks and tiles they place a layer of coal dust between each tier in succession up to the very top of the kiln so that when the furnaces below are lighted, the fire spreads throughout, and when the coal is consumed the kiln remains seasoned, without the superintendence of anyone, while the bottom becomes cool long before the extinction of the flames above ".[1]

Although this process could arouse enthusiasm in a foreign diplomat, who must have been a novice in technique, an English inventor like Rovenzon was far less sanguine over the achievement. Newcastle coal, " as it is now used ", he wrote, " doth many times spoyle. much of the bricke-clampe, by making it run together in a lumpe ".[2] In 1663, Sir Balthazar Gerbier, the well-known courtier, who attempted to present himself to the public as an expert on every subject from the principles of banking to the art of rhetoric, estimated in a book entitled *Counsel and Advise to all Builders*, that, out of every 20,000 bricks " burnt with Seacoles ", at least 5,000 were unfit for use.[3] Possible improvements in the process engaged the attention of inventors and men of science. They observed considerable differences in the damage done by different kinds of coal. Rovenzon pointed out that " bricks may bee made of pit-coal or stone-coal, or any other of the priviledged fewels, better than with New-castle sea-cole ". It was seen that some grades gave off a larger quantity of gas than others. " By lighting a pipe with the several coals ", Houghton remarked, " a vast difference may be perceiv'd ; that of soft coal giving a taste of the mineral, the other is pleasant as if lighted with paper ".[4] The problem was to find a grade relatively free from gas, or to adopt some coal from which a large percentage of the gas could be removed without destroying the burning properties. Towards the end of the seventeenth century a mixture of " coal-ashes " with street sweepings or clay, known as " Spanish ", often replaced ordinary coal in the brick kilns—and evidently to good purpose.[5] For, in considering causes for the great expansion in the Newcastle trade between 1634 and 1674, Petty lays special stress upon the increase in the number of bricks burnt with coal.[6]

According to Plot, seven tons were needed to burn a clamp of 16,000 bricks ; and, as it required 9,000 bricks, and nearly two tons of lime, to make one square rod of a wall two feet thick, it is evident that building operations were bound to involve a heavy consumption

[1] *Cal. S.P. Venetian*, 1617–19, pp. 320–1.
[2] *Treatise of Metallica*, 1613, p. 13.
[3] Gerbier, *Counsel and Advise to all Builders*, 1663, pp. 52–3.
[4] Houghton, *op. cit.*, vol. i, p. 119.
[5] *Ibid.*, vol. i, pp. 186 sqq. " By that shift ", wrote Defoe (*The Complete English Tradesman*, 1726, vol. i, p. 35), the brick makers " save eight chaldrons of coal out of eleven ".
[6] Petty, *Political Arithmetic*, 1690, p. 99.

of coal.[1] The quantity used in brick making varied greatly as between kilns. The stiffer the clay, the greater the quantity of fuel. In general one ton of coal or " Spanish " could replace two wagonloads of wood. There is little reason to doubt that, before the end of the seventeenth century, most English brick making, and doubtless the baking of nearly all pottery and earthenware, was undertaken with the help of a coal or cinder fire.

Tobacco pipe makers are known to have employed coal early in the seventeenth century. Nearly a ton was consumed in baking 24 clay pipes by the method used in this manufacture at Broseley. The process undoubtedly involved a good deal of breakage.[2] Some time before 1639 the London Company of Pipe-makers began to pay an annual salary of £40 (by no means poor remuneration in those days) to a man who instructed the members " how to burn with sea-coal ; whereby a great quantity of wood was preserved."[3] Charles II incorporated the company on the understanding that the members should use only mineral fuel.[4]

That was the basis upon which Mansell and his partners received in 1615 their well-known patent for the exclusive manufacture of glass in England. The problem of substituting coal for firewood in this manufacture had apparently concerned King James himself, and a sum of £5,000 is said to have been spent on preliminary experiments.[5] Coal was first successfully burned in a glass works before 1612. " Very lately ", wrote Sturtevant, in his *Metallica*, which was published in that year, " by a wind furnace greene glass is made as well by pit-coale at Winchester House in Southwark as it is done in other places with much wast and consuming of infinite stores of billets and other wood-fuell ".[6] The change from wood had been made possible by closing the clay pots, or crucibles, in which the potash and sand were melted down, so that the glass need not come into direct contact with the

---

[1] Plot, *op. cit.*, p. 128 sqq. ; Gerbier, *op. cit.*, pp. 55–61. In Staffordshire the brick makers continued to burn raw coal as it came from the pit. Defoe (*loc. cit.*) implies that 11 chaldrons served to burn 100,000 bricks. Gerbier (*op. cit.*, pp. 52–3) says that 15 loads of wood were required to burn a clamp of 20,000 bricks. According to Houghton, " sometimes they have been better burnt with two chalder and a half [3¾ tons] than at other times they have been burnt with ten chalder [13¾ tons] ".

[2] As Professor Gay suggests to me. See *V.C.H. Salop*, pp. 440–1. Twenty-four was the maximum number of pipes that could then be baked at once in a single kiln.

[3] *Cal. S.P.D.*, 1639, p. 384.   *Exch. Spec. Com.*, no. 4242, deals with pipe making in London.

[4] *Cal. S.D.P.*, 1663–4, p. 126. As in brick making, about 20 per cent of the product was unfit for use (*V.C.H. Salop*, pp. 440–1). Towards the end of the seventeenth century, tobacco pipes, like bricks, were usually baked in a " cinder ", rather than in a raw coal, fire. " I presume ", wrote Houghton, (*op. cit.*, vol. i, p. 205), " coals that smoke will spoil colour ".

[5] *Pat. Roll* 11 James I, pt. xvi, no. 4 ; cited *V.C.H. Surrey*, vol. ii, p. 301.

[6] S. Sturtevant, *A Treatise of Metallica*, 1658 ed., p. 4. Dud Dudley, writing fifty years later, claimed that glass melting with coal had been first successfully accomplished at Dudley (*Mettallum Martis*, 1665, p. 35). Hartshorne thinks some coal must have been used in the glass furnaces as early as 1608 (*Old English Glasses*, p. 182). The first patent which mentions the use of coal, is that granted to Slingsby in 1610 (see above), and it is certain that the process could have had little commercial importance before that date.

fire in the furnace. This invention was attributed to Thomas Percivall, one of the partners in the Mansell patent of 1615.[1] " We could not heretofore be induced to believe ", one reads in the preamble to a preliminary patent granted in 1614, " that it would ever have been brought to pass as we are assured thereof by plain and manifest demonstration, several furnaces . . . now being at work ".[2]

The first successful experiments were made with Scotch coal, which, as has been explained already, was relatively free from noxious gases.[3] For several years the glaziers regarded Newcastle " seacoal " as altogether too sulphurous to be of service. Owing to the high cost of importing fuel from Scotland, Mansell attempted to melt glass with Welsh anthracite in a furnace which he had erected near Milford Haven, and he also tried to carry on a manufacture near the river Trent in Nottinghamshire, where the mines yielded a mineral somewhat similar to that found near the Firth of Forth. But the cost of transporting the finished glass from Wales or Nottingham to London outweighed all the advantage which could be obtained from lower fuel costs near the collieries. " For his last refuge ", if we may credit a story inspired by Mansell himself, he " was enforced . . . , contrary to all men's opinions, to make triall at New-castle upon Tyne, where, after the expense of many thousand pounds, that worke for Window-glasse was affected with New-castle Cole ".[4]   Before 1624 it had been found feasible to make white glass, crystal glass, and all other kinds of glass, with common " seacoal ".

Not all the glaziers, of course, abandoned wood fires immediately after the proclamation of 1615.[5] Eventually, however, economic law came to the aid of statute law : the rising price of timber made the costs of the older fuel prohibitive, while it was found that better glass for daily use could be produced with coal than with wood. Before 1700 coal must have been almost the only fuel burned in all the British glass works. At a furnace in Nottinghamshire belonging to Sir Percival Willoughby, from 500 to 600 tons of coal had been thought necessary in 1615 for the production of 800 cases, or 80 tons, of glass—one year's output. As in boiling down salt and baking bricks, the amount of coal consumed varied according to the nature of the raw product. " Broad glass spendeth both more coales and asse [ash], quantitie for quantitie, then drinking glass dothe ".[6] Perhaps one would not be far wrong in estimating the annual consumption of coal in glass manufacture just before 1700 (when there were 90 furnaces

---

[1] *Pat. Roll* 12 James I, pt. iii, no. 9; as cited *V.C.H. Surrey*, vol. ii, p. 301, where a good account of these patents may be found.

[2] *Pat. Roll* 11 James I, pt. xvi, no. 4 ; cited as above.

[3] See above, pp. 117–20.

[4] *S.P.D., James I*, vol. clxii, no. 63 (printed Hartshorne, *op. cit.*, pp. 426–31). See also the answer to Mansell as given in *S.P.D., James I*, vol. clxii, no. 64 (printed Hartshorne, pp. 423–6).

[5] Cf. *Acts of the Privy Council*, 1615–16, pp. 469–70, 471–2.

[6] *H.M.C., Report on the MSS. of Lord Middleton*, pp. 500–1. It is estimated that the cost of coal for the furnace will be £125 for the year. As " everie rooke of coales " cost 5s. 6d., and as a " rooke " was slightly more than a ton (see Appendix C), I have concluded that from 500 to 600 tons was probably burned at such a furnace.

in England and probably a score or more in Scotland [1]) at more than 50,000 tons.

When we remember that at least 300,000 tons more were probably burned in the manufacture of salt, that metal work, lime burning, textiles, brick making, and brewing each required tens of thousands of tons every year, that the firing of tobacco pipes, earthenware and pottery of all kinds, the making of soap, starch, and candles, the public baking of bread, the production of alum, copperas, saltpetre, and gunpowder, all involved heavy expenditures for coal, it hardly seems an exaggeration to estimate the coal annually burned in manufactures at nearly a million tons before the eighteenth century. Thus approximately a third of all the fuel mined in Great Britain was consumed in native industries, at the very time to which Macaulay referred when he said that coal was little used in manufactures.

Judged by modern experience, one-third of the output of coal seems but a small proportion to use for industrial purposes. Of the 183,000,000 tons consumed in Great Britain in 1913, more than four-fifths was burned in industry. If a survey shows that, in the late seventeenth century the proportion was vastly smaller, this must not be taken to indicate that at that time manufacturers generally burned firewood, turf, or peat rather than coal. What it shows is that the demand for fuel of all kinds in industry was relatively unimportant in comparison with the demand to-day. England was not yet the home of the steam engine and the express train. Gas and electricity had still to be invented. Nearly half of all the coal consumed in industry in 1913 went into channels—railways, coasting steamers, pumping engines, gas works, etc.—for which there was no equivalent in the seventeenth century. Were we able to carry our calculations back to the reign of Henry VIII, and to estimate the quantity of wood, turf, and peat then consumed, we should find that manufacturing processes absorbed a far smaller proportion of all fuel produced than in the reign of Charles II. If the country used two-thirds of its fuel for household purposes in 1688, it may well have used five-sixths in 1550. What is astonishing is the great quantity of coal burned in industry a century before the Industrial Revolution is supposed to have taken place.

Once the technical obstacles had been overcome, coal might be substituted for wood in an industrial process even more rapidly than in the domestic hearth. For where, as in malting or glass making, the product from a kiln or furnace heated by coal was found to be superior in quality to that from one heated by wood, an actual premium was put on the use of mineral fuel. No amount of rhetoric, on the other hand, could persuade our ancestors, any more than it can persuade us, that coals make as pleasant a fire in the drawing room or the bed chamber as do logs. While there were Londoners who maintained that seacoal fumes counteracted the city smells of sewage, these fumes remained, as Smollett said, " a pernicious nuisance to lungs of any delicacy of texture ". So long as men could afford the luxury, they clung to wood fires. The difference between the price of wood and coal was generally great enough to oblige the majority of householders to burn coal ; but it is almost certain that, in trades where the substitution involved no technical problem, or where the technical problem had been completely solved, it required a smaller advantage in the price to induce the artisan than to induce the householder to change his firing. After the Restoration a brick maker had to pay about twenty shillings to purchase enough firewood to give him the equivalent in heat of one ton of coal.[1] As soon, therefore, as mechanical improvements made it possible to produce satisfactory bricks in coal-heated kilns, the brick maker was forced by competition from his rivals to purchase mineral fuel (or " Spanish "), wherever he could get it for less than twenty shillings per ton, as he could in London, Dublin, Cambridge, Leicester, York, Birmingham, Edinburgh, Glasgow, and in nearly every district where there was any considerable concentration of population or of industry.

Before 1700 the substitution of coal for firewood had relieved the pressure upon timber supplies in England, except in those localities where metallic ores were smelted. While Evelyn could warn the nation that, to strengthen its naval power, it must look to the planting of trees,[2] while the Crown continued to show concern over the inroads made upon the woodlands,[3] the danger of exhaustion, as reflected in the writings of persons concerned with economic affairs and in the actual price of timber, was an issue far less ominous in the decade from 1690 to 1700 than in the decade from 1600 to 1610. A French visitor, well informed on many questions, could write in 1698 of the abundant forests, and could argue that the use of coal by the English could not be ascribed to any lack of wood.[4] " I do not find logs especially dear ", he added. While the price of firewood had increased so much more rapidly than the price of commodities generally during the century preceding the Civil War, that, according to Wiebe's index numbers, the ratio of the former to the latter had risen from about one to about 2·6[5], firewood was certainly no dearer relative to the general price level at the end of the seventeenth century than on the eve of the struggle between Crown and Parliament.[6] Houghton and others found it possible to advocate the destruction of the forests

---

[1] Two loads at 10s. the load (Gerbier, *Counsel and Advise to all Builders*, 1663, pp. 52–3).

[2] *Sylva*, 1664, p. 108.

[3] Cf. R. G. Albion, *Forests and Sea Power*, 1926, esp. chap. v.

[4] H. M. de V., *Mémoires . . . par un voyageur en Angleterre*, p. 196.

[5] See above, p. 158.

[6] G. Wiebe, *Zur Geschichte der Preisrevolution des 16 und 17 Jahrhunderts*, 1892, p. 375.

	1451–1500	1633–42	1643–52	1653–62	1663–72	1673–82	1683–92	1693–1702
General Prices .	100	291	331	308	324	348	319	339
Price of Firewood	100	780	490	662	577	679	683	683

These figures are naturally very imperfect, and the fluctuations in timber prices from decade to decade appear to be impossibly wide, but they may be taken to show that the price of firewood remained fairly stable after 1642.

as being useless ornaments to a country that had begun sometimes to talk in terms, so common in our day, of pure economic efficiency.[1] Their advice apparently did not go unheeded ; the timber shortage had been so much eased by the adoption of the coal fire that Defoe, in commenting upon its use in the upper rooms of London taverns, could write : " It is not immaterial to observe what an Alteration it makes in the Value of those Woods in Kent, and how many more of them, than usual, are yearly grubbed up, and the Land made fit for the Plough ".[2]

Gregory King, in 1696, estimated the value of all timber annually felled in England at £1,000,000, less than half of which was consumed for firing.[3] If we assume that it cost on the average about 20 shillings to buy enough firewood to obtain the same heat value as could be obtained from one ton of coal,[4] then it is clear that the timber annually burned for fuel in England was equivalent to about 500,000 tons of coal. As nearly 3,000,000 tons of the latter were probably burned in Great Britain at this time,[5] it is evident that coal must already have been of overwhelmingly greater importance than firewood in the economy of the British people.[6]

" Coal is one of the greatest sources of English wealth and plenty ", writes Monsieur Ticquet in his letter to a Frenchman concerning the industrial progress which he observed across the Channel. Although written in 1738, some time after the period with which we are specially concerned had ended, this report would appear, from what we know of the use of coal in industry, to apply almost equally well to conditions at the end of the seventeenth century. " I regard it ", he continues, " as the soul of English manufactures. . . . Here coal serves for all the domestic uses for which we employ wood—whether for the hearth, the kitchen, or the laundry. Bakers and pastry cooks heat their ovens with it. Brewers, sugar refiners, dyers, hatters, confectioners, distillers, who are here very numerous, vinegar and soap makers and bleachers of linen cloth heat their boilers with it. In a vinegar factory in London there are £40,000 worth of boilers and other vessels. The window panes, which are more beautiful here than in France, the whitest cut glass, the drinking goblets, and the mirrors are melted with no other fuel than coal. Carpenters never employ planks, even when they were cut ten years before, unless they have been dried in a sort of stove by this mineral. Coal is burned to bend not only planks but also the largest beams used in shipbuilding. Coopers have no other fire. Ordinary pottery and earthenware are baked with coal. You know that they bake their bricks with coal in Flanders and even in the neighbourhood of Marly. But, as the use of bricks is far more extensive in this country than in Flanders, they make different kinds, which

---

[1] See above, p. 190.
[2] Defoe, *Tour*, 1748 ed., vol. i, p. 138.
[3] King, *Natural and Political Observations*, in Chalmers, *Comparative Strength of Great Britain*, 1804 ed., appx., p. 53.
[4] See above, p. 221.
[5] See above, p. 20. Not much more than 200,000 tons was shipped to Ireland and abroad (see above, p. 79).
[6] King's estimate leaves Scotland out of account. There can be little doubt that in the more populous parts of that kingdom coal was quite as generally used for fuel as in England (see above, pp. 45, 94–5, 105, 107–8).

they bake in ovens of different forms, of which I will bring you the designs.   All workers in iron and in copper make use of coal. . . .   I explored the shires between Monmouth and Warwick, filled as they are with iron and copper manufactures, in the company of my [English] friend, who was well content to show me the wealth of the workers in this district.   I observed with surprise the skill of these artisans and the comforts they enjoy.   Their villages seemed to me as well built as the finest towns in Flanders, and I think they are richer. The prodigious consumption of provisions brought into the markets astonished me greatly . . . I noticed . . . a large number of processes which are unknown to our workers, notes of which I will show you in Paris.   It is astonishing with what ease they make their ironmongery and the small jewelry which they send to India" [1]

[1] *Archives Nationales*, O¹1293.

CHAPTER IV

THE ROLE OF COAL IN BRITAIN'S ECONOMIC SUPREMACY

" CHARCOAL causes many ore mines to lie unwrought for want of fuell ",
a petitioner wrote in 1623, in urging the Privy Council to support a
process which some were sanguine enough to think would make it
possible to smelt and to refine lead with seacoal or pitcoal.[1] Not
until the very end of the seventeenth century, however, did such a
process prove a commercial success,[2] and not until the American
Revolution had coal generally replaced charcoal in smelting.[3] Meantime, where iron stone was concerned, woods remained, to quote
Yarranton, " as the Breast is to the Child ; let that cease, all dies ".[4]
Long before wood actually " ceased ", the activities of the smelters
diminished.   Just as to-day in the case of coal it is less the ultimate
exhaustion of the seams, than the immediate increase in mining costs,
that handicaps British industrialists in competing with the products
from other countries where the mines have been less intensively worked,
so in the seventeenth century it was primarily the high price of timber
that tended to check the conversion of iron ore into metal.   Firewood
for a furnace in Wyre Forest sold in 1630 for about three times the
price it had brought at a furnace in Cannock Chase, in 1588.[5] Since
the purchase and carriage of wood, and its conversion into charcoal,
had represented more than half the total cost of producing a ton of
iron in 1588,[6] it is evident that the increasing charge for fuel must
have pushed the price of iron so high that small metal workers had
to economize on their supplies.   Iron imported from Sweden, Flanders,
and Spain could undersell the product from the furnaces in south-
eastern England.[7] Even had there been a market for metals at any
price, the State could not have allowed smelters to exhaust indefinitely
timber deemed necessary to supply the navy.   As it was, Parliament
passed legislation to protect the forests from the ravages of the iron
masters as early as 1558 ;[8] and, from that time onward, statesmen,
by statutes, proclamations, and administrative action, limited, in
nearly every district of England and Scotland, their right to take
timber.[9] The failure to smelt with coal alone explains the British

---

[1] S.P.D., James I, vol. cxlv, no. 66.
[2] Lewis, The Stannaries, p. 24, n. 4.   [3] See below, p. 251.
[4] Yarranton, England's Improvement, pp. 147–9.
[5] V.C.H. Shropshire, vol. i, p. 460. The smelters paid 2s. 4d. and 3s. 2d.
per cord in 1630, as against 1s. in 1588. And the price in 1588 was much higher
than in 1550 (cf. above, pp. 158–61).
[6] See above, p. 193, n. 7.   [7] Yarranton, op. cit., p. 149.   [8] 1 Eliz., cap. 15.
[9] V.C.H. Worcestershire, p. 267 ; Harleian MSS., 6838, ff. 152–6, 165–6,
171–6 ; R. R. Steele, Bibliography of Tudor and Stuart Proclamations, no. 1750 ;
P. Hume Brown, Scotland before 1700, pp. 273–4 ; W. I. Macadam, " Notes
on the Ancient Iron Industry of Scotland ", in Proceeds. of Soc. Antiq. of Scotland,
N.S., vol. ix, 1886–7, pp. 89–90 ; Ehrenberg, Hamburg und England, 1896,
p. 4 ; S.P.D., Charles I, vol. cccxxi, no. 42 ; Cal. S.P.D., 1637–8, p. 239 ; 1653–4,
pp. 142, 152, etc.

policy of encouraging imports of pig and bar iron from the American colonies, while rigidly excluding imports of finished metal goods.[1] A continental writer on commerce remarked, in the eighteenth century, that the destruction of wood had cost the English all their advantage in mines of lead and iron.[2]

This destruction must have cost them such advantages as they possessed in other industries as well, had they not been able to substitute coal for wood. Manufactures which consumed relatively little fuel would, no doubt, have suffered less than others, but there was no manufacture that did not benefit by cheap firing material. If all tradesmen had been forced to depend entirely on firewood, the price of timber must have risen far higher than it did. There could have been only one remedy—short of depending on imports of foreign timber—the intensive planting of trees. Such a step would, in its turn, have involved continual encroachments upon pasture lands and harvest fields, and therefore a decline in the output of textiles and of grain.[3]

Even had there been enough timber to keep down the price, would the history of British manufactures during the seventeenth century have been the same ? There is good reason to think not, and to believe that before 1700 coal had already a positive, if not always an easily discernible, influence on the development of industry, an influence extending far beyond its use as a substitute for wood. Coal being, as we have had occasion to remark, so much cheaper in the immediate neighbourhood of the mines than at a distance, it is clear that, if other considerations were equal, manufactures were more likely to thrive in the mining districts than elsewhere. The evidence suggests that the beginning of the momentous economic movements, which to-day have concentrated the heavy industry both of western Europe and of the United States round the coal-fields, may be traced farther back than their traditional starting point in the late eighteenth century into Stuart times.

For industrial processes in which the expense of coal was small in comparison with the total costs of production, the advantages of setting up business in colliery districts were hardly likely to weigh much against other considerations. At Keswick, during Elizabeth's reign, copper makers, who needed a little mineral fuel in addition to large quantities of charcoal, thought it worth their while to advance money towards a search for coal seams in the immediate neighbourhood of the deposits of copper, but there is no evidence that they even weighed the possible advantages of carrying either the ore or the smelted copper to a coal mine already discovered at a distance of some twenty miles from their plant.[4] During the seventeenth century, however, those who worked ores into finished products sometimes thought it a decisive advantage to be as near as possible to a supply of mineral fuel. A skilful smith, wrote Colbert in 1673, when discussing means

---

[1] Beer, *Commercial Policy of England towards the American Colonies*, pp. 84–5, 88–9.
[2] [? Accarias de Serione] *La richesse de l'Angleterre*, Vienna, 1771, p. 41.
[3] Cf. above, pp. 190–1, 221–2.
[4] Collingwood, *Elizabethan Keswick*, pp. 33, 36, 44, 73.

of provisioning the French navy, could make anchors more cheaply and easily at Rochefort than anywhere else, since he could have coal better and cheaper there than in the heart of the kingdom, and could select his iron from those places where it is of the best quality, and it was easier to transport from the heart of the kingdom iron suitable for making anchors than the anchors themselves, owing to their great weight.[1] It would be stretching conjecture too far to suppose that the movement of the English iron industry away from Surrey and Sussex into the western counties along the Welsh border [2]—a movement not yet completed at the beginning of the eighteenth century—can be ascribed to the absence of coal in the former, and the presence of coal in the latter, district. This movement was due primarily to the exhaustion of timber supplies in the south-eastern counties, where the State made its first serious attempt to protect the forests from the ravages of glass makers and iron masters, and, whenever possible, forced them to move away. Yet it is certain that these masters, having migrated to the counties on both sides of the Severn, found at least one reason for remaining there in the abundance of mineral fuel, which, together with the deposits of iron ore, attracted every kind of smith and metal worker to whom they could sell the product from their furnaces.[3] " In all these [western] Countries ", wrote Yarranton, " there is an infinite of Pit Coals, . . . and the Iron Stone growing with the Coals, there it is manufactured very cheap, and sent all England over, and to most parts of the World ".[4]

Cheap coal supplies cannot be regarded as an important factor in the westward movement of the textile industry during the fifteenth and sixteenth centuries, from York and Beverley into the West Riding and southern Lancashire. That movement has been explained as an escape from the gild restrictions of the older towns, on the part of cloth makers who wished to meet a new demand for the export of woollens.[5] Yet there can be no doubt that during the seventeenth century these cloth makers, especially the scourers and dyers, found a considerable convenience in the quantity of surface coal available all through the district in which they settled. It is tempting to believe that there was a connection between the success of the textile manufactures—

---

[1] Clément, *Lettres, instruction et mémoires de Colbert*, vol. iii, p. 501.

[2] In 1720, when the total annual production of iron in England and Wales was thought by one writer to have been 17,350 tons (Scrivenor, *History of the Iron Trade*, p. 59 ; but cf. Ashton, *Iron and Steel*, p. 235), only 2,000 tons were produced in Sussex, Kent, and Hampshire, as against 10,600 tons in Gloucestershire, Monmouthshire, Herefordshire, Shropshire, Cheshire, Staffordshire, and Worcestershire. The remainder came from Wales, Yorkshire, Derbyshire, Warwickshire, and Nottinghamshire.

[3] Another probable reason was the difficulty experienced by the masters of the south-eastern counties in competing with foreign iron smelted with cheaper firewood. At inland towns, foreign iron must have been more expensive than near the coasts.

[4] *England's Improvement*, pp. 58–9. Yarranton is referring to Worcestershire, Staffordshire, Shropshire, Derbyshire, Warwickshire, Monmouthshire, and the Forest of Dean. Cf. Dud Dudley's remark (*Mettallum Martis*, 1665, p. 53) that, " had not these parts abounded with cole, it would have been a great deal worse with [the smiths] then it is ".

[5] Heaton, *Yorkshire Woollen and Worsted Industries*, pp. 45–7.

which clustered about Leeds, Wakefield, Bradford, Halifax, Burnley, Bolton, Rochdale, and Manchester—and the scores of small colliery enterprises which sprang up in almost every parish. " Such, it seems ", wrote Defoe, when he visited Halifax early in the eighteenth century, " has been the Bounty of Nature to this Country that two Things essential to Life, and more particularly to the Business followed here, are found in it, and in such a Situation as is not to be met with in any Part of England, if in the World beside ; I mean Coals, and running Water on the Tops of the highest Hills. I doubt not but there are both Springs and Coals lower in these Hills ; but were they to fetch them thence, it is probable the Pits would be too full of Water ; it is easy, however, to fetch them from the upper Parts, the Horses going light up, and coming down loaden ".[1]

For industrial processes in which the expense of buying fuel represented a large portion of the total cost of production, the early importance of coal as a factor in determining the place of manufacture is incontestible. Crown monopolists who controlled the alum works near Whitby found constant difficulty during the years before the Civil War in making a profit on their sales, owing to the high charge of shipping coal from the Wear and hauling it overland to the pans. More than £150 was spent in one year (1613–14) in a futile search for coal mines near the Yorkshire alum houses.[2] Manufacture remained there, nevertheless, for the monopolists did not believe they could gain any appreciable advantage by carrying the stone to the collieries. In 1612 William Phillips of Newcastle came forward with the announcement that a mine of alum contiguous to a mine of coal had been found within seven miles of his native town.[3] He estimated the probable saving in fuel costs, if the works could be transferred from Yorkshire to the Tyne, at 20s. per Newcastle chaldron. We do not know whether he succeeded in his project of starting a local manufacture, but, if he failed, it must have been because the alum stone was either not sufficiently abundant or of poor quality.[4]

It was cheap coal, of course, that led to the concentration of the salt manufacture at Shields, on the Firth of Forth, and at other points along the coasts where seams outcropped.[5] Pan owners at Yarmouth, King's Lynn, or in the south of England, having, like the alum monopolists, to bring their fuel supplies some distance by sea, or to buy wood at dear prices, could not long compete with pan owners in the coal fields. With the rising price of firewood, the brine spring

---

[1] *Tour*, 1769 ed., vol. iii, p. 145.

[2] *V.C.H. Yorkshire*, vol. ii, p. 384. Coal could be had cheaper in Yorkshire than in the Isle of Wight, which had probably been the first home of the English alum industry (cf. above, p. 184).

[3] *Lansdowne MSS.*, 152, no. 6, f. 111.

[4] Perhaps the project succeeded. We know that small quantities of alum were exported from Newcastle during the seventeenth century (*Exch. K.R. Port Bks.*, bdl. 191).

[5] Notably at Whitehaven, Blyth, Cullercoates, Seaton Delaval, at Saltcoats, in Ayrshire, and, for a short period, at Brora in Sutherland (*V.C.H. Cumberland*, vol. ii, p. 359 ; Sinclair, *Statistical Account of Scotland*, vol. vii, p. 173 ; Chambers, *Domestic Annals of Scotland*, vol. i. p. 302 ; *Acts of the Parliaments of Scotland*, vol. viii, p. 609).

operators in Worcestershire and Cheshire gradually lost the advantage which their more saliferous liquid gave them over the sea salt makers, who had to consume larger quantities of fuel in producing a ton of salt, for at the " wiches " it became necessary either to buy ever dearer firewood, or to pay for a long haulage of coal.[1] To carry tubs of brine to the collieries would have been even more expensive than to carry sacks of coal to the springs. Once rock salt was discovered, the principal seats of manufacture changed again, because it was easy to bring the rock to the midland coal pits before dissolving it into brine. Rock salt, therefore, could undersell salt made at the wiches, as the operators of brine pits complained, " just so much as the Coals are Cheaper . . . viz. 12s. 9d. or 13s. per ton ".[2] The sea-salt industries at Shields and in Scotland also felt this competition ; and the decline of the salt works in those districts dates from the end of the seventeenth century.[3] Thus we find the chief centres of the salt industry shifting, at the end of the sixteenth century, to the colliery districts of the north of England and the Firth of Forth, and then shifting again, at the end of the seventeenth century, to the colliery districts of Derbyshire, Staffordshire, Shropshire, Flintshire, and Lancashire.

We have seen already how Mansell moved his glass manufacture from one coal field to another, in an effort to combine the advantage of a cheap and suitable fuel with ready access to the market for his output. It occurred to some English and Scottish merchants that the high charge of carrying Scotch coal to London might be overcome by erecting furnaces near the Scottish collieries ; and it was on this basis that glass making was successfully established at Leith and Wemyss.[4] Of the four principal centres of the English glass industry at the end of the seventeenth century, three—Stourbridge, Newcastle, and Bristol—owed their situation chiefly to their proximity to the coal seams, while the fourth—London—retained its importance largely because of the advantages for marketing glass in the thickly settled, commercialized Thames valley.[5]

It would be idle to pretend that the location of the coal seams

---

[1] As late as 1580, the " wiches " were still the principal seat of the English salt manufacture (Robert Hitchcock's " Pollitique Platt for the Development of Fisheries ", printed Tawney and Power, *Tudor Economic Documents*, vol. iii, p. 241), but they soon lost their supremacy in the face of the growing competition from the industry in Durham and Northumberland (see above, pp. 176–9).

[2] *The Case of the Brine-Pits Truly Stated*, (?) 1699. The brine spring operators admitted that it cost their rivals something to carry the rock to the mines, but contended that this extra cost was offset by the fact that it was more expensive to sink a brine pit than a rock salt pit.

[3] See above, pp. 177–8.

[4] *Privy Council Register of Scotland*, 1st Series, vol. xii, pp. xv–xvi, 374, 428, 439, 451–2, 481, 760, 772. There is a possibility that a manufacture of glass was begun in Ayrshire at this time. One of the English capitalists who was helping to finance the Scottish glass venture refers to a trip he made from Edinburgh to " Eyare " (Ayr ?), a town near certain glassworks (*Chanc. Proc.*, *James I*, C. 26/44).

[5] Of the 90 English glasshouses listed by Houghton (*Husbandry and Trade*, vol. ii, pp. 48–9) more than half were situated at places within a few miles of coal pits.

determined the main movement of population in Great Britain during the sixteenth and seventeenth centuries. The rapid growth in the size of London cannot be ascribed to coal.[1] Counties rich in that mineral still ranked among the more sparsely settled shires, when, at the end of the seventeenth century, Houghton used the hearth tax returns to make his estimate of the number of houses in England.[2] Population centred in the Thames valley and in the lower basin of the Severn, neither of which areas possessed substantial mineral resources of their own ; although it is important to remember that, as a result of facilities for transport by water, coal could be had at a far lower price than firewood in nearly all the towns in these valleys, especially along the Severn.[3]

The direct influence of coal can be seen in a number of small movements of population within the mining counties themselves. These movements may be compared to those which are going on to-day in Siberia. Siberia is still the most thinly populated part of Russia, just as the region north of Cambridge and Coventry and west of the Severn remained, on the whole, the most thinly populated part of Great Britain in 1700 ; yet in Siberia there is a tendency for many small industries to concentrate in the neighbourhood of coal mines, which have been of late more intensively exploited. It is possible to observe a similar tendency in seventeenth-century Britain.

Local husbandmen, thrown out of employment by the spread of pasture farming, evicted manorial tenants, and wandering vagrants, some of them uprooted, perhaps, by the social changes which followed the Reformation, began to settle in villages clustered about the more important colliery enterprises of their native counties or regions. During the last half of the seventeenth century, advantage was taken of the cheap coal at Neath to develop the lead and copper industries there, and, as a consequence, the place was transformed from a tiny hamlet into a substantial town. " These Coal-Works and Work-Houses ", boasted Sir Humphrey Mackworth, the well-known midland capitalist who had gained control of the local colliery by marrying the owner's daughter, " employ a great number of men, women, and children, to whom several thousand pounds are paid every year, which circulates in this neighbourhood, and other trades are thereby increased, the market much improved, and the rents better paid ".[4] Thus, if modern phraseology may be permitted, the new employees brought purchasing power, which encouraged the advent of more tradesmen.

Wherever coal seams outcropped near the coast, we find the story repeated. Whitehaven in 1566 is said to have been a hamlet of six houses, with one bark the size of a large row boat, and a feeble export trade in " Herrings and Codfish, nor anything imported but salt ".

[1] For an explanation of the growth of London at this period, see N. S. B. Gras, *The Evolution of the English Corn Market*, pp. 73–7, and *passim*.
[2] See above, p. 81 n. Cf. Mantoux, *The Industrial Revolution in the Eighteenth Century*, revised ed., 1928, pp. 357–9.
[3] Cf. above, pp. 80–1, 96–7.
[4] Quoted G. G. Francis, *Copper Smelting in the Swansea District*, 1867, pp. 82–6.

By means of Sir John Lowther's well-managed exploitation of coal mines in the hinterland, " great numbers of people [had] been drawn to bring their effects, settle their Families and build Houses at White-haven to carry on a trade there " ; and at the end of the seventeenth century the town had a fleet of eighty merchant vessels, substantial exports of salt, as well as of coal, and local manufactures of lime and iron wares.[1]   Sunderland must have been nearly as small a place as Whitehaven in Elizabeth's reign, for Camden makes no mention of either in his *Britannia*, which appeared in 1586, and which seems to give a fairly complete record of English towns and villages.   In 1700 Sunderland had as many inhabitants as Liverpool,[2] and this growth may be ascribed almost entirely to the trade in coal and salt, based on the increasing importance of collieries up the river Wear.[3]   Blyth-nook, according to Defoe, as well as Cullercoates and Seaton Sluice, " derives its Origin from the Coal-Trade, having some advantage from its Situation, which brought it first to be regarded, and has since preserved it in Esteem ".[4]

Newcastle had a different origin ; yet in no district was commercial and industrial expansion, based on mining, carried so far during the seventeenth century as in the Tyne valley.   There were, to continue quoting Defoe, " two Articles of Trade here, which are particularly owing to the Coals, viz. Glasshouses and Salt-pans ".   Limeburning, brewing, shipbuilding and the attendant smith work, hardware " after the manner of Sheffield ", and all the other manufactures which came to line both banks of the river, grew more rapidly because cheap mineral fuel reduced the costs of production.[5]   Besides, as William Phillips pointed out in 1612, in urging the establishment of alum works near Newcastle, the Tyne provided an especially advantageous channel for the export of any commodity " at easie rate, to every porte as well within the realme as without, by shipps which came daylie thither for Coles ". [6]

This concentration of population and of trade in the neighbour-hood of the coal measures was by no means limited to districts near the coasts.   In the hills which slope up from both banks of the Severn at Broseley, there developed during the seventeenth century several industries, lured by the almost unrivalled opportunities for mining coal above the level of free drainage, and by the facilities for shipping goods up or down the wide slow-flowing river.   As early as the reign

---

[1] Fletcher, *Archæology of the West Cumberland Coal Trade*, pp. 271–4 ; cf. Defoe, *Tour*, 1769 ed., vol. iii, pp. 312–15.
[2] About 5,000.   See J. W. Summers, *The History and Antiquities of Sunderland*, 1858, pp. 4–6.
[3] In the eighteenth century the abundant supplies of cheap coal also attracted to Sunderland important glass works and lime kilns (Surtees, *History of Durham*, vol. i, p. 13).
[4] Defoe, *op. cit.*, vol. iii, p. 241.
[5] *Ibid.*, pp. 237, 239.   The Winlaton iron works, which employed 1,500 men, were founded in 1690 (Green, *Chronicles and Records of the Northern Coal Trade*, p. 197).   A collection of letters concerning trade, written from Newcastle between 1677 and 1683 (*Cambridge Univ. Library*, D.d. vii, 26), shows that the town already had considerable dealings in iron.
[6] *Lansdowne MSS.*, 152, no. 6, f. 111.

of James I the town was famous for its tobacco pipes. Smokers all over England habitually asked each other: " Is yours a Broseley? "[1] In connection with the colliery there (as we can see on a map drawn for exhibition at a meeting of the Royal Society), some enterprising traders had begun, before 1700, to extract tar, pitch, and oil from shale, taken from about the coal mines to be distilled in great cauldrons, set up near the rails laid from the pits to the river.[2] At Broseley also bricks, tiles, and all kinds of earthenware were made ; while at Madeley and Coalbrookdale, across the Severn, were " severall Iron Works, Forges and Furnaces ", and a number of lime kilns.[3] Evidence of a most active industrial community can be found long before Abraham Darby brought historical fame to the district by his successful experiment in smelting with coke.

An early eighteenth-century traveller wandering from Birmingham to Wolverhampton, and thence to Stourbridge by way of Walsall, must have been amazed to find the inhabitants everywhere employed in industry ; for the " Black Country " had already taken on its characteristic aspect. Here were collieries, glass works, lime, brick, and malt kilns in abundance ; [4] but the fame of the district rested upon the manufacture of every sort of iron, copper, and brass ware. " We cannot travel far in any direction out of the sound of the nail hammer ".[5] " Every Farm ", wrote Defoe, " has one Forge or more ; so that the Farmers carry on two very different Businesses, working at their Forges as Smiths, when they are not employed in the Fields as Farmers. And all they work they bring to market, where the great Tradesmen buy it up, and send to London ".[6]

The trade of the smith was not a new one, of course. Camden had found Birmingham " swarming with Inhabitants, and echoing with the noise of Anvils ".[7] Yet there is reason to believe that between 1650 and 1700 the population of the town increased from 5,472 to 15,032,[8] a larger number of inhabitants than could then be found in any town of Lancashire.[9] In an area ten miles round Dudley, Yarranton, who knew the country well, found " more people inhabiting and more money returned in a year than is in these four rich, fat counties "—Warwickshire, Leicestershire, Northamptonshire, and Oxfordshire.[10] His statement suggests that a large portion of the population of Warwickshire and Worcestershire, which ranked among the most densely settled English shires in 1700, may have been massed along the Staffordshire county line close to the collieries. For the concentration of industry there must have been due, in part at least,

[1] *V.C.H. Salop*, pp. 440–1.
[2] *Royal Soc., Philosophical Transactions*, vol. xix, 1695–7, p. 544.
[3] *V.C.H. Salop*, pp. 422, 447–9, 458–61.
[4] Plot, *Natural History of Staffordshire*, p. 128.
[5] Quoted, *V.C.H. Worcestershire*, vol. i, p. 272.
[6] Defoe, *op. cit.*, vol. iii, p. 412.
[7] *Britannia*, 1772 ed., vol. i, p. 452.
[8] Hamilton, *Brass and Copper Industries*, 1926, pp. 124–5, 139.
[9] Manchester is supposed to have had less than 10,000 inhabitants at the beginning of the eighteenth century (Mantoux, *The Industrial Revolution in the Eighteenth Century*, revised ed., 1928, p. 365).
[10] Yarranton, *England's Improvement*, p. 52.

to the frequent outcrops of exceptionally thick coal seams. Such an advantage did the artisans derive from cheap fuel, and so famous were they for their skilful workmanship, that, in addition to local supplies of ore, iron was brought from the Forest of Dean and from the American colonies, copper from Sweden and brass from Holland, to be made into finished goods, occasionally destined for export to the very countries whence the ore had come.[1] At the same time, small industrial enterprises developed in increasing numbers about the towns of Sheffield and Derby. The cutlery work produced in the smoking forges at Sheffield, with the aid of the coal supplies furnished by pits all round the town, was carried by packhorse down the valley of the Humber, to be distributed among the inhabitants of all parts of England.[2]

Nowhere, perhaps, is the connection between mineral wealth and the early growth of manufactures more plain than in the case of Scotland. It was an event of fundamental importance for Scottish economic history when James I, after his coronation as King of England, ordered a supply of " great coal " from mines on the Firth of Forth to be sent to the palace at Westminster, for it called the attention of English statesmen and merchants to the special qualities of Scotch coal.[3] When, a few years later, the first success in making glass without firewood was achieved in London with this same Scotch coal, a few Englishmen and Scotchmen saw that the possession of so valuable a commodity opened new possibilities for a country which had been until this time an industrial desert. " Nature ", wrote in 1618 one of the bitterest critics of the Scottish people, " hath only discovered unto them some mynes of coales to shew to what end shee created them." [4] They had already taken her lesson to heart. It is obviously no accident that the manufacture of glass in a fire of Scotch coal was followed immediately by the granting of a series of patents for starting " new " industries in the northern kingdom, patents which are, in the opinion of one of the most scholarly of Scottish historians,[5] certain evidence of the first industrial awakening of Scotland. Besides the setting up of glass furnaces and the rapid development of salt manufacture along the Firth of Forth, attempts were made between 1610 and 1630, partly with English capital, to introduce, at various places in Fife and in the Lothians, soap boiling, sugar refining, tile making, and the manufacture of alum, saltpetre, and gunpowder.[6]

---

[1] Hamilton, *op. cit.*, pp. 130, 139.

[2] *V.C.H. Yorkshire*, vol. ii, pp. 393–4 ; Lloyd, *The Cutlery Trades*, pp. 124, 139 ;   Defoe, *Tour*, 1748 ed., vol. iii, p. 101 ;   Jars, *Voyages métallurgiques*, vol. i, pp. 253–5.   The proximity of Sheffield to good iron mines also made it a natural centre for the growth of the metal trade.   The local grades of iron could be mixed with other grades, in which the region was deficient, but which could be imported from Sweden by way of the Humber.

[3] *Denmilne MSS.*, vol. ix, no. 23.

[4] *H.M.C., Report on the MSS. of Lord Middleton*, p. 184.

[5] P. Hume Brown. See his introduction to *Privy Council Register of Scotland*, 1st Series, vol. xii, pp. xiii–xv.

[6] *Ibid.*, 1st Series, vol. ix, p. 599 ; vol. xi, pp. 275, 306, 322 ; vol. xii, pp. 91–2, 104, 106–7, 232–3, 500, 503, 505–6, 508, 516–19, 771–2 ; 2nd Series, vol. i, p. 377 ; vol. ii, pp. xxxii–iii, 333–4, 424–5 ; M. Mackintosh, *History of Civilization in Scotland*, 1892–6, vol. iii, pp. 313–14, 319.

Before the Restoration, all these trades had secured a firm foothold in Scotland, and, in addition, the Scotch were trying to produce their own tobacco pipes, and their own wool cards to serve the expanding native textile industry.[1] " Manufactures " had become a common word in the Scottish vocabulary, and there were merchants in Edinburgh sufficiently rich to lend money to the Government,[2] and to help finance colliery enterprises.[3] The substitution of coal for wood as firing had transformed Scotland from a country exceedingly poor to a country exceedingly rich in fuel, and therefore in potential industrial power.

Just as the presence of collieries in certain districts of Scotland and of England provides one explanation for the growth of manufactures there, so the more general exploitation of British, as compared with continental, coal mines provides one explanation of the more rapid development of manufactures in Great Britain than in foreign countries, between the accession of Elizabeth and the reign of William III. The advantages she derived from her mineral wealth tended steadily to increase during the seventeenth century, with every rise in the price of timber abroad, with every fresh substitution of coal for firewood in native industry, with every movement of native industry in the direction of the mining districts.

Abroad, too, there were movements of manufactures towards the coal fields. At Liége the armament industry had become so important by the beginning of the seventeenth century that European statesmen regarded an alliance with the Principality as a first-rate strategic move.[4] The Liégeois developed an extensive manufacture of potash, glass, soap, iron wares, and tobacco pipes ; distilleries flourished in the town, and Liége alum competed with English alum on the market at Amsterdam.[5] It might be possible to trace a similar grouping of trades round the mines at Namur, Charleroi, and Mons.

Except in the Belgian provinces, there is no evidence, until after 1700, of any concentration of industry about the coal seams comparable to that which can be observed in many British fields.[6] And the Belgian

[1] Mackintosh, *op. cit.*, pp. 294–336 ; Scott, *Joint-Stock Companies*, vol. i, p. 282, and *passim*.

[2] *Calendar of Suppl. Parl. Papers* (General Register House, Edinburgh), vols. ii and iii, *passim*.

[3] See below, vol. ii, p. 31.

[4] Pirenne, *Histoire de Belgique*, vol. iii, p. 249 ; F. Henaux, *Fabrique d'armes de Liége*.

[5] F. Pholien, *La verrerie au Pays de Liége* ; Pirenne, *op. cit.*, vol. iv, pp. 427–30. Professor Pirenne states (*ibid.*, vol. v, pp. 358–9) that the Liége artisans were unable to use coal in alum-making until well into the eighteenth century. This seems to me doubtful—unless he has in mind the " calcining " of the stone (see above, p. 209), rather than the process of boiling in pans. It is unlikely that the Liégeois could have been ignorant of the use of coal in the latter process, when its use for this purpose went back at least to the beginning of the seventeenth century in England. In spite of its high price, wood was still burned for calcining alum stone in the Liége district at the beginning of the nineteenth century (L. F. Thomassin, *Mémoire statistique du département de l'Ourthe*, 1879, p. 419).

[6] One might possibly except from this statement the St. Etienne district, where the extensive production of arms and ironmongery was due partly to the supply of cheap coal (Boislisle, *Correspondance des Contrôleurs Généraux*,

mines were made pawns in the political struggles of the seventeenth century. None of the Great Powers secured a firm hold on any one of the colliery districts. Situated within the boundaries of small independent or semi-independent states, the collieries of Hainaut passed from the control of the Spanish to the control of the French, and back again to the control of the Spanish ; while the entire mining basin from Mons to Liége was invaded and ravaged again and again by the armies of Richelieu in their struggle with the Spanish, by the armies of Louis XIV, and by the armies which first the Dutch and then the Continental Alliance raised to oppose him. It is no wonder that trade and commerce suffered.

As the cost of firewood and of other surface fuels rose, France and Holland, which were—after the decline of Spain and Venice—the two chief commercial states on the Continent, had either to produce their manufactured goods at an ever increasing expense, or else to depend to an even greater extent on imported coal. The latter alternative, which the Dutch appear to have adopted, had serious drawbacks. To begin with, the prices paid for fuel were almost sure to be higher than in England,[1] where there was more than enough coal to supply the native demand. We must not suppose that this proved, as it might in our own industrialized age, an almost fatal handicap to certain Dutch manufactures ; yet, unless the artisans and textile workers, who made up a large part of the population,[2] could offset the higher price which they paid for fuel by securing other raw materials more cheaply than the English, or by better workmanship, they were likely eventually to lose trade. This dependence on imported fuel, moreover, put the Dutch partly at the mercy of the commercial policy of foreign states, who, if they so wished, could (as England did) impose heavier and heavier taxes on their own exports of coal.[3] In time of war, exports from enemy states might be suspended almost entirely.[4]

Colbert's letters reveal his desire to make France independent of foreign nations for her supplies of mineral fuel as well as for many manufactures. It was an age of mercantilism ; statesmen had become possessed by the strange complex that all that is needed in order to secure the production of goods of any sort at home is to forbid their importation from abroad. To judge by the tariffs which they laid successively upon coal from Great Britain, coal from Liége, and coal from Mons, it was upon the weapon of exclusion that Colbert

---

vol. iii, no. 496 ; A. Des Cilleuls, *Histoire et régime de la grande industrie*, 1900, p. 132), but one could hardly except the Ruhr district (Beck, *Geschichte des Eisen*, vol. ii, p. 1196 ; Hue, *Die Bergarbeiter*, vol. i, p. 351).

[1] This statement has to be qualified, for at the very end of the seventeenth century, the English Government imposed taxes upon the coastwise shipment of coal, comparable to those placed upon exports (see below, vol. ii, pp. 310–4).

[2] O. Pringsheim, *Beiträge zur wirtschaftlichen Entwickelungsgeschichte . . . Niederlande*, 1890, p. 12.

[3] See below, pt. v, ch. ii.

[4] Towards the end of the sixteenth century Holland was prevented by the Spanish from having its usual supply of Liége coal ; and it was this blockade which led the Dutch to develop a more extensive trade in Ruhr coal on the Rhine (Hue, *Die Bergarbeiter*, vol. i, pp. 347–8). During her wars with England, Holland was deprived almost entirely of English coal (see above, p. 84).

and his successors largely depended for the development of the mineral wealth of France.[1]   Seldom has any economic weapon proved so obvious a boomerang.   Foreign imports were, indeed, effectively checked, but there was no increasing output of domestic coal to replace them.   Sugar refiners, glaziers, smiths, and other artisans along the north-eastern frontier and near the north and west coast of France abandoned the use of mineral fuel, which some of them had only recently adopted, and either shut down their works or purchased dear supplies of firewood, to the despair of *intendants* who were already convinced of the necessity for forest conservation.[2]   In a *mémoire* which he submitted to the *Contrôleur Général* in 1699, Vauban, the famous fortress builder, characterized by Voltaire as " perhaps the only general who preferred the welfare of the State to his own ", pointed out that, if it were the aim of French statesmen to encourage native industry, it was a suicidal policy to deprive the country of the means of carrying on that industry by imposing high duties on coal :—

" The honour which the King has done me in entrusting me with the chief management of his fortifications, and the complaints which I receive on every hand from the undertakers of these works as a result of the new tariff on coal imported from Mons, have obliged me to examine with care what the King might derive from this latest move ; but, after having wracked my brains about it, I have at last come to the conclusion that, taking all the factors into consideration, it can result only in an illusory increase in his Majesties revenues, and may subsequently turn into a great loss, even endangering all the other advantages that the King gets out of coal, and depriving the people of its use altogether, or at least reducing them to the hard necessity of buying it at prices set by his enemies. . . . Since . . . this tariff falls almost entirely upon the common people, the soldiers, and the manufacturers, as well as on the fortifications, I flattered myself that you would not take it amiss were I to inform you of the prejudice which the King's works will sustain. . . .

" According to my calculation, every cubic fathom of masonry will cost 46 *sols* more . . . for mortar alone, whether it be the works of the King or those of private persons ; brick work, taking into account the burning of the lime as well as the baking of the bricks, will cost 5*ll*.12*s*. 9*d*. more, which new charges will fall upon all the King's works now under construction at Condé, Quesnoy, Landrecies, Valenciennes, Bouchain, Cambray, Tournai, Lille, Menin, Douay, and Arras,—upon barracks and magazines as well as upon fortifications. . . .

" One must reckon besides that all the iron work done for the King . . . whether for the artillery or the buildings of his Majesty, will cost considerably more.

" To which one must add that in the Tournai district, wood having doubled in price since the last war, the common people of Tourna

---

[1] See below, vol. ii, pp. 229–31.
[2] Boislisle, *Correspondance des Contrôleurs Généraux*, vol. ii, p. 498.   From 1670 to 1700 the French market was practically closed to Liége coal.   For the concern of the *intendants* over the wood shortage, see above, p. 162.

and the soldiers garrisoned there use nothing but coal for heating ; which fact, combined with their sufferings from the high price of bread and beer, constrains them to raise their charge for a day's work and therefore for its produce, and this charge falls in part on the fortifications, in part on the troops, who desert because of their sufferings, and the people, who are already tormented by hunger.

" All brewers, dyers, lime burners, brick, tile, and earthenware makers, potters and founders of every kind, wax chandlers and refiners of salt suffer from this tariff, and have raised the price of their produce in proportion to the tax, so that everyone is inconvenienced by it. . . .

" From Tournai there used to be a trade in lime, worth 300,000 florins or more, which was sold in the Catholic Low Countries and in Holland, a trade which we have lost, or which we shall lose, as a result of this last increase in the tariff, because the traders of Ath[1] will be able to supply lime cheaper than those of Tournai. . . .

" We shall not only lose the trade in lime from Tournai, but Heaven only knows at what price they will then sell us coal ! for by that time they will have us at their mercy and will be able to do without us in the matter of lime. Moreover, it is certain that far less coal for the Catholic Low Countries will pass through Tournai,[2] and that the citizens of Ghent, Bruges, and Ostend will supply themselves with English coal ; in which case we shall have a considerable diminution in the King's revenue from the re-export of coal ".[3]

These carefully thought-out arguments made little impression on the directors of public finance. A tariff is a contagious malady ; the artificial wall which British and French statesmen had erected against the passage of commodities across the Channel differed only in its height from walls erected along every European frontier. Every colliery district in the Belgian provinces was surrounded by such a wall.[4] As a consequence, most continental countries added voluntarily to the advantages which Great Britain enjoyed, when, towards the end of the seventeenth century, coal came into more general use among manufacturers.

---

[1] Ath is a small town about 20 miles north of Mons on the river Dendre, which flows into the Scheldt, east of Ghent. A project had been on foot for several years to build a canal from the river Haine to the Dendre at Ath (E. Bidaut, *Mines de houille de . . . Charleroi*, 1845), in order to bring Mons coal to the Low Countries by water without passing through France. This canal had not yet been dug, however, and coal for the lime kilns at Ath must have been brought most of the distance over ground.

[2] In the absence of a canal from the Haine to the Dendre, Mons coal for the Low Countries had to be carried down the Haine to Condé, and thence to Tournai on the Scheldt. (Cf. above, pp. 128–9). At the time Vauban wrote, this passage involved the crossing of two frontiers. His predictions concerning the inevitable result of a high tariff upon the coal traffic through Condé and Tournai were justified by subsequent developments, as is shown in a *Mémoire sur le commerce des charbons de terre venant de Mons pour Tournay*, printed by the coal merchants of north-eastern France about 1720 (*Bibliothèque de l'Arsenal*, Paris, MSS. 4018, vol. i, f. 109).

[3] Boislisle, *op. cit.*, vol. i, no. 1874.

[4] L. Devillers, *Inventaire analytique des archives de . . . Mons*, 1882, vol. ii, pp. 41 sqq.

The advantages which Great Britain derived from her coal mines before the Industrial Revolution were not limited to manufactures. According to Monsieur Ticquet, whom we have already had occasion to quote, coal was the basis for improvements in farming, which, he tells us, the English had carried much farther than the French. These improvements were due, he thought, to the use that husbandmen throughout England and Scotland made of lime, which could be burned cheaply everywhere in the neighbourhood of collieries. " Since they have applied lime to agriculture ", Monsieur Ticquet proceeds, " the lands in those districts where it is to be had for a reasonable price have tripled their revenues. Lime seals up the pores of sandy soils and enables them to retain enough humidity to nourish plants; it ferments cold earth. The English put more or less lime into all the fertilizers which they compound for the improvement of different soils ".[1] Before the end of the seventeenth century husbandmen had begun to employ coal-ashes and soot, besides lime, for fertilization, and we have the word of a Swedish traveller that they made " a matchless manure ".[2] They could be procured wherever mineral fuel was burned. Hertfordshire farmers, living as far as thirty miles from London, " think it good Husbandry "—to quote one of the greatest early authorities on agriculture—" to give ten, eleven, and twelve pence for a single bushel of Coal-soot, to lay on Clover, St. Foyne, Wheat, Barley, natural Grass, etc." [3] Owing to the wide area under which coal could be mined, the English and Scotch could buy both lime and soot more cheaply than foreign husbandmen; and this may explain partly why Great Britain, between 1550 and 1700, was able to feed an increasing population, despite the increase in pasture farming and in land required for industry in general. Notwithstanding the multiplication of manufactures on both banks of the Tyne, the exacting Matt Bramble found that " the country . . . yields a delightful prospect of agriculture and plantation ".[4]

In time of war, the English derived from coal advantages not only directly, in the production of war materials [5]—ammunition, artillery, muskets, lime for fortifications, pitch and tar for caulking, and metal for sheathing the hulls of battleships—but also indirectly. The task of transporting every year larger quantities of mineral fuel from one British port to another led, no less than the colonial trade, to the

---

[1] *Archives Nationales*, O¹1293.
[2] Kalm, *Account of his Visit to England . . . in 1748*, trans. by J. Lucas, 1892, p. 139.
[3] William Ellis, *The Modern Husbandman* (4th ed.), pp. 91–4.
[4] Smollett, *Humphry Clinker*, letter from Tweedmouth to Dr. Lewis.
[5] Colbert writes (1677) to M. de Moyenneville, *Intendant des Fortifications*, who is preparing for war with the Dutch : " Je vous enverray incessamment l'arrest que vous avez demandé pour l'exemption des droits d'entrée du charbon de terre que les entrepreneurs de Calais feront venir. Tenez la main qu'ils en fassent de grands magazins tant pour cette année que pour la prochaine, et travaillez de vostre costé à en amaser quantité, suivant les ordres que je vous ay donnés " (Clément, *Lettres, instructions et mémoires de Colbert*, vol. v, pp. 198–9). Thus we see that, under the stress of war, Colbert himself recognized that his tariff weapon was double-edged.

development of "a goodlie fleete of tawle and serviceable Shipps
. . . together with a great multitude of skilfull marriners and sea-
faringe men used and imployed about the same to the generall and
great good and safety of this Realme ".[1] Admirals and statesmen
looked more and more to the colliers, which sailed between Newcastle
and London, to provide the navy with trained seamen always available
to serve within a few days' notice, and this traffic took on, therefore,
a very special strategic importance.[2] " I doubt not ", Houghton
remarks, " but it might be the foundation of our being too hard at sea
for all opposers ".[3] Before 1700 the Irish coal trade is said to have
employed a fleet of nearly 200 ships, and the west-coast trade must
have employed as many more. " It's evident ", wrote in 1691 Walter
Harris, an authority on commercial relations between England and
Ireland, " that whatever Trade constantly employs them [the colliers]
tends to their encrease, and the encrease of our Seamen ".[4]

Modern writers who deal with the economics of transport have
laid due stress upon the part played during the nineteenth century
by coal exports, and by cheap fuel for bunkers, in encouraging the
British to become the carriers for half the world. But they have
rarely commented upon the relation of the early trade in mineral
fuel to the predominant position that England had achieved as a
maritime power before the Industrial Revolution, though the connection seemed obvious to many contemporaries. Historical students
need hardly be reminded that, until the reign of Henry VII, the
country had almost no navy and far fewer merchantmen than the
principal continental states. If Admiral Mahan had gathered round
him a group of Crown officials, prior to the accession of the Tudors,
he might have found it difficult to persuade them that their nation's
destiny lay on the sea. It is significant that the first period of rapid
expansion in English shipping should correspond with the first period
of rapid expansion in the coal trade.[5] This trade provided, according
to many seventeenth-century writers, the "chief nursery" for
English seamen.[6] An anonymous petitioner of the time of James I
called upon the Crown to consider "what an infinite number of shipps
and people are now sett on worke in England by coles onely ; more
than by all other kinds of merchandize els ".[7] "The Colliery-Trade",

---

[1] *Exchequer Decrees*, Series IV, vol. iii, f. 106. The fact that coal had relieved
pressure on the forests must have made it easier to get timber for building ships.

[2] Cf. Mantoux, *The Industrial Revolution*, pp. 289–90, note.

[3] *Husbandry and Trade*, vol. ii, pp. 155–6.

[4] Walter Harris, *Remarks on the Affairs and Trade of England and Ireland*,
1691.

[5] W. Sombart, *Krieg und Kapitalismus*, vol. ii, pp. 175 sqq. ; R. G. Marsden,
" English Ships in the Reign of James I ", in *Trans. Roy. Hist. Soc.*, N.S.,
vol. xix, 1905, pp. 309–42 ; and see above, pp. 172–3.

[6] Cunningham, *Growth of English Industry and Commerce, Modern Times*,
1921 ed., p. 319 ; Evelyn, *Sylva*, vol. ii, pp. 108–9 ; *Remembrancia MSS.*, vol. iv,
no. 52 ; *Reasons humbly offered to . . . Parliament for their not laying any further
Imposition upon Coals imported to London*, 1689. See also citations by E. R.
Turner, " English Coal Industry in the Seventeenth and Eighteenth Centuries ",
in *Amer. Hist. Rev.*, vol. xxvii, 1921–2, p. 6 ; and cf. below, Appendix H.

[7] *S.P.D., Charles I*, vol. xiv, no. 9. This document is, I believe, wrongly
assigned to the reign of Charles I (see below, vol. ii, p. 221 n.).

wrote in 1700 Charles Povey, a London coal merchant, later prominent for his part in organizing the postal service, " brings up a greater number of Seamen, than all our Navigation elsewhere, and it must be acknowledg'd that when Owners and Masters of Vessels gain by Trading to Newcastle, then, and then only, our Trade to foreign Parts is in a flourishing Condition ".[1]

Contemporary opinions such as these are confirmed by other evidence. More vessels may have been engaged in the fishing industry than in the carriage of coal, but they were of a smaller size.[2] If, in value of goods shipped, the commerce in coal was inferior to the export of cloth, or the import of French wine, in actual bulk of cargo, and consequently in ship's space required, it was many times more important than both together. Never had so bulky a commodity been carried by sea in such vast quantities. Before the end of the sixteenth century nearly, if not quite, as many vessels were engaged in the carriage of coal from Newcastle to foreign countries, as were engaged in the carriage of all commodities from foreign countries to London, then by far the richest English trading port.[3] And early in the seventeenth century the coastwise coal traffic from Newcastle required from five to ten times as much shipping tonnage as the overseas coal traffic. When we add, to the ships employed by the coal trade from Durham and Northumberland, the ships employed by that from Scottish and west-coast ports, it seems likely that, at the time of the Restoration, the tonnage of colliers had come to exceed the tonnage of all other British merchantmen. River barges and lighters, as well as sea-going vessels, provided an excellent school for seamen, and the coal trade led to the building of large numbers

---

[1] *A Discovery of Indirect Practices in the Coal-Trade*, 1700, p. 43. In opposing restrictions upon the export of Scotch salt, the representatives of the Scottish burghs pointed out in 1630 that one-half of all Scottish shipping was employed in carrying coal and salt, and that it would be a fatal blow to the commercial prosperity of the burghs if the ships engaged in these two trades were left idle (*Privy Council Register of Scotland*, 2nd Series, vol. iv, p. xvii). The Scottish coal trade also came to be regarded as an important " Nursery of many Seamen " (J. Spruel, *An Accompt Current betwixt Scotland and England Ballanced*, 1705, p. 17).

[2] See above, p. 173 n.

[3] During the year ending Michaelmas, 1594, 852 ships carried 35,934 tons of coal from Newcastle to foreign ports (*Exch. K.R. Port Bks.*, 185/6). During a period of nine months, from Michaelmas, 1601, through June, 1602, 714 ships entered the port of London with cargoes from abroad (L. R. Miller, " New Evidence on the Shipping and Imports of London ", in *Quart. Journ. Econ.*, vol. xli, 1926–7, no. 4, pp. 740 sqq.). The combined tonnage of these 714 ships was 32,050 tons. During the three summer months, when trade was at its height, 300 or more additional ships probably came to London. Even so, the tonnage of the ships engaged in this trade would not be greatly in excess of that of the colliers leaving Newcastle for foreign parts, since approximately one ton of shipping was required to carry one ton of coal. We must remember, too, that not all Tyne valley coal consumed abroad was carried overseas directly from Newcastle ; some went first to other English ports, whence it was reshipped to foreign countries (see below, vol. ii, p. 363). If we added the vessels employed in these reshipments to those employed directly in exporting coal we might find that during the last fifteen years of Elizabeth's reign the Newcastle coal exports actually required more shipping tonnage than the London imports of all commodities.

of such barges and lighters.[1] The coal trade from Newcastle to London was relatively no less important in the late seventeenth than in the late eighteenth century, when, Adam Smith observes, it " employs more shipping than all the carrying trades of England ".[2]

The coal trade may be regarded, in short, as a magnet which helped to draw Englishmen to seek their profit and their livelihood in ocean commerce. Neither the Spaniards, nor the Portuguese, nor the French, nor the Dutch felt the same pull, because none of these peoples had any comparable ocean traffic in coal.

" On explique comment Francois II succède à Henri III, Charles IX, à Francois II et Henri III à Charles IX, mais personne n'enseigne comment Watt succède à Papin et Fulton à Watt ". Perhaps no one will ever be able satisfactorily to explain the complicated inter-mingling of economic and social and intellectual forces which underlie the great inventions ; yet much thought has been given to the question since Hugo's criticism of the subject matter of history. Only two years after his words had appeared in print, Jevons gave reasons for supposing that problems connected with coal mining had provided the incentive for certain British inventions.[3] This theory has been further developed by other thinkers, notably by Kulischer and by Max Weber ; while the more general problem of the causes behind inventive achievement has received some attention from very diverse quarters.[4] Neither the implications nor the limitations of Jevons' theory have been examined as closely as they might be ; nor has it been realized how important a bearing upon the theory has a history of the coal industry in the seventeenth century. Jevons himself pushed his study of origins back beyond 1700 only in one case—that of the steam engine. It is a part of his thesis, in fact, that Great Britain contributed nothing essential to the inventive power of mankind before the Industrial Revolution.

Since Jevons wrote, it has been frequently suggested that until the eighteenth century the English owed practically all their technical skill to foreign teaching. That they did borrow from abroad is evident enough. When Elizabeth ascended the throne, England was unques-tionably one of the most backward among the nations of Europe in industrial technique and in scientific skill. Cecil imported Lorrainers to teach glass making, Germans to teach salt making, Nürnbergers to teach battery work ; the first English paper mill was erected in 1588 by a German ;[5] Eustachius Roche, a Fleming, was empowered in the 'nineties by the Scottish Parliament to employ throughout the northern kingdom his new method of making salt, and his invention

---

[1] *V.C.H. Salop*, p. 426 ; Hylton Dale, *The Fellowship of Woodmongers*, 1923, p. 73.

[2] *Wealth of Nations*, bk. ii, chap. v.

[3] Jevons, *The Coal Question*, 1865, chap. vi.

[4] A summary of the literature on this subject may be found in R. C. Epstein, " Industrial Invention: Heroic or Systematic ? ", in *Quart. Journ. of Econ.*, vol. xl, 1926, pp. 232–72.

[5] Scott, *Joint-Stock Companies*, vol. i, p. 116 ; vol. ii, pp. 416–17.

for supplying a better draught in kilns and stoves.[1] Nor did British borrowing from foreigners cease after the death of Elizabeth. Yarranton travelled to Leipzig to learn the process of tin and latten plating. The continental influence is apparent in the proposals of many inventors. It was Paule Timmerman, "marchant straunger", who appealed in 1615 to the Privy Council for support in his effort to introduce a "new method" of sugar refining, the London manufacture having been hitherto in the hands of some few Dutch and English refiners.[2] King James I seriously considered the granting of special privileges to a group of "French Undertakers", who offered to transmute iron into steel "with pitt Coles onely", and who are said to have received exclusive rights in France, though there is no reason to suppose that their method proved more successful than the methods of the great majority of patent seekers, who promised much but accomplished little.[3] In the list of patentees one finds occasional foreign names—there is James Vanderbrooke, "Commise-Generall of Holland", who proposed to reduce the consumption of fuel in boiling down sea water to salt, there is Frederick Wagoner, "borne beyond the seas in Zeland", who had a "new furnace" for smelting lead, there is Sir John Christopher van Berg, a Moravian knight, there are Sir Thomas de Mayenne and Francis Fandell de Fresue, natives of France, and many others [4]—and the absence of a foreign name on the actual patent is no proof, as Hyde Price has shown, that the project is of English origin.[5] English patentees were in the habit of referring to their travel, study, and observation abroad, and this would seem to imply an acknowledgment that technical achievement had been pushed farther on the Continent than at home. It is well known that in canal building Great Britain lagged far behind France and Holland.[6]

In mining technique the obligations of the British to foreigners, and to the Germans in particular, have been a favourite theme of investigators. These obligations are easy to establish. One can point to the group of Almaynes, who came to work in the copper mines at Keswick in 1565 [7], to the use of gunpowder for blasting rocks in sinking shafts at Chemnitz some time before the British used it for that purpose,[8] and to the fact that the method of ventilating coal mines with the aid of a furnace erected at the mouth of adits to draw out the bad air originated at Liége.[9] One can also point out

---

[1] *Acts of the Parliaments of Scotland*, vol. iv, pp. 182–4, 187–8 ; *Privy Council Register of Scotland*, 1st Series, vol. vi, pp. 17–18, 837.

[2] *Acts of the Privy Council*, 1615–16, pp. 284, 305, 579, 655.

[3] *Cotton MSS. Titus B.V.*, f. 165 ; *Cal. S.P. Venetian*, 1610–13, p. 3. These Frenchmen were rivals of Sir William Slingsby and his partners (*Harleian MSS.*, 7009, no. 4 ; and see above, p. 215) for the grant of such a patent, and it was Slingsby who received the grant in the end.

[4] *Patent Specifications*, nos. 80, 81, 92a, 119, 211.

[5] Wm. Hyde Price, *The English Patents of Monopoly*, 1906, pp. 62–4.

[6] Cf. below, pp. 258–9.

[7] *Acts of the Privy Council*, N.S., vol. vii, p. 229.

[8] Granville Poole, *Historical Review of Coal Mining*, 1924, p. 82 ; *Royal Soc., Philosophical Transactions*, vol. i, 1665, pp. 82 sqq., 109.

[9] *Ibid.*, vol. i, pp. 79–81.

that the pumping devices used in the British collieries during the seventeenth century are all described more or less adequately in Agricola's famous compendium of metallurgical knowledge, published in 1556.[1] At Wollaton colliery in Nottinghamshire, " there are ", wrote an " inginer " in 1610, " models to be seene of all the water-workes that are of any worth or value in Italye, Germanye or the Low Contryes ".[2]

In view of the evidence of the indebtedness of English technique to that of the Continent, there is a temptation to dismiss the question of inventions in England under the Tudors and Stuarts without further examination. Yet to do so is to neglect a period of fundamental importance in the history of those scientific and engineering triumphs which are in large measure responsible for the manner in which western Europe and America live to-day. The study of foreign methods implies something besides ignorance : it implies an eager desire to learn. There is no great movement, either in art or in science, that has not begun by turning its face towards the past. If there be truth in the commonly accepted view that science is an accumulation of knowledge, and that each individual's contribution is but one further clearing in the forest of the unknown, then it is difficult to see how any investigator could start out unless by following the trails blazed by his predecessors until those trails end. The attempt to trace invention, or even a particular invention, back to its origins, is limited only by our lack of knowledge concerning the past. English technique in the seventeenth century was profoundly influenced by that of Holland, France, Germany, Spain, and Italy ; and continental technique owed much, in its turn, to the slow infiltration of improved processes traceable to the more advanced civilizations of the East.[3] Such considerations do not justify the belief that the work of ancient peoples explains all the advances in mechanical technique made in Europe before the eighteenth century. At the beginning of the sixteenth century the West had already added something of its own to the teachings of the East ; during the seventeenth century it was destined to add much more, and the chief centre of experiment and development was destined to shift from the Continent to Great Britain.

For the technique borrowed from the Continent proved inadequate. Try as they would, the British found it impossible, with the pumping devices which had served the copper and silver miners of Bohemia, Hungary, the Tyrol, or the Harz, to force a considerable volume of water high enough successfully to drain the deeper pits. Scores of new drainage projects were " noised abroad " ; but the persistent failure of their sponsors (who usually claimed hitherto undiscovered powers for

---

[1] Agricola, *De Re Metallica*, ed. H. C. and L. H. Hoover, 1912, bk. vi.
[2] *H.M.C.*, *Report on the MSS. of Lord Middleton*, pp. 173-4.
[3] E. W. Hulme, " The History of the Patent System under the Prerogative and at Common Law ", in *Law Quart. Review*, vol. xii, 1896, pp. 141-2. See also Beckmann, *History of Inventions*. The mechanical principles behind most of the pumping devices used by the medieval miners had been known to the Greeks and Romans (cf. the Hoovers' note on machinery in their edition of *De Re Metallica*, p. 149).

some well-worn device) to perform the feats which they promised caused mine managers to adopt a sceptical attitude.[1] Nevertheless these attempts are symptoms of a healthy state of mental activity. For every successful inventor, as for every great painter, there must be hundreds of failures. Behind the superficial contrivances, of which we have so much evidence in the patents granted by the Crown, there was gradually developed a solid foundation of experiments destined to culminate in the discovery of the steam engine.

Between the first known attempt in 1631 " to raise water from coal pitts by fire ",[2] and the erection in 1712 [3] at a Staffordshire colliery of a Newcomen steam engine, which made the raising of water from a mine " by fire " a commercial success, there is a long and intricate story, still only partially known, of efforts to embody an increasing amount of scientific experience in a practical form.[4] It is by no means a story of purely English achievement. At least three foreigners—De Caus, Huyghens, and Papin—made outstanding contributions to the final result. Into the dispute as to which nation deserves the largest measure of credit we need not enter. A scientific concept which is purely national is, perhaps, as rare as a pure race. All students of early inventions can agree, at least, upon three propositions : that the problem of colliery drainage concerned the British more than the French ; that Huyghens and Papin spent more of their time in England in contact with English " natural philosophers " than their English colleagues—David Ramsay, court clockmaker, the Marquis of Worcester, Samuel Morland, Savery, and Newcomen—spent in France ; and, finally, that the first engines were erected at English not at continental mines.[5]

Hitherto, when we have spoken of British borrowing, we have referred to the adoption of processes which had been previously used on the Continent. It is clear that British borrowing in the case of the steam engine—if borrowing it can be called—was of an altogether

---

[1] J. C., *The Compleat Collier*, 1708, p. 22 ; Stephen Primatt, *The City and Country Purchaser and Builder*, 1680 ed., pp. 28–9. The source material dealing with the early pumping devices for draining mines is voluminous. Mention may be made of the following : *Pat. Spec.*, nos. 8, 9, 21, 29, 37, 48, 49, 50, 55, 57, 76, 84, 110, 117, 135, 136, 142, 174, 175, 179, 186, 208 ; *Commons Journ.*, vol. ix, p. 300 ; Rymer, *Foedera*, 1704, etc., vol. xvii, p. 102, and vol. xix, p. 236 ; *S.P.D.*, *Charles I*, vol. clxxvi, no. 3, and vol. clxxxvii, no. 46 ; *Sloane MSS.*, 4812, f. 6 ; *Acts of the Parliaments of Scot.*, vol. iv, p. 176 ; *Privy Council Register of Scotland*, 1st Series, vol. xii, pp. 258, 277, and 3rd Series, vol. vi, p. 406.

[2] Rymer, *Foedera* (Hague edit.), vol. viii, pt. 3, p. 153. The patent was granted to David Ramsay, who also obtained patents for a number of other industrial processes. Although this was the first attempt to use steam in raising water from a coal pit, it was not the first investigation into the forcible effects of steam (cf. J. P. Muirhead, *The Life of James Watt*, 1859, pp. 88–9).

[3] Newcomen is said to have built an engine in 1705, but it was not tried at a colliery until 1712 (Galloway, *Annals of Coal Mining*, pp. 238–9).

[4] There are already many versions of the story (cf. Muirhead, *op. cit.* ; R. H. Thurston, *History of the Growth of the Steam Engine*, chaps. i–iii ; Galloway, *op. cit.*, pp. 236–41 ; Mantoux, *The Industrial Revolution*, pp. 318 sqq.).

[5] T. Birch, *History of the Royal Society*, 1756, esp. vol. iv, *passim* ; *Royal Soc., Philosophical Transactions*, *passim* ; Muirhead, *op. cit.*, pp. 97–122 ; John Buddle, *The Marquis of Worcester's Century of Inventions*, 1813 ; James Halliwell, *A Brief Account of the Life, Writings, and Inventions of Sir Samuel Morland*, 1838.

different sort. In this case a definite advance in the technical skill of mankind was made on English soil. And this is not the only instance of such an advance to be found in seventeenth-century history.

On the occasion of the hundredth anniversary of the first passenger railway, M. Charles Foley suggested that it would be appropriate to commemorate, not only George Stephenson, but also the Frenchmen whose work had contributed towards the invention of the railway, among them Papin, and Beaumont who, in 1630, "established at Newcastle roads of parallel wooden rails ".[1] Beaumont came of a family prominent in Leicestershire for centuries, so that his French blood must have been diluted, to say the least ; he died in 1624 ; and, though he introduced these rails into the north of England,[2] he was almost certainly not their inventor. Notwithstanding its errors, M. Foley's statement serves to call attention to the fact that the modern railway developed out of the efforts to solve two separate problems : that of draining collieries more thoroughly and more economically, and that of moving coal overground more cheaply. Towards a solution of this second problem English experience made, during the seventeenth century, a contribution scarcely less important than its contribution towards a solution of the first.

Doubtless the principle behind the "wagon-way", as it came to be called,[3] may be traced much farther back ; but our first description of such a way, as laid in 1606 between Broseley colliery and the Severn, with the express approval of the Privy Council,[4] suggests that decided improvements had been made on the devices previously used by the German metal miners. This is not the first English wagonway, for we read of "railes" in an account book kept at Wollaton colliery in 1597–8.[5] As Beaumont became one of the owners of adjoining collieries at Strelley and Bilborough a few years later, it was undoubtedly at this colliery, and not at Broseley, that he learned about the railed way, which he later introduced into the north of England.[6] There is some reason to think that an attempt was made to introduce a similar device at Scottish mines, in 1606, when Thomas Tulloch, of Inveresk, petitioned for a patent "for making . . . ingynis . . . for the . . . aisie transporting of coillis betuix colpotis, sey and salt pans of this realme ".[7]

At Broseley in 1606, the colliery owner, James Clifford, had set down "tylting railes", "which weare made fast to certen of the grownde". Along these a "very artificiall engine" of timber conveyed coal downhill to the river.[8] The expression "tylting railes" is significant : it could hardly have been applied to the thick wooden

[1] *Echo de Paris*, April 24th, 1925.
[2] See below, vol. ii, pp. 14–6, 66–7.
[3] It was also called a "Newcastle road" (T. J. Taylor, *Archaeology of the Coal Trade*, p. 150).
[4] *Star Chamb. Proc.*, *James I*, 109/8, 310/16.
[5] *H.M.C.*, *Report on the MSS. of Lord Middleton*, p. 169.
[6] See below, vol. ii, pp. 15–6.
[7] *Privy Council Register of Scotland*, 1st Series, vol. vii, pp. 278–9.
[8] *Star Chamb. Proc.*, *James I*, 109/8, 310/16. Until the eighteenth century the rails were always made of wood.

planks mentioned half a century earlier by Agricola—planks a foot or more wide, laid in parallel pairs, so that ordinary cart wheels rolled loosely over the surface, an iron pin underneath the cart extending down between the planks to prevent the vehicle from running off its specially constructed path. We cannot be sure that the " artificial engine " at Broseley was equipped with modern flanges, or rowlets, to fit the rails, but we know from Roger North's description [1] of the wagonways near Newcastle that such rowlets were common before 1676. All doubt concerning the novelty of the English wagonway should be removed by the knowledge that it was introduced into the Ruhr district at the end of the eighteenth century as an " englischer Kohlenweg ",[2] although the passages described by Agricola had been common in connection with German mining at least from the beginning of the sixteenth century.

Before the eighteenth century other technical advances were made as the result of English experience. There is the invention of boring rods to determine the nature of mineral seams without digging pits.[3] There is the discovery of a method of obtaining pitch and tar from shale.[4] There are numberless developments connected with the substitution of coal for firewood.

These developments have been almost entirely overlooked, owing to the tendency of investigators to focus their attention upon the single problem of smelting iron with mineral fuel. The smelting of iron was only a part of the larger problem of smelting and refining all metals, and that again was only a part of the still larger problem of making coal a suitable substitute for wood in industry generally, and of reducing the total consumption of fuel of all kinds in a given process. Experiments towards solving the problem of substitution may be divided into two broad classes : those directed towards finding improved furnaces or kilns to protect the material burned from the noxious properties of coal, and those directed towards ridding the coal itself of those noxious properties.[5]

It was the first quest which led to the invention of the closed pots for melting glass, and, at the very end of the seventeenth century,

---

[1] Quoted Galloway, *op. cit.*, p. 156.

[2] Beck, *Geschichte des Eisen*, vol. iii, p. 960.

[3] See Appendix N.

[4] The process is said to have been discovered by Martin Ele (Thomson, *Royal Society*, p. 518 ; *Royal Soc., Philos. Trans.*, vol. xix, 1695–7, p. 544). The pitch and tar used in England for caulking the hulls of ships had been previously made from the bark of trees ; and the new process must, therefore, have effected some small saving in the supply of timber. Attempts were made to distil tar and pitch from coal (Beck, *op. cit.*, vol. iii, pp. 304–5 ; *Pat. Spec.*, nos. 214, 259, 329 ; Evelyn, *Sylva*, p. 86), but the process was not apparently a commercial success in Great Britain until the end of the eighteenth century, when the Earl of Dundonald started the manufacture in Midlothian (Sinclair, *Statistical Account of Scotland*, vol. ix, pp. 278–9). In the time of Agricola, the Germans extracted from coal some sort of a tar which they found useful for medicinal purposes, and for protecting the vines from worms (Beck, *op. cit.*, vol. i, pp. 104–5). The use of this tar for medicinal purposes continued at least until the end of the seventeenth century (Bünting, *Sylva Subterranea*, bk. xx).

[5] The problem of finding better furnaces presented itself in connection with the burning of coal for domestic, as well as for industrial, use. Many were the seventeenth-century projects for " curing " smoky chimneys.

to the invention of a reverberatory furnace,[1] which made possible the smelting of lead and copper, and the transmutation of copper into brass, with a pit-coal fire. Sir Nicholas Crisp, one of the wealthiest royalist financiers, and a farmer of the coal taxes at the Restoration, is said to have "much facilitated" the work of making copperas, by an "ingenious contrivance" that enabled the heat from the fires to be "conveyed to all parts of the bottom and sides of the furnace".[2] Technical improvements in other furnaces appear to have made it possible also to burn coal in extracting silver from lead ore and in the manufacture of steel. In 1623 a certain Lewyn van Hack had undertaken to extract silver from lead in Cardiganshire, using seacoal in the smelting and charcoal "only in the refyning".[3] We know nothing about his methods or his success, but in an article written in 1678, and devoted to the methods of separating silver and other bodies from lead ore, Dr. Christopher Merret remarks that the "latest invention is a new furnace. The conveniency . . . is, that a little fire, and that of New Castle coals, will do the work".[4] In 1610 a group of Frenchmen, in asking for the privilege of making steel in England, promised to do the work "with pitt cole onely".[5] While this particular project does not appear to have succeeded, the new fuel was almost certainly used soon after in the process of "cementation".[6] The furnace in which the iron bars were embedded in charcoal could, like the new glass furnaces, be heated to harden the metal into steel without direct contact between the flames and the iron, for, as in glass making, the sensitive materials were protected by being put in crucibles.[7] The work of adapting industrial processes to the burning of the new fuel was often accomplished piecemeal, and the quest for a method of performing the actual smelting of the ore must not be allowed to obscure other important mechanical experiments concerned with the substitution of coal for wood.[8]

There are essentially two very different methods of treating coal to rid it of its noxious properties : the first is to combine it with certain extraneous matter in briquettes, the second is to reduce it to coke by burning. Briquette making did not originate with the English;

---

[1] A patent for a reverberatory furnace was granted to Viscount Grandison in 1678 (*Pat. Spec.*, no. 206), but a satisfactory furnace was not invented until just before 1697 (Houghton, *Husbandry and Trade*, vol. ii, pp. 187 sqq.).

[2] *Royal Soc., Philosophical Transactions*, vol. xii, 1678, pp. 1056–9. On Crisp, see below, vol. ii, p. 307.

[3] *S.P.D., James I*, vol. cxlv, no. 66, and also the account of the project in *Sackville MSS.*, in a document dated May 30th, 1623. An attempt had been made in 1613 to extract silver from lead ore in a coal fire (*Privy Council Register of Scotland*, 1st Series, vol. x, pp. 95–6).

[4] *Philos. Trans.*, vol. xii, pp. 1046–52.

[5] *Cotton MSS., Titus*, B.V., f. 265.

[6] *Pat. Spec.*, no. 33, and Dud Dudley, *Mettallum Martis*, 1855 ed., p. 35, both refer to the use of coal in making steel.

[7] This process of making steel with a raw coal fire at Birmingham was described by Monsieur Ticquet in 1738 (*Archives Nationales*, 0¹1293), and I think it is a fair assumption that the method was not then a new one.

[8] The cases cited in the text are probably nothing more than samples of the general improvement made in boilers, furnaces, and kilns during the seventeenth century.

ARTIFICIALL FIRE, *21.*
O R,
COALE for Rich and Poore.

This being the offer of an Excellent new Invention, by Mr. RICHARD GESLING Ingineer, (late deceased) but now thought fit to be put in Practise.

Read,   Practice,   Judge.

F I R E and *Water* be two of the most excellent Creatures that ever God created:if this be wanting,there is no subsistence to *Man*,if Water be wanting,no subsistence to *Beast*; where these two predominate,there is neither life for *Man* or *Beast*,and where they *master* they leave no *servant*; of these two we shall need to speake at this time but of *Fire* only; and what hath been left to us by that painfull *Ingineer* Mr. *Gesling*, who was the first *Inventor* of this, I here have thought good to put to publike view, that all both Rich and Poore may be provided for in some competent way; had hee lived, he might have put it to Presse, but I seeing none to stirre in it, I having a Writ from him, have here set it forth in these times of scarcity,which if thou practise and make experiment, it shall be to thy profit and comfort. It may bee some foolish lasie persons may say, as some fine Nosed City *Dames* used to tell their Husbands; O Husband! we shall never bee well, wee nor our Children, whilst wee live in the smell of this *Cities Seacoale smoke*; Pray,a Countrey house for our health, that we may get out of this *stinking Seacoale smell.* —— But —— how many of these fine Nosed *Dames* now cry, Would to *God* we had Seacoale, O the want of Fire undoes us!, O the sweet *Seacoale* fire we used to have, how we want them now, no fire to your *Seacoale*! Thus now they see the want of that they slighted in times past; This for the *rich*,à word for the *poore*.

The great want of *Fewell* for *fire*,makes many a poore Creature cast about how to passe over this cold *Winter* to come, but finding small redresse for so cruell an enemy, as the cold makes some turne Thieves that never stole before, steale *Posts*, *Seats*, *Benches* from *doores*, *Railes*, nay, the very *Stocks* that should punish them, and all to keep cold *Winter* away: Now that all may be provided for, and the better furnish't before *Winter* comes, take this *Modell* to help thee at the cheapest and easiest rate that *Experience* can find out; there bee many wayes,of all which thou shalt find here both the *sweetest*,*wholesomest*, and *cheapest*,and most *usefullest* and *beneficiall* both to *rich* and *poore*, I tell you this *Secret* is worth the looking after, and by it many thousands may be set at worke,and yet before Summer be over,store may bee made: It may bee you will say, Why was it not begun sooner this yeere? But I tell you two Reasons; one,the want of *Seacoale* to help; the other it is never too late to seek profit: do thou practise it,and thou shalt see profit to proceed suddenly; and so God prosper thee and thy labours.

Take three Load of *red Morter*, such as you make Bricks with double Loads, half a Chaldron of good *Seacoales* of the smallest and best, three Sack-fulls of the best *Smalcoale*, foure bushells of *Saw-dust*, foure trusses of *Straw* chopped; worke all these together with water stiffe as Bricks, then when it is worked all together very well,take foure sacks of the dust of *Smalcoale*,and with that used as they do the Sand for casting of Brick, then cast the Ingredients as you cast Brick, but halfe so thick, and dry it as Brick is dryed; or you may make it up in round Balls not too big, with Charcoale, or Smalcoale dust, on the outside, and so laid to dry, when they be thorow dry, burn them with a little Scotch-coale, or Wood, or any Combustible matter to fire it, or with two or three wooden chips to kindle your fire withall, and to keepe in the life of the fire, and these cast a most excellent heat, and keepe fire for any use, to Rost, Boile, or Bake, for the richer sort; but bee sure you lay them not too close on the fire, but as you see your pattern upon this Paper, mingled with a Scotch-coale or two.

For the Poorer sort, Cow-dung mingled with Saw-dust and Smalcoale, made up into Bals,'or in a square like a Tile, not too thick, and dry ed,make a very good Fewell, but something noisome. Also that which comes out of the paunches of Beasts killed, it being dryed is excellent fire.

Horse-dung in Balls with Saw-dust, or the duft of Smalcoale, or Charcoale dust, dryed, is good Fewell, but the smell is offensive.

Greenwich Heath, or *Hounslow* Heath Turfe well dryed,is very good fewell, with a little Scotch-coale burnt with it.

Peate if well dryed,but well fatted with seggy or flagge roots from Fenny places, is a very good firing,mingled with Coale when it is burnt.

Some make an Oven with Kennell durt,with a hole at the top for the heat and smoke to ascend up in the chimny, and with six or seven bricks raise the bottome and make up the shed,and then dawbe up the oven, leaving the hole at top,and before put in a few Scotch-coale,and after it is kindled will keepe fire a weeke, every two dayes putting addition of the kennell durt to keep it whole, and putting Seacoale, or Scotch-coale as the fire declines; this is a fire which casts a good heat at the mouth and top,but not commendable nor fit to rost.

Above all things let mee perswade all men to sweepe their Chimneys cleane against winter, for with Scotch-coale, or Coale minged with wood, or with any of these *Chimneys* grow extraordinary foule, and he that meanes to keepe his house from firing, let him bee sure to keepe his Chimney cleane: Thus if thou makest use, thou shalt see thy labour worth thy paines, and bee thankfull to him that hath been the Instructer, for this Artificiall Fire.

LONDON,
Printed by *Richard Cotes* for *Michael Spark Senior,* 1644.

A RECIPE FOR MAKING BRIQUETTES, 1644
(British Museum, 669, f. 10(11))
Reproduced through the courtesy of the officials of the British Museum

they learned the process from the Liégeois,[1] and briquettes were probably common in all the Belgian colliery districts [2] before any effort was made to introduce them into England. John Thornburgh, Dean of York, received in 1590 a grant from the Crown for " refining " seacoal, " to corecte the sulphurous nature ".[3] It may be assumed that he proposed to treat mineral fuel after the manner of Liége, but we have found no sufficiently definite description of his method to feel certain of this.[4] Sir Hugh Platt's project, published in 1603, leaves us in no doubt. The third son of a wealthy London brewer, Platt, besides being well known in his time as the author of a tract (which passed through many editions) on the art of preserving woman's beauty, took a serious interest in problems of scientific farming, and demonstrated the power contained in a jet of steam. Like most English scientists of his day, but unlike those of two generations later, he did little original thinking. He is frank enough to admit this in describing the process of making briquettes, or " Coleballes ", as he called them. To " knit up " a bushel of seacoal, he explains to house-holders, take half a peck of loam and dissolve it in a small tub of water. Then take the best quality seacoal, strew it on a stone or paved floor and break it into small pieces either with a hammer, a mallet, or under the feet. Spread the small pieces about one handful thick upon the floor, and sprinkle some thin " pap " all over the heap. Then turn with a shovel or spade, and repeat the process until the mixture is soft enough to make balls in the hands, " according to the . . . making of snowbals ". Place these balls one by one out of contact with each other until they have had a few days to dry. " And so," concludes Platt, " you have seacoles wrought up into bals simply of themselves, according to the maner of Lukeland [Liége] in Germanie : which forme of firing hath been in use with them for many years past ".[5] This coating of loam would, Platt thought, temper the noxious fumes which so annoyed the London domestic consumer. But, despite the efforts of several subsequent " inventors " to improve upon Platt's recipe,[6] briquettes never became a success in the capital. Apparently the coal of Newcastle, like that of St. Etienne (where similar projects failed later on),[7] did not lend itself readily to this treatment. In Pembrokeshire, however, the " culm ", or small coal

---

[1] The Liégeois may have learned it from the Chinese (cf. above, p. 13 n.).

[2] They were no novelty in the Mons colliery district in 1620 (Devillers, op. cit., vol. ii, p. 41 ; see also Hue, Die Bergarbeiter, vol. i, p. 351).

[3] S.P.D., Eliz., vol. ccxxxiii (October 10th, 1590) ; see also Lansdowne MSS., 67, no. 20, and cf. below, vol. ii, pp. 214–5.

[4] I do not know of a case, however, in which the word " refining " is applied to the coking of coal, so perhaps the word itself may be taken as evidence that Thornburgh was concerned with making briquettes.

[5] A new, cheape, and delicate Fire of Cole-balles, wherein Seacole is by mixture of other combustible bodies, both sweetened and multiplied, 1603. Shortly before Platt printed his " receipt ", an attempt had been made to use coal for domestic fuel in Bavaria by treating it " auf niederländische Art " (Hue, op. cit., vol. i, p. 344). For a description of the manner of making " Coal Balls " at Liége at the beginning of the eighteenth century see, Royal Soc., Philosophical Transactions, vol. xli, 1741, p. 672.

[6] Artificiall Fire, or Coal for Riche and Poore, 1644 ; Birch, History of the Royal Society, vol. ii, p. 117.

[7] Aléon Dulac, Observations générales sur le charbon minéral, f. 2.

of anthracite, was mixed with mud " and work'd up into Balls " for use in the domestic hearth, where they made a much better fire than raw culm.[1]

The idea that briquette making might purify coal of "all malignant proprieties which are averse to the nature of metallique substances ", and therefore permit its use in smelting, seems to have been at the back of Sturtevant's mind when he published his well-known treatise in 1612.[2] Octavious de Strada, a Bohemian gentleman, who is said to have been the first to attempt to smelt iron with coal in the Liége district, may have worked with briquettes.[3] He secured, with the aid of an English courtier, Sir John Hacket, patents for his process both in England and in Scotland ;[4] but, like so many other projects of a similar sort, it failed to accomplish its purpose.

It is difficult to understand why the making of briquettes should have preceded the making of coke. When the problem of ridding mineral fuel of its noxious properties presented itself, it might be thought that the first solution likely to occur to the mind would be to treat coal as logs were treated in order to obtain charcoal. A smith, it might be supposed, would have observed that some of the débris from his coal fire could be made to burn again ; and this might have led him to try to char mineral fuel. We have not been able to trace the use of " cinders " of coal (which Houghton defined as " the smaller sort sifted from the ashes ")[5] back farther than the middle of the seventeenth century ; but their adoption by a number of " tradesmen " (brick and tobacco-pipe makers, lime burners, salt and alum manufacturers)[6] between 1660 and 1700 indicates that the principle of a second burning came to be well understood during this period. In 1620 the Crown granted a patent for " charking earth fuel " to be used in smelting, and this is the first unmistakable reference to such an attempt that has been found.[7] The patentees attributed the

---

[1] *A New Present State of England*, 1727 (?), vol. i, p. 313.

[2] *Metallica*. Lewis (*Stannaries*, p. 24, note) thinks that Sturtevant may have conceived the idea of coking the coal ; but this is difficult to reconcile with Sturtevant's reference to "Lenicks ", a pressing or moulding instrument made "for the tempering, stamping, and commixing of Sea-coale or Stone-coale ". This sounds like a development of the briquette-making process: Sturtevant evidently hoped that a blending of the better stone coal with seacoal might add those " deficient properties which, as they are in charcoale, so ought they to be found in pit-coale " (*op. cit.*, pp. 90–1, 98, 106). Cf. above, pp. 115 n., 116–7.

[3] Beck, *Geschichte des Eisen*, vol. ii, p. 1213. Galloway (*History of Coal Mining*, pp. 45 sqq.) thinks de Strada coked his coal. At least until the end of the seventeenth century scientists in Germany, discussed the feasibility of smelting with coal moulded (*pochen*) " nach Lüttischer und Brabändischer Art " (Bünting, *Sylva Subterranea*, bk. xxx).

[4] Rymer, *Foedera*, 1704, etc., vol. xviii, p. 870 ; *Privy Council Register of Scotland*, 2nd Series, vol. i, pp. xvi, 61, 62.

[5] Houghton, *Husbandry and Trade*, vol. ii, p. 81. Jars (*Voyages métallurgiques*, vol. i, pp. 210–12) thought that " cinders " was a synonym for coke in the New-castle district.

[6] Houghton, *op. cit.*, vol. i, pp. 185, 205 ; Collins, *Salt and Fishery*, p. 51 ; *Royal Soc., Phil. Trans.*, vol. xii, 1677–8, p. 1053.

[7] *Patent Spec.*, no. 15. There is, however, a record of one Michell, " a collier ", who came to Downhead, a village in Somerset, and there undertook in 1616 to make " charcke coale " (E. H. Bates, *Quarter Sessions Records for* . . .

invention to one of their number, a certain Hugh Grundy.   We find the name of John Gasper Wolffen on a subsequent patent, to remind us that foreigners may also have contributed to the discovery of a satisfactory coke ; [1] but beside Wolffen we must put the English names of Grundy, Thomas Payton of Deptford, " gentleman inventor ", Sir John Winter, the prominent colliery owner and landlord from Gloucestershire, and many others.   It is not known who first made the Derbyshire " Coaks " which proved successful in drying malt. There are records between 1620 and 1660 of a great number of efforts to produce " charked coals " ; [2] and, with two exceptions, [3] all these efforts were directed towards providing a more pleasant fuel, not for the smelter, but for the maltster, the brewer, or the domestic consumer. [4] Winter's method, of which Evelyn has left us a full description, was to place the coal to be " charred " in pots, like the pots for glass-making, so that the flames of the fire did not come into direct contact with the mineral. [5]   The method successfully adopted in Derbyshire resembled more the age-old manner of making charcoal.

" The collier ", to quote Houghton, " sets six or eight waggon loads of coal in a round heap upon the ends, and as pyramidal (large at bottom and small at top) as they will stand.   If it be a wind he sets Fleaks to shelter from it, and then into a hole left in the middle to the top of the heap (or pit as they call it) he throws a shovel full or two of fire, which by spreading itself each way fires the pit round ; this burns and blazes till the smoke and flame ceases, and it's all of a red fire, then he covers all the heap with dust, and that side first which by help of the wind burns most, or where the fire breaks out, which immediately damps it and makes them dead coals, which thus stands till next morning, or longer according to their occasion, and then with a rake like a gardiners with six or eight teeth, he pulls them down round the heap, and the dust falls to the bottom, which is thrown up on a heap to damp the next pit ". [6]

With coke made in much the same way, Abraham Darby solved the problem of iron smelting, which, unlike the smelting of lead and

Somerset (Somerset Record Society, vol. xxiii, 1907), p. 216).   The word " collier " was given many meanings during the seventeenth century.   It had been previously used most commonly to denote persons engaged in converting wood into charcoal, and this may well be the sense in this record.   On the other hand, the fact that Downhead is situated within a mile or two of coal pits in the Mendip district, where the new mining industry took on a considerable importance at the beginning of the seventeenth century, suggests that Michell may have experimented with mineral fuel.   On the Continent we hear of attempts to make coke, for use in the domestic hearth, before the end of the sixteenth century (Helen Boyce, The Mines of the Upper Harz, 1920, p. 79).

[1] Pat. Spec., no. 65.   See also S.P.D., Charles I, vol. ccclxxvii, no. 61.

[2] Pat. Spec., nos. 51, 78, 93, 94 ; Cal. S.P.D., 1655, pp. 171, 191 ; 1660–1, p. 507 ; Evelyn, Sylva, vol. ii, pp. 108–9.   According to Monsieur Ticquet (Archives Nationales, $0^1$1293), copper founders used coke for making buttons, rings, and other jewelry.

[3] Grundy and John Copley, who received a grant in 1655 to make iron with "charked pit coal " (Pat. Spec., no. 15 ; Cal. S.P.D., 1655, p. 191).

[4] In one case (Pat. Spec., no. 51) it was proposed to " reduce " peat or turf to be used in smelting, as well as in boiling salt, burning bricks, tiles, etc.

[5] Diary, July 11th, 1656.   Winter's coke apparently proved unsatisfactory for domestic consumption.

[6] Houghton, op. cit., vol. i, p. 109.

tin, could not be accomplished with raw coal, even after the invention of the reverberatory furnace. The story of the successive efforts to smelt with mineral fuel has been told by a number of modern investigators.[1] In these accounts, the attempts of Rovenzon and Sturtevant, and above all that of Dud Dudley, have often received an attention altogether disproportionate to their importance. Dudley's was but one of hundreds of such attempts, and the chief reason why it has made so great an impression on modern investigators is that Dudley published a full account of his process,[2] and claimed—with little warrant, Mr. Ashton thinks[3]—to have been successful. Much confusion concerning the success of various attempts to smelt with coal prevailed in the seventeenth century, and this confusion has been sometimes reflected in modern histories, owing to the fact that iron masters, like lead and copper masters, often made purchases of mineral fuel in addition to their purchases of charcoal. For instance, when in 1597 Sir Percival Willoughby took over certain iron works at Codnor, in Derbyshire, it is provided in the agreement that he shall have " soe manie Rooks of Cole at Harther Pytts as shalbe employed for the making of Iron at Harther fordge and the Iron works there ".[4] This coal the masters used commonly for purposes other than the actual smelting—while efforts to produce pig iron with it, according to an account of 1678 of iron works in the Forest of Dean, had proved " hitherto . . . ineffectual ", it was burned in the " finery " for casting pig into bar iron.[5] Yet in some instances the masters probably attempted—and perhaps with partial success—to use certain of the less sulphurous coals in their blast furnaces. That would explain why Malynes thought iron could be melted with " flaming sea coale " or " Scots coale ".[6] Thomas Proctor, one of the first suitors for a patent to smelt with mineral fuel, recognized " that all Seacoale is not of lyke goodness nor all Iron Stoane of lyke quallytie to bee cast blowne or wrought ". He believed that " in some places Iron may be made very well with Seacoale onlye . . . and in some other places cannot be performed but with a mixture of seacoale, turffe or peate and some quantitye of woode coale ".[7] In practice neither the efforts of Proctor, nor those of his seventeenth-century successors, enabled the iron masters to produce a satisfactory iron at a reasonable cost, even with the best grade of coal and the most susceptible of stones.[8]

---

[1] Cf. Galloway, *History of Coal Mining*, pp. 39 sqq. ; Beck, *Geschichte des Eisen*, vol. iii, pp. 160 sqq. ; Hyde Price, *The English Patents of Monopoly*, pp. 107 sqq. ; Lewis, *Stannaries*, pp. 20 sqq. ; Mantoux, *The Industrial Revolution*, pp. 289 sqq. ; T. S. Ashton, *Iron and Steel in the Industrial Revolution*, chaps. i, ii.

[2] *Mettallum Martis*, 1665.   [3] Ashton, *op. cit.*, pp. 10–12.

[4] *Chanc. Proc., Eliz.*, Z./z./12 (Zouch v. Willoughby and Cranewell, Nov., 1597).

[5] *Roy. Soc., Philos. Trans.*, vol. xii, pp. 931–5. Cf. *H.M.C., Report on the MSS. of Lord Middleton*, p. 495.

[6] *Lex Mercatoria*, p. 270.   [7] *Lansdowne MSS.*, 59, no. 73.

[8] The remarks of Pettus in his *Fleta Minor* (1683), to the effect that metals could not be smelted without wood fuel, are too well known to need citation here. It would, of course, be a mistake to interpret these remarks as proving that no coal whatever was used in the smelting process during the seventeenth century.

But the ultimate discovery by the elder Darby at the beginning of the eighteenth century [1] of a successful process depended on innumerable experiments, made for the most part in seventeenth-century England (which Jevons considered almost barren of inventive skill), experiments often directed towards the smelting of metals other than iron, and even more often not directed towards a solution of the smelting problem at all.

This is not the place to discuss the causes which held back the general use of coal in smelting until the second half of the eighteenth century,[2] so that it was still possible, in 1750, for an English writer on economic subjects to think that " iron-ore is not converted into malleable iron with any fire but wood or charcoal ".[3] It is clear that, before the accession of George I, in 1714, most of the technical principles, apart from those relating to textiles, upon which the Industrial Revolution was to be based, had been more or less fully worked out, and that the Revolution itself had become inevitable. Even before the time when that Revolution is supposed to have begun, English industrial technique served as a model for the continental nations. Steam engines for draining mines, patterned after those first used in Great Britain, were set up early in the eighteenth century at collieries in the Low Countries and in France. All the uses made of mineral fuel by artisans and domestic consumers in France had been introduced, we learn from a *mémoire* written in 1742, " à l'imitation de l'Angleterre ".[4] The French studied and adopted English methods of fertilizing farm lands. An Englishman named Sutton set up at a mine in Normandy in 1752 a " ventilator ", said to have been the best yet contrived.[5] William Wilkinson, a brother of the famous iron-master, came to le Creusot to teach iron-making.[6] The leadership of the seventeenth-century British school of mineralogy—whose observations concerning the nature of minerals, and the lie of seams in the earth, may be studied in such works as Plot's *Natural History of Staffordshire* and Sinclair's *Hydrostaticks* [7]—is reflected by the common use in

---

[1] Ashton, *op. cit.*, pp. 28 sqq.

[2] There are probably three main causes : (i) the effort to keep the process a secret in order to prevent knowledge of it from reaching foreigners (Jars, *Sur le charbon*, f. 405) ; (ii) imperfections in the process as tried ; it was apparently on this account that Mr. Wood's " design of supplying the whole nation with Pit-Coal " broke down in 1730 (*Tracts relating to various Trades*, B.M., 816, m. 13, nos. 9–14) ; (iii) local vested interests in the old charcoal-burning furnaces (Lewis, *Stannaries*, p. 26). It had been opposition from such interests that had prevented Dudley—so he alleged—from having his process adopted commercially. Vested interests also held back the construction of canals in this same period (see below, pp. 258–9).

[3] *The State of the Trade and Manufactory of Iron in Great Britain*, 1750. Coal was little used in Great Britain for smelting either lead (*V.C.H. Derbyshire*, vol. ii, p. 345) or tin (*Lewis, Stannaries*, p. 26) for many years after the necessary technical improvements in the processes had been discovered.

[4] *Détail de differents moyens . . . pour diminuer la consommation du bois en France*, Sept., 1742 (*Archives Nationales*, O¹1293).

[5] Aléon Dulac, *op. cit.*, f. 39.

[6] Beck, *op. cit.*, vol. iii, pp. 1032–3.

[7] See also Charles Leigh, *National History of Lancashire and Cheshire*, 1700 ; Christopher Merret, *Pinax Rerum Naturalium Britannicarum*, 1666 ; and the writings of Martin Lister (1638–1711), John Woodward (1655–1722), etc.

eighteenth-century France of English technical terms, such as " flat-broad coal ", " hanging coal ", " wildfire ", " bad air ", " coaks ", and by the careful attention which Frenchmen gave to all the publications of the Royal Society. " Having read their books in France ", wrote Monsieur Ticquet, one of the great number of Frenchmen who came to study English technical skill at first hand, " I observed here with great pleasure their practice. . . . I have observed a great many processes unknown to our workmen ".[1]

At the foundation of the new skill was the work of the English " natural philosophers ". Seldom has there been effected a more perfect marriage between pure science and practical achievement than in Restoration England, through the efforts of the group of men who founded and carried on the Royal Society. It had been the intention of the founders " the more to communicate to each other their Discoveries ".[2] They did not content themselves with carrying out an infinite number of laboratory experiments designed to shed light on every possible aspect of the organic and the inorganic world. They sought foreign collaboration. Throughout the length and breadth of Britain they aimed to stir up interest in, and elicit contributions on, every conceivable subject. " All places and Comers ", they note with satisfaction, " are now busy and warm about this Work and we find many noble Rarities to be every Day given in not only by the hands of learned and professed Philosophers ; but from the Shops of Mechanicks ; from the voyages of Merchants ; from the Ploughs of Husbandmen ; from the Sports, the Fishponds, the Parks, the Gardens of Gentlemen ".[3] Their own members came from every walk of life, and among them were most of the distinguished men of the time. Under one roof in London one might have met Boyle and Newton and Leibnitz, Evelyn and Pepys and Christopher Wren, Petty and Southwell, Lowther and Brereton, Papin and Huyghens and Becher.

To understand the England of the Restoration, one must turn to these men, just as one turns to Michel Angelo and Leonardo to find the meaning of Italy in the Renaissance, or to Montesquieu, Voltaire, and Rousseau to find the meaning of eighteenth-century France. Restoration England has neither the eternal beauty of Renaissance Italy nor the moral grandeur of 1789 ; nor did it make an English gift to the world comparable to the gift made by the age of Shakespeare. Its achievements and its limitations are summed up most completely in the lives of Boyle and Newton.

It was an age concerned above all with the conquest of the knowable. " Your Majesty ", began the dedication of the Royal Society to its patron, " will certainly obtain immortal fame, for having established a perpetual Succession of Inventors ". Complete confidence prevailed as to the power of man eventually to master the physical world. " That perchance may be easy for the next, which seems impossible to this generation ", wrote Fuller, in referring to the unsuccessful attempts

---

[1] *Archives Nationales*, 0¹1293. Cf. Des Cilleuls, *Histoire et régime de la grande industrie*, p. 30.
[2] Thos. Sprat, *History of the Royal Society*, 1667, p. 56.
[3] *Ibid.*, pp. 71–2.

to smelt iron with mineral fuel.[1]   Religion, nationality, profession—
all were subordinated to this ultimate aim of technical mastery,
subordinated to a degree perhaps never before found in history.
For the efforts of her " natural philosophers " England had a respect
which alone makes possible the growth of a great school of science or of
art.   To a contemporary like Addison, Newton and Boyle were giants
of the stature of Leonardo da Vinci, and their fellow members in the
Royal Society enjoyed a prestige for which it is difficult to find any
parallel in our own age.   Newton, Voltaire tells us, " a vécu honoré de
ses compatriotes, et a été enterré comme un roi qui aurait fait du bien
à ses sujets ".[2]   Upon finding that De Caus had been put in a mad-
house by Richelieu for pestering the Cardinal with his revelations
concerning the power of steam, Worcester remarked :   " In my
country, instead of shutting him up, he would have been rewarded ".[3]
Ticquet expressed surprise that in England those who devoted themselves to a study of the technique of agriculture were held in esteem,
and that " philosophers, men of quality and rich merchants " took
part in this study.   With respect to science and invention, wrote
Voltaire in his *Siècle de Louis XIV*, the seventeenth may be better
called " le Siècle des Anglais ".

It is no wonder, therefore, that Papin and Huyghens, Leibnitz
and Becher should have found the atmosphere of London particularly
sympathetic, and should have been frequent guests at the meetings
of the Royal Society.   Foreign scientists at the end of the seventeenth
century valued highly their contact with the chiefs of the English
school ; one of Becher's first acts upon his arrival in England was to
demonstrate his method of purifying coal " before Mr. Boyle at
Windsor ".[4]

Boyle's interest in experiments of this nature, and the constant
discussion of the Royal Society concerning the means of increasing
the output of various commodities, or of heightening the efficiency of
different technical processes, show a trend of the utmost importance.
These " natural philosophers " of the English Restoration were
economists as well as scientists ; [5] and, beside their old religion,
which they resolved should not stand in the way of their achievements, they had begun to erect a new religion—the religion of production.   It was one of Boyle's principal propositions, to which he reverted
again and again, " that the Goods of Mankind may be much increased
by the Naturalist's Insight into Trades ".[6]   There is no doubt that,
if challenged, this is the philosophy by which Boyle would have justified
his own activities and those of his colleagues.   It is not, perhaps, too
much to say that this state of mind, which first found complete
expression in Restoration England, and the state of mind that has

[1] Quoted Galloway, *History of Coal Mining*, p. 50.
[2] *Lettres Philosophiques*, xiv.
[3] Quoted Muirhead, *Life of Watt*, p. 99.
[4] Beck, *op. cit.*, vol. iii, p. 297.
[5] For the influence of practical interests on the scientific work of Boyle and his
school, see Hermann Kopp, *Geschichte der Chemie*, 1843–7, vol. i, pp. 146 sqq. ;
Emil Kander, *Johann Joachim Becher als Wirtschafts- und Sozialpolitiker*, 1922
(a thesis submitted at the University of Berlin).
[6] *Works*, vol. iii, p. 167. See also pp. 141 sqq.

made of the personal accumulation of wealth a moral virtue,[1] are the intellectual foundations upon which modern capitalism rests.

There is a clear relation between the appearance of the extremely influential British school of " natural philosophers ", and the growth of the British coal industry. The proceedings of the Royal Society for the first thirty or forty years following the grant of its charter in 1660, are full of discussions which have a bearing, often direct and perhaps even more frequently indirect, upon problems connected with mineralogy, with the mining and the use of coal. Colliery owners like Lowther, who was a member, or Sir Roger Mostyn, who was not, send in accounts of explosions or of new methods of finding coal. Local naturalists send in samples of mineral fuel, or of strata resembling it, to be analysed and examined. Sir Robert Southwell reads a paper on the advantages of digging canals to supply London with cheaper coal. Boyle conducts experiments in order to determine the difference between coal and wood. The members meet to consider the projects on foot for smelting with coal. They puzzle over the causes for the strange fires which occur in the coal seams.

It would be a mistake, of course, to suppose that they talked predominantly of coal. To get the proper perspective it is necessary to go through Boyle's suggestions of " General Heads for a Natural History of a Countrey ". The problem of minerals, and of mineral fuel in particular, occupies only a small fraction of the whole.[2] Natural philosophers were concerned with the circulation of the blood, with disease, with physiology, with the climate, with astronomy, chemistry, physics, and pure mathematics ; yet again and again they reverted to trade and industry as if these subjects were the point of departure for most of their investigations. If they were not in the strict sense of the word inventors, there is no doubt that they were eager to throw all their learning into the service of invention. Soon after the incorporation of the Royal Society the King signified his pleasure that no patent should be granted for any " philosophical or mechanical invention ", until it had been approved by the Society.[3]

And the problem of invention in the seventeenth century was to a considerable degree directly related to coal. " Through necessitie, which is the mother of all artes ", wrote Howe, in 1631, " they have of very late yeeres devised the making of iron, the making of all sorts of glasse, and the burning of bricke, with sea coal or pitcoale ".[4] In this case the necessity was the substitution of coal for wood ; in other cases it was the more adequate drainage of the mines, or the cheaper carriage of fuel over ground. It is possible to gain some idea of the extent to which inventive effort was inspired to exert itself by problems connected directly or indirectly with the coal industry, by presenting in tabular form, according to subject matter, the patents for inventions taken out between 1561 and 1688. These patents do not give us, by any means, a complete survey of inventive

---

[1] Cf. R. H. Tawney, *Religion and the Rise of Capitalism*, 1926.
[2] *Royal Soc., Phil. Trans.*, vol. i, pp. 186–9. See also his " Articles of Inquiry Touching Mines ", *ibid.*, vol. ii, pp. 329–43.
[3] Birch, *op. cit.*, vol. i, p. 116.
[4] Stow's *Annals*, ed. E. Howe, 1631, p. 1025.

effort during the period under review,[1] as is made plain by the fact
that nearly all the successful inventions had no patent. But they
perhaps offer a satisfactory sample of the processes with which
inventors concerned themselves.

TABLE XII

Patents for Inventions, 1561–1688 [2]

Patents [2]	1561-70	1571-90	1611-40	1660-88	Total for 1571-1688
(i) *in which the relation to the coal industry is certain—*					
Drainage of mines	3	3	14	23	
Sounding for mines	—	—	1	—	
Lighting in mines	—	—	—	1	
Better furnaces, ovens, etc.	2	3	21	29	}136
Special treatment of coal	—	1	7	3	
Smelting	—	1	16	8	
(ii) *in which there is strong reason to suspect at least an indirect relation to coal—*					
Better means of communication (canals, harbour dredging, etc.)	1	3	15	16	}99
Manufacturing processes	8	13	14	29	
(iii) *in which a relation to coal is unlikely—*					
Manufacturing processes	{3	{6	{39	28	}82
Agriculture				6	

[1] There is no record, for instance, of the efforts to invent new methods
of ventilating the mines,—efforts of which we get plenty of evidence from other
sources (cf. below, pp. 363–4).

[2] The records of patents granted between 1561 and 1570 and between
1571 and 1590 are taken from E. W. Hulme, " The History of the Patent System
under the Prerogative and at Common Law " in *Law Quart. Rev.*, vol. xii,
1896, pp. 141 sqq., and vol. xvi, 1900, pp. 44 sqq.), the remainder from the
Patent Specification files in the Patent Office, London. This leaves a gap after
1590 for which we have no evidence.

[3] In the case of patents for draining mines, " coal pits " are generally
mentioned specifically. Collieries were not, however, the only objective for
which pumping devices were designed. They were designed also to drain other
mines, to drain the fens, and to dredge rivers and harbours ; but there can be no
doubt that the most insistent call for such devices came from the owners of the
collieries. And coal had played an important part in bringing about the demand
for dredging engines ; the harbours of the Tyne and of the Firth of Forth were
choked with sand bars, making navigation hazardous, chiefly because the
colliers sailed in ballasted with sand-bags, which the crew tossed into the river
before loading a cargo of coal (*Privy Counc. Reg. of Scotland*, 1st Series, vol. xiii,
pp. 262–4, 786–7 ; 2nd Series, vol. iv, pp. 294, 298, 303, and vol. vi, pp. 57–8).
After many administrative efforts to check this practice, it was formally prohibited by Act of Parliament, 19 George II, cap. 22 (see Cunningham, *Growth
of English History and Commerce, Mod. Times*, pp. 488–9). The harbours at
Sunderland and Easter Cockenzie had to be deepened to be made accessible
at all to the larger colliers (*Privy Counc. Reg. of Scotland*, 3rd Series, vol. vi,
pp. 383–5). Again, inventions designed to facilitate transport—devices such
as locks for canals, better carriage wheels, river dredging engines—cannot be

Problems arising out of the extraction, the carriage, and the use of coal apparently provided a capital motive for inventive effort in the seventeenth century ; and it is hardly an exaggeration to suggest that the rapid expansion of the British coal industry prior to 1700 was one principal cause for the development of that scientific spirit in Great Britain without which the Industrial Revolution itself would have been impossible. A striking confirmation of this thesis is to be found in the fact that, on the Continent, it was precisely in the district around Liége, where the coal mines had been most intensively exploited, that the most notable strides in inventive effort and technical skill were made during the seventeenth and early eighteenth centuries.[1]

It has been shown that the development of British coal mining in the period from 1550 to 1700 made possible the growth of manufactures ; that indirectly this development encouraged more scientific agriculture, the growth of a great merchant marine, and the founding of a school of natural philosophers and inventors, whose discoveries have contributed so largely to the building up of the modern industrialism of western Europe and America. It remains to draw attention to the part played by the coal industry in increasing the size of markets and the economic interdependence of different sections of the country.

Wood can be made to grow almost anywhere, and most societies depending on wood fuel can therefore obtain their supplies of timber in large measure near the place of consumption. But coal resources are normally localized in a limited number of regions, and the adoption of coal in place of wood necessarily involves a greater economic interdependence between districts, because the number of persons who are obliged to secure fuel from a distance is immensely increased. A country like England may be divided into a number of small self-sufficient groups, each able to nourish and clothe and warm itself, so long as it depends on wood for fuel, for then all the products essential to life may be acquired from any soil that is reasonably fertile. But coal, unlike corn, or wool, or wood, cannot

divorced from the coal industry altogether, for the problem of transport was becoming more and more a problem of the transportation of coal (see below, pp. 258–9).

I have made a somewhat arbitrary division between manufactures in which one may suspect a relation to coal, and manufactures in which such a relation is unlikely. In general, I have included patents for new methods of salt and soap boiling, etc., when the nature of the method is not described, under the former heading, since the chief economies which could be made in these manufactures were in fuel costs ; patents for mills to grind corn, for spinning engines (there are three of these), saw mills, etc., under the latter heading. Some of the patents are for carrying out several different operations, and there is therefore great difficulty in deciding under what heading to include them. Where this is the case, I have classified the patent according to what appears to be the chief object in the minds of the promoters.

[1] *Royal Soc., Philos. Trans.*, vol. i, pp. 79–80 ; Hue, *Die Bergarbeiter*, vol. i, p. 344 ; Léopold de Genneté, *Nouvelle construction de cheminée qui garantie du feu et de la fumée*, Paris, 1759 ; Clément, *Lettres, instructions et mémoires de Colbert*, vol. iv, pp. 595–6, etc. Natives of Liége, as well as Englishmen, were sought after by the French to teach them the new improvements in mining technique (Thomassin, *Mémoire statistique*, p. 414).

be made to grow, or even to reproduce itself, and consequently the society which turns to coal adopts as a necessity of life a product which can be had only where the forces of nature have seen fit to provide it.   Hence the general use of coal means that some areas must produce a surplus to supply others and must therefore specialize in the production of mineral fuel.

England before 1500 had been made up of a number of separate market areas, each supplying almost entirely its own simple wants. Judged by modern standards, the London of the later Middle Ages was still a small town depending for its necessities upon the surrounding counties.   As Professor Gras and others have shown, the two succeeding centuries brought enormous changes in the size and nature of English markets.   To an ever increasing extent one region exchanged its surplus products for those of another region, and the London area had elaborate trading relations with every other district in England.   By the accession of Charles I, London had already become a modern city—a city of traders, money lenders, and middlemen, nourished with the surplus products of a nation.

The fact that the country as a whole was adopting coal as fuel in the sixteenth and seventeenth centuries encouraged the population of regions rich in minerals to specialize in coal mining and in manufactures dependent for their success on cheap supplies of coal.   England had not yet begun to use her coal as a means of purchasing the food and raw materials of other countries, but there were already certain small areas, miniature Englands of to-day, which, because of their advantages in coal, had begun to depend upon other areas for their corn. The Tyne valley affords the most outstanding example.   The economic position of this area is well described in a petition of the Newcastle Merchant Adventurers to Parliament in 1652.   " By Providence ", they point out, " they are seated in a barren and poore cuntry, which commonly requires great supply of corne and other necessaryes ", and they have been " very serviceable " in bringing these commodities by sea and supplying them at reasonable rates, " which they were better able to doe by reason of the advantage and opportunity they have above other places in the commodity of cole, by which they commonly make their voyages outwards and bring home their returns for small freight ".[1]   A principal source of their grain supply was the fertile region of Bedfordshire, Cambridgeshire, Huntingdonshire, and parts of Northamptonshire, in the upper Ouse valley.   Professor Gras has found that the quantities of corn sent to Newcastle were considerable.[2] Nor was the Tyne the only mining area dependent upon other districts for a part of its grain supplies.   In 1620, the justices of Nottinghamshire argue that there is no need of a storehouse for corn, as counties

---

[1] F. W. Dendy, *Extracts from the Records of the Merchant Adventurers of Newcastle-upon-Tyne* (*Surtees Soc.*, vol. xciii, 1895), p. 179.

[2] Gras, *op. cit.*, pp. 53–4, 104, 277, 321–2, Appendix A and D.   In the years 1663–4, 1671–2, 1684–5, 1688–9 more shipments of corn were sent from King's Lynn to Newcastle than to any other port, not excepting London (*ibid.*, pp. 308–9).   On the other hand, we have evidence that the husbandry in the neighbourhood of Newcastle was greatly improved in the eighteenth century (see above, p. 237).

which send up the Trent for coals bring in corn whenever it is needed.[1] The owners of Bedworth colliery in 1684, in soliciting financial support from the citizens of Coventry, emphasize the advantages derived from the resort of persons from Northamptonshire and adjacent parts to Bedworth. In return for fuel, they brought " quantities of wheat, barley and malt to Coventry, thus lessening the price of these commodities ".[2] The districts round Derby obtained their barley in exchange for coal carried overland into Leicestershire and Northamptonshire. Already in 1603 Scotland exported coal and salt and imported grain.

" The more we study . . . the history of communication by water in England ", writes Mantoux in his history of the Industrial Revolution, " the more do we realize how closely it was interwoven with the history of coal ".[3] The economic interdependence of various districts, which had resulted in part from the substitution of coal for wood as fuel, greatly increased the traffic by sea and land. A very large proportion of that traffic, measured by bulk, was in coal.[4] The fact that this was of all commodities the bulkiest in proportion to its value put a special premium on inventions for reducing the costs of transportation. The relation of coal to the eventual discovery of the steam railroad has already been explained. Efforts to introduce sailing wagons, and to remodel ships in order to reduce the number of seamen, were especially associated even in the seventeenth century with the carriage of coal. It has long been recognized that a principal cause for canal building and river dredging in Great Britain, during the late eighteenth and early nineteenth centuries, was the need for transporting coal inland at cheaper rates. This need had begun to be felt already by the middle of the seventeenth century. Projects were " noised abroad " for digging canals, for dredging rivers, and for making them navigable by installing locks. Yarranton sponsored an attempt in 1661 to make a stream in Worcestershire navigable from Stourbridge to the Severn, in order to provide an outlet by water for the produce from the Staffordshire collieries. A flood destroyed the works, and the undertakers could not raise enough capital to revive their project, owing perhaps to the determined opposition of Shropshire interests, who appealed to Parliament in fear lest the proposal, if carried into effect, should ruin the prosperity of the mines and the allied industries at Madeley, Broseley, and Benthall.[5] By a similar procedure the Corporation of York discouraged an attempt to make the rivers Aire and Calder navigable up to Wakefield and Leeds ; for, although the people of York stood to gain from the coal

---

[1] *Cal. S.P.D.*, 1619–23, p. 130.

[2] *Exch. Deps. by Com.*, 36 Charles II, Mich. 43 (Answer of John Dearen of Coventry, clothier).

[3] Mantoux, *op. cit.*, p. 111.

[4] See above, pp. 238–40 ; and cf. pp. 100–19.

[5] J. Priestley, *Historical Account of the Navigable Rivers, Canals and Railways throughout Great Britain*, 1831, p. 632 ; Yarranton, *England's Improvement*, pp 65–6 ; *H.M.C., 5th Report*, pt. i, appx., p. 160b ; *ibid., 15th Report*, pt. x, p. 65 (*Report on the MSS. of the Shrewsbury Corporation*). It was not until 1782 that Staffordshire coal actually found an outlet by canal down the river Severn (Galloway, *Annals of Coal Mining*, p. 331).

which would thus find an outlet by water, they stood to lose—as they thought—because a navigable river would enable the textile workers of the West Riding to sell their produce more cheaply in competition with the textile workers of York, who had already a market by water. It was not until 1699 that an Act of Parliament was passed to make these rivers navigable.[1] Again, the Corporation of Nottingham opposed all projects for dredging the Trent above the town through fear of the competition of Derbyshire coal in the lower Trent valley, a market hitherto served almost exclusively by the collieries about Wollaton. In spite of such opposition, adventurers succeeded in deepening the river Ouse for the passages of barges up to St. Neots, and the river Thames for the passage of barges up to Oxford. One of the chief objects in both cases was to cheapen the cost of carrying Newcastle coal farther inland.

In the light of our knowledge that the French were backward in exploiting their mineral resources, we might assume that the considerable development of canal building in France during the second half of the seventeenth century bore no relation to the development of the coal industry. It is interesting, therefore, to find a tract published in 1638 advocating the construction of canals to connect the principal river systems of France, and particularly to connect the Loire with the Seine, on the ground that such waterways will make possible the cheap transport of the turf and coal found near St. Etienne and elsewhere in central France, and will therefore diminish the consumption of timber and the import of British coal.[2] Obviously we cannot leave the question of mineral fuel altogether out of account in searching for the origins of the project to build the Seine-Loire canal, which was begun in 1675. Six years before, Colbert had referred to the advantages for traders in coal that would result from the proposed deepening of the river Allier from Brioude to Pont-du-Château.[3] The relation of coal to the development of inland waterways is older than has been suspected.

Already before the end of the seventeenth century something had been done in England, under the stimulus provided by the urgent need for cheap coal, to reduce the cost of transport by land and water. Not only had several rivers been deepened, making possible the carriage of commodities by barges for greater distances, wagonways had, as we shall see, materially cheapened the expense of land carriage in certain regions,[4] and improvements in shipbuilding had reduced the number of sailors required for sailing a freight vessel.[5] These advances in the technique of transport were tending further to increase the economic interdependence of different parts of Great Britain, which the localization of the coal resources did much to encourage. Thus coal was at the root of the main developments

---

[1] Priestley, op. cit., pp. 7 sqq. ; Defoe, Tour, 1748 ed., vol. iii, pp. 121–3 A New Present State of England, vol. i, pp. 281–2, 289.

[2] Charles Lamberville, Alphabet des Terres à brusler et à Charbon de forge, 1638.

[3] Clément, Lettres, instructions et mémoires de Colbert, vol. iv, p. 437.

[4] See below, p. 385.

[5] See below, pp. 390–1.

which lead, as Adam Smith was perhaps the first to point out, to the increasing division of labour—another of the foundations upon which modern industrialism rests.

To trace the influence of coal in stimulating heavy industry, in promoting English agriculture and sea power and invention, and in increasing the size of markets and the division of labour, farther back into history than has been hitherto attempted is something beyond a mere exercise in research. If it can be shown that Great Britain owed her very extensive commercial and industrial development during the sixteenth and seventeenth centuries in some measure to her advantages over other countries in the matter of coal supplies, the case of those who argue that the economic strength of a nation rests mainly upon the policy pursued by its statesmen is materially weakened. Frederick List and his successors of the Protectionist School have looked to the growth of England during the sixteenth and seventeenth centuries for confirmation of their theories ; Seeley made this growth the basis of his argument concerning the influence of British policy upon economic progress. But the late Professor Unwin has done much to undermine the theories both of Seeley and the neo-Mercantilists,[1] and it can now be seen that coal offers an important alternative explanation for the development of British economic power in this period.

It is not suggested, of course, that the mineral wealth of England was the sole, or even the principal cause underlying this remarkable economic development. Obviously the fact that she remained largely aloof from continental quarrels, especially during the reigns of Elizabeth and James I, gave her an enormous economic advantage. While large parts of the Continent were constantly drenched with blood, the English countryside, except for the brief period of the Civil War, remained undisturbed. The passage of goods from one part of England to another could be effected free of heavy tolls ; whereas on the Continent, as we have seen in the case of coal, high tariffs imposed damaging restrictions on the movements of commodities. And while the Continent was busy with warfare, England had the opportunity to gain control over a vast and wealthy colonial empire. The importance of all these factors in establishing British commercial supremacy before the Industrial Revolution has long been recognized.

What has not been recognized is that commercial supremacy was based to no small extent upon the growth of native industry, and that the growth both of industry and commerce was based to no small extent upon coal. If this is true at so early a period in the history of western Europe, the part played by coal in producing our modern economic civilization is greater than has been realized, and there is no exaggeration in the verses composed about the new fuel not long after it had come into common use.

---

[1] See especially Unwin's *Studies in Economic History*, edited by R. H. Tawney, 1927.

" For 'tis not only Fire commends this Spring ;
A Coale pit is a Mine of ev'rything.
Wee sinke, a Jack of all Trades shop, and found
An invers'd Burse, an Exchange under-ground.
This Proteus Earth converts to what you'd ha't ;
Now you may weav't to silke, then coyn't to Plate ;
Or (what's a Metamorphosis more deere)
Dissolve it, and twill melt to London Beere :
For whatsoer that gawdy Citty boasts,
Each Month derives to these attractive Coasts.
Wee shall exhaust their Chamber, and devoure
The Treasures of Guild-hall, the Mint, the Tower.
Our Staiths their morgag'd streets will soon divide,
Blathon own Cornewall, Stella share Cheapside,
Thus will our Coale-pit's Charity and Pitty,
At distance, under-Mine and Fire the Cittie.[1]

[1] *News from Newcastle*, 1651.

PART III

COAL AND THE OWNERSHIP OF NATURAL RESOURCES

PART III

COAL AND THE OWNERSHIP OF NATURAL RESOURCES

THE working of minerals has given rise to conflict between rival claimants ever since the value of such deposits first came to be known. Even if the broad question of the ownership of seams of lead, iron, or coal in the earth is settled, a large number of other problems still remain to be solved. How much may the owner of the minerals ask for conceding the right to exploit them ? In order to turn his riches to account, it is necessary to dig in the soil, to the detriment of the surface value. Considerable tracts of arable or pasture land may be sacrificed for sinking shafts, providing space for machines or workmen, or for the carriage of the mineral from the pit head. Sometimes owing to negligence, but often unavoidably, mining operations cause the earth's surface to give way and damage is done to farms and houses. What shall be paid for these losses ? When two mining companies are working in the same seam, how shall it be decided exactly where the rights of each begin and end ? Supposing the operations of one aid or hinder the other—what is to be given as compensation ? Once the guiding principles have been worked out, there remains the problem of applying them, which, in turn, involves the discovery of the facts in each particular case, by means of elaborate measurements underground and complex investigations on the part of persons expert in geology or natural science. The history of the slow development of the law governing mineral property shows a dissipation of human energies in endless legal struggles, often accompanied by illegal sabotage and acts of violence.

We may consider, in the first place, the actual ownership of mineral wealth as it lies in the earth, without touching on the questions of surface damage and wayleaves, which arise only after exploitation has begun. As soon as the treasures of the earth's crust become accessible to man's labour, they are claimed as property, and a payment of " royalty " is asked for their use in the same way as a rent is charged by landowners for the use of their land. Investigators are agreed that, historically, royalty owners may be divided into four categories : the overlord, whether he be king, prince, or bishop ; the community, in the person of the State or sovereign power ; the owner of the surface ; the concessionnaire or mine owner.[1] " Their relative rights ", as Mr. and Mrs. Hoover have put it, " have been the cause of never-ending contentions ever since a record of civil and economic contentions began ".[2]

---

[1] Sometimes the first and second groups are classed under a single heading : the State. There is, however, an immense difference between the conception of personal ownership by a feudal lord and the modern conception of community ownership by the sovereign power as trustee for all the people.

[2] H. C. and L. H. Hoover, note on mining law in their edition of Agricola's *De Re Metallica*, p. 82.

CHAPTER I

THE OWNERSHIP OF MINERALS

Such knowledge as we have of ancient law suggests that in the Greek states, and probably also in Egypt, Lydia, Assyria, Persia, India, and China, the ownership of minerals was frequently vested in the State.[1] In Roman law it is not possible to find any general principle of invariable application, but when mineral wealth was not worked directly by the State, it appears often to have been leased to concessionnaires, who paid a tithe to the central government. Only minerals discovered in Roman provinces after the conquest were ever in private hands. During the early Middle Ages it was common for European princes, in their efforts to establish a claim to minerals, to construe certain passages of Latin commentators as proof of the assumption of regalian rights by the Roman State. Modern scholars have questioned this interpretation, but it may have aided German feudal and ecclesiastical lords in establishing a title to such minerals as were found within their small jurisdictions.[2] The strength of national feeling, and the growing power of the Crown in France,[3] made it possible for King Charles VI to assert, by royal edict in 1413, a hypothetical title to all minerals, but, in general, until the eighteenth century owners of land had an important voice in, and shared in the fruits of, the exploitation of minerals under their holdings. In both Germany and France, as elsewhere on the Continent, the view that the rights both of landowner and concessionnaire were limited by the sovereignty of the overlord was maintained, though it was frequently ignored in practice.

Only in Great Britain were the rights of the landowner to all minerals, except gold and silver, made absolute. This principle was determined during the sixteenth and seventeenth centuries, when the opposition to the interference of the Crown in economic affairs found expression in legal decisions and statutes. Before this period the law as to the ownership of mineral wealth was not clearly defined. There

---

[1] See the Hoovers' note on mining law in their edition of Agricola, *op. cit.*, pp. 82 sqq., and G. R. Lewis, *The Stannaries*, pp. 65–7.

[2] The literature on the history of German mining law is voluminous. See K. T. von Inama-Sternegg, *Deutsche Wirthschaftsgeschichte*; Schmoller, " Die geschichtliche Entwickelung des Unternehmungs", in *Jahrbuch für Gesetzgebung*, vol. xv, 1891, pp. 660–710, 963–1029 ; Herman Achenbach, *Das gemeine deutsche Bergrecht*, Bonn, 1869 ; Adolf Arndt, *Bergbau und Bergbaupolitik*, Leipzig, 1894. A list of authorities on the subject is given by A. Zycha in *Vierteljahrschrift für Sozial- und Wirthschaftsgeschichte*, vol. v, 1907, pp. 238–92, " Zur neuesten Literatur über die Wirthschafts und Rechtsgeschichte des deutschen Bergbaues ".

[3] The best historical treatment of French mining law is that of Lamé-Fleury, *De la législation minérale sous l'ancienne monarchie*, Paris, 1857–80. See also H. Achenbach, *Das französische Bergrecht*, Bonn, 1869. A further list of authorities is given by Rouff, *Les mines de charbon en France au XVIIIe siècle*, pp. xlvi–vii.

was no English statute like the edict by which Charles VI of France proclaimed the right of the Crown to all minerals. Apart, however, from the theory of Norman law by which all subjects were presumed ultimately to derive their title, not only to minerals but to land, from the King, and apart also from special mining areas like the Forest of Dean, the Mendip Hills, or the High Peak of Derbyshire, where exceptional customs, said to be of extremely ancient origin, hold sway to this day, records show that the Crown made several early attempts to establish its right over base metals in the lands of subjects.[1] " A reasonable interpretation", says Lewis, speaking of two such attempts made in the twelfth and thirteenth centuries, " would be that the Sovereign did claim all metallic mines, and enforced his claim where practicable ", but never with general success.[2] Such a conception of the royal prerogative was still alive under the Stuarts, as is shown in England by various projects for declaring coal or lead to be mines royal,[3] and in Scotland, by the terms of an Act of 1592, which declared it lawful for the King " for reasonable composition to set in feu farm to every Earl, Lord, Baron, and other freeholder . . . all and whatsoever mines of gold, silver, copper, lead, tin and whatsoever other metals or minerals which is or may be found within their own lands and heritages ".[4] In the same period the Crown claimed special rights in connection with saltpetre both in England and in Scotland, by virtue of which its agents might search the property of any subject in quest of earth from which to extract it. The claim was regarded as an infringement upon popular rights, and served to strengthen the distrust of all similar pretensions.[5] Projects for extending the *regale* were rendered increasingly unpopular by the growing opposition to all projects which aimed to strengthen the royal prerogative. The Act of 1592 was left to accumulate dust among the parliamentary papers in Edinburgh. It was not printed, and is not mentioned by leading Scottish commentators of the seventeenth century.[6]

In England, the opportunity for the Crown to claim the ownership

---

[1] See Lewis, *op. cit.*, pp. 79 sqq.; Hoover, *op. cit.*, pp. 82–6; *V.C.H. Somerset*, vol. ii, pp. 362 sqq.; *V.C.H. Derby*, pp. 325 sqq.; H. G. Nicholls, *The Forest of Dean*, 1858; Thomas Sopwith, *An Account of the Mining Districts of Alston Moor, Weardale, and Teesdale*, 1833; *Additional MSS.*, 32465 (" Liberties of Miners of the Peak, Co. Derby "); *Royal Commission on Mining Royalties*, 1890–3, *Final Report*, pp. 6–7, 82, and *1st* to *4th Reports*, *passim*.

[2] *Op. cit.*, p. 75.

[3] See below, pp. 283–4.

[4] *Act Anent the Mines*, 1592, cap. 31. This Act was apparently designed not to establish a title of the Crown to these mines, which was taken for granted, but to enable freeholders to work minerals in their own lands on payment of a fixed portion of the product to the Crown. See the statement in 1704 of John Binning (*Letter to a Member of Parliament*), who urged that the hitherto unprinted Act of 1592 be published as an encouragement to the search for metal mines. (See also *Privy Council Register of Scotland*, 1st Series, vol. ix, pp. 333–4). Under the terms of the Act, landowners were to be forced to exploit minerals found in their land.

[5] J. Beckmann, *History of Inventions*, vol. ii, p. 509.

[6] See the note by the Lord Advocate, submitted to the Coal Industry Commission, 1919 (*Report*, vol. iii, Appendix 57, pp. 150–1).

of base minerals in the lands of its subjects [1] was compromised, at the beginning of the great expansion in coal mining, by a decision given by the Court of Exchequer in 1566. The famous case of the *Queen v. the Earl of Northumberland*, in which most of the legal talent of the realm was engaged on one side or the other, arose out of the entry of agents of the Society of Mines Royal into a vein of copper ore, containing some silver, at Newlands in Cumberland, under land owned by the Earl.[2] All twelve judges were agreed that mines of gold or silver, whether or not in the lands of a subject, belonged by prerogative to the Crown, and they gave unanimous judgment for the Queen. Nine judges held that ore of any kind came within the *regale* if it could be proved to contain a trace of the precious metals. The three remaining judges, in dissenting, argued that all minerals belonged to the subject, provided that the value of the precious metal in them did not exceed that of the base. All agreed that a mine which contained no gold or silver was not to be deemed a mine royal. Apparently the dissenting judges persuaded their colleagues to reopen the general question of what proportion of precious metal should constitute a mine royal, for, in the concluding portion of the judgment, it was stated that, if the proportion of gold or silver were negligible, the minerals should not belong to the Crown. Obviously this last clause in the decision was incapable of clear interpretation. While Malynes in 1622 considered a copper mine a mine royal,[3] and while Pettus believed that the Crown might search for treasure in anyone's ground,[4] the general trend of opinion during the seventeenth century seems to have been opposed to a rigorous enforcement of the doctrine of the majority. In 1622 the Privy Council was urged to encourage private lead mining, on the ground that " all lead ore in England and Wales has silver in it although not to the proportion of a royal mine ",[5] and Pettus in 1670 held that this proportion was not reached "where the Oar . . . digged . . . doth not yield . . . so much Gold or Silver, as that the value thereof doth exceed the charge of Refining ".[6] These ambiguities were cleared up by statutes of 1689 and 1693, which in part overruled the decision of 1566 by declaring " that no mine of copper, tin, iron or lead, shall hereafter be adjudged, reputed or taken to be a Royal Mine, although gold or silver may be extracted out of the same ", provided that the Crown have a right of pre-emption of these precious metals, paying the mine owners a certain fixed rate per ton of base ore.[7] From that time until the present day, both in

---

[1] See, for instance, the warrant " for to search for moynes " in any land, granted in 1562 to Thomas Thurland, David Höchstetter, and others (*Lansdowne MSS.*, 5, no. 47).

[2] Edmund Plowden, *The Commentaries or Reports*, 1818 ed., pp. 310–40. A lengthy summary of Plowden's report of this case is given in W. Bainbridge, *The Law of Mines and Minerals*, 1878 ed., pp. 122–8. A briefer synopsis is that of W. R. Scott, *Joint-Stock Companies*, vol. ii, pp. 385–6.

[3] *Lex Mercatoria*, p. 259.

[4] Sir John Pettus, *Fodinae Regales*, 1670, p. 20.

[5] *S.P.D., James I*, vol. cxlv, no. 66 : *Reasons why the Council should further Levine Van Hack's work of smelting and refining lead ore* (? May, 1622).

[6] Pettus, *op. cit.*, p. 9.

[7] 1 Wm. and Mary, cap. 30 ; 5 Wm. and Mary, cap. 6. The quotation is from section 4 of the first Act.

England and in Scotland,[1] the courts have held that the landowner has an absolute right to any mine in his own lands, except mines of gold and silver, which are scarcely ever found in the British Isles, and except in special areas, like the Forest of Dean and the High Peak, where ancient custom takes precedence over statute law.[2] Thus was it settled that the mining law of Great Britain and her colonies—except in so far as it has been modified by statute—should differ from that both of ancient civilizations and of other modern nations.

Between the early twelfth and the late sixteenth centuries, when overlords throughout Europe were laying their hands on the mineral wealth within their respective jurisdictions, we find little interest taken in coal compared with that taken in the precious metals, and in tin, copper, lead, and even iron. Coal generally escaped the claims of princes simply because it was thought to be worth less than other minerals.[3] The title of the English Crown to precious metals rested in large measure upon their " excellence " and upon their alleged importance for national defence.[4] Coal was regarded, even as late

[1] In Scotland, so far as I know, we have no statute defining a mine royal after that of 1424, cap. 12, but the Scottish courts in the eighteenth century acted as if the Acts of William and Mary were applicable to their country, and interpreted the Scottish Act of 1592 as entitling the freeholder to demand a grant, and not merely authorizing the King to make one, and (although this is certainly contrary to the spirit in which the law was enacted) they interpreted the Act as applicable only to mines of gold and silver (*Report of Coal Industry Commission*, 1919, vol. iii, pp. 150–1). In the nineteenth century it was held in the courts of common law that the Act of 1592 was controlled by the earlier Scottish statute of 1424, which claimed only gold and silver as mines royal (*Roy. Com. on Mining Royalties, 2nd Report*, Qs. 7340–3). See also below, pp. 282–3.

[2] The exceptions also include the tin mines of Devon and Cornwall, and certain areas of North Wales. Minerals (of whatever nature) under the seashore or sea bottom or tidal navigable rivers are also vested in the Crown. A good statement of the existing law was given in 1893 by R. W. Cooper, a sol citor of Newcastle-on-Tyne (*Roy. Com. on Mining Royalties, 4th Report*, Qs. 19466–8). See also Bainbridge, *op. cit.*, pp. 117–19, and Redmayne and Stone, *The Ownership and Valuation of Mineral Property*, 1920, pp. 8–12.

[3] On the relative lack of interest taken during the Middle Ages in coal as compared with other minerals both in Great Britain and on the Continent, all authorities appear to agree. See, for instance, R. M. Garnier, *History of the English Landed Interest, Modern Period*, 1893, p. 33 ; Lewis, *Stannaries*, p. 97 ; R. L. Galloway, "Earliest Records of the Working of Coal on the Banks of the Tyne", in *Arch. Aeliana*, N.S., vol. viii, 1880, pp. 168–9. On the early importance of lead mining see W. H. Pulsifer, *Notes for a History of Lead*, 1888, pp. 116, 126, etc. In Germany, where we find the greatest development of mining in the Middle Ages, the chief quest was for copper and silver, and we hear next to nothing of coal workings.

[4] Plowden, *Reports*, pp. 310 sqq. Baron Barham argued on behalf of the Crown that the most excellent things should belong to the most excellent person, and further that the Crown could not fulfil its duty to defend the realm unless it enjoyed a natural provision of treasure. See also Pettus, *op. cit.*, pp. 28–9 ; Scott, *Joint-Stock Companies*, vol. ii, p. 383. The saltpetre *regale* was defended on the ground that a supply of native saltpetre, to be used in the manufacture of gunpowder, was necessary for the defence of the realm (Bainbridge, *Law of Mines*, p. 133). Note also Adam Smith's remarks concerning treasure trove : " It was put upon the same footing with gold and silver mines, which, . . . were never supposed to be comprehended in the general grant of the lands, though mines of lead, copper, tin, and coal were, *as things of smaller consequence* " (*Wealth of Nations*, 1920 ed., p. 268). (The italics are mine.)

as Elizabeth's reign, as noxious and disgusting, and there was still no jurist far-sighted enough to urge that one day wars would be won by the nation which possessed the largest supply of this mineral. When such workings as existed were at shallow depths, and men of learning lived in ignorance of the nature of the seams (some imagining that they grew like surface vegetation), doubt even prevailed whether coal could properly be classed with the more esteemed minerals.[1] On manors where it served the tenants for firing, it was subject to the same manorial regulations as peat, turf, and wood, which were frequently allowed the copyholder as a customary right.[2] The charter of Beauchief Abbey in the late thirteenth century gave the canons liberty to take and carry away seacoal both for their own and their tenants' use.[3] In Lancashire such was the custom as late as the sixteenth century. When an action was brought, in the reign of Elizabeth, against a tenant of Ightenhill manor, the defendant alleged that his ancestors " since memory of man " had dug coal " for the necessarie fyre burned on this holding ".[4] Along weathered outcrops it was scarcely more trouble to get coal than turf, and it is not an accident that the early documents refer to " taking " and " digging " rather than mining. Indeed, it was sometimes believed that mineral fuel might more properly be classed with what the French call *carrières*, or quarries, than with mines.[5] In a discussion of the true meaning of the Scottish " Act anent the Mines ", John Lyndesay wrote in 1592, " I deny that tawb [turf ?], coal quarrell or lyme may justlie be called metals ". Like legal opinion in the nineteenth century, though for a different reason, he held that coal could not be subject to the provisions of the Act.[6]

On the Continent the landowner was nowhere as strongly entrenched in his title to minerals as in Great Britain. But there, too, for similar reasons, princes and other overlords during the Middle Ages

---

[1] In medieval times, the idea was generally held both in England and Belgium that coal was to be found only at short depths, and was a form of vegetation which grew like other forms (Decamps, *Mémoire historique sur l'industrie houillère de Mons*, pp. 58 sqq.). This view was still held by M. Trudaine, *intendant* at Lyons, concerning the coal mines round St. Etienne, as late as 1709 (Boislisle, *Correspondance des Contrôleurs Généraux*, vol. iii, no. 496).

[2] *Duchy of Lancs., Deps.*, 50/V/1 ; *ibid., Pleadings*, 78/B/10 ; *Cal. of Duchy of Lancs. Pleadings*, vol. ii, p. 95 ; R. H. Tawney, *The Agrarian Problem in the Sixteenth Century*, pp. 127, 129.

[3] *V.C.H. Derbyshire*, vol. ii, pp. 350, 352.

[4] *Duchy of Lancs. Pleadings*, 109/A/9. See *Duchy of Lancs., Deps.* 77/B/6, for the same custom in Haigh manor near Wigan. Also *V.C.H. Lancs.*, vol. iv, p. 108 ; Whitaker, *Whalley*, 1818 ed., p. 255, cited Crofton, *Lancashire and Cheshire Coal Mining Records*, p. 36. For the same custom in South Wales, see G. G. Francis, *Charters granted to Swansea*, 1867, p. 7 ; and in Scotland, A. S. Cunningham, *Mining in the Kingdom of Fife*, 1913, pp. 3–4.

[5] *Roy. Com. on Mining Royalties, Final Report*, p. 59. The word " mine ", as first used, applied to the ore or mineral, and not to the place from which it was taken. The dual meaning first appeared at the end of the fourteenth century (see Galloway, *Annals of Coal Mining*, pp. 25–6).

[6] *Balcarres MSS.*, vol. ix, f. 26 (" Mr. John Lyndesay's answers to the laird of Merchinston's observations upon the act of parlament of the metals ", August 1st, 1592).

seldom asserted with vigour their claim to the ownership of coal. In the Belgian provinces, where coal mining could already be spoken of in the sixteenth century as an industry of some importance, two main systems of ownership prevailed : that of Hainaut,[1] and that of the Pays de Liége, which applied also in essentials to the province of Limbourg.[2] In Hainaut surface owners had no rights to the coal under their holdings, not even the right of priority in exploitation. On the other hand, the title of the ultimate overlord (the Archduke of Hainaut at the beginning of the seventeenth century), which extended to mines of lead and iron, did not apply to coal and stones.[3] Deposits of mineral fuel were first found in what was then the domain of various ecclesiastical houses, such as the Chapter of St. Wandru and the Abbey of St. Ghislain, and these foundations combined in the thirteenth century to lay down rules for the exploitation of coal.[4] In their capacity of *seigneurs haut-justiciers*, these lords alone had the right to grant concessions. Their functions, in so far as they concerned the *droit de charbonnage*, were definitely set forth in 1619 by an archducal charter. The ownership of coal was completely separated from that of the surface, the former being, in theory at least, vested in the community, for which the *seigneurs haut-justiciers* acted as trustees. They granted what are known as " apex " concessions to a seam or part of a seam, regardless of the surface boundaries of landholders. We have here what is perhaps the only instance of this kind of grant in early coal mining history on the Continent, though it was not uncommonly found in other kinds of mining, being a logical outcome of the legal doctrine that the ownership of the minerals was separate from that of the surface. Concessions could pass from father to son, but were subject to the payment of two dues (*redevances*) to the *seigneurs haut-justiciers* : the *cens*, a fixed sum given in return for the power to exploit a mine, and the *entrecens*, a percentage of the produce, ranging from one-fourth to one-fiftieth, and paid either in money or in deliveries of coal at the pit head. The *droit de charbonnage*, once granted, could be withdrawn only if it were proved in court that the concession involved grave dangers to the public interest.[5] Here, then, we find the machinery for something that approaches state control of concessions.

In the Pays de Liége all coal mines, except those under royal roads, which belonged to the prince-bishop of Liége, were owned by the landlord. Many landlords sold their titles to the surface while reserving the minerals, and this led to a distinction between the *terrageur* or *rendeur* (owner of the subsoil) and the *hurtier* (owner of

[1] Hainaut included within its territorial boundaries the coal mines about Mons and Charleroi. The small coalfield of Namur was, it seems, considered a part of the *regale* of the counts of Namur (G. Arnould, *Mémoire sur le bassin houiller du couchant de Mons*, 1877, p. 28).

[2] See below, p. 288.

[3] Arnould, *op. cit.*, p. 22.

[4] Decamps, *op. cit.*, pp. 58 sqq.

[5] Arnould, *op. cit.*, pp. 23–7. According to the report of M. Harzé, made to the Royal Commission on Mining Royalties (*4th Report*, p. 130), the *entrecens* amounted to from one-tenth to one-twentieth of the product extracted.

the surface). A grant from the *terrageur* was all that the miner needed to begin operations. This grant could be either a *rendage*, which entitled the recipient to open as many pits as he liked, or a *permission* to open only one pit. In the second case the concessionnaire's title ended when the first pit was filled in. The *terrageur* was entitled to a royalty paid either in money or in coal delivered, usually one-eighty-first of the product if the mine was below the level of free drainage, one-forty-first if above that level. His right as grantee was not absolute, like that of the lord of the manor in England or the *seigneurs haut-justiciers* in Hainaut, and it was possible to obtain a concession without his consent, by means of either the so-called *droit de conquête* or the *permission de quarante jours*. The *droit de conquête* permitted the entrepreneur, if refused a grant by the *terrageur*, to prove by witnesses and experts that he possessed the means and the technical skill necessary to drain the mine. If he could satisfy the court of his capacity, and if the *terrageur* himself did not at once begin to work the coal, then the entrepreneur received a grant in law subject to payment of the usual royalty. By the *permission de quarante jours*, an entrepreneur (or mining company) had the right to dig a pit without grant, and, if he could prove that the *terrageur* had been aware of his operations and had not brought proceedings against him within forty days, he secured the title to the particular seam worked, subject again to payment of royalty. Besides the *droit de terrage*, as this royalty was called, the entrepreneur, if he depended upon an adit not his own for drainage, must pay the *cens d'areine* (another one-eighty-first of the product) to the owner of the adit.[1] The adit owner further acquired such seams as had first been rendered workable by virtue of his operations, and although he was obliged to pay *terrage* to the *terrageur*, it has been questioned whether he did not in fact become the proprietor of all the reclaimed coal.[2] In the Pays de Liége, it is clear, the power of landowners over their mineral wealth was reduced to small dimensions.

The common view is that, in Germany, coal was everywhere outside the rights over minerals established by sovereign princes of the Empire.[3] Certainly in the Saar district coal deposits were left at the free disposal of surface owners.[4] The most important seams in the Wurm field, near Aachen, were all under four large manors (*Herrschaften*), in each of which the lord enjoyed an absolute right to the coal. He usually worked it himself through special officials.[5]

In French law no distinction was made between coal and other

---

[1] Jars, *Voyages métallurgiques*, vol. i, pp. 371–80 ; G. de Louvrex, *Recueil contenant les edits et règlements fait pour le Pais de Liége et Comte de Looz*, vol. ii, pp. 218–28. M. Harzé (see above) says the *terrage* amounted to one-eightieth of the product.

[2] Lewis, *The Stannaries*, p. 72.

[3] J. P. Bünting, *Sylva Subterranea*, 1693, chap. v ; L. Beck, *Geschichte des Eisen*, vol. ii, p. 101; Inama-Sternegg, *Deutsche Wirthschaftsgeschichte*, vol. iii, pt. ii, p. 145.

[4] A. Hasslacher, *Geschichtliche Entwickelung des Steinkohlenbergbaues in Saargebiete*, 1884, pp. 406–11.

[5] Käntzeler und Michel, in *Aachener Echo der Gegenwart*, 1873, nos. 126, 207–11.

minerals until 1601, when Henry IV, partly in order to awaken interest in the new fuel, set free from his sovereignty coal and certain other mines, and relinquished his royalty of one-tenth of the produce " par grâce spéciale en faveur de sa noblesse et pour gratifier nos bons sujets propriétaires des lieux ".[1] As the King relinquished only by " special grace " what he tacitly affirmed were his just rights, he reserved the power to resume them at some future date. Minerals were by no means fully vested in the landowner, and that the Crown still felt free to lease coal in the lands of its subjects is shown by a concession made in 1670 to a certain Sieur Jaer, a native of Liége, of all seams of mineral fuel in Rethélois, without payment of any dues whatsoever, except compensation for damage done, to surface owners.[2] A more general concession was given to the Duc de Montausier in 1689 by the Council of State. He was granted permission to exploit for forty years all coal mines in the kingdom, except those in Nivernais, which had been conceded a few months before to the Duc de Nevers, and those already opened by landed proprietors.[3] The Duchesse d'Uzès, who succeeded to the concession of her father, Montausier, in 1692, was strongly opposed, in attempting to exercise her rights under it, by various proprietors of coal pits in Anjou, but in 1695 the courts ruled that in the future no new mines might be opened by those proprietors without her permission.[4] The professed purpose of such Crown grants had been to increase the supply of fuel, but the efforts of the duchess in this direction were generally regarded as unsatisfactory, and this may have forced the Council of State in 1698 to issue a decree reversing the decision of 1695, and making it lawful for any landowner to exploit coal mines whether already open or not.[5] Within six years, however, new monopolies were granted to members of the nobility.[6] In practice these concessions concerned such large areas and were so general in their terminology that it was very difficult to enforce them, and they had little effect on the diggings of the local peasantry along the coal outcrops. Small landholders continued to work seams in their own holdings, generally, no doubt, in complete ignorance of contemporary legislation.[7]

Nowhere on the Continent did the landowner exercise so extensive a control over the production of coal as in Great Britain, and many even of such rights as he had enjoyed in early France and Germany he was destined to lose before the nineteenth century. The *régime*

[1] Rouff, *Les mines de charbon en France au XVIIIe siècle*, p. xvi. In 1604 the Crown claimed one-thirtieth of the product of coal and iron mines, to be used as a benefit fund for miners (*Roy. Com. on Mining Royalties, Final Report*, p. 58 ; cf. below, vol. ii, p. 172). See also A. des Cilleuls, *Histoire et régime de la grande industrie en France au XVIIe et XVIIIe siècles*, p. 131.
[2] Clément, *Lettres, instructions et mémoires de Colbert*, vol. iv, pp. 595–6.
[3] E. Lamé-Fleury, *De la législation minérale sous l'ancienne monarchie*, 1857, pp. 137–8.
[4] *Ibid.*, p. 138 ; O. Couffon, *Les mines de charbon en Anjou*, 1911, p. 14.
[5] *Mémoire du Sieur des Casaux du Hallay, député de Nantes, sur l'état du commerce en général*, March 4th, 1701, in Boislisle, *Correspondance des Contrôleurs Généraux*, vol. ii, appx., p. 198.
[6] Lamé-Fleury, *op. cit.*, pp. 138–9.
[7] Rouff, *op. cit.*, chap. iv.

of free exploitation in France was brought definitely to an end in
1744, by a decree which prohibited all persons from working coal
in the future, except under a concession from the State.[1] Although
the provisions of this decree were stubbornly resisted by surface
owners, the legal power of the Crown to revive a right which it had
alienated only by " special grace " was not questioned. As in the
case of the special concessions previously granted, the professed object
was a greater output of coal, which had become by general
acknowledgment a national necessity. All other systems, it was
urged, had been tried without success, and, in reply to critics like
Turgot, who saw in the new law only a means of adding to the revenue
of the Crown,[2] the government could point to an increasing production
of coal, which, however, probably began some years before the
issue of the decree. Mirabeau succeeded by his eloquence in persuading
the National Assembly, which had just sanctified the rights of private
property, to retain the power of the State over minerals, and not to
hand them back to the proprietors of the soil. Those who drafted
the Act of 1791, which is to this day the basis of the French law of
mineral property, were careful that the actual ownership of minerals
should not be given to the State.[3] They are at the disposal of the
nation in this sense only : that such substances cannot be exploited
without its consent and except under its surveillance ; but no royalties
are in fact taken by the government. The landowner or his assignee
is preferred to other applicants, and the act does not apply to quarries
(carrières) or seams within a hundred feet of the soil, which are
deemed the property of the surface owner. It is clear that, on terms
such as these, the shallow medieval coal mines would have belonged
to the landowners.

French mining legislation in the eighteenth century had a wide
influence on that of other continental countries.[4] The Act of 1791
(replaced nineteen years later by that of April 21st, 1810, which
did not modify it in essentials) was applied to all mines in the Belgian
provinces ; though existing concessions by rendage, droit de conquête
and permission de quarante jours in the Pays de Liége and grants of
the seigneurs haut-justiciers in Hainaut were recognized, provided
that the concessionnaire immediately registered his grant. A proportional rent of from $2\frac{1}{2}$ to 3 per cent of the gross profits was henceforth taken by the State. The royalty formerly paid to the seigneurs
haut-justiciers was abolished, but in the Pays de Liége terrage, cens
d'areine, and other dues were still exacted.[5]

In Germany we need concern ourselves with only two systems
of mineral law, the Prussian and the Saxon. Ten years after the

[1] The decree is printed by Rouff, op. cit., pp. 597–600, and analysed by him,
pp. 63–82. The power of granting coal mines was given to the Contrôleur
Général des Finances.
[2] Turgot, Mémoire sur les mines et carrières, 1764 (?), pt. ii.
[3] Rouff, op. cit., pp. 545–82.
[4] Roy. Com. on Mining Royalties, 4th Report, pp. 200–3 ; Final Report,
p. 58.
[5] Ibid., 4th Report, pp. 130–1 ; E. Bidaut, Mines de houille de l'arrondissement
de Charleroi, 1845, p. 20.

issue of the French decree of 1744, Prince Wilhelm Heinrich, as sovereign of the Principality of Nassau-Saarbrücken, claimed the *regale* over coal mines in part of the Saar district, and proceeded to work them as a state enterprise.[1] Eventually the Saar mines came into the hands of the Prussian Crown, which took a direct part in the development of the German coal industry during the second half of the eighteenth century. When Jars visited the Duchy of Magdeburg in 1768, he found that the chief shares in the most important colliery near Halle had been bought by the King of Prussia, who received on coal leased elsewhere in the Duchy a royalty of one-tenth of the value of the produce.[2] The modern Prussian law has followed very closely that of France.[3] Coal and base metals are placed in the same class as gold and silver, and belong to the first finder or claimant, whose right, however, must be confirmed by the government. Landowners have no claim to any royalty, except in the case of a few mines on the left bank of the Rhine, where ancient customs, resembling the *terrage* at Liége, were not at once abolished. Two per cent of the gross profits of all mining enterprises are due to the government. According to Saxon law, on the other hand, coal and lignite, as distinguished from other minerals, are vested in the surface owner, who is entitled to a royalty if he leases them to others.[4]

Though French law does not give the State full ownership of minerals, they cannot be disposed of at the discretion of the surface owners, as in Great Britain. Nor is the landowner on the Continent, apart from a few instances founded on early customs, able to exact any tribute other than what is paid for actual interference with his surface rights.[5] In England and Scotland, as is well known, his royalties amount to-day to approximately £6,000,000 a year.[6]

How is it that the British landlord has so long enjoyed these privileges unmolested ? In some districts of England immemorial customs were strong enough to prevent, and in a few cases still prevent, surface owners exercising to the full those rights to which they eventually became entitled under common law. Tin miners could

---

[1] Hasslacher, *Geschichtliche Entwickelung des Steinkohlenbergbaues in Saargebiete*, pp. 410–11, 419 ; Beck, *Geschichte des Eisen*, vol. iii, p. 984.

[2] Jars, *Voyages métallurgiques*, vol. i, pp. 314–18.

[3] *Roy. Com. on Mining Royalties*, *4th Report*, pp. 150–3 ; *Final Report*, pp. 105–7.

[4] At the time of Jars' visit (*op. cit.*, vol. i, pp. 323–4), the landowner in Saxony might exploit coal under his land, but he could sell only a certain fixed amount at a fixed price, and was obliged to pay a percentage of his product to the King. He has apparently strengthened his position during the nineteenth century.

[5] In the St. Etienne region, in France, royalties nearly as high as those in England have been paid to surface owners for coal. These royalties are the result of old regulations, and apply only to a very small proportion of the French coal annually raised (*Roy. Com. on Mining Royalties*, *4th Report*, Qs. 19317, 19320, 19322, 19325). Elsewhere on the Continent royalties, whether paid to the State or to the surface owner, are but a fraction of those levied in Great Britain (*ibid.*, *Final Report*, p. 44).

[6] See the evidence of Sir Richard Redmayne in *Report of Coal Industry Commission*, 1919, vol. ii, p.1158.

prospect freely on wastes and demesne lands in Cornwall, and in Devon on enclosed lands as well, and were entitled to possession of the mine once they found the ore, subject to the payment of a portion of their product to the lord of the manor.[1] In the Mendip lead mining district of Somerset, as we learn from a code of privileges drawn up in 1470,[2] it was lawful for miners to dig pits anywhere, upon application for a licence, which could not be refused, and on the understanding that every tenth pound of lead should go " to the Lord of the Soil where it was landed ". Licences to dig were granted by, and royalties paid to, the " Four Lords Royal of Mendip "—the Bishop of Bath and Wells, the Abbot of Glastonbury, Lord Bonville, and the Lord of the Honour of Richmond—who alone were entitled to royalties. In approving a concession granted during Elizabeth's reign to Sir Bevis Bulmer of a lead mine under the manor of Chewton Mendip, the Privy Council was careful not to infringe on the privileges of those who had worked there before the coming of Bulmer, according to the " laws and orders for the minerals ".[3] Another lead mining district with special customs consisted of the wapentake of Wirksworth and seven small liberties in the High Peak, in Derbyshire. Here administration was in the hands of " barmasters " who represented the Crown, and to whom miners, on discovering ore, applied for a concession or " meer " —a portion of the seam twenty-nine yards in length—which was granted without payment of any compensation to the owner of the soil. A certain percentage of the product was due the Crown.[4]

In other districts local lords granted the metal miners special privileges which resembled those of the High Peak. The lord of Hope manor in Flintshire, for instance, had a code of " laws and customs of the Minorie ", set forth in a document of 1353, which apparently applied also to several neighbouring manors, and which may be very nearly matched point by point by the Derbyshire code.[5]

---

[1] H. C. and L. H. Hoover, note on mining law in their edition of Agricolas' *De Re Metallica*, pp. 82-6 ; *Roy. Com. on Mining Royalties, Final Report*, pp. 25-9. The whole question of Stannary mining law is discussed by Lewis in his book on the Stannaries (see especially chap. iv).

[2] As printed by the *Roy. Com. on Mining Royalties, Final Report*, p. 82. This written code was drawn up as the result of a dispute between the people and Lord Bonville, over the rights and customs claimed by the miners. The document is also to be found in Thos. Houghton, *The Compleat Miner*, 1688 (pt. iii), which contains valuable information concerning the mines and miners of Derby, Gloucester, and Somerset. On the lead miners' privileges in the Mendips see also *V.C.H. Somerset*, vol. ii, pp. 362 sqq. ; *Star Chamb. Proc., James I*, 49/18, 153/17 (the latter deals with the minery court of the manor of Chewton) ; *Chanc. Proc., Eliz.*, H. 2/15 ; Lewis, *op. cit.*, p. 80 ; J. W. Gough, *The Mines of Mendip*, 1930.

[3] *Acts of the Privy Council*, N.S., vol. xiv, pp. 353–5 (February 28th, 1587).

[4] *Roy. Com. on Mining Royalties, 3rd Report*, Qs. 15256–15503 (testimony of A. M. Alsop, barmaster for the wapentake of Wirksworth, and Thos. Shimwell, barmaster of the High Peak). According to the Hoovers, a royalty was paid to the landlord as well as to the Crown. Mining concessions here, as well as in the coal district of Hainaut, were granted according to the " apex " law. See also *V.C.H. Derbyshire*, vol. ii, pp. 325 sq. ; Lewis, *op. cit.*, p. 81.

[5] *Exch. T.R. Misc. Bks.*, vol. ccxcvii, ff. 42 sq. For this manuscript I am indebted to Mr. D. L. Evans, of the Public Record Office, who tells me that he proposes soon to print it. This step should be awaited with interest by the historical student, for it has not heretofore been generally realized that special

The "marchants" and miners, upon finding the mine, chose a " Barmastre ", who delivered to the " Finder " two " Fyndemeeres ", each meer being in this case twenty-two yards (10 fathoms 7 feet) in length of seam. The lord of the field was invested with a third " meer " adjacent, just as in the High Peak the barmaster claimed a half meer for the King. In both districts miners were to be tried for offences in a special court, and in both we find the curiously barbarous punishment by which a miner convicted a third time (for bloodshed in Flintshire,[1] for theft of ore in Derbyshire) [2] should have the palm of his right hand pierced with a knife, and be pinned to the stocks until he either die or tear himself away. Lords of manors are said to have established similar barmotes at Eyam and Litton in Derbyshire, and it seems possible that at one time areas of special mining customs were much more common in England than has been generally supposed. Such special privileges, even when they were originally granted by the landlord himself, imposed restrictions on his rights over minerals in the same way that manorial customs limited his rights over surface land.

But, before the sixteenth century, mining customs, like manorial customs, had already begun to lose their strength. The decisions of the common law judges sometimes replaced, or overruled, the decisions of the mine courts. Even in those few regions where ancient privileges have been maintained down to the present day, they are to be regarded as survivals, whose influence on economic and social life belongs to the past.

So far as we know, there is only one district in England where coal mining was ever directly subject to such special privileges. When and how the customs originated which have made the laws of mineral property in the Forest of Dean unique will perhaps always remain a mystery. They were, no doubt, first applied to iron mining, which was the chief industry of the Forest until the seventeenth century, but in an early charter, ascribed to the reign of Edward I,[3] it was set down that " sea coal mine is as free in all points as the oare mine ".[4] Only adults born in the Hundred of St. Briavels who had

mining customs existed in Flintshire. The document reveals features common to all the better-known special mining districts—Devon and Cornwall, Mendip, High Peak, Forest of Dean, and Alston Moor in Cumberland. The " meers " granted by the barmaster were inheritable, subject to continuous working of the mine and payment of a " lot " to the lord of the minery. The miner's rights included the space and necessary timber for building his house, and permission to put his cattle to graze in the pasture of Hope lordship. From the reference to " marchants ", it appears that capital was furnished by traders in the ore, a common manner of financing the early mining partnerships both in England and on the Continent. No reference is made to the kind of ore extracted, but it was probably lead.

[1] *Ibid.*

[2] L. F. Salzman, *English Industries of the Middle Ages*, pp. 44–6. In Derbyshire the culprit's hand was pinned to the uprights of his windlass, not to the stocks.

[3] First printed as the *Laws and Customs of the Miners in the Forest of Dean*, ed. T. Houghton, 1687. See also H. G. Nicholls, *Iron Making in . . . the Forest of Dean*, 1866, pp. 71–82, and G. R. Lewis' article in *V.C.H. Gloucestershire*, vol. ii, p. 221.

[4] As quoted Galloway, *Annals of Coal Mining*, p. 28.

laboured as miners a year and a day were eligible to become owners of minerals. The " free miners ", as they have always been called, had the right to search for ore and coal in all lands within the Forest bounds, " wether . . . in his majesties soyle or in the soyle of any other ".[1] Ownership was conferred upon them (as in France during the last fifty years of the *ancien régime*) by the Crown, through a representative called the " Gaveller ".[2] " In case any coales or Oare have been discovered in any pits . . . by the . . . industrie . . . and charge of the . . . mynors who digged . . . the same . . . then the . . . Gaveler . . . hath been allwayes accustomed to . . . deliver every such mynor . . . his . . . share . . . out of the same pitts ".[3] Concessions were made, not of parts of the seam as in Derbyshire, but of plots of ground called gales, under which the minerals became the property of the free miner who had found them. These gales were designed to provide sufficient space for a single colliery enterprise.[4] Under the exigencies of deeper mining, requiring wider surface grants, the minimum area for a gale was increased several times during the seventeenth and eighteenth centuries.[5] Free miners became owners of mineral property in perpetuity, and might will, sell, or lease their gales entirely at their own discretion.[6] The Crown, though never in a strict sense owner of the mines, received certain payments which were in the nature of a royalty rent.[7] Each miner gave the gaveller a penny when he came to " gale " the pit—that is to confer ownership by entering the concession, together with the names of the partners, in his books. Another penny was due from each miner on the Tuesday of every week in which his profits exceeded nine pence. Finally, the Crown had a right, once the pit had been sunk, to add for every four free miners one partner, known as the " kingsman ", who might dig on behalf of the King without sharing in the capital expenses.[8] If the pit was sunk in the soil of a subject, he too was entitled to a share in the enterprise.[9]

---

[1] *Exch. Deps. by Com.*, 13 Charles I, Mich. 42.

[2] In the seventeenth century there were two gavellers, or " gailors ", the gaveller " above the wood ", and the gaveller " below the wood "—each evidently having jurisdiction over half the forest. (*Ibid.*, 27 Charles II, Mich. 28).

[3] *Ibid.*, 13 Charles I, Mich. 42.

[4] Evidence of Thos. Forster Brown before Select Committee of the House of Commons on Woods and Forests and Land Revenues of the Crown. (Printed in *Roy. Com. on Mining Royalties, 2nd Report*, pp. 416–21).

[5] See above, p. 75, and *V.C.H. Gloucestershire*, vol. ii, pp. 227–8.

[6] To-day no free miner may hold more than three gales.

[7] Perhaps technically the miners did not own the minerals either, since they were obliged to pay royalties for them. It is a difficult legal problem to define the owner, precisely because the customs prevailing in the Forest of Dean are survivals from a time before the sense of private property either in land or minerals existed in the form in which we know it to-day.

[8] *Exch. Deps. by Com.*, 27 Charles II, Mich. 28. How often in early times advantage was taken of the right to put in a fifth man is uncertain. To-day the payments due the Crown are all lumped together in a single royalty, estimated in advance by the gaveller on the assumption that a fifth partner has a right to his share of the gross profits. The royalty may be reassessed every 21 years at the request of either party. The burden of this extra share upon the enterprise obviously becomes greater with every increase in capital expense.

[9] *V.C.H. Gloucestershire*, vol. ii, p. 221.

Under the Stuarts, when coal mining replaced iron mining as the foremost industry of the Forest, we find the Crown attempting to curtail the rights of the free miners, and to grant wealthy petitioners general mineral concessions, such as were commonly accorded in other districts by lords of the manor. It is true that the old customs had become somewhat out of date, as even the colliers themselves were aware. They were suited to a state of things in which mining required little capital, and in which each partner was at once adventurer, foreman, and hewer. Already in the days of the Commonwealth more money was needed to start a colliery than poor miners could muster,[1] and they had begun to hand over their shares to capitalists who were foreign to the district. A case in point is the bargain, referred to in 1652, by which Christopher Tucker and five other native miners conveyed their " parts " in a pit called " Maplepitt or Yellowshaft " to John Brayne, Esquire, who was to pay annually £20 to Tucker, and £10 to each of the others, out of the profits issuing from the mine.[2]

It was one thing, however, for the free miner to recognize his financial limitations and to sell his rights for cash ; it was quite another for him to see an outside power act as though his rights had ceased to exist. When Christopher Morgan of St. Briavels told a body of commissioners in 1637 that, although he had not worked in a coal mine for thirty-five years, yet he hoped his right and interest " in the same is always capable . . . accordinge to the auncient custome of mynes there ",[3] he was expressing the united answer of free miners to the pretensions of the Crown. Morgan was a witness on behalf of a great number of natives of the Forest, defendants, together with Sir John Winter, in a suit brought on behalf of Edward Terringham, " gentleman of his majesties Privy Chamber ", who had just received a Crown lease of " all the mines of coal and quarries of grindstones within the Forest of Dean ".[4] The Attorney-General, Sir John Banks, contended that Terringham might prevent all others from digging coal in the district. He complained that the defendants, both on their own initiative and as the servants of Winter, were mining to the prejudice of Terringham's lease. The free miners answered that they were entitled by immemorial privileges to mine on their own initiative, and denied that they had worked at Winter's colliery in a wood called " Norchards ", which they believed was outside the Forest bounds, and where he had employed Staffordshire men and other strangers.[5] I have not found the judgment in this

---

[1] Cf. below, pp. 401, 412 sqq.

[2] *Exch. Decrees*, Series IV, vol. v, ff. 272, 279.

[3] *Exch. Deps. by Com.*, 13 Charles I, Mich. 42, dated October 9th, 1637, at Little Dean (Attorney-General, on behalf of Edward Terringham *v.* Sir John Winter and others).

[4] Galloway, *Annals*, p. 208.

[5] Such was the testimony of Christopher Tucker (*Exch. Deps. by Com.*, 13 Charles I, Mich. 42). John Teckoll, another miner, witness in the same suit, and Wm. Yearworth, of Newland, witness in another suit between the same parties (*ibid.*, 13 and 14 Charles I, Hilary 16), said that Winter employed " both forest men and strangers ".

Sir John Winter was one of the most important coal capitalists and

case, but it is clear that Terringham soon tired of spending money
in support of his pretensions, as was only natural when the terms
of his lease entitled him to a £300 annuity if he failed to derive
a profit from the mines.[1] He is said to have given up working on his
own behalf after six months, because coals " were not of soe greate
value neyther did the worke turne to soe greate accompt " as he had
expected. The free miners apparently did not challenge his right
to mine in the Forest, but only his desire to interfere with their own
concessions. A number of them actually worked under him for
wages, and also after he departed, under one Tucker (perhaps the
man already referred to), who claimed to have a title from Terringham.[2] After the Interregnum, Terringham's descendants revived
his claims,[3] and, in 1668, an Act of Parliament gave the Crown
permission to lease coal mines and stone quarries in the Forest of Dean
for a period not exceeding thirty-one years. By another clause,
however, the Act provided for the continuance of all the lawful
privileges of the miners, and, in spite of a lease in 1667 to Captain
Francis Terringham, the family seem to have ceased to urge their
claim before the accession of James II.[4] The miners succeeded in
holding on to most of their special privileges, and henceforth outside
capitalists were able to enter the field only by buying up free shares.[5]
For the freedom of British miners generally, however, this victory
was as empty as that of a general who wins a skirmish with a few of
his shock troops, only to be routed by the enemy in the real battle.

Why was the Forest of Dean the only district in Great Britain
where coal mining was subject to special customs ? The early use of
coal in working up iron,[6] in a region where iron ore and mineral fuel
were extracted side by side from shallow diggings, explains why
coal should have come under the same regulations as iron. But
how shall we explain why those regulations should have been applied

---

industrialists of his day. His connection with the Forest of Dean was not
limited to this suit. He purchased the Crown's interest in mines and lands
within the Forest in 1640, after Terringham's lease was already in being, and
he also assumed the Crown's obligation under that lease to pay Terringham
an annuity of £300, if he failed to make a profit from his coal works. Winter's
relations with the free miners after 1640 are uncertain, but his descendants
were apparently preparing in 1684 to restore their interest in the Forest to the
Crown. On Winter's interests in collieries, see below, vol. ii, pp. 11–12.

[1] He is said to have spent £2,500 in lawsuits (Galloway, *op. cit.*, p. 208).
[2] *Exch. Deps. by Com.*, 27 Charles II, Mich. 28.
[3] *Ibid.*, also 27 and 28 Charles II, Hilary 21 (Katherine Terringham, widow
of Francis Terringham v. Jas. Yarworth and others, free miners).
[4] Galloway, *op. cit.*, p. 208 ; *Cal. Treasury Books*, vol. ii, p. 205.
[5] The Terringhams failed in their effort to prevent colliers from selling
coal on their own account (*Exch. Deps. by Com.*, 27 Charles II, Mich. 28), and
in 1711 the right to carry coal out of the Forest was limited to the inhabitants
of the Hundred of St. Briavels, a limitation which increased the difficulties of
outside concessionnaires (*V.C.H. Gloucestershire*, vol. ii, p. 229). Lewis (*op.
cit.*, p. 175) holds that by their acknowledgment in 1613 that they held
their privileges merely by the grace and charity of the Crown, the free miners
definitely mortgaged their rights for the future. It is true that eventually
many of the special customs were set aside, but the principle that mineral
ownership is in the free miners has been maintained to this day.
[6] See above, pp. 11–12, 201–2.

in the Forest of Dean to iron, which elsewhere, like coal, escaped them ? Perhaps the great antiquity of the iron industry in this district enabled the miners to establish for themselves a preferential position at a time when miners generally received special privileges from princes in many parts of Europe.[1] The Forest of Dean was famous for its iron as early as the twelfth century, and continued until the fourteenth century to be the only considerable iron-producing area in England.[2] The customs of the district present an intriguing field for study and speculation, but for the student of general coal-mining legislation their importance must not be exaggerated. After all, they offer only a local exception to the general rule that coal is everywhere in the hands of the landowners. It may be doubted whether the Forest of Dean accounted for more than one per cent of the total British coal output in the seventeenth century. To-day its output is proportionately even smaller. So far as can be judged, the customs of this district, isolated as it is from the other coalfields of Great Britain, have had little direct influence on the general law of mineral ownership.

The British coal industry was of small consequence before the sixteenth century, and the Crown had no thought in early times of claiming special rights over a commodity which was regarded both as unimportant and as belonging to the same species as wood and turf. Even had the Crown been disposed to make such a claim, its power to do so had been compromised in advance, when, in 1217, the Forest Charter was granted at the dictation of the barons. Henceforth manorial lords could, in most parts of the realm, without fear of question, dig or break ground in their own holdings. At a time when mineral fuel was taken from weathered outcrops with scarcely more effort than was required to get peat or turf, and when men lived in ignorance of the depths of the seams, it is easy to understand how the terms of the Forest Charter could be invoked to cover the digging of coal.[3] The first serious test of such an interpretation came, it seems, at the beginning of the fourteenth century in the County Palatine of Durham, over which the Bishop of Durham enjoyed virtual sovereignty. It has been suggested that, in respect to the proprietorship of mines, his privileges somewhat exceeded those of the Crown, and that he claimed the ownership, not only of the precious metals, but of other minerals, including coals.[4] His bailiffs had been interfering with the lords of towns and his other free tenants who were digging coal and iron ore in their manors, and in 1303 Bishop Anthony Beck, under pressure from them, admitted that any manorial lord " may mine for coal and iron in his several lands and also where others have common, in the same manner as the bishop does in his several lands and where others have common ".[5]

[1] Cf. below, p. 289.
[2] Salzman, *English Industries of the Middle Ages*, pp. 24–6.
[3] This point is well made by Galloway (*op. cit.*, pp. 18–19).
[4] G. T. Lapsley, *The County Palatine of Durham*, 1900, p. 58.
[5] *Cal. Close Rolls*, 1302–7, pp. 100–1.

There is no record of any other attempt to establish a *regale* over coal during the Middle Ages.[1] While mineral fuel was not mentioned in the case of 1566, it was definitely excluded from the minerals over which the Crown had special rights by the judgment then given,[2] for we have no example in England of a seam in which coal was found in conjunction with a precious metal, even though, according to the Frenchman, Aléon Dulac, who wrote in 1786, an argentiferous coal seam had been discovered near Chemnitz in Saxony.[3] The judges in 1566 might have reconsidered their decision, had they known that it would form one link in the process by which the Crown was eventually deprived of any substantial rights over minerals in Great Britain. Had they desired this issue, they would hardly have found unanimously in favour of the Queen in the case at trial. They did not realize, of course, that there were practically no gold or silver mines in the land, nor could they foresee the rôle which coal was destined to play in British history. Their judgment preceded by some years the great expansion in the coal industry ; and it is conceivable that they might have guarded themselves against giving coal to the landowner, had they been aware of its nature and potential importance. The legal principle that those resources which were excellent and of value for national defence belonged to the sovereign [4] might have been extended to cover coal. But, in 1566, England was only on the threshold of an age when men could speak of ordering the trade in coal so as to turn the coal mines " into Mynes of Golde " to the Crown.[5] To study the evolution of science and industry is to distrust the wisdom of sanctifying any legal decision.

It is not easy to discover a principle defining the status of coal in early Scottish law. Although it has been held in modern times by a Lord Advocate that an Act of 1424, which established the Crown right to gold and silver, tacitly acknowledged that coal and, indeed, all minerals, except the precious metals, were outside the *regale*,[6] it is to be noted, nevertheless, that the Queen in 1565 granted permission to John Stewart and his son William, " if in their search for mines between the water of Tay and the sherifdom of Orkney . . . they find any coalheughs, . . . to work them, provided they are not within 10 miles of any of the Queen's dwelling places ".[7] Again, in 1583, when all the mines and minerals in Scotland were leased by the Crown to Eustachius Roche, he was given leave to take as much coal as he needed for refining and melting his metal.[8] Such terms suggest that mineral fuel may not have been considered strictly private property, even if found in the lands of a subject. The failure of these general monopolies to yield any considerable supply of metallic ore led to the

[1] R. L. Galloway, " Earliest Records of the Working of Coal on the Banks of the Tyne", in *Arch. Aeliana*, N.S., vol. viii, 1880, pp. 168–9.

[2] See above, p. 268.

[3] *Observations générales sur le charbon minéral*, f. 23.

[4] See above, p. 269.

[5] See below, vol. ii, p. 217.

[6] *Report of Coal Industry Commission*, 1919, vol. iii, pp. 150–1.

[7] *Privy Council Register of Scotland*, 1st Series, vol. i, pp. 330–1.

[8] *Acts of the Parliaments of Scotland*, vol. iii, p. 369b ; R. W. Cochran-Patrick, *Early Records relating to Mining in Scotland*, p. 16.

passage of the " Act anent the Mines " in 1592, which encouraged
landowners to work metals and minerals within their own holdings,
but expressly reserved a tenth of the produce as due to the Crown.
Coal is not named in the Act and the framers did not intend to include
it.[1] But, during the late sixteenth and seventeenth centuries, most
of the grants of lands " in feu " refer to " coal-heughs " as if they
were not included in the concession without a specific clause.[2] Sometimes a small special rent was asked for the coal, separate from the
ground rent, as in a charter of 1643 to Edward Bruce, who was bound
to pay £22 annually for coals in the bounds of the " Common Muir
of Culross ".   The charter also reserved " the privilege to the Crown
of winning coals for the King's fire ".[3]   Whether or not these grants
were to be made the basis of a claim by the Crown to ultimate
ownership of mineral fuel depended on the interpretation of future
generations.   It is not inconceivable that in Scotland a legal case,
similar to that which the French king advanced in favour of the
decree of 1744, might have been made in favour of a lapsed *regale*
over coal.

Even in England, projects for the Crown to take over the coal
mines were heard of during the seventeenth century, but it is not
always clear whether they envisage actual ownership of the minerals
or merely the management of the trade as a State monopoly.   One
such project, presented in 1628, suggests that the Crown shall take
" the Benifett of all Coale mynes to be fownde in the Bishoprick of
Durisme, uppon remoove of that Bishop, alowinge him Coales for his
own expense " ; [4] but no reference is made in it to the acquisition
by the Crown of all coal mines under privately-owned lands.   A
proposal for " taking coale pits into the hands of the state " was gossip
in London taverns in 1653 at the time of the first Dutch war,[5]
and after the Restoration we hear of the same scheme in a modified
form.[6]   Towards the end of the reign of William III, Moses Stringer
suggested that mines and minerals, in lands where they were not
being properly exploited, should be let to the Crown for a small rental,
and a right of priority in exploitation, such as is allowed in France,
granted to the landowner.   Mineral riches, he urged, ought to be in
State hands and managed by a national stock, because " the Ignorant
Country Man knows not more of these when his Plough . . . turns
up any of them, than Æsop's Cock did of the Gem he found on the
Dunghill ".[7]

---

[1] See above, p. 270.
[2] *Royal Commission on Mining Royalties*, 1893, 2nd Report, Qs. 7325 ; *MS.
Cal. of Scottish State Papers*, 1292–1761, *passim*.   Sometimes we find grants
of coal without the land above it (*Acts of the Parliaments of Scotland*, vol. iv,
pt. i, p. 606).
[3] *Report of Coal Industry Commission*, 1919, vol. iii, p. 232 (Appx. no. 80),
[4] *Stowe MSS.*, 326, ff. 7 sqq., printed below, Appendix J (i) ; and see below
vol. ii, pp. 273 sqq.
[5] *The Two Grand Ingrossers of Coles : viz., the Woodmonger and the
Chandler*, 1653.
[6] *S.P.D., Charles II*, vol. cxix, no. 24, I, printed below, Appendix J (ii).
[7] *English and Welsh Mines and Minerals discovered in some Proposals
to the . . . House of Commons*, 1699, pp. 11–12.

In France this alleged inability of small landholders to develop their own mineral wealth furnished the principal argument used in the eighteenth century in support of the decree of 1744.[1] We may ask therefore why none of the projects for depriving surface owners of their absolute power ever met with success in England. We should be neglecting more subtle, and probably more important, reasons if we saw a complete explanation in the growing antagonism in Stuart times to all schemes of State control.[2] No doubt any project to turn coal seams into " mines royal ", associated as it often was with making the King " sole merchant of coles ",[3] must have aroused suspicions of a Crown monopoly ; and must have been subjected to the scorn of the pamphleteer, who covered with epithets each new economic octopus. But, even if proposals that the State should take over the ownership of the coal seams had been successfully divorced from the efforts of the Crown to interfere with private initiative in industry and trade, they would probably not have seriously engaged the attention of Parliament or of any considerable party within the nation.

We may perhaps help to explain why no such proposal ever became practical politics by showing that the chief causes which led to the resumption of mineral ownership by the State in France did not exist in England or Scotland. Whereas the feeble output of the French collieries at the beginning of the eighteenth century—altogether insufficient to meet the needs of that country—demanded radical reforms, the British coalfields were by far the most productive in the world, and agreements were frequently entered into by colliery proprietors to limit the supply of mineral fuel, which, unless controlled, was bound to outstrip demand.[4] It could not effectively be urged in England, as it was urged in France, that large concessions, such as were necessary for the establishment of mining enterprises of the most up-to-date kind, could be had only if mineral ownership were taken out of private hands. For, as we shall proceed to explain, the agrarian changes of the sixteenth and seventeenth centuries were accompanied by a great number of legal decisions, the general trend of which was in the direction of limiting rights over minerals to the lord of the manor, and to a few of the most powerful freeholders. At the same time, the number of customary tenants was being reduced.[5] Thus the vesting of royalty rights over coal and other base minerals in the landowner in England did not involve the division of the *regale* between anything like as many persons as it would have involved in France, which remained a country of small landowners, many of whom claimed a right to the coal beneath their holdings. In England, therefore, it was possible for mining adventurers to obtain comparatively large concessions to exploit coal, without being obliged

[1] See Rouff, *Les mines de charbon en France au XVIIIe siècle*, chap. iv.

[2] Dr. Levy (*Monopoly and Competition*, 1911, see esp. pt. i, chs. iii and iv) has shown that the anti-monopoly feeling, aroused by the Stuart patents, came earlier in England, and was much more violent there, than in any continental country.

[3] Cf. below, pt. v, ch. iv.

[4] See above, pt. i, ch. iii, below, pt. iv, ch. iii, (iii) and (v).

[5] Cf. above, p. 156 ; below, p. 318, and vol. ii. p. 327.

to bargain with scores of peasant proprietors, as the early adventurers in France were frequently obliged to do before the central government resumed the ownership of minerals. The decline in the number of small landholders in England is a fact of capital importance to the student of mining legislation. In 1918, the income from more than forty per cent of all British coal royalties was enjoyed by fifty persons.[1] The spread of great landed estates, in short, is connected with the power of the English landowner, successfully to absorb the *regale*, which in the case of France reverted to the State. In the chapters which follow an attempt is made to explain the nature of this connection.

[1] *Report of Coal Industry Commission*, 1919, vol. iii, pp. 229–30. Of a total of £5,960,365 paid in coal royalties for the year ending September 30th, 1918, £2,340,684 was received by 50 persons. More than five-sixths of the royalties were in the hands of 854 persons.

CHAPTER II

THE JUDICIAL PROCESS FOR SETTLING MINERAL RIGHTS

In order to understand the evolution of mineral law in Great Britain, it is not enough to show, as we have tried to do in the last chapter, that coal was left outside the *regale* of the Crown because of legal decisions given before its value had been recognized. We must also explain why no movement developed in England in favour of the assertion of rights over coal on the part of the Crown, when, during the seventeenth century, the importance of the new fuel became apparent. One explanation, we have suggested, is to be found in the decisions of the courts concerning the mineral rights attaching to various kinds of tenure. It is necessary briefly to discuss, therefore, the nature of the judicial process which resulted in these decisions.

The influence of manorial custom and authority was already seriously on the wane before the expansion of the coal industry began. Although within certain manors, where coal had been dug before the reign of Elizabeth, definite mining regulations existed, they were not comprehensive enough to cover all the cases which arose with the development of colliery enterprise, and, in the few instances where they could be applied to new coal workings, they were tending to lose their force. In the absence of a body of accepted principles, it was natural that interested parties should advance extravagant claims. As a consequence, a large number of disputes arose. If one side could stretch its interpretation of the law so could the other. Coal mining in all ages appears to have been a peculiarly fertile source of litigation ; [1] but there is some reason to suppose that, in proportion to the number of persons interested in the industry, there have never been so many disputes as in England and Scotland under Elizabeth and the first two Stuarts. It seems to have been almost impossible at that time to have had a connection with coal, either as lord of the manor, copyholder, promoter, entrepreneur, foreman, or miner, without being dragged into a suit, at least as a witness. Some persons were engaged in a great number of cases at once. James Clifford, who worked coal in his own manor of Broseley in Shropshire, is said to have had between forty and fifty suits waiting trial in 1606.[2] Two decades later Arthur Player was engaged in at least a score of cases concerning his collieries in the Bristol field.[3] Patrick Edmonstone, an active Scottish mine owner, was either suing

---

[1] See Jars, *Voyages métallurgiques*, vol. i, p. 384, on the disputes round Liége in the eighteenth century. " Il est peu d'entreprises qui soient sujettes à tant de procès que celles des mines de charbon ; il en voit chaque jour de nouveaux ".

[2] *Star Chamb. Proc.*, *James I*, 310/16.

[3] *Chanc. Deps.*, P. 18/H (testimony of William Symonds).

or sued by all his neighbours in Midlothian in 1637.[1] Christopher Anderson, farmer of Crown mines near Leeds, constantly appeared in pleadings in the Duchy of Lancaster between 1585 and 1600.[2] He is only one of fifty mine owners at inland collieries of whom the same might be said. An important mine like that at Bedworth in Warwickshire was scarcely ever free of lawsuits.[3] In Durham and Northumberland the litigation was disproportionate even to the importance of what was then the most productive field in the world. Months of research could be spent in classifying the parties in cases still extant. To these must be added the disputes (and doubtless they were extremely numerous) of which no record has been left us.

By what procedure, and on what authority, were these cases to be settled ? The continental fields in which the production of coal was of consequence during the sixteenth and seventeenth centuries all had legislation, or well defined customs, or arbitral bodies, to deal with problems raised by colliery enterprise. Even in France, where mining of all kinds was still backward, royal decrees were issued to regulate the manner of sinking shafts, to prevent accidents, and to provide medical aid in case of need.[4] Ecclesiastical regulations as early as 1248 and 1252 laid down detailed rules for the exploitation of coal in the Mons basin, limiting the time of year when pits could be worked, and prescribing the methods of working them.[5] Many of the more important rules were included in the charters granted to the Estates of Hainaut, particularly in the charter of March 5th, 1619, which fixed the nature of the *droit de charbonnage*.[6] There was no special mining court, but the *Cour Souveraine des Etats du Hainaut* had jurisdiction in questions of fact, and in the interpretation of long established customs.[7] For the district round Aachen, special working rules were promulgated in writing by the authorities of the four principal lordships where coal was dug. Officials known as *Kohlwieger*, for whom no exact equivalent appears to exist in other districts, represented the lord in each *Herrschaft*, and supervised work at the pits. They seem to have had full power to settle all disputes.[8]

The Liége district offers the most complete and interesting example of a special colliery jurisdiction. From the fourteenth century

---

[1] *Privy Council Register of Scotland*, 2nd Series, vol. vi, p. 525.

[2] *Duchy of Lancs. Pleadings*, 144/A/9, 149/A/35, 152/A/6, etc.

[3] *V.C.H. Warwickshire*, vol. ii, pp. 221–3 ; *S.P.D.*, *James I*, vol. cxxxiii, nos. 67–9 ; vol. cxxxiv, nos. 31–2 ; vol. clviii, no. 23 ; *Charles I*, vol. cciii, no. 92 ; vol. cciv, nos. 82–7 ; vol. cccclxxx, no. 50 ; *P.C.R.*, vol. xxxi, pp. 507, 533, 541 ; vol. xxxii, pp. 271–2 ; vol. xxxviii, p. 424 ; vol. xxxix, p. 659 ; vol. xlii, pp. 65–6, 84–5, 112–14 ; vol. xlix, p. 355 ; *Exch. Deps. by Com.*, 36 Charles II, Mich. 43

[4] Des Cilleuls, *Histoire et régime de la grande industrie en France au XVIIe et XVIIIe siècles*, p. 131.

[5] Decamps, *Mémoire historique sur l'industrie houillère du couchant de Mons*, pp. 58 sqq., and Appendices.

[6] G. Arnould, *Mémoire sur le bassin houiller du couchant de Mons*, pp. 22–3.

[7] *Ibid.*, p. 29.

[8] Käntzeler and Michel in *Aachener Echo der Gegenwart*, 1873, no. 207. Though we never hear of a *Kohlwieger* except in the Aachen region, his duties are said to have resembled those of the *Markscheider* at early German metal mines (*ibid.*).

onwards the Liégeois obtained their water supply from certain adits constructed to drain the miners' shafts. It became, therefore, the concern of all citizens that mining should be carefully carried on. Only four adits—*areines franches*—furnished pure water, and it was essential that they should not be contaminated by the water from other adits—*areines batardes*.[1] We are inclined to think that the famous Liége miners' court, the *Voir Jurés des Charbonnages*,[2] may originally have been created to supervise the water supply. Certainly it remained one of the chief functions of the court to inspect these *areines franches* every fifteen days. Each inspection gave rise to a written report, filed for later reference. The members of the miners' court (at first four, later seven in number) were all chosen from the profession of *houilleurs*, persons skilled in coal mining, who were members of one of the thirty-two town gilds (*corps de metier*).[3] They were subjected to an examination by the *Echevins de la Justice Souveraine de la cité et pays de Liége*, to prove that they possessed sufficient knowledge and capacity to serve. They concerned themselves from early times, not only with the adits, but with all questions raised by colliery operations, and became a special court, confirmed in its power by the Emperor Charles V, to settle all disputes and give judgments—*recors*.[4] Appeals from its decisions could be made only to the ordinary local courts, not to judges of the Empire. Although its jurisdiction was limited to mines in the Pays de Liége, cases arising in the neighbouring collieries of the Wurmrevier or the Province of Limbourg were sometimes voluntarily submitted to it by the parties engaged. As the members of the miners' court continued to be trained miners, their reports and judgments, which have been preserved in the state archives at Liége, were based on expert knowledge, and are said to be highly esteemed by modern engineers who have read them.[5] Some of the more important decisions, together with the ancient customs of the colliers, were embodied as early as 1487 in a code— *La Paix de St. Jacques* [6]—which served as a model for the code of the Province of Limbourg, enacted in 1694, and remained the mining law of the district until the French Revolution, when the court was abolished.[7]

Miners' courts for settling disputes were common in medieval

---

[1] To connect an *areine franche* with an *areine batarde* was a capital crime (Jars, *op. cit.*, p. 378).

[2] *Ibid.*, pp. 371–2 ; Thomassin, *Mémoire statistique*, p. 419.

[3] The members of the *Voir Jurés* were paid 2 écus a day when actively employed (Jars, *op. cit.*, p. 372).

[4] For some of the *recors* see M. G. de Louvrex, *Recueil contenant les édits et réglements faits pour le pays de Liége et comté de Looz*, vol. i, p. 346 ; vol. ii, pp. 202–3, 205–7, 216–18, 220–8, 229–30, etc.

[5] H. Pirenne, *Histoire de Belgique*, vol. v, pp. 357–8. According to Professor Pirenne (*ibid.*, vol. iii, pp. 248–9), " les archives des 'voir jurés des charbonnages ' fourniraient sans doute le sujet d'une des études les plus attachantes que puisse fournir l'histoire économique, et d'une de celles aussi qui montreraient le plus glorieusement tout ce que peuvent l'énergie et l'ingéniosité humaine ".

[6] Printed Louvrex, *op. cit.*, vol. ii, pp. 190–9.

[7] *Réglement général en matière de houillerie pour la Province de Limbourg*, March 1st, 1694, printed Jars, *op. cit.*, vol. i, pp. 382–90. The code does not differ in essentials from *La Paix de St. Jacques*.

England in the lead and tin mining districts, where special customs prevailed. Cases were submitted to juries of miners, which met at regularly appointed intervals, in the Stannaries of Devon and Cornwall, in the Mendip Hills, Alston Moor, and the High Peak.[1] Even the lordship of Hope in Flintshire had a yearly " Court of the Minorie ".[2] But, except in the Forest of Dean, where juries of twelve, twenty-four, or forty-eight free miners were chosen to decide suits concerning the working or sale of all minerals in the Forest, we find no record of special mining courts in connection with the coal industry.

There are probably two principal reasons why all the important British coalfields remained free from those special customs and miners' juries which were a feature of the coal mines of Liége, the lead mines of Mendip, and the tin mines of Devon and Cornwall. Like the practices of chivalry, miners' privileges originated in the early Middle Ages ; like the craft gilds, they were in full vigour between the twelfth and the fourteenth centuries, when coal was of small importance compared with iron, lead, tin, or the precious metals, and when it was commonly associated with surface fuels like timber or brushwood, peat or turf, rather than with minerals, which could be more easily divorced from the manorial economy.[3] The second reason relates to the geographical distribution of the British coal seams. Special mining customs could be applied most easily in small local areas of intensive production, like the early colliery districts of Mons or Liége, or the lead district of the High Peak, where the pits were all concentrated in an area of a few square miles. They had less chance to get a foothold in British coal mining, which was carried on in hundreds of different parts of the island.

Special mining jurisdictions, then, appear to have had little influence upon the evolution of British law concerning the coal mines. Nor do we find in England an attempt to guide the development of mineral law by legislative enactment, as in France. Except for the Acts of William and Mary defining a " mine royal ",[4] one may search in vain through the statutes of Parliament and the proclamations of the Crown, during the sixteenth and seventeenth centuries, for any regulations for the working of coal.[5] Just as in other fields of legal

---

[1] Lewis, *Stannaries*, pp. 79–81, 88.

[2] *Exch. T.R. Misc. Books*, vol. ccxcvii, f. 42.

[3] Garnier (*History of the Landed Interest*, 1893, p. 33) advances the interesting hypothesis that coal was never the object of special mining customs, because it was first dug " at a later period than lead or tin, when the manorial economy was fully established, and when everything belonged to the lord of the manor that could not be proved by ancient custom to belong to the people ".

[4] See above, p. 268.

[5] There were Acts concerning taxation and monopolies, but none regulating coal mining proper. Not until the eighteenth century do we find statutes dealing directly with mineral law. For instance, a statute of George II (10 Geo. II, cap. 32 s. 6) made it felony, punishable by death, to set collieries on fire ; and by another statute of the same reign (13 Geo. II, cap. 21) attempts of a proprietor

procedure the English have steered clear of written codes, so in the law of their coal mines they have depended upon the accumulated decisions of the law courts for the settlement of disputes.

There is, as has already been stated, one exception. In the Forest of Dean the rights and privileges connected with the working and sale of coal and iron were shaped by a miners' parliament, an institution similar to the tinners' parliament in Cornwall and Devon—though there the membership appears to have been more aristocratic [1]— but without any parallel on the Continent. Free miners chose forty-eight representatives to sit in what was called a Mine Law Court.[2] The first of these Courts of which we have record, though probably not the first in fact, was held in 1663 ; and between 1663 and 1741 there were sixteen meetings at somewhat irregular intervals. The enactments of the Mine Law Court fixed and modified the mining laws for the local community. As the origin of this institution is still uncertain, it is impossible to say whether the special customs of the Forest of Dean, which existed long before the seventeenth century, were in the beginning the work of assemblies of free miners, or whether the earliest parliaments merely confirmed customs which had a different origin. Like the *Voir Jurés* at Liége, the Mine Law Court was eventually abolished, together with some of its more uncompromising enactments, which were thought to interfere with the capitalistic exploitation of coal seams. To this day, however, in important features, mining law in the Forest of Dean is still that formerly sanctioned by the miners' parliament, and it differs in many respects from what is enforced elsewhere in Great Britain.[3]

Outside the Forest of Dean it was left to the ordinary law courts to contribute, by their decisions, towards the formation of those principles which are now set forth in treatises on mineral law, and in some measure to determine who should get royalties, wayleave rents, and compensation for damage due to subsidence, and how much. The decisions given between the accession of Elizabeth and the revolution of 1688 are of great importance. Before 1560 there was very little litigation dealing with the coal industry, and, though we have a record of cases tried in the King's Bench as early as 1306,[4] the amount of collective legal opinion on coal mining was small. For the entire reign of Henry VIII only three cases relating to coal mines are preserved in the records of the Star Chamber. For the reign of James I there are scores. Out of judgments in thousands

to drown his rival's colliery were to be compensated by the payment of treble damages together with the costs of the suit. A bill to settle rents on coal wharves and wayleaves in the counties of Durham and Northumberland was introduced in Parliament in 1696 (*Commons' Journals*, vol. xi, pp. 456, 617, 623), but was never passed (see below, p. 335 n.). In a case like that of 1632 in Warwickshire, in which the colliery owners had undermined the public highway to the danger of traffic, the Privy Council ordered the local justices to stop them (*P.C.R.*, vol. xlii, pp. 65–6), but there was no written law on the subject.

[1] Lewis, *op. cit.*, pp. 128–30.

[2] *V.C.H. Gloucestershire*, vol. ii, pp. 222–3 ; Nicholls, *Iron-making in . . . the Forest of Dean*, p. 78.

[3] *Roy. Com. on Mining Royalties*, 1893, *2nd Report*, pp. 416–21.

[4] *V.C.H. Derbyshire*, vol. ii, p. 350.

of suits submitted for trial during the century after the accession
of Elizabeth, precedents were established which have been followed
by jurists ever since, and which have given British mining law a special
character unlike that of any continental system, or, so far as we know,
of the system of any ancient civilization.

The complexity of the English administration of justice must
impress all who attempt to unravel the tangled thread of logic by
which cases dealing with the same colliery were brought now under
one jurisdiction, now under another. The effort is liable, indeed,
to be wasted, for the problem is sometimes insoluble. Even the
most learned legal writers of the early seventeenth century could not
always have told just where came the dividing line between the
authority of the Chancery Court of the Palatinate of Durham and
that of the Council of the North, between the Lancashire Assizes,
the Chancery Court of the Palatinate of Lancaster, and the Court of
the Duchy Chamber of Lancaster. Cases were constantly being passed
from one judicial authority to another. A suit between tenants of
Broseley in Shropshire and the lord of the manor, over their respective
mineral rights, was fought out in at least five separate courts—
the Shrewsbury Assizes, the Council of Wales, the King's Bench, the
Court of Chancery, and the Star Chamber.[1]

The particular court in which an action was brought depended
primarily upon the choice of the parties who joined issue, particularly
the plaintiff, and only secondarily upon the nature of the suit.
Sometimes a court rejected a plea on the ground that it could
more fittingly be judged in another jurisdiction. A case brought
in the courts of the Duchy of Lancaster was dismissed as a mere
matter of "trespass", to be settled at common law.[2] Another,
in the Star Chamber, was dismissed because it dealt with a question
of title.[3] For the same reason a dispute over the coal "heugh"
of Wolmet in Midlothian was referred by the Scottish Privy Council
to the Lords of Session.[4] Instances could be multiplied. But, though
the nature of a suit sometimes prevented its trial in some particular
court, this seldom determined within exactly what jurisdiction it
should fall. It was easy for clever lawyers to arrange their client's
bill of complaint to fit the requirements of almost any court. By
alleging that bodies of husbandmen had taken violent possession of a
colliery in their client's land, they could turn a simple issue over title
into a case of riot, in order to insure trial by the Star Chamber. Thomas
Bagshawe, of Ridge Hall, Derbyshire, for instance, filed in 1606
a bill of complaint in which he alleged that Ralph Cooke, "yeoman",
and twenty-two other persons had assembled, "armed with pitch-

---

[1] *Star Chamb. Proc., James I*, 86/18; 109/8, 9; 294/25; 310/16.
[2] *Duchy of Lancs. Pleadings*, 156/B/5; *ibid., Decrees and Orders*, vol. xx,
ff. 527, 543 (Wm. Barcrofte v. John Higgins, Chris. Jackson, and others, 1592).
[3] *Star Chamb. Proc., James I*, 4/3, and 227/1–7 (Attorney-General v.
Fabian Heywood and others; Sir Stephen Proctor v. Sir Wm. Ingelbye and
others, 1603).
[4] *Privy Council Register of Scotland*, 1st Series, vol. vii, pp. 156–7 (Sir James
Sandilands v. Francis and George Wauchop, 1605).

forks, bows and arrows, pistols, guns and other weapons ", and threatened Bagshawe's servants with death if they did not cease mining in Fernilee manor. The title to minerals there was in dispute between Bagshawe, who had purchased the manor, and one John Hibbert, who held a lease still in being from the preceding lord. The assailants, according to Bagshawe, ordered his servants to leave behind the coal they had extracted, and beat a certain John Davie, " a collier expert and experienced in such workes ", in the hope of disabling him and of depriving Bagshawe of skilled advice.   All the charges were denied by the defendants.[1]

We may say, in general, that the tendency was more and more away from settlement of disputes in manorial courts, where the lord was frequently an interested party. Customs of the manor were frequently submitted for confirmation or modification to other courts.[2] Sometimes, as in the decision given by the Court of the Duchy Chamber of Lancaster concerning the manor of Shelton in north Staffordshire,[3] old privileges were revoked. Whenever possible, advantage was taken of special jurisdictions.   Most of the important collieries in Durham could be brought under the authority of the County Palatine, and one can scarcely run through a dozen of the bills of complaint filed with the Chancery Court of the Palatinate, without meeting with a dispute arising out of the local coal industry. A suit over a small colliery in the manor of Codden, in the Tyne valley, was tried as late as 1590 in the court of the Archbishop at York.[4] The Council of the North, we learn indirectly,[5] tried colliery cases arising on manors in Durham, Northumberland, and Yorkshire, before it was abolished in 1641, along with the other special organs of Tudor administration. The Council of Wales and the Marches tried colliery cases arising not only in Welsh counties but also in Shropshire and Gloucestershire.[6] As a great number of Crown manors in several counties were administered through the Duchy of Lancaster, many suits, particularly in Lancashire and Yorkshire, were heard in the Court of the Duchy Chamber.[7] These suits were usually brought within the Duchy jurisdiction on the ground that Crown interests in mineral property were involved. Other cases dealing with coal in Lancashire were tried in the courts of the County Palatine of Lancaster, and

---

[1] *Star Chamb. Proc., James I*, 50/23.
[2] See, for instance, *Duchy of Lancs. Pleadings*, 87/M/7 ; 162/A/35.
[3] *Ibid.*, 144/B/11 (Sir Ralph Bagnall *v.* James Burne and others).
[4] *A History of Northumberland*, vol. iv, p. 173, note.
[5] *Chanc. Proc., James I, C.* 20/74, 154/12 ; *Star Chamb. Proc., James I*, 154/12 ; *Court of Augm. Proc.*, 9/99 ; *Duchy of Lancs. Pleadings*, 146/F/20 ; S.P.D., *Eliz.*, vol. cclxiii, no. 72, I.   The records of the Council of the North, which doubtless contained much valuable material with reference to the coal mines, particularly in Durham, have been lost (see R. R. Reid, *The King's Council in the North*, 1921).
[6] *Star Chamb. Proc., James I*, 310/16 ; *Chanc. Deps., James I*, P. 18/H ; *Court of Requests Proc.* 25/103.
[7] For instance :   *Duchy of Lancs. Pleading*, 25/B/1 ; 87/M/4, 7 ;   106/A/1 ; 144/A/9 ; 149/A/35 ; 159/T/7 ; 162/A/35 ; 189/L/7 ; 189/H/8, 15 ; 193/J/1 ; 195/S/10 ;   204/O/1 ;   205/S/27 ;   211/G/6;   *Duchy of Lancs. Deps.*, 22/L/3 ; 73/R/15.

occasionally in the Court Leet of Manchester.[1] Mineral wealth in lands formerly belonging to the monasteries and confiscated by the Crown was subject to the jurisdiction of the Court of Augmentations, until its dissolution in 1554.[2] Thereafter similar cases were usually tried by the Court of Exchequer, which concerned itself with Crown rights in the coal industry throughout the country.[3]

Colliery disputes of all kinds were frequently brought directly, or on appeal from other jurisdictions, to be settled in the higher courts. One can find little difference in the type of case before the Star Chamber,[4] while that institution still existed, and the Court of Chancery.[5] In general these courts refused to handle questions merely concerning title, but accepted disputes over alleged damage caused by rival collieries, the collection of debts contracted by partners, or the enforcement of the terms of leases or working contracts. They dealt with cases originating in Durham, Northumberland, Cumberland, Yorkshire, Lancashire, Derbyshire, Nottinghamshire, Warwickshire, Staffordshire, Shropshire, Gloucestershire, Somerset, Glamorgan, Carmarthenshire, Denbighshire, and Flintshire—in short, in practically every region where the coal industry was of any importance. Some disputes were also submitted to the Court of Requests,[6] though the number was small, partly, no doubt, because the particular advantages which were there offered to poor men were counterbalanced by the cost of bringing a plea in so distant a city as London, more than a hundred miles from the nearest coalfield. Still others were tried in the King's Bench.[7] And, before the Civil War, the Privy Council sometimes stepped in to settle a case. Apart from three special instances,[8] the Council appears to have intervened

[1] *Duchy of Lancs. Deps.*, 19/T/3 ; Crofton, *Lancs. and Cheshire Coal Mining Records*, p. 53.
[2] For instance : *Court of Augm. Proc.*, 3/45 ; 6/37 ; 7/11 ; 14/67, 76 ; 18/95 ; 23/5 ; 31/46 ; 32/34 ; *Court of Augm. Misc. Bks.*, vol. cxi, ff. 16–19 ; vol. cxii, ff. 17–24 ; vol. cxvi, ff. 55–65.
[3] For instance : *Exch. Deps. by Com.*, 12 Eliz., East. 1 ; 21 Eliz., Hil. 8 ; 29 Eliz., East. 4 ; 8 James I, Trin. 1 ; 11 James I, Trin. 6 ; 3 Charles I, East. 19 ; 3 and 4 Charles I, Hil. 2 ; 11 Charles I, Mich. 46 ; 12 Charles I, East. 25 ; 13 Charles I, East. 14 ; 16 Charles I, Mich. 22 ; 14 Charles II, East. 26 ; 29 Charles II, East. 20 ; 36 Charles II, Mich. 43 ; *Exch. Spec. Com.*, nos. 2621, 4355, 5996 ; *Exch. Decrees*, Series iv, vol. ii, f. 243*d* ; vol. iii, f. 141 *i*.
[4] For instance : *Star Chamb. Proc., Henry VIII*, 5/138 ; 22/94 ; *ibid., James I*, 4/3, 69/18 ; 86/18 ; 92/6 ; 106/7 ; 109/8, 9 ; 154/12 ; 155/5 ; 167/17 ; 224/19, 20 ; 227/1–4 ; 228/13 ; 245/6 ; 288/5 ; 294/25 ; 310/16, 33.
[5] For instance : *Chanc. Proc., Eliz.*, C.c. 6/40 ; F.f. 5/31 ; O.o. 2/46 ; U.u. 1/50 ; *ibid., James I*, A/9/24 ; B/20/72 ; C/9/69 ; C/24/5 ; D/1/29 ; D/9/10 ; F/4/53 ; G/1/76 ; *Chanc. Proc.*, Series ii, 371/4 ; 425/47 ; *Chanc. Deps.*, P/7/T ; P/18/H.
[6] *Court of Requests Proc.*, 25/103, 28/56, 40/106.
[7] *Exch. Spec. Com.*, no. 4689 ; *Exch. Decrees*, Series iv, vol. i, f. 129 ; *Star Chamb. Proc., James I*, 294/25 ; *Chanc. Proc., James I*, C.20/74.
[8] One exception was the dispute between rival factions at Newcastle over the management of the " Grand Lease " (see below, vol. ii, pp. 121–5) ; another was the struggle between the rival collieries of Griff and Bedworth concerning the coal supply for the town of Coventry (*P.C.R.*, vol. xxxi, pp. 507, 533, 541 ; vol. xxxii, pp. 271–2 ; vol. xxxviii, p. 424 ; vol. xxxix, p. 659 ; vol. xlii, pp. 65–6, 84–5, 112–14 ; vol. xlix, p. 355) ; the third was an appeal by one Hewet Osborne, Esq., for an injunction to prevent interference with a " sough " which drained

only in suits dealing with rights of wayleave for carriage of coal from the pits to the water, on the Tyne, on the Wear, in Cumberland, and in Pembrokeshire.[1] After the Restoration such questions were left for the regular law courts.

In Scotland, as in England, there were no special courts for mineral law, and the proper jurisdiction for coal mining cases was no less uncertain. The greater number of suits, especially in the counties of Fife, Clackmannan, Stirling, Lanark, and Ayr, were probably settled by the local justices or by the Lords of Session, whose early proceedings fill hundreds of unindexed volumes in the General Register House in Edinburgh.[2] In the Lothians the parties very frequently made an effort to bring their disputes before the Scottish Privy Council,[3] whose sittings at Holyrood made it simple and inexpensive for litigants from collieries along the southern shore of the Forth to submit their differences there. Notwithstanding the principle that the Privy Council should not consider questions of disputed title, the number of suits which were brought before the chief Scottish governing body nearly every year was considerable, though many were returned to the lower courts.[4]

With so many different courts available, it was natural that suitors should instruct their lawyers to plead in those where their chances of a verdict seemed most favourable, and there is ample evidence that the latter often spent more time in manœuvring over matters of jurisdiction than on any other point connected with the case. One of the Strelleys complained that his neighbour, Willoughby, had brought an action against him in the Star Chamber in order that " the poor men ", his miners, might be " dragged up to Westminster ", and their master " put to dyvers and menfold other vexacious trowbyle

---

his coal pit, until such time " as the matter maie be tryed by the due course of the common lawe " (*Acts of the Privy Council*, N.S., vol. xxix, pp. 657–8).

[1] *P.C.R.*, vol. iii, pp. 622, 718, vol. xxxii, pp. 499, 679–80 ; vol. xxxiii, p. 66 ; vol. xliv, pp. 544–5, 647–8 ; vol. xlv, pp. 248, 487–8. An action brought by the Earl of Northumberland in 1623 concerning the mining of coal, by a Newcastle partnership, under a place called " Bird's Nest ", which the Earl insisted was part of his manor of Newburn, was eventually referred to the Court of Chancery as a private cause not fit to be tried by the Council (*ibid.*, vol. xxxi, pp. 561, 592 ; vol. xxxiii, pp. 558–60 ; *S.P.D., James I*, vol. cxxxvii, no. 134). A similar suit over a waste called Sugley had been brought in the King's Bench and the defendants had appealed to the Court of Exchequer (*Exch. Decrees*, Series IV, vol. i, f. 129 ; *Exch. Deps. by Com.*, 8 James I, Hil. 1 ; 9 James I, East. 16).

[2] Time has not permitted me to examine these valuable manuscript volumes, but Mr. Paton, searcher of records in Edinburgh, informs me that they contain reports of many coal mining disputes. Cases brought before the Scottish Privy Council were constantly being referred to the Lords of Session (*Privy Council Register of Scotland*, 1st Series, vol. vi, p. 236 ; vol. vii, pp. 156–7 ; 2nd Series, vol. iii, p. 563). The Privy Council persistently refused to decide " the heretable right to the coale ".

[3] Cases from Fife and Lanarkshire and perhaps other countries, were also brought before the Scottish Privy Council.

[4] It is worth noting that two Englishmen who were being sued by John Nicholson, of Lasswade, from whom they had leased coal mines, denied that, as foreigners, they could plead before the Court of Session, where Nicholson wished to have his case tried, and forced him to join issue with them in the Council. (*Privy Council Register of Scotland*, 1st Series, vol. xiii, pp. 668–9.)

and costys ".[1] Oliver Selby, who worked coal in Kyloe manor near the Scottish border, protested in 1603 that Thomas Gray had filed a bill against him in the Star Chamber, instead of with the Council of the North, simply to inconvenience him and put him to the expense of the long journey to London.[2] According to Huntingdon Beaumont, a certain Roger Fenwick, his rival, in an effort to prevent the success of Beaumont's colliery near Blyth, surreptitiously secured a decree against him in the Chancery Court of the Palatinate for non-payment of debt. Fenwick and others then kidnapped and carried him by force into the jurisdiction of the County Palatine in Durham, where they exhibited the decree and threatened to put him in jail unless he paid at once £52, which, if we are to believe his own story, he had never owed.[3]

Various reasons prompted litigants to bring their actions in the national courts. London merchants, who were creditors of colliery proprietors, were naturally eager to sue for recovery of debts in the capital, where their wealth and legal connections gave them special power. For the lords and gentry and provincial traders, who owned or leased the greater number of coal mines, it was not always such good policy to bring actions in London, particularly when their object was to curtail the mineral rights of manorial tenants. The Court of Star Chamber and the Councils of Wales and of the North, after the bridling of the Court of Requests, offered some protection to small tenants and copyholders in their struggle to maintain rights once theirs by custom of the manor. On the other hand, the owners of mineral property and their lessees were sometimes able to make facile tools of local juries.[4] Many of them preferred, therefore, to bring actions at common law ; while the local tenants frequently sought refuge in the royal courts. The suit in which in 1588 the freeholders and copyholders of the manor of Kippax, near Leeds, succeeded, by an appeal from common law to the Council of the North (whose decision was afterwards challenged in the Court of the Duchy Chamber of Lancaster), in protecting their rights against the digging of John Freston, of Altofts, farmer of the queen's mines in Kippax,[5] is typical of a considerable number of cases in every part of the country where coal was worked. They strengthen the ground of those who argue that the suppression, at the time of the Civil War, of the Courts of Star Chamber and Requests and of the Council of the North weakened the position of manorial tenants.

It is clear that the proper jurisdiction for mining cases was scarcely better defined, in the sixteenth and seventeenth centuries, than the state of mineral law itself. This uncertainty over jurisdiction added to the time and energy spent in litigation concerning coal mines. Courts would accept a suit, deliberate over it, even appoint commissioners to take depositions, and finally refuse jurisdiction. Under these circumstances the force of judgments was

---

[1] V.C.H. Nottinghamshire, vol. ii, p. 326.
[2] Star Chamb. Proc., James I, 154/12.
[3] Ibid., 69/18.
[4] R. H. Tawney, The Agrarian Problem in the Sixteenth Century, pp. 397–400.
[5] Duchy of Lancs. Pleadings, 146/F/20.

weakened, the number of appeals or new suits increased, and the way thrown open for every sort of intrigue. Obviously, in such a contest, the victory was likely to go to the litigant who had the longer purse, and could wear out the resources of his rival in a protracted legal battle, which frequently became a war of financial exhaustion. Thomas Killigrew and Cicely Crofts, two servants of the Crown, are said to have spent £1,700, the equivalent of £15,000 or more in modern currency, in prosecuting the King's title to a coal mine in Benwell manor on the Tyne.[1] Five thousand pounds, a larger sum than was sometimes required to start a mining enterprise, is said to have been spent by one side in a dispute between the Tyne colliery owners and certain suitors for a patent to " survey " coal, with a view to permitting the sale of only the superior grades.[2]

No matter in what court a case was finally settled, the judges were liable to be without any special knowledge of the technicalities of mining. They had to depend largely upon the investigations of commissions and the depositions of witnesses to provide themselves with the material for a decision.[3] The distances to be travelled by witnesses—often poor coal hewers—were frequently too great to permit of testimony being given in the presence of the judges. Even when it was possible for witnesses to appear in person, supposedly impartial commissioners were still needed to settle questions of fact, and to determine the sums due for damage to surface property or for wayleave rent. For instance, in a case brought in 1634–5 in the Chancery Court of the Palatinate of Durham concerning the wayleave to be charged by copyholders of Whickham, the judges left it to a commission to settle the rent due.[4] The Privy Council, in deciding similar cases, always appointed commissioners. In the judgments given by the Courts of Exchequer, of Chancery, of Augmentations, and of the Duchy Chamber of Lancaster, and by the Privy Council of Scotland, these reports of commissioners often played a dominant rôle in the settlement of questions of fact. Findings of commissioners were never questioned by the court unless it could be shown that they had acted from self-interest. Commissioners often became, therefore, arbiters in fact and, sometimes, also in name.

Generally it was only unimportant cases which were actually settled by " arbiters ", independently of the court in whose jurisdiction the suit had fallen. In 1665 we find the Bishop of Durham's Halmote Court submitting a dispute over deliveries of coal to arbiters, as " too small a matter to concern this court ".[5] Sometimes the parties

[1] Cal. S.P.D., 1637–8, pp. 247, 419.

[2] S.P.D., James I, vol. cxxvi, no. 20.

[3] We have an instance of a coal mining dispute settled by commissioners at least as early as the fourteenth century. In 1357–8 the Court of Chancery tried a suit brought by the Prior of Tynemouth against the burgesses of Newcastle, whom he charged with digging for mineral fuel in his moor of Elswick (R. L. Galloway, Annals of Coal Mining, p. 42).

[4] Palat. of Durham, Decrees and Orders, vol. i, ff. 107 sqq., 228 sqq. (Thos. Liddell v. Richard Jackson and others).

[5] Green, " Chronicles and Records of the Northern Coal Trade ", in Trans. No. of Eng. Inst. of Mining Engineers, vol. xv, 1865–6, pp. 267–8 (Blakeston v. Dikes).

themselves chose arbiters before the cause had come to law, as in the case of a " delphe " of coal in Woodsetts (Derbyshire) where both sides agreed upon four mediators.[1]

While courts seldom delegated their power of giving a judgment, the real authority in settling many disputes, and in fixing the payments due for damage to the earth's surface, and for wayleaves, rested with commissioners, whether they were arbiters in name or merely in fact. Though occasionally, as in the dispute between John Nicholson of Lasswade, in Midlothian, and James Tulloch, " tackesman " of his colliery,[2] the commissioners were chosen by the parties to the suit, they were usually named by the court within whose jurisdiction the case had fallen, subject of course to the approval of plaintiff and defendant. Once appointed, they acted in very much the same capacity as the members of the miners' court at Liége: examining witnesses, visiting the colliery, often descending into the pits to make tests and measurements. But the miners' court was a standing body of technical experts, while the English commissions were made up of men chosen to facilitate the settlement of one particular dispute. They might never again be called upon to serve. They were nearly always local gentry or members of the trading classes. It was common for the Privy Council to call upon the county justices to assess wayleave rents, but frequently it simply selected men of prominence in the county, who occupied no official position.

In Durham and Northumberland, as in the Lothians, there was seldom a commission which did not include at least one member who was a prominent coal owner, and sometimes all the commissioners had extensive interests in the coal industry. Sir Thomas Riddell, Sir George Selby, and Thomas Liddell, three Newcastle merchants whose investments in mines in the north of England equalled, if they did not exceed, those of any other adventurers in the reign of James I, constantly appeared as commissioners in important mining cases.[3] In Scotland prominent mine owners such as the Earl of Mar, Lord Sinclair, Lord Elphinstone, Sir Walter Seton, and Sir John Wauchope served no less frequently as commissioners than did the leading Newcastle merchants.[4] It follows that the point of view of the great colliery owner must have found expression in most of the mining cases at least in these two districts.[5]

It is not an accident that " chancery suit " has become a synonym for delay, and when the time spent in filing bill and answer, replication and rejoinder, in appointing a commission and taking depositions is considered, it is obvious that, even with all possible expedition, it must often have taken many months to get a judgment. To

---

[1] *Chanc. Proc., Eliz.*, O.o. 2/46 (Rich. Ogden *v.* Robert Strongfellow and Frances his wife, 1588–9).

[2] *Privy Council Register of Scotland*, 1st Series, vol. xiii, pp. 368–9.

[3] *P.C.R.*, vol. xxxi, pp. 561, 622, 718; vol. xxxii, p. 499; vol. xxxiii, p. 66; *Exch. Spec. Com.*, nos. 4355, 5996.

[4] *Privy Council Register of Scotland*, 2nd Series, vol. vi, pp. 525–6; 3rd Series, vol. i, p. 388, and vol. ii, pp. 274–5; *H.M.C., Report on MSS. of Earl of Mar and Kellie*, p. 280.

[5] Cf. below, vol. ii, pp. 119 sqq., 157 sqq.

wait for the wheels of justice to turn was often to lose your case ; if your neighbour was digging coal under your holding, what you wanted was to stop him at once ; if he obstructed an adit by which you drained your own colliery, it was cold comfort to be upheld in the courts a year after your mine had been lost by flood. When summary action was demanded, therefore, appeal was made for an injunction, either to some local justice or to one of the higher courts. Thus, the Steward of the Honor of Pomfret, at the request of Roger Mallett, farmer of Crown mines in the manor of Rothwell, which was within the Honor, forbad William Lucas to win coal in the manor pending trial of his title ; [1] the Court of Augmentations enjoined Sir Francis Leeke from interfering with the sale of coal mined by William Bolles, at Swanwick in Derbyshire ; [2] the Scottish Privy Council bound John Gaw for the safety and indemnity of George Bruce's colliery at Culross, which was endangered, according to the allegations of Bruce, because Gaw had broken down the " water-gang ", or dam, used to supply water to turn the wheel of the pumping engine at the pit head.[3] In Scotland one or both parties to a suit were commonly required to " find caution ", i.e. to make a deposit subject to confiscation if they did not comply with the rulings made by the court to govern their activities pending trial of the case.[4]

Judges and commissioners in the sixteenth century were not entirely without tradition to guide their decisions. Such precedents as were available were derived from two chief sources : from the special usages of medieval mining districts, and from the customs of particular manors. Though it is possible to trace the influence of the medieval lead and tin miner in certain colliery practices—to see, in the habit, at mines near the Forth and in South Wales, of giving the lessee a piece of coal as a token of possession,[5] a relic of

[1] *Duchy of Lancs. Pleadings*, 87/M/4, 7.

[2] *Court of Augm. Proc.*, 31/46. See also the injunction served on the copy-holders of Colne manor (*Duchy of Lancs. Pleadings*, 159/T/7 ; *ibid.*, *Decrees and Orders*, vol. xx, ff. 248, 438).

[3] *Privy Council Register of Scotland*, 2nd Series, vol. viii, p. 267. Another interesting case is that in which Wm. Wallat of Tough disputed the title of Duncan Tailzeour, burgess of Dunfermline, then in possession of the " coal heugh " of Tough. Tailzeour agrees to pay for the right to continue working during the period pending trial, provided the decision goes against him. He also agrees not to put in more colliers than he has had in the past, in order that he may not exhaust the mine, and not to work the pillars so as to endanger the supports ; all subject to a forfeit of 500 marks Scots (*ibid.*, 2nd Series, vol. i, pp. 73–4).

[4] See, for instance, bonds of caution furnished in a suit over the colliery of Wolmet, in Midlothian. The principal parties each deposited 5,000 marks Scots, their servants 1,000 marks Scots (*ibid.*, 3rd Series, vol. ii, pp. 665–8).

[5] In 1660 Robt. Rose received possession on behalf of the Duchess of Hamilton of a colliery at Kinneil, " by delivery of a great Coall at Langhauch heugh by Alex. Burges, overman " (*Hamilton MSS.* : " Possession of the Lands and Works of Kinneil", July 20th, 1660). At the same time, a handful of salt was presented as a symbol that the Duchess had taken over the pans. At Llanelly in Carmarthenshire, when one Thomas Brent leased a mine, he was given " a piece of mony " in earnest of the bargain (*Star Chamb. Proc., James I*, 155/5).

the old convention whereby the lord's bailiff received a dish of ore when he invested the finder with ownership ; to discover, in the fixing of a royalty rent at a certain proportion of the profits or product of the mine,[1] a connection with the medieval compact by which the owner was entitled to a definite percentage of the metal extracted ; to refer back a clause in a sixteenth or seventeenth-century lease, requiring continuous mining,[2] to the ancient rule that the miner forfeited his title if he abandoned work for longer than a specified period—still, these are survivals largely of academic interest. The delivery of a piece of coal was simply a picturesque act, having lost its original significance as a symbol that the giver was the finder of the mine. Clauses in a lease fixing the royalty in proportion to the product, or requiring the lessees to work continuously, are not to be taken as evidence that medieval mining traditions had any real power in the coal industry. These clauses were imposed to fit the needs of the time rather than at the dictation of ancient usage.

Of far greater importance for the decisions of Tudor and Stuart judges and commissioners was the influence of manorial custom. For the settlement of a suit dealing with coal mining one subject was examined by nearly every commission, was on the lips of nearly every witness, was cited by plaintiff and defendant alike—the customs of the manor. What then were these customs and how did they arise ? Wherever coal was discovered and worked there was a need to define the terms on which it might be dug and sold, the respective rights of tenant and lord. And since, in medieval England, coal was wanted chiefly for use as fuel within the manor, it was natural that regulations concerning this mineral should come within the scope of manorial law. Each manor laid down its own rules. In proportion to the importance of mineral fuel in the manorial economy, and in proportion to the length of time that coal had been worked, these regulations

---

[1] It is possible that this practice was derived from the customs in some of the ancient lead mining districts. In the manor of Nidderdale, in Yorkshire, the same percentage was fixed both for lead and coal. We find that a fixed proportion of the coal produced, or of the profit realized, was sometimes taken as a royalty in the manor of Stratton-on-the-Fosse, near the Mendip lead district, and in the hundred of Prestatlyn, in Flintshire, near to other ancient lead mines (*Court of Requests Proc.*, 22/4 ; *Cal. Treas. Papers*, vol. i, p. 242 ; vol. ii, p. 10 ; *Parl. Surveys, Somerset*, no. 39). In Scotland, the King sometimes claimed the tenth load of gold or silver, and, in a licence of 1555 granted to tenants by the Arch-dean of St. Andrew's to dig coals, the ninth load which came on the " coal hill " (pithead) was due the convent of Dunfermline (*H.M.C.*, *4th Report*, appx., p. 499). On the manor of Westerleigh in Kingswood Chase, freeholders and copyholders might mine for coal within their lands upon payment of one-tenth of their profits to the lord, a payment known as the " Lords parte " (*Chanc. Proc.*, Series II, 371/4).

[2] The lead and coal mines of Nidderdale (Yorks.) were leased in 1547 for forty years, with the proviso that the lease should be void if the lessee ceased work for two years (*Court of Requests Proc.*, 22/4). During the Civil War, Ralph Cole, a royalist and prominent Newcastle coal owner and merchant, complained that, having been unable to work the " Wilson's Field " colliery in Gateshead, because of the wars and his imprisonment, he had been thrust out by the owner under a clause in the lease permitting the lessor to re-enter if mining was discontinued for a year (Welford, *Records of the Committees for Compounding*, p. 165).

were more or less complete.   At Cowpen and Elswick in Northumberland, at Ewloe in Flintshire, and Stratton in Somerset,[1] there existed
a whole series of well understood customs, approaching an unwritten
code, for in all these manors coal digging was an old story.   Within
other manors, where coal had only recently been discovered, such
rules, if they existed, did not carry the authority of age.

When the authority of manorial custom varied so greatly from
manor to manor, no coherent body of principles capable of being
applied to coal mining cases in general was likely to emerge from it.
The emergence of such principles was rendered still more difficult
because, in manors where complete rules were established, they
frequently differed in important respects from those in other manors
in the same county.   Elswick manor, for example, continued to have
its own peculiar system of assessing wayleave rent for carriage of
coal from pit to staith, a system different from that followed across
the Tyne at Ryton and Whickham, and even from that in the adjoining
manor of Benwell.   The system in Elswick was based upon an ancient
right of manorial tenants to enjoy, in return for all damage that they
might sustain from mining operations, a monopoly over the transport
of coal to the river.   When, in the seventeenth century, Elswick
colliery became one of the principal enterprises in the north, tenants
leased their privileges to the owners of wagons, who were able to
make profits over and above their rent, because, owing to the freedom
from direct payments for wayleave and surface damage, a wain,
according to the testimony of one witness,[2] was able to earn 18*d*.
a day more at Elswick than at other places in Northumberland.
Most of this sum was probably paid as wayleave rent to the tenants of
the manor.

More interesting still were the customs of the manor of Tunstall,
in north Staffordshire, where the lord granted leases or copyholds
of a " row ", or seam, of coal,[3] instead of leasing that part of the seam
which lay under a particular division of the manor.   This is perhaps
the only instance in British coal mining of an " apex " concession,[4]
according to which leases are made of a seam or fraction of a seam,
without reference to the ownership of the surface.   Possibly this
manner of leasing was derived from the customs of the lead mining

---

[1] *Exch. Deps. by Com.*, 41 Eliz., East. 19, and *A History of Northumberland*, vol. ix, pp. 225-7, and Appendix 5 (for Cowpen) ; *Exch. Deps. by
Com.*, 32 Charles II, Mich. 30 (for Elswick) ; *Court of Requests Proc.*, 25/103
(for Ewloe) ; *Minis. Accts.*, 1123/3, cited in *V.C.H. Somerset*, vol. ii, p. 371
(for Stratton).   See also *Exch. Deps. by Com.*, 11 James I, Trin. 6, for customs
concerning coal in the manor of Eckington (Derbyshire) ; *Star Chamb. Proc.*,
James I, 92/6, 228/13, for the manor of Tunstall (Staffs.) ; *Duchy of Lancs.
Pleadings*, 87/M/7, for Rothwell (Yorks.) ; *Chanc. Proc.*, Eliz., C.c. 6/40, for
Wednesbury (Staffs.) ; Bainbridge, *Law of Mines*, 1878 ed., pp. 56-8, for
Bolsover (Derbyshire).   These are only a few of the cases in which we find evidence
of special manorial customs concerning coal mining.

[2] Thos. Sootheran, of Elswick, yeoman (*Exch. Deps. by Com.*, 32 Charles II,
Mich. 30).

[3] *Star Chamb. Proc.*, *James I*, 228/13 (John Podmore *v.* John Colclough
and others, 1609).

[4] Possibly there was a similar custom within the adjoining manor of
Newcastle-under-Lyme (*Duchy of Lancs. Pleadings*, 90/B/41 ; 144/B/11).

district of the High Peak in Derbyshire, where " apex " concessions appear to have been the rule. The fact that such a practice was apparently peculiar to one manor in the case of coal mining serves to emphasize the local differences which prevailed in the colliery districts. In spite of the tendency for the courts to modify manorial custom and to evolve common mining laws for the whole country, not all local peculiarities were ironed out ; until the passage of the Real Property Act of 1922, the rights of copyholders with respect to minerals differed materially from manor to manor.[1] To understand the special historical development of the English law of mines, one must be conscious of the process by which judicial decisions have been superimposed upon ancient manorial customs, which could be applied only with some difficulty to the problems raised by coal mining, but which might, nevertheless, influence the judicial decisions.

Manorial customs left their trace upon mining law to a greater extent than the special usages of early mining communities ; but the chief characteristic of both was their diversity ; while what was needed to facilitate the expansion of the coal industry was uniformity. Towards the end of the sixteenth century there began to develop, partly as a result of the growing demand for coal, what might be called a national policy towards coal mining ; and this policy had an increasingly important influence on judicial decisions. Before the end of the seventeenth century it had already begun to establish a body of general principles which could be applied throughout the country. All districts were in need of fuel ; life in London was becoming impossible without ever larger supplies of Newcastle sea-coal ; and the significance of the new industry was impressed on the minds of pamphleteers and statesmen at a time when they were peculiarly conscious of the opportunity for the State to take part in the direction of economic affairs. A successful colliery became in the eyes of politicians a public benefit. It added to the supply of fuel, provided work for the unemployed, and increased the revenue from taxation.[2] Judges pointed to " the working, venting and transporting of coales " as " very requisite and . . . usefull and commodious for all his Majesty's people ".[3] In a case brought in 1578 against Charles Jackson, farmer of the Queen's mines in the manor of Houghton in Yorkshire, the defendant sought to prove that his colliery was of value to the whole community, " for that those quarters are utterly destitute of all woodds and other fewell ".[4] Some years later, the Court of the Duchy Chamber of Lancaster protected Katherine Sherrington's pit in Orrell, just east of Wigan, from the violence of her competitor, William Orrell, in spite of his plea that she was

---

[1] Bainbridge, op. cit., pp. 42 sqq., 50, 53 sqq. ; Redmayne and Stone, Ownership and Valuation of Mineral Property, pp. 57 sqq. In Scotland feudal or universal tenure corresponds with English copyhold tenure, but the customs are uniform—the estate of superior and vassal being distinct and separate, and the vassal having a freehold estate including the minerals, with full power to work them (Bainbridge, op. cit., p. 49).

[2] See below, pt. v, esp. ch. i.

[3] Palat. of Durham, Decrees and Orders, vol. i, f. 109.

[4] Exch. Deps. by Com., 21 Eliz., Hil. 8.

encroaching on his rights, because, in the judgment of the Court, her mine was " a great good and benefitt to the neighbours and cuntrey thereabouts ", and its failure " a loss to divers poor people and others ".[1]

At the same time, it was recognized that collieries might be so carelessly run as to be definitely prejudicial to the public welfare. The Scottish Privy Council permitted the Edmonstones to continue mining after 1600 at Wolmet in Midlothian, only on the understanding that they " sould not inutile waist and destroy the saidis coillis and colhewis aganis the commone weill of the cuntrie ".[2] In 1623, commissioners appointed by the Privy Council to survey a coal mine near Newcastle, called the " Bird's Nest ", stopped the works in the public interest, on the ground that they were likely to be drowned out if the mining methods of the present entrepreneurs were continued.[3] Here, again, the courts look beyond the legal merits of the case, and consider the public need for continued supplies of fuel. Scarcely a plea was brought before the Scottish Privy Council in which that aspect was not urged by one side or the other. Frequently the suitor in Scotland would insist upon the dual social benefit conferred by his colliery. As an example, we may cite the argument of Major John Biggar, in 1667. The success of his colliery at Wolmet is, he says, for the general good, for " many poor people and ther families are maintained and employed and the toune of Edinburgh and the countrey about served with the product of ther labours . . . the mantenance of coalyiers [being] . . . of publick concernment for the intertainment of the poor people ".[4] On the Tyne and Wear, also, and wherever there were large mining enterprises, judges on occasion took into account the service alleged to have been rendered by the owner in providing work for the needy. The success of the Grand Lease colliery at Whickham, according to its owners, who explain its importance in a document of 1618, is of great benefit to London, and " other Citties, Townes and Countries in the Sowth " ; it is of value to the King's revenue from customs, and it affords relief to " multitudes of people who are dailie set on worke in the . . . colemynes, whereby they, there wives, children and famylies are norished and mayntened ".[5] It was recognized that, if landholders asked a high price for granting colliery owners the right to sink pits or to carry fuel, this was, as the Newcastle Company of Hostmen pointed out in a petition to Parliament in 1690, " no small prejudice to trade", as it tended to " raise the price of coal ".[6] The English Privy Council had already established a formula, adopted later by

---

[1] *Duchy of Lancs., Decrees and Orders*, vol. xxiii, ff. 236, 303–4. For a statement of the case, see *Duchy of Lancs. Pleadings*, 195/S/10 ; 204/O/1 ; 205/S/27.

[2] *Privy Council Register of Scotland*, 1st Series, vol. vi, p. 170 (Petition by Patrick Edmonston and his son, Archibald, November 4th, 1600).

[3] *S.P.D., James I*, vol. cxxxvii, no. 34.

[4] *Privy Council Register of Scotland*, 3rd Series, vol. ii, pp. 371–2.

[5] *Star Chamb. Proc., James I*, 245/6 (Sir Peter Riddell and others *v.* Francis Liddell and others).

[6] *Commons' Journals*, vol. x, p. 385.

the Chancery Court of the Palatinate of Durham in similar cases,[1] for settling disputes over wayleaves and colliery drainage. Commissioners, whether appointed for Northumberland, Durham, Cumberland, or Pembrokeshire, were told " to procure [for the landowner] such recompense . . . as shalbe thought just and reasonable ", having first an eye to the public interest and remembering that " the incourrageing and advanceing of coalworks may importe as well the comonwealth in . . . fewell, as . . . his majestie in his customes ".[2] When, however, it was a choice between the King getting his tax and the people more coal, the Council was guided by motives of doubtful public interest, as is shown by its refusal in 1625 to aid certain mine owners on the Wear to secure better terms for wayleave, " until the masters and owners of coales in generall to be vented at Sunderland . . . shall agree among themselves, to allow the king such profit ", i.e. a tax of a shilling per chaldron, paid on the Tyne, from which the Wear mines had been exempted by Parliament, " as upon the coales of Newcastle ". The Council contended " that the increase of the vent of coales from Sunderland would be in divers respects more to the prejudice . . . of his majesties revenew aryseing of the Newcastle coales then to the benefitt of the comonwealth otherwise ".[3] Such a conflict of interests probably interfered only on rare occasions with the principle established concerning wayleave rent. Judges found it increasingly possible to adopt formulas like that invented by the Privy Council as a means of overruling local customs, when they were inimical to the development of the mineral wealth of the country.

Evidently the conditions under which disputes were settled in the coalfields during the sixteenth and seventeenth centuries were increasingly favourable to the owners of large collieries, and increasingly unfavourable to manorial tenants who wished either to work coal in a small way within their own holdings, or to obtain compensation for the use of their land by the owners of large collieries. The customs of special mining jurisdictions, of medieval origin, which afforded the small enterprise some protection, had practically no influence upon English decisions in coal mining cases. The suppression, in the seventeenth century, of the Star Chamber and other royal courts deprived the manorial tenants of those jurisdictions in which they stood the best chance of obtaining a sympathetic hearing. The complicated nature of judicial procedure in coal mining cases, involving as it did the expenditure of large sums, naturally gave an enormous advantage

---

[1] *Palat. of Durham, Decrees and Orders*, vol. i, ff. 107 sqq., 228 sqq.

[2] *P.C.R.*, vol. xxxii, p. 499. The same principle of public interest was also invoked by the Council in *ibid.*, vol. xxxi, p. 718 (Durham) ; vol. xxxiii, p. 66 (Durham) ; vol. xliv, pp. 544–5 (Northumb.), and pp. 647–8 (Pembrokeshire) ; vol. xlv, pp. 248, 487–8 (Cumberland). Although the Crown derived no revenue from the midland coal trade, the Privy Council used the same formula concerning the " public interest " in dealing with a case concerning collieries in Warwickshire (*ibid.*, vol. xxxi, p. 541).

[3] *Ibid.*, vol. xxxii, pp. 679–80 (Sir Rich. Lumley and Sir Wm. Lampton *v.* Geo. Lilburne and Humfrey Wharton, 1625). Cf. below, pt. v, *passim*.

to rich mining adventurers, and the frequent selection of these adventurers as commissioners in important cases further strengthened their position. Finally, the supervention, at the expense of manorial customs, of a national policy which aimed at getting the minerals of the country developed as extensively and economically as possible, could not fail to affect adversely the interests of the yeomanry. In view of the weakness of the small landholder's position before the courts, it is not surprising that we should find the trend of judicial decisions unfavourable to him.

CHAPTER III

THE CONCENTRATION OF MINERAL RIGHTS

LAND in Elizabethan England was divided into hundreds of thousands of separate holdings : freeholds, leaseholds for lives or for terms of years, copyholds of various kinds, tenancies at will. These holdings were grouped in thousands of manors, large and small, rich and poor, including woodlands, pastures, and cultivated fields, and dividing the country into units each with a separate legal, a separate economic, personality. Though large holdings were, no doubt, fairly numerous, there is reason to believe that, even in sparsely populated counties, the vast majority of holdings were probably under seventy-five, in some other counties under twenty, acres.[1] As these divisions had been established without any relation to the mineral seams which ran in every direction under the soil—dipping in one place, breaking off in another, rising in a third to the surface—it was a matter of accident under what holdings coal was found. For the mining concessionnaire it was important to know what mineral rights went with each tenure. He had to ascertain whether it would be necessary to come to an agreement with every tenant in whose ground he proposed to sink shafts, to build hovels for his workmen, to carry fuel, or to make passage for the water drained from his pits ; or whether, since even a small manor usually offered an ample area for one colliery enterprise on the sixteenth or seventeenth century scale,[2] he might make a single settlement with the lord for all these rights.

(i) Rights of Freeholders

In the case of land in the demesne there was seldom any ambiguity as to rights.[3] Both surface and minerals were in the lord, and mining

---

[1] R. H. Tawney, *The Agrarian Problem in the Sixteenth Century*, 1912, pp. 32–3, 64–5, Tables II and IV.

[2] Cf. below, pt. iv, ch. i (i).

[3] Unless, of course, the coal or a part of the coal was specifically excepted when the grant or sale of the manor was made. For instance, in 1630, when Sir W. Hewitt sold the townships of Amble and Hauxley in Northumberland, he reserved all mines of coal, with liberty to work them, and with wayleave and staithleave (Bainbridge, *op. cit.*, p. 219). Such exceptions were rare before the end of the sixteenth century, for mineral fuel was generally regarded as of too small importance to be worth reserving. An interesting case in point is the action brought by Sir Thomas Leghe, of Hoddesdon, against Robert Ryshwood, of Crofton in Yorkshire. Leghe claimed the coal in the lordship of Crofton by lease from the King, but Ryshwood had been able to satisfy the Council of the North that Leghe's claim could have no force, because he (Ryshwood) had been seized in the lordship of which the mine was parcel before the date of the alleged lease, and consequently all grants of the minerals therein were void unless made by him (*Court of Augm. Proc.*, 9/99). Evidently the Crown officials had not thought of reserving the mines when they granted to Ryshwood these lands, recently acquired by the dissolution of Nostell Priory (cf. Appendix M (i)).

was subject entirely to such conditions as he chose to impose. Free-holders under the lord of the manor were not in the same unassailable position. Sometimes, by a clause in their grants, as in the Bishop of Durham's lordships of Ryton, Stella, and Winlaton on the south bank of the Tyne,[1] they renounced all claim to the ownership of the coal mines. Usually, no doubt, they were better off than other tenants. Sometimes they had as full rights over their own holdings as the lord himself over his demesne. On the manors of Kippax in Yorkshire, Colne in Lancashire, Bromfield in Denbighshire, and Halesowen in Worcestershire, they might dig for coal without asking anyone's permission.[2] A copyholder of Halesowen is reported in 1608 as saying " that he feared to seeke or gett coale in his then customary lands, but if it had byn his freehold lands he said he would have sought for coales ".[3] In the lordship of Alnwick, in Northumberland, neither the lord nor anyone in his name could take the minerals from under a freehold without the consent of the freeholder.[4] At Whickham in Durham, apart from the Grand Lease colliery, held of the Bishop as overlord, there were several mining companies, each operating in separate freeholds under different leases ;[5] and in the case of the Brinkburne freehold, formerly owned by Brinkburne Priory, there was a curious ancient custom that the landowner or his lessees might follow the seam into his neighbour's holding (thereby establishing what was in effect an apex concession), unless this neighbour sank a pit of his own to countermine, or " thurl ", his rival.[6] Across the river, on the Crown manor of Benwell, where the nature of the holdings is revealed in the map reproduced facing page 305, it was decided by the Court of Exchequer in 1637 that " the mynes found within the severall freehould lands . . . wholly belonge to the freeholders ", and that in a part of the manor where freeholds lay so mingled with customary land that they could not be plainly distinguished, " a thirde parte of the mynes found therein . . . might by usage belonge to the freeholders of the lands and the other two partes belonge to the king ".[7]

The lord frequently made good his claim to ownership of the

---

[1] *Palat. of Durham, Decrees and Orders*, vol. i, f. 497.
[2] *Duchy of Lancs. Pleadings*, 146/F/20 (1588) ; *Exch. Deps. by Com.*, 1659, East. 20 ; *Palat. of Lancs. Bills*, 57/101 ; *Court. of Augm. Proc.*, 18/95.
[3] *Exch. Deps. by Com.*, 5 James I, Hilary 17 (Littleton v. Low).
[4] *A History of Northumberland*, vol. ii, p. 455.
[5] See below, p. 361 n.
[6] *Parl. Surveys, Durham*, no. 8. The practice of sinking a pit near the boundary of one's land, if a neighbour was known to be mining, was called having " an eye on " his operations (*H.M.C., Report on MSS. of the Duke of Portland*, vol. vi, p. 100), and was a fairly common custom. The lord of Colne manor in Lancashire, when his freeholders, digging in their own lands, got under his waste, sunk " unto them . . . and soe [did] spoyle theire work " (*Exch. Deps. by Com.*, 1654, East. 20), but it is unlikely that there was any custom, as was the case in Brinkburne freehold in Whickham, by which a freeholder's encroach-ments could be judged legal unless they were prevented by direct action on the part of his neighbours.
[7] *Exch. Decrees*, Series IV, vol. iv, ff. 260, 265. The surface bounds of holdings in Benwell had long been in doubt (*Court of Augm. Proc.*, 38/2 ; *Augm. Misc. Bks.*, vol. cxvi, ff. 55–65 ; *Exch. Spec. Com.*, no. 1730 ; *Cal. S.P.D.*, 1631–3, p. 241).

minerals even against his freehold tenants. If he was occasionally unsuccessful, as in Wednesbury manor in Staffordshire, where husbandmen dug coal in their freeholds without licence of the lord in spite of his attempts to prevent them,[1] he often succeeded in keeping freeholders on the same footing as his other tenants, as in Ewloe lordship in Flintshire, where any coal dug without licence was forfeit to the king or his farmer,[2] or in Haigh manor in Lancashire, where freeholders had to pay " certaine bonuses, presents and averages " for digging,[3] or, finally, as at Westerleigh near Bristol, where, like the copyholders, they might " break the soil, dig for coales . . . dig levels, . . . take timber, . . . in respect whereof the Lord of the . . . Mannor . . . is to have a tenth parte or some other part of the coales, . . . all charges and reprisals touching the same being first deducted ".[4]

Nowhere, perhaps, was the lot of a freeholder so hard as at Cowpen in Northumberland. Not satisfied with citing old records to prove that he was forbidden to dig without the consent of the " chief lord ", the farmers of the Queen's coal sought in 1598-9 to prevent him from asking any compensation when they mined in his lands. Since eight of the sixteen pits which they had sunk were in freeholds, as is stated on the diagram facing this page, they made much of the issue. While they were able to point to docile tenants like John Smith, who testified that the farmers had mined in his holding without any compensation—" nor did he ever crave it or think it was due him,"—others combined against them to show that on one occasion at least they had been forced to compound. In the heat of their resentment two of the tenants are said to have beaten and cut with a sword a hewer employed by the farmers, when he was at work in a pit within a certain Cuthbert Watson's freehold. No record of the result of this dispute is to be found, but probably the freeholders failed to make good their claim.[5]

In cases where the freeholder's title was not directly challenged, concessionnaires and lords of manors could sometimes play upon his ignorance or inferior social position to trick him out of his rights. A group of adventurers who planned, at the beginning of the reign of Charles I, to extract coal from a seam in the manor of Lumley, one-third under the demesne, two-thirds under the freeholds of Thomas Fotherley and other yeomen, promised to aid the tenants—" being but simple and weak men "—to establish against the lord of the manor, Sir Richard Lumley, their title to the coal under their freeholds, and thereby induced them to sign away their minerals on a

---

[1] *Chanc. Proc., Eliz.*, C. 6/40 (Wm. Comberford *v.* Eliz. Nicholas and others, April 17th and 21st, 1589).

[2] *Court of Requests Proc.*, 25/103 (Holcroft and others *v.* Edw. Stanley, reign of Philip and Mary). Later, however, the rights of freeholders to mine coal were apparently confirmed (see below, p. 317).

[3] *Duchy of Lancs. Deps.*, 77/B/6 (Nicholas Butler and others *v.* Roger Bradshawe, Esq., and others, 1554).

[4] *Chanc. Proc.*, Series II, 371/4 (Wm. Player *v.* Geo. Bullock and others, 1624).

[5] *Exch. Deps. by Com.*, 41 Eliz., East. 19 ; *A History of Northumberland*, vol. ix, pp. 225-7.

twenty-one year lease, in return only for fire coal and a lump payment of 13s. 4d. for every surface acre actually spoiled. Having in fact connived with Lumley, the adventurers in question, Jeffrey Walker [1] and his partners, Thomas Liddell of Newcastle and William Errington, subsequently forgot their promise. They left the yeomen to discover that the lord was receiving £160 in annual rent for his minerals under the demesne, though Fotherley and his friends believed that their title to the minerals under their freeholds was equally good.[2] Six husbandmen of Cumberland, freeholders of the Crown manor of Distington, were no less effectively deceived in 1620, not only into giving away their mineral rights in perpetuity, but into paying £20 to boot, when Henry Fletcher, a prominent colliery adventurer, engaged himself to free them from their " lyveryes and wardshippes then claymed " by the King.[3] As Fletcher was careful, in taking over the mines, to keep the only copy of the agreement, the tenants were left without means of holding him to its terms.

On Broseley manor in Shropshire, during the first years of the reign of James I, the lord of the manor, James Clifford, met with a more highly organized opposition to his schemes for preventing freeholders from profiting by the coal under their grounds. While not denying their ownership, he had refused to grant them wayleave across the manorial waste separating their holdings from the river Severn, thinking thus himself to monopolize the trade from Broseley. His tenants challenged him on two counts. First of all, they objected to his building cottages on the waste for his " colliers and minerall men ", on the ground that they did not want to have for their neighbours the kind of labourer he hired.[4] Without waiting to get a decision in the courts, the tenants tore down the cottages and the hedges.[5] In the second place, they denied Clifford's right to sink pits in or set foot upon a certain pasture called " Calcott ", in the possession of one of the tenants, Richard Wilcox, a pasture in which there was known to be coal deep under the soil. Clifford believed that the winning of this seam was essential to the success of his whole enterprise. The tenants, though they did not propose to dig here themselves, needed a passage over " Calcott " for the coal from a parcel of their own freeholds, called " Birchleasowe ". To gain control of the pasture became, therefore, the aim of both sides. Clifford, by compounding with Wilcox some years before the freeholders had recognized the

[1] Jeffrey Walker was himself a " yeoman " of Lumley manor (*Palat. of Durham, Bills and Answers*, bdl. 29 : Wharton v. Liddell), and probably played a nefarious part in persuading the freeholders that his partners, whom they did not know, were acting in good faith. Whether Thomas Fotherley was as " weak " a man as he chose to declare himself in this case may be doubted. In 1623 he had been appointed one of two subcommissioners for the sale of various Crown manors in the north of England (*Sackville MSS.* : Warrant appointing Wm. Hill and Thos. Fotherley, 1623).

[2] *Palat. of Durham, Bills and Answers*, bdl. 24 (Fotherley and others v. Walker and others).

[3] *Chanc. Proc., James I.,* C. 24/5.

[4] *Star Chamb. Proc., James I,* 310/16 (Rich. Wilcox, gent., and others v. James Clifford, Esq., and others). Cf. below, vol. ii, pp. 147, 150–1.

[5] *Star Chamb. Proc., James I,* 86/18 (Clifford v. Rowland Lacon, Esq., Wilcox, and others).

importance of the coal industry, had been able to sink pits in the pasture.[1] Early in the year 1606, after having unsuccessfully brought an action in Chancery to oust the lord, a large number of tenants assembled with Wilcox at Calcott colliery, seized various machinery used at the pit, attempted to kill one of Clifford's workmen, and carried off his "tylting railes", along which his wagons ran down to the Severn.[2] They then laid wooden rails of their own to bring coal from "Birchleasowe", but Clifford's men lost no time in hacking these rails to pieces and in destroying also "the gates, highewaies, and the passage of and for the . . . Engin".[3] Henceforth, as quickly as one side laid down rails, they were torn up by the other. In November, 1607, the freeholders are said to have carried out another fierce raid on Calcott colliery. They stoned Clifford's labourers above ground, took away the ropes "of the wyndles . . . which did drawe upp the . . . workmen", kept them imprisoned in the mine for four days without food, and later beat two of the colliers so that they could do no work.[4] We do not know how the issue was finally settled, but we suspect that the power of the freeholders in this case, not only persistently to oppose the lord by force, but to sustain many pleas against him in various courts, may have been due in large measure to the fact that among their number were several "gentlemen" and "esquires".[5]

(ii) *Rights of Leaseholders and Customary Tenants*

Freeholds, after all, made up only a small portion of all manorial tenures in the sixteenth century, and they were tending to split off into independent units, bound only by a formal subservience to the authority of the manor. Far the greater number of holdings belonged neither to freeholders, nor to leaseholders, but to customary tenants, among whom copyholders predominated.[6]

It may be taken for granted that the leaseholder seldom had a title to minerals, unless they were specifically included in his lease. Nevertheless, leaseholders sometimes disputed the right of the lord to work coal in their holdings. In 1685 the lord of the manor of Atherton, near Wigan, was obliged to bring an action against several of his tenants, who had resisted his "getting, bankeing and takeing of coales", on the ground that coal had not been excepted from their original deeds, nor had the lord reserved any "freedom of passage, egress or regress". The lord was unable to produce the deeds, but he

---

[1] *Ibid.*, 109/8 (Clifford *v.* Wilcox and others).

[2] Cf. above, p. 244 ; below, p. 384.

[3] *Star Chamb. Proc., James I*, 310/16.

[4] *Ibid.*, 109/9 (Clifford *v.* Wilcox) ; see also *ibid.*, 294/25 (Wm. Wells, gent. *v.* Clifford and others).

[5] Struggles between the freeholders and their lords over mineral rights were common all through the coalfields. In 1585 Sir John Byron was engaged in pressing his title as lessee of the mines in the large manor of Rochdale (H. Fishwick, *The History of Rochdale*, 1889, pp. 24–5), and there the freeholders seem eventually to have made good their claim to the right to work coal in their own lands (*ibid.*, pp. 28–30 ; *Palat. of Lancs. Bills*, 41/144).

[6] R. H. Tawney, *The Agrarian Problem in the Sixteenth Century*, pp. 25 (Table I), 40–1, 47–9, 287–8.

said that he was informed by ancient persons that all mines in the lands were excepted from the leases.[1] While it appears that lords had some difficulty in asserting their right to mines in the case of lease-holders who held their land under long-term leases, unless the mines were specifically reserved, the more wary soon learned to except coal in making new grants of any land that might contain minerals.[2]

In the case of copyhold tenures, the rights concerning coal differed greatly, both according to the terms of the tenure and the customs of particular manors. During the late sixteenth and the seventeenth centuries, these rights of copyholders were subjected to judicial interpretation in innumerable cases. Unlike the free-holders, customary tenants never appear to have claimed for them-selves full ownership of minerals underneath their ground.[3] Their pretensions were twofold. On the one hand, they claimed a right, often based on an ancient manorial custom allowing them to take fuel for their own use,[4] to dig surface coal in their own holdings in return for certain obligations to the lord. On the other hand, they claimed power to refuse access to minerals beneath their soil, asserting that such access damaged the surface, in which they believed their title to be absolute. As the day of small unobtrusive diggings was giving way to that of large colliery enterprises, whose operations reduced the amount of land available for farming, by taking up space in sinking pits, erecting machinery, and carrying away coal, the tenants might claim that the lord, by permitting mining, was breaking the terms of his original agreement with them.

For his part, the lord hoped to find in his coal deposits a means of increasing his income, which had suffered from the fall in the real value of his tenants' rents, occasioned by the rise in general prices. Not only must the tenants be prevented from digging themselves, they must be stripped of their power to refuse access to minerals under their holdings, or to demand excessive compensation. Seldom did a copyholder succeed in retaining, after the sixteenth or seventeenth century, the privilege of digging in his holding. In place of his ancient claim to get his own fire coal, he was usually prevailed upon to accept the lord's promise that he might have fuel from the lord's mine, either

---

[1] *Palat. of Lancs. Bills,* 38/27. Cf. *ibid.,* 39/53.

[2] Reservations of the coal in leases of farms in Little Hulton, Lancs., began in 1575 (*V.C.H. Lancashire,* vol. ii, p. 357 ; Crofton, *Lancashire and Cheshire Coal Mining Records,* pp. 42–7). For reservations of the coal in Black Callerton, Hartley, and Horton Grange, Northumberland, by the Delaval family in 1612, see *Waterford MSS.,* no. 16 J : " A Booke of severall things touching the Estate of Seaton Delaval ". See also *ibid.,* no. 146, bdl. 2 ; *Additional MSS.,* 6687, f. 183 ; Bainbridge, *op. cit.,* p. 219, for seventeenth-century reservations of coal in Northumberland and Derbyshire. Cf. Lister, *Coal Mining in Halifax,* p. 274.

[3] The copyholder and lord may, of course, by mutual agreement, " enfranchise " the land, i.e. convert it from copyhold to freehold, thus vesting ownership of the minerals in the new freeholder (Bainbridge, *Law of Mines,* 1878 ed., p. 64).

[4] If custom had permitted the tenant to gather fallen branches, chop wood, or take turf and peat for his fire, he had generally been able in early times to extend this customary right to cover the digging of coal (cf. *Duchy of Lancs. Pleadings,* 77/B/6 ; 109/A/9 ; 189/L/7 ; *V.C.H. Lancashire,* vol. iv, p. 108). Cf. above, p. 270.

free or at a price below that charged to other customers.[1] This often seemed an equitable arrangement, because it relieved the tenant from labour in a pit, but, in reality, it meant that he relinquished in perpetuity his claim to a share in the coal under the manor ; and eventually the lord was able to plead some new capital expense, or the competition from a neighbouring colliery, as an excuse for raising the price paid by his tenants, until they were placed on the same footing as other buyers. As late as 1599, one Henry Barcroft clung to the literal interpretation that copyhold tenure on Colne manor in Lancashire carried with it " liberty of fyrebote of cole mynes ". An action was brought against him by Lawrence Lister, farmer of the Queen's mines in the manor.[2] We are unfortunately left without the decision of the court. But we know that the trend of legal opinion throughout Lancashire and Yorkshire was against permitting a copyholder to break the soil without special licence from the lord.[3] Although, in 1572, John Nutter was able to show a permit from the steward, " according to . . . ancient custome and usage " of Rothwell manor in the Honor of Pomfret, authorizing him to get coal in his copyhold, the Court of the Duchy Chamber gave judgment against him, holding that a grant by the steward did not bind the Queen as overlord.[4] Within twenty-five manors in Durham and Northumberland, in which the Crown had important rights as lord, surveyors in 1611 found that all coal mines under customary lands held of the King, as well as under leaseholds, could be worked by the tenants only if they obtained a specific grant.[5]

In a few, and probably exceptional, cases the customary tenants were better treated. Several copyholders in Halesowen manor in Worcestershire in 1607 were known to have won coal without licence of the lord, though it was perhaps illegal to do so there,[6] and in the lordship of Bromfield and Yale in Denbighshire the tenants said that they " doe usually digg and gett in there owne severall fields

---

[1] *Exch. Deps. by Com.*, 32 Charles II, Mich. 30 ; *Waterford MSS.*, no. 16 J, as cited above ; Francis, *Charters Granted to Swansea*, p. 7. These are only a few instances of the special privileges granted to manorial tenants in buying fire coal. Cf. above, p. 105. An early instance of such an agreement is that of Selston, in Nottinghamshire, where during the reign of Edward I the tenants gave up " all mines and diggings " of coal and iron, and in return were to receive annually thirty cartloads of coal (*V.C.H. Nottinghamshire*, vol. ii, p. 324).

[2] *Duchy of Lancs. Pleadings*, 189/L/7 ; *ibid., Decrees and Orders*, vol. xxii, ff. 516, 526, 573, 762, 804, 883.

[3] To this day the tenant has a right to work open unoccupied mines (Redmayne and Stone, *Ownership and Valuation of Mineral Property*, p. 40).

[4] *Duchy of Lancs. Pleadings*, 87/M/7 ; *ibid., Decrees and Orders*, vol. xv, f. 182. For the trend of opinion in Lancashire and Yorkshire see also *Duchy of Lancs. Pleadings*, 87/M/4 ; 109/A/5 ; 119/A/11 ; 146/F/20, 185/T/12 ; *Palat. of Lancs. Bills*, 36/159.

[5] *Exch. Spec. Com.*, no. 5037. The same principle applied within the Bishop's manor of Whickham, where we find that Robert Fawdon, though seized of a tenement in Greenlaw, was obliged to take a lease in order to work the coal in his holding (*Palat. of Durham, Bills and Answers*, bdl. 9 : Hardinge v. Barlaw, 1610).

[6] *Exch. Deps. by Com.*, 5 James I, Hilary 17, as cited *V.C.H. Shropshire*, pp. 459–60.

. . . stones, leade, Coales, or any other mettall ".[1] A number of copyholders—many holding only for life—within Westerleigh manor, near Bristol, claimed as good a right as the freeholders, and dug and sold coal in defiance of their lord's concessionnaire, Arthur Player, protesting that never should he strip them of their privileges in his effort to monopolize the local trade.[2] The greatest freedom of all prevailed within the manor of Eckington, in the northern extremity of Derbyshire, not far from Sheffield, where it was said that " aney coppiehoulder . . . may sinke pittes for gettinge of Coales in aney of theire coppiehould landes, for theire owne uses and expenses and may also make sale of theire coales which they gett . . . soe that it be noe hindraunce to the lord's sale ". It was left to a jury of the manor to decide when there was a hindrance, in which case the offending tenant might " forfeite his . . . land for gettinge and sellinge of coales " for a period of not more than one year.[3] In Durham the copyholders in the Bishop's manors claimed the right to mine coal during and after the seventeenth century, but it is exceedingly unlikely that they were able to make good this claim without a licence from the Bishop,[4] although, even as late as the end of the nineteenth century, they retained a right in one small manor to work the coal of the upper seam, and maintained a sort of village pit—apparently the only case in the county in which a copyholder worked minerals.[5]

What the customary tenant most often dreaded was not that he would be forbidden to dig himself, but that the lord or a concessionnaire would carry on mining operations in his holding, thereby spoiling the crops or interfering with the pasture. No sum could console the thrifty husbandman for the desecration of his soil. Against this he put up as stubborn a resistance as he could command, and many were the obscure battles fought with pitchfork against pick and shovel to prevent what all tenants united in branding as a mighty abuse. Modern treatises on mineral legislation tell us that coal under copyholds can be worked only with the copyholder's consent.[6] Their authors might have difficulty in persuading some seventeenth-century tenants, were they resurrected, that they could have taken refuge under such a rule. As it was, the copyholders lived in constant fear of the discovery of coal under their land. A searcher for coal in a manor belonging to the Earl of Hertford in Wiltshire, in 1655,

---

[1] *Exch. Deps. by Com.*, 16 Charles I, Mich. 22.

[2] *Chanc. Proc.*, Series II, 371/4.

[3] *Exch. Deps. by Com.*, 11 James I, Trinity 6. For this reference I am indebted to Mr. W. Romaine Newbold, of Philadelphia, who first suggested to me that the Exchequer Depositions might contain much valuable information concerning the coal mines.

[4] The claim itself was apparently somewhat ambiguous, for while the copyholders, in stating their position, say in one place that they can " not only use the Lime-Stone, and other Quarries within [their] . . . Grounds, but work the other Mines also ", they admit that this can be done only " where there is [a] . . . Custom or Usage to warrant " it, and they suggest in another place that it can be done only " upon making some reasonable Acknowledgment to the Bishop, [and] upon obtaining his Licence to work the same " (*The Case of the Copyhold and Leasehold Tenants of the Bishoprick of Durham*, n.d.).

[5] *Royal Commission on Mining Royalties, 4th Report*, 1893, Qu. 19,563.

[6] Bainbridge, *op. cit.*, p. 37 ; Redmayne and Stone, *op. cit.*, pp. 57–8.

was at first assisted by the tenants ; but they later refused to help him, " fearing that he may dig in their grounds ".[1] There must have been many tenants throughout the country like one Lydd, who, when he heard that a workman digging a " sawepitt " had found a seam under his copyhold in Halesowen, ordered it to be filled up again, saying that, if Sir John Lyttelton, lord of the manor, knew thereof, he would mine for the " said coles " ; [2] or like John Hartley, Stephen Hargraves, and Edmund Robinson, who offered to pay for leases of the minerals beneath their copyholds in Great Marsden in Lancashire, " which offers they made to stop spoiling their grounds [and] rather than strangers should get leases to search for coal, than for any benefit they were likely to get ".[3]

When the lord himself worked the coal, the copyholder stood a poor chance.[4] Dorothy Selkane, one of a group of adventurers holding the lease of coal mines within a Crown manor, petitioned Salisbury in 1609 for royal support against the copyholders, insisting in the course of her letter that " In this Country [? Cumberland] the Lords of other Mannors where Coles are have and doe digge notwithstandinge any Coppyholders or leassees estate ".[5] As against the concessionnaire the tenant was scarcely in a better position, unless he had the support of his lord. A case in point is that of the copyholder Thomas Hobbes, who charged Arthur Player, at the end of the reign of James I, with trespass for digging in his meadow in Westerleigh, though Player had already agreed to pay considerable compensation. When Player sent a " coleminer ", William Symonds, with two of his fellows, to sink a pit in the meadow, they were prevented by Hobbes' tenant, " who threw in the earth as fast as the colliers dug it out ".[6] A copyholder's right to claim trespass in such cases was upheld in court,[7] but it is worth noting that Thomas Roberts, lord of Westerleigh, was himself engaged in suits against Player, his concessionnaire, and that, according to the testimony of the same Symonds, Roberts had combined with Hobbes to prevent Player from getting at the rich seam under the meadow, which they planned to work for their own profit. Player was locally unpopular because of his attempts to monopolize the trade, attempts that were supposed to have raised the price of coal in Bristol, and it is not certain that the customs on manors in this region were generally so favourable to the tenants as at Westerleigh.

When the tenant was not supported by the lord in his struggle against the concessionnaire, the latter was almost certain to prevail. On the manor of Stratton, in the Mendip coalfield, a witness deposed that " if any tenant or occupier of any enclosed ground . . . did

---

[1] H.M.C., *Report on MSS. in Various Collections*, vol. i, p. 131.

[2] *Exch. Deps. by Com.*, 5 James I, Hil. 17, as cited *V.C.H. Shropshire*, p. 459.

[3] *Duchy of Lancs. Spec. Com.*, no. 648.

[4] See, for instance, the dispute between Thomas Pacy, lord of Babington manor, and two of his copyholders, in which, we are told, " the said Mr. Pacy had the best of it " (*V.C.H. Somerset*, vol. ii, pp. 381–2).

[5] *S.P.D., James I*, vol. xlviii, nos. 119, 119 I : Dorothy Selkane to Salisbury, October 22nd, 1609.

[6] *Chanc. Deps.*, P/18/H.

[7] Bainbridge, *op. cit.*, p. 39 (Player *v.* Roberts).

refuse to permit and suffer any grantee or lessee of the colemynes to work the same . . . it hath been taken and reputed to be the custome . . . that such grantee or lessee might notwithstanding beginn and proceed in the said work ", subject to the payment of reasonable damages.[1] Throughout Somerset, adventurers seem always in the end to have gained access to the minerals under copyholds, although in one instance, in Stratton, a group of mining partners were obliged to wait for the death of a tenant named Nicholas Everett, after which one of their number, by marrying his daughter, came to an agreement with the widow.[2] In the manor of Houghton, in the West Riding of Yorkshire, Charles Jackson, farmer of Crown mines, claimed in 1577 a right " to digg . . . coles as well in Errable grounds of the copiholders . . and in their crofts and on the backsides of ther dwellinge houses, . . . whether . . . copiholders of inheritance or of any other estate, without yielding any Recompence . . . for the same ".[3] Commissioners decided, however, that Jackson's pits must be " dug in convenient places, . . . filled up afterwards, and [the] rubbish taken away, the tenant to be recompensed for forebearing his land which is good for corn ".[4] Similar decisions were given by the Court of the Duchy Chamber of Lancaster concerning the rights of copyholders in Leeds and Rothwell manors.[5] In Lancashire, too, if we may judge by the action of a jury of Ightenhill manor, Crown farmers were permitted to dig in any copyhold or leasehold, but had to pay for damage done to the surface.[6] Such was also the rule in Durham and Northumberland on manors owned by the Crown [7] or the Bishop, although in the Bishop's manors the copyholders sought to maintain a right to refuse the use of their lands for mining operations, except " upon such Terms and Agreements " as they thought fit.[8] Generally

---

[1] *V.C.H. Somerset*, vol. ii, p. 381. See *Chanc. Proc., James I*, B. 20/72, for customs on the neighbouring manors of Benter [? Binegar] and Midsomer Norton.

[2] *V.C.H. Somerset*, vol. ii, p. 381.

[3] *Exch. Deps. by Com.*, 21 Eliz., Hil. 8.

[4] *Exch. Spec. Com.*, no. 2621. According to one witness, William Prince, a copyholder in the neighbouring manor of Kippax, the claim of the Crown farmer there to refuse to pay any damages to the copyholders was actually sustained.

[5] *Duchy of Lancs. Pleadings*, 144/A/9, 149/A/35, 152/A/6, and *Decrees and Orders*, vol. xx, ff. 196, 427 (Chris. Anderson, Esq., *v.* Thos. Casson, Alice Calbeck and others, 1587–90) ; *Duchy of Lancs. Pleadings*, 193/J/1, and *Decrees and Orders*, vol. xxii, f. 785 (Henry Johnston, copyholder, *v.* Richard Bland, gent., Queen's lessee of coal mines, 1599–1600).

[6] *Star Chamb. Proc., James I*, 310/33 (Nich. Waddesworth, yeoman, *v.* Henry Jackson and others, November, 1621).

[7] In this connection see the petition of Thomas Castelin, in 1576, soliciting license to search for and work coal mines in Crown lands north of the Trent, and agreeing to compound with copyholders and leaseholders (*S.P.D., Eliz.*, vol. cx, no. 18).

[8] *The Case of the Copyhold and Leasehold Tenants of the Bishoprick of Durham*, n.d. To-day the right of the Bishop (in the person of the Ecclesiastical Commissioners), as overlord, to work coal in the copyholds is not questioned, but in 1888 his right to let down the surface was successfully challenged by a copyholder, since which time it has been necessary for his lessees to pay underground wayleaves to the surface holders (*Royal Commission on Mining Royalties, 4th Report*, Qs. 19,564–74).

speaking, while the right of customary tenants to compensation was upheld—and a loophole thus provided through which they might harry the unwelcome adventurer over the exact sum he must pay—they were in practice rarely able to refuse access to their grounds when faced by a powerful mining concessionnaire, and the compensation they received was often totally inadequate.[1]

(iii) *Rights to Minerals under the Manorial Waste*

We have still to consider one class of land, the waste, often an extremely important part of the manor, both for the surface and for the minerals underneath. Except in a few instances, such as that of the territory in Nottinghamshire and Derbyshire known as the Honor of Peverell, where the " soil of all waste grounds . . . and the benefit of all mines therein belong to the Crown unless the Lord of the manor can show a grant under the great seal ",[2] the wastes were the lord's, subject to customary rights on the part of the tenant, which varied from manor to manor. Two privileges in particular encumbered the lord who wished to benefit by the coal under his wastes. The first was the right of pasture. From early times tenants had insisted that the digging, and even the carriage, of coal was a menace to their grazing, because it destroyed the grass and endangered the lives of their cattle, who frequently stumbled into unfenced workings.[3] The second was the right to fuel. Freeholders and copyholders had been allowed to take wood, turf, and peat, not only from their own lands, but from the commons, and in some districts the decline of other kinds of firing had caused an extension of this privilege to cover the getting of coal. Thus, as we learn from a dispute of 1527, all tenants of the Crown manor of Ightenhill, in Lancashire, and all inhabitants of the town of Burnley had " enjoyed time out of mind sufficient cole for their fewell to be had from the coal pits of [Burnley] waste ", and over a hundred of them had just taken advantage of this custom to dig their winter supply.[4] By the " customary " of Bolsover manor

---

[1] *Cal. Treas. Bks.*, vol. iv, p. 808 ; *V.C.H. Somerset*, vol. ii, p. 381. Cf. below, p. 333.

[2] *Parl. Surveys, Notts.*, no. 21. During the reign of Charles I, the Gorings, who farmed the Crown mines in the Honor, had some difficulty in enforcing their right to minerals in the wastes, as is shown by the actions which they brought against Sir Percival Willoughby, owner of Trowell manor, who resisted their digging in Trowell moor (*Exch. Decrees*, Series IV, vol. ii, f. 234d ; vol. iii, f. 141d), and against the " Mayor, Commonalty and citizens of London ", owners of Tibshelf manor (in right of the " late Hospital of Savoy "), who claimed the coal under Tibshelf moor (*Exch. Deps. by Com.*, 13 Charles I, East. 14). The King's coal mines in the Honor were leased again as a unit in 1663 (*Cal. S.P.D.*, 1663-4, p. 160). Within the Crown chase of Kingswood, in Gloucestershire, the lords of the surrounding manors had, by successive encroachments, almost entirely deprived the King of his mineral rights (*S.P.D., James I*, vol. liv, no. 8 ; *Chanc. Proc., James I*, C. 13/25 ; *Parl. Surveys, Gloucester*, no. 12 ; *Exch. Deps. by Com.*, 27 Charles II, Mich. 29 ; *Exch. Decrees*, Series IV, vol. i, ff. 190, 223, 229, 231).

[3] *V.C.H. Yorkshire*, vol. ii, p. 336.

[4] *Duchy of Lancs. Deps.* 19/T/3 (Richard Townley, gent., farmer of the King's " Coal Beds ", *v.* Hugh Habergham, gent., and others).

it was " lawful for sokemen to dig for sea coal . . . to their proper uses, without the view and delivery of the bailiff of foresters ".[1]

This privilege, granted at a time when the commercial value of coal was slight, came to be regarded by the lord at the end of the sixteenth century as a means whereby tenants could take his minerals, and thereby " disinherit " him. At the same time the right of pasture was apt to interfere with his power to lease the coal, for though in many—probably in the great majority of—wastes the lord's ownership of and right to work minerals were recognized,[2] in others, like the wastes of Rotherham and Greasbrough in south Yorkshire,[3] the freeholders and customary tenants claimed that coal could be mined only with their consent. Hence the lord was eager to rid himself of both customary privileges. Here it was that a clash occurred. On the one hand the lord fought his tenants' pretensions to dig coal in the waste,[4] on the other the tenants challenged the mining concessionnaires who held leases from the lord, and sometimes forced them to guard the shafts night and day, to prevent indignant tenants from carrying off the coal or doing violence to the workmen and the pits.[5]

The lord often sought to enforce what he regarded as his rights, by enclosing parts of the waste where he supposed minerals could be obtained. Some of the famous enclosures of the early seventeenth century were certainly made to prevent tenants from interfering with the working or leasing of coal. Such enclosures aroused a bitter resistance. In 1605 the freeholders and customary tenants of the Earl of Derby's manor of Kirkby Malzeard, in the North Riding, assembled at the pits of his farmer in Thorp moor, common land which the earl had tried to enclose, carried away in wains much coal from the bank, sprinkled the remainder about in the bushes and hollow places, and filled up the pits, which they thought a menace to their cattle.[6] Richard Bold, Esq., who worked coal in his waste in Sutton manor near St. Helens, had his enclosures pulled down sixteen

---

[1] *V.C.H. Derbyshire*, vol. ii, p. 352. Sometimes, as within Wakefield manor (Yorks.) and Whickham manor (Durham), freeholders enjoyed by custom the right of wayleave through the wastes for the carriage of coal dug in their own land (*Duchy of Lancs. Pleadings*, 162/A/35 ; *Palat. of Durham, Bills and Answers*, bdl. 2 : Thos. Surtees, gent., and others *v.* Jno. Hedworth).

[2] *Duchy of Lancs. Pleadings*, 108/W/12, 189/L/7 ; *Duchy of Lancs. Spec. Com.*, no. 418 ; *Parl. Surveys, Derby*, no. 10 ; *Exch. Deps. by Com.*, 14 Charles II, East. 26 ; Garnier, *History of the Landed Interest*, vol. ii, pp. 31–2.

[3] *Exch. Spec. Com.*, no. 2759. When the coal mines of Alnwick lordship were leased in 1567, the " free common moore " was excepted (*A History of Northumberland*, vol. ii, p. 455).

[4] *Duchy of Lancs. Pleadings*, 102/T/13, 208/A/27 ; Crofton, *Lancs. and Cheshire Coal Mining Records*, p. 40.

[5] *Duchy of Lancs. Pleadings*, 96/T/15 (Francis Tunstall, esq., *v.* Myles Hudleston, gent., and others). Tenants hindered the King's farmers from mining on a common in the bailiwick of Estchisham in Denbighshire (*Court of Augm. Proc.*, 6/36), on the wastes of Haslingden, Rosendale, Tottington, and other manors in Lancashire (*Duchy of Lancs. Pleadings*, 32/K/2), and on the waste of Bradford manor in Yorkshire (*ibid.*, 156/A/10).

[6] *Star Chamb. Proc.*, James I, 4/3.

times by the freeholders who claimed " common of pasture ".[1] The inhabitants of Duffield Forest took advantage of the Civil War to throw open the commons, which the King had recently enclosed under an order of the Duchy of Lancaster, awarding him one-third of this Crown waste, 3,468 acres in total area, he having chosen " all the places where the coal delfes now are sunk ".[2] In spite of the tenants' resistance, much valuable mineral property was successfully enclosed.

Sometimes an agreement was reached. The lord had little difficulty in overruling the alleged right of his tenants to dig for fuel free of charge. Already in 1527 Sir Thomas More, as Chancellor of the Duchy of Lancaster, ordered the tenants who had taken coal from Burnley waste [3] to pay 3d. for every second fother to the farmer of the King's mines. By the well-known " Bolsover decree " of 1578, tenants of this Derbyshire manor could have coal from the lord's pit at 15d. the wainload (roughly one ton), but could dig themselves only if the lord failed to provide them with sufficient fuel for their own use.[4] Where the tenants did not claim a right to dig, it was frequently possible to reconcile their other privileges with the lord's desire to lease his coal, without recourse either to lawsuits or violence. For instance, the freeholders and commoners of the manor of Shirlett in Shropshire confirmed, by a written agreement,[5] the lord's full right to mine in the wastes, in return for a promise that he would protect their cattle and construct roads and gateways for their convenience. On Westerleigh manor in Gloucestershire, by assigning to the copyholders and free-holders a part of the common " in which . . . there was to be no spoil of the herbage by making or working any coal pits ", the lord was permitted to enclose the remainder.[6] Similarly on Ewloe manor in Flintshire, where the freeholders had interfered with the operations of Sir John North, the King's farmer of coal mines, an agreement was signed in 1626 by which the Crown gave up 480 acres of demesne lands to the freeholders, who in return renounced all their rights in the commons.[7] A somewhat similar arrangement appears to have been made at about the same time between the freeholders and Sir John Byron, lord of the manor of Butterworth, in the parish of Roch-dale. Byron sold the freeholders a portion of the manor, and it was understood that they should share in the profit made from coal mines within this part, according to the value of their freehold lands.[8] At the end of the seventeenth century there remained few wastes in which the lord's power to mine, or to lease his coal, was seriously restricted.

---

[1] *Duchy of Lancs. Decrees and Orders*, vol. xxviii, f. 310b; cited Crofton, *op. cit.*, pp. 57-8. The court sanctioned the enclosures pending further trial, but ordered Bold to cause " the pits and holes to be carefully looked to and filled up after the work is ended to prevent any danger to the plaintiffs' cattle ".

[2] *Parl. Surveys, Derby*, no. 16. See also *ibid.*, no. 3, and *Exch. Deps. by Com.*, 1659, East. 27.

[3] See above, p. 315.

[4] Bainbridge, *op. cit.*, p. 56.

[5] Printed *V.C.H. Shropshire*, p. 459.

[6] *Chanc. Proc.*, Series II, 371/4.

[7] *Exch. Decrees*, Series IV, vol. iii, f. 131d.

[8] *Palat. of Lancs. Bills*, 41/144.

As a result largely of judicial decisions made during the sixteenth and seventeenth centuries, the right to work and to lease minerals had come to be concentrated in the lords of manors. No tenants except freeholders retained a claim to the ownership of minerals, and not all the freeholders found it possible to make their claim effective. The enclosure movement steadily diminished the number of separate holdings, and curtailed the rights of tenants in the wastes and commons, while the system of primogeniture tended to keep great landed estates intact. All these factors combined to reduce mineral ownership in Great Britain into the hands of a relatively small number of persons. Towards the end of the eighteenth century, Jars, the famous French engineer and metallurgist, remarked that in the Newcastle district practically all royalty rights rested in a few rich men.[1] They were then spoken of as royalty owners, the word royalty being applied, apparently for the first time, to the payments which landowners received for the working of minerals within their own lands and the holdings of their tenants. In Great Britain, the meaning of the word has undergone, in fact, a curious inversion. An attribute of sovereignty in feudal times, when sovereignty was decentralized, the *regale* has been absorbed, not by the sovereign State, as in France and most other continental countries, but by the landowner. Thus a word originally applied to the rights of the sovereign as against the subject, is now applied to the rights of the subject as against the sovereign.

Whatever may be thought of the ethics of despoiling small tenants, it is clear that the concentration of mineral rights in the hands of the lords of manors was a principal factor in making tolerable the British system of privately-owned royalties. In France the peasants generally appear to have been much more successful in maintaining rights to dig for coal in their own holdings and to deny the lords and their concessionnaires access to minerals beneath these holdings, until the decree of 1744 was made effective.[2] The prevalence of small, inefficiently-worked collieries which resulted from these rights, was, as we have indicated, the principal cause for that decree, by which the French State resumed its title to minerals. The limitations placed in England upon the power of manorial tenants to work coal and to share in the returns when the lords or their concessionnaires worked it, and the decline of peasant proprietorship, do much to explain why the private ownership of minerals became a principle of English law.

---

[1] *Voyages métallurgiques*, vol. i, p. 181.
[2] Cf. Rouff *Les mines de charbon en France*, esp. chs. iv and v.

CHAPTER IV

THE LANDOWNER'S REVENUE FROM COAL MINES

(i) Royalties

WHAT share of the selling price of coal did those fortunate landowners, who were able to establish a clear title to the ownership of the subsoil, receive by virtue of it ? During the sixteenth and seventeenth centuries, it was usual to call the whole return which a landlord received from leasing his coal a " rent ". This " rent " commonly included not simply the element which we now term a royalty, but also payments for wayleave to carry fuel from the pithead, for using the lord's ground to build colliers' cottages, for taking timber from the wastes ; in short, for all the damage to property involved in carrying on a colliery enterprise.[1] In addition, the " rent " sometimes covered less passive services rendered by the landlord. There was no sharp line separating a contractor, who worked coal for the landlord, from a lessee.[2] Although the " Articles of Agreement " entered into by John Farrer, Esq., and Abraham Shaw, yeoman, in 1663, might possibly be called a lease, because Shaw was to bear all preliminary capital expenditures, he was " to have allowance of the fourth part of the said charges, except of boring with wimble for searching, out of the half of the profit accrueing to . . . Farrer forth of the premises, which one half . . . profit is to be paid to . . . Farrer by . . . Shaw, so soon as there shal bee any sale or profit made of any coles ".[3] As Farrer shared to some extent in the profit and risk, he cannot be regarded as a simple lessor, nor can the entire return that he received be called a royalty. Wherever the landlord took any share—as frequently he did—in directing, encouraging, or financing the work, a part of his " rent " must be considered—to use the terminology of the economists—as interest, profit, or wages of management.[4]

---

[1] The word " royalty ", in the sense of a payment made by the lessee of the coal to the landlord, became common in the colliery districts in the eighteenth century (cf. Jars, *Voyages métallurgiques*, vol. i, p. 181), but until after Mill's time economists used the term " rent " interchangeably for the landlord's return from his leases of land and for that from his leases of minerals. (Cf. below, p. 328.)

[2] Cf. below, pp. 423–4.

[3] Lister, " Coal Mining in Halifax ", in Wm. Wheeler, *Old Yorkshire*, 2nd Series, 1885, pp. 274–5.

[4] Again, in cases where coal was leased together with land (*Duchy of Lancs. Pleadings*, 90/B/41 ; *Cal. S.P.D.*, 1591–4, p. 530 ; Crofton, *Lancs. and Cheshire Coal Mining Records*, p. 57), with lead mines (*Augm. Off., Misc. Bks.*, vol. ccxxx, f. 67), with corn mills (*Duchy of Lancs. Pleadings*, 51/B/10), or with salt pans (*Chanc. Proc., James I*, D. 9/10 ; *Cal. S.P.D.*, 1635–6, p. 305), the share of the total payment exacted in respect of coal is never stated. Such cases are rare, however.

With these qualifications in mind, let us consider the return received by the landlord, as it appears in coal leases of the sixteenth and seventeenth centuries. An exact analysis of the different elements composing it is impossible, but it is probable that the larger share of the " rent " charged consisted of what we now call a royalty.[1]

Royalties before the sixteenth century, to judge from our examination of the Bishop's leases in Durham, tended to be remarkably high, in some cases perhaps as much as a third of the selling price at the pithead.[2] Early leases of coal rarely followed the method commonly adopted at mines of metallic ore, of assessing the royalty as a fixed percentage of the profits, or, more frequently, of the output. It was usual in letting coal mines to ask for a certain annual rent, and then to limit the number of pits that could be worked, and the number of hewers (sometimes called " picks ") to be employed in each pit.[3] Thus a maximum limit was set upon output, while the mineral owner was generally assured of a definite return, whether the enterprise yielded a profit or not. Royalties were only fixed in this way at the small number of mines where coal was actually dug for sale. When, in the sixteenth century, the new industry became important throughout the country, it was not this ancient type of lease which was adopted by the greater number of landlords. For the most part, they proceeded to lease their coal, as they leased their land, for a fine and a fixed annual " rent ",[4] based on a very superficial investigation of the possible value of the minerals to the concessionnaire. These landlords acted as if coal were in the same class as the products of the soil. Nor was this remarkable when some scientific authorities were not yet sure whether or not coal, like trees and surface vegetation, could reproduce itself.[5]

We have already referred to the part played by the Crown in softening the terms of leases.[6] Even on manors where coal had been sold for a profit before the Reformation, ancient clauses limiting output tended to disappear. When the coal of the manor of Denton

---

[1] In addition to the money " rent ", the lessee usually bound himself to make free deliveries of coal to the lord, and often to supply some of the manorial tenants with fuel at a price which might not cover the expenses of production. Rents paid purely in kind were never common in English coal mining; but, before the sixteenth century, some Scottish owners of " heughs " paid their dues to the Crown in coal (*Exch. Rolls of Scotland*, vol. vii, pp. 319–20, 403, 405, 535, 628; vol. viii, pp. 61–2, 332, 402, 511, 602; vol. ix, pp. 15, 104, 172, 242, 398, 465; vol. x, pp. 32, 89, 175, 276, 330, 408, 493, 560, 754).

[2] See above, p. 139.

[3] In leases of mines in Belper Ward, Derbyshire, in 1314–15, the " rent " was determined, " as so often in similar circumstances in the northern counties ", by the number of picks (*Duchy of Lancs. Minis. Accts.*, 1/3, as cited *V.C.H. Derbyshire*, vol. ii, p. 350). See also *H.M.C.*, *Report on the MSS. of Lord Middleton*, p. 100.

[4] Occasionally leases were drawn so that the rent payable increased during the term of the lease. The lessees of a mine at Hawkesbury in Warwickshire paid only a " pepper corn " a year until they had extracted 100 tons of coal; thereafter they paid £40 annually for the next five years, and £60 for the remainder of the term of their 21 year lease (*Exch. Deps. by Com.*, 36 Charles II, Mich. 43).

[5] See above, p. 270.

[6] See above, pp. 144–8.

on the Tyne was rented in 1569, the lessee was still limited to an output of 20 chaldrons (of 8 bowls) per working day ; but when, in 1601, the same mines were again leased—and to the same family, the Erringtons—the only restriction was that an extra " rent " should be paid for every additional pit worked.[1] Even that restriction was often abandoned. A lease made in 1574 of mines in Gateshead stipulated that the rent of £1 13s. 4d. should cover only the two pits then producing coal, and that £3 6s. 8d. should be paid for every additional pit worked ; but in 1597 the Crown accepted an annual " rent " of £16 plus £10 fine, in return for the right to mine, without any limit on the output.[2] While there remained some lands [3] where the royalties were still levied according to the number of pits worked, the common lease during the period from 1580 to 1640 stipulated for a fixed payment, regardless of the quantity of minerals extracted.[4] Except that he had to fill up old shafts, leave barriers of unworked coal underground to prevent him from mining outside his concession, and hand over active pits in a condition to be worked by his successor,[5] the lessee was generally free to mine as he chose.[6] In addition to the grants of coal, his lease almost always included the right to erect machinery upon, and to carry fuel over, the landlord's holding ; in fact all privileges necessary for operating a mine, in so far as it

---

[1] *Augm. Partics. for Leases*, 109/61, 54.

[2] *Ibid.*, 35/61 ; 37/85.

[3] Such as Chopwell manor, Greenlaw freehold in Whickham, Blackborne and High Lambton in Durham, Stratton in Somerset (*ibid.*, 38/43 ; *Cal. S.P.D.*, 1603–10, p. 384 ; *S.P.D.*, *Charles I*, vol. ɒii, no. 77 ; *Exch. Spec. Com.*, no. 3758 ; *Palat. of Durham, Bills and Answers*, bdl. 9 (Harding v. Barlawe, April 11th, 1610) ; Welford, *Records of the Committees for Compounding*, p. 165 ; *V.C.H. Somerset*, vol. ii, pp. 371, 380).

[4] See Appendix L for a sample. A full list of references to coal leases which apparently did not contain any clause limiting output would take up too much space. The following should be considered as only a few examples : *Duchy of Lancs. Pleadings*, 90/B/41 ; 108/W/12 ; 109/A/5 ; 119/A/11 ; 144/B/11, 189/L/7 ; *Duchy of Lancs. Spec. Coms.*, nos. 418, 433, 648 ; *Augm. Partics. for Leases*, 25/25 ; 29/78 ; 121/18 ; 149/18, 34 ; 159/12 ; 171/46 ; 178/1 ; 217/1, 14 ; 223/14 ; 225/3, 98 ; *Parl. Surveys, Derby*, nos. 3, 10, 18 ; *Durham*, no. 6 ; *Flint*, no. 3 ; *Lancs.*, nos. 8, 19 ; *Northumb.*, no. 5 ; *Court. of Augm. Proc.*, 23/6 ; *Palat. of Durham, Bills and Answers*, bdl. 5 (Pinshon v. Todd) ; *ibid.*, bdl. 24 (Stevenson v. Liddell and others) ; *Palat. of Durham, Decrees and Orders*, vol. i, ff. 19, 226 ; Green, "The Chronicles and Records of the Northern Coal Trade", in *Trans. North of Eng. Inst. of Mining Eng.*, vol. xv, 1865–6, pp. 196–7 ; *S.P.D.*, *Charles I*, vol. ɒii, no. 78 ; *Cal. S.P.D.*, 1598–1601, p. 15 ; *Chanc. Proc., James I*, G. 1/76 ; Fishwick, *History of Rochdale*, p. 80 ; *Privy Council Register of Scotland*, 1st Series, vol. xiii, pp. 668–9 ; *H.M.C.*, *Report of the MSS. of Lord Middleton*, p. 173.

[5] By a clause in his lease of coal mines in Bramcote (Notts.) in 1609, Sir Percival Willoughby agreed to deliver up, at the expiration of his term, " such and soe many pitt and pittes open and chandrable and fitt for gettinge of cooles therein as shall be wrought and cooles gotten in at any time within three years before th' end . . . of the said terme [of 21 years] ". (*H.M.C.*, *Report on the MSS. of Lord Middleton*, p. 173.) At Strelley (Notts.) in the same year (1609) the lessee covenanted " to worke all the said workes orderlie as by the Arte of Collierie might be done for the benefitt of future times " (*Chanc. Proc., James I*, B. 40/70).

[6] Frequently he had also to conform to local customs (see above, pp. 298–301), but this was apparently not stipulated in his lease.

lay within the power of the lessor to grant them.[1] In some cases the lessee could " pull down and take away " such " engines and buildings " as he had erected,[2] much as the modern tenant of a house may remove the fixtures which he himself installs. Often there was a provision to release the lessee from his annual payments if the mine proved unworkable.[3] As was their custom with farm lands and pastures, landlords at times granted coal by copy of court roll ; [4] but, as they generally desired to replace customary tenure by lease-hold tenure, such grants were rare and it is usual to find leases for a term of years, most frequently for twenty-one years.[5] This term survived longer than did the method of assessing royalties, and in 1765 Jars tells us that concessionnaires still found it the normal period for a return on the capital expended.[6]

The profits arising from the revolutionary expansion of the coal industry at the end of Elizabeth's reign were not at once shared by the landlord. Generally he had leased his mines for fixed annual payments, adjusted sometimes to a small output, sometimes, on manors where there had as yet been no sale of coal, in the hope of a future output. Then came a greatly increased demand for coal, accompanied by deeper mining and much larger collieries. Where £5 or £10 a year had been considered a high valuation for the landlord to put upon his share in the produce of one pit in the northern counties from 1575 to 1600, concessionnaires were found after 1630 who were ready to pay £100.[7] Since the " rents " had generally been fixed regardless of output, the landlord had no means of bettering his position until the lease expired.[8] In some cases he received, no

---

[1] The lessor might reserve the right to grant wayleave through his grounds to other colliery companies, provided the grant did not prejudice the mining operations of the lessee.

[2] *Exch. Deps. by Com.*, 36 Charles II, Mich. 43.

[3] See, for example : *Augm. Partics. for Leases*, 38/32 ; 39/59 ; 109/11 ; *Lansdowne MSS.*, 71, no. 13 ; Welford, *op. cit.*, p. 287.

[4] *Chanc. Proc.*, Series II, 371/4.

[5] Leases for much longer terms were not unknown. The " grand lease " of the Bishop of Durham's coal in Whickham manor was granted for 99 years, but this can hardly be regarded as a case in which the lessor entered into the agreement of his own free will (cf. above, pp. 151–2). We have even found a case in which the coal under a plot of land in Great Marsden, Lancashire, was granted for 980 years (*Palat. of Lancs. Bills*, 66/14). On the other hand, we find examples of leases for very short terms. The coal under a plot of land in Rochdale was leased in 1658 for a five year period (*ibid.*, 32/44). But such short terms usually indicate that the agreement was more in the nature of a working contract than a lease (see above, p. 319 ; below, pp. 423–4). Sometimes the lessees were allowed a few extra months in which they might market their coal. A lease of the colliery at Lumley (Durham) expired July 6th, 1645, but " for carriage, in winter 1645 " (*S.P.D., Charles I*, vol. Dii, no. 78).

[6] *Voyages métallurgiques*, vol. i, p. 181. To-day, although the 21 year term has not been entirely abandoned, the common term for a lease is longer— 42 to 60 years (Redmayne and Stone, *Ownership and Valuation of Mineral Property*, pp. 65–7). This is natural because the capital invested is many times greater, and the life of a colliery much longer, than in the seventeenth and eighteenth centuries.

[7] *S.P.D., Charles I*, vol. Dii, no. 77.

[8] His position was made still worse by the continued fall in the value of money.

doubt, less than half of one per cent of the selling price of a ton.[1] In others he might be comparatively well off. Wide variations were the rule, and it was the most fertile, or best situated, mines which tended temporarily to pay the smallest royalties per unit of coal extracted.

Conditions so unsatisfactory to the landlord naturally could endure only until the lease fell in, when he demanded a higher " rent ". Everywhere royalties increased ; and the value of coal-bearing land rose rapidly, selling for as much as £150 an acre in south Staffordshire after the Restoration.[2] This increase was the swifter because of the emergence of a new economic scourge—the speculator in mineral property, who capitalized his superior knowledge of the future of the coal industry by taking leases of coal from unsuspecting landlords, and immediately subletting at a much higher rent. It was certainly without any intention of working the minerals himself that Henry Rosse, merchant and gentleman of London, took in 1596 a twenty-one year lease from the Crown of coal mines in the wastes of Cockfield moor in Raby lordship, at an annual rent of £4. A few months later Rosse sublet the property for sixteen years to Anthony Felton, who later conveyed most of his interest to two Newcastle merchants— for £22 a year over and above the Queen's " rent ".[3] In a similar way, George Lilburne of Sunderland and George Grey of Southwick, on November 20th, 1625, were able to get rid of their twenty-one year lease of Lumley colliery, dating only from July 6th, 1624, for the whole of the unexpired term at an annual profit of £90.[4] Such speculation was not limited to the great enterprises of Durham county. Throughout Yorkshire and Lancashire we find many coal mines sublet several times over before the original lease has expired.[5] Some men, like a certain Robert Leigh,[6] carried on a business in buying and selling mineral property, which did not differ in principle from that of the modern dealer in oil shares. Their activities not only forced up royalties more rapidly than they would otherwise have risen, but probably also settled them at a higher level, by demonstrating to the landlords how much could be got. By high rents,[7] frequently by even higher fines, the owners sought to recoup themselves for

[1] Cf. above, p. 147.

[2] Plot, *Natural History of Staffordshire*, p. 127. At about the same time a witness from the manor of Stratton in Somerset believed that "coles raised out of one acre of land may be worth more than the inheritance thereof" (*V.C.H. Somerset*, vol. ii, p. 382). At Newcastle, a part of the town moor (known as "Nun Moor"), rich in coal, that had been rented for 23s. 4d. a year in 1489, was sold in 1678 for £800 (*Documents of the Newcastle Corporation* : Typewritten abstract of a case of 1820 concerning the town moor).

[3] *Star Chamb. Proc., James I*, 265/3.

[4] *S.P.D., Charles I*, vol. Dii, no. 78.

[5] *Ibid., James I*, vol. clxxi, no. 67 ; *Duchy of Lancs. Pleadings*, 100/G/4 ; 109/A/5 ; 116/F/2.

[6] *Duchy of Lancs. Pleadings*, 183/L/13.

[7] In 1617 the coal mines of Strelley manor, near Nottingham, yielded £206 in annual rent (*Chanc. Proc., James I*, B. 40/70) ; in 1611 those of Mangotsfield manor, near Bristol, were leased for 21 years at £250 annually for the first 4¾ years, and £300 for the remainder of the term (*ibid.*, C. 9/69). These rents were very much larger than those which were being paid for more important mines leased by the Crown towards the end of Elizabeth's reign (see above, pp. 146–7).

what they no doubt regarded as their losses during the period when the industry expanded so rapidly. In the period of relative depression that followed the boom of the late sixteenth and early seventeenth centuries,[1] some mine owners found themselves unable to make sufficient profits to pay the high royalties.[2]

Both landlord and concessionnaire, having found that the uncertain terms of the Elizabethan lease might turn to their disadvantage, were in a mood to adopt another method of reckoning, which would establish some definite relation between the royalties paid and the value of the coal extracted. At many of the English metal mines, and at the coal mines in the Liége and Mons districts, in Saxony, the Saar district, and the Duchy of Magdeburg,[3] royalties were always taken at some fixed proportion—for example, a tenth—of the pithead price of the mineral. One might have expected the English landowners to adopt the same method of assessment. Curiously enough, however, sliding scales for coal " rents " have never been in favour in the British Isles, and even to-day, although they are on the increase, coal royalties are usually fixed on some other basis.[4] I have been able to find only two instances before 1725 of what could properly be called a sliding scale:[5] one on the manor of Chopwell in Durham, where, in 1649, Sir Henry Vane, by an old lease of the reign of Charles I, was accountable to the Crown for half of all the coals raised, if they exceeded the yearly value of £5; the other in Derbyshire, where a mine was let in 1720 by one of the Sitwells of Renishaw, the payments to vary according as the profit was over or under £50 annually.[6]

The system of assessing royalties at a fixed sum per unit of coal produced—still the most common system—apparently originated in the English coal mines. Before 1700 many landlords throughout the country had adopted it. Ralph Delaval's lease, made in 1611, of the mines of Seaton Delaval to Sir William Slingsby of York, whereby the lessee contracted to pay 8d. for every 24 bowls of coal extracted, is certainly an early example.[7] At about the same time the mines of Llanelly in South Wales were let for 2s. 2d. per " way " of coal (approximately four tons) taken from the " lower vayne ", and 1s. 10d. per " way " taken from the " upper vayne ".[8] In 1631 Sir John Hedworth leased the famous colliery of Harraton on the Wear, reserving to himself 16d. on every chaldron of nine-quarters coal (between two and two and a half tons), and 10d. on every chaldron of five-quarters coal.[9]

[1] See below, vol. ii, pp. 74–8.
[2] See, for instance, Ralph Cole's petition of 1646 concerning his collieries in Durham (Welford, op. cit., p. 165).
[3] Cf. Jars, op. cit., vol. i, pp. 314–19, 323–4; and Hasslacher, Steinkohlenbergbaues in Saargebiete, p. 407.
[4] Redmayne and Stone, Ownership and Valuation of Mineral Property, pp. 71–2, 77 sqq.
[5] I use this term in the same sense as do Redmayne and Stone, who do not regard a fixed rent per ton as constituting a " sliding scale " (op. cit., pp. 71–2, 78).
[6] Parl. Surveys, Durham, no. 3; V.C.H. Derbyshire, vol. ii, p. 354.
[7] Waterford MSS., no. 16 J (" A Booke of severall things touching the Estate of Seaton Delaval ").
[8] Star Chamb. Proc., James I, 155/5.
[9] Palat. of Durham, Decrees and Orders, vol. i, ff. 533–4. Nine-quarters indicates a seven-foot seam, five-quarters a four-foot seam.

As time went on, the common unit for assessing royalties in the northern counties became the " ten " (of approximately fifty-two tons). But whether reckoned by the "ten", the "bowl", the "way", the "stack", (as in Derbyshire), the " quarter ", the " pitt load ", the " basket " (as in Lancashire) or the ton (as at Oakengates in Shropshire),[1] the principle of assessment was everywhere the same. And, while the new system was not general at the end of the seventeenth century, it was everywhere gaining ground. According to one authority, who shows a special knowledge of conditions in the north of England, it had become scarcely less common for landowners to ask for " a certain rent out of every chaldron without fine " than to ask for a fine and a fixed annual rent.[2] With the new method of assessment came most of those provisions typical of the modern colliery lease [3] : the exemption from royalties of coal consumed at the mine, the payment of royalties only after extraction had begun, the inclusion of the so-called " certain rent " representing a minimum below which royalties could not fall.[4]

While it is clear that " rents " rose rapidly after 1610, it is difficult, owing to the still common method of leasing coal for a fixed sum irrespective of output, to estimate what charge they represented upon each ton raised. From a few instances in which rent was paid according to the quantity of fuel extracted, it appears that landlords expected more after 1660 than they had expected during the reign of James I. At Llanelly in 1612 the lessor received 2s. 2d. and 1s. 10d. per " way ".[5] In the neighbouring lordship of Gower, during the Commonwealth period, the rate had risen to 4s., or about 1s. per ton.[6] A similar increase took place in Durham and Northumberland. Ralph Delaval asked only 8d. on every chaldron in 1611,[7] but in 1631 the thick coal at Harraton was leased for 16d. per chaldron,[8] and about 1665 the coal of Great Lumley brought 15d. per twenty-one corves,[9] or about 8d. per ton.

This evidence of an increase in the amount of the royalty per ton between 1610 and 1660 cannot be explained by a rise in the pithead price of coal, for the indications are that during this half century the return received by the colliery owner per unit of coal produced increased very little.[10] Moreover, the increase in royalties was certainly much greater than the figures just given would indicate. It was only

---

[1] *V.C.H. Derbyshire*, vol. ii, p. 355 ; *V.C.H. Shropshire*, p. 455 ; Crofton, *op. cit.*, pp. 46, 61 ; *Duchy of Lancs. Pleadings*, 183/L/13.

[2] S. Primatt, *The City and Country Purchaser and Builder*, 1680 ed., p. 29.

[3] A typical modern colliery lease is given by Redmayne and Stone, *op. cit.*, pp. 54–7.

[4] A lease by the Newcastle Corporation of Walker colliery in 1757 included a provision of " certain rent ". One finds an early example in a lease of 1675 of mines at Whitley in Northumberland (Granville Poole, *Historical Review of Coal Mining*, 1924, pp. 325–6).

[5] *Star Chamb. Proc., James I*, 155/5.

[6] C. Wilkins, *South Wales Coal Trade*, 1888, p. 14.

[7] *Waterford MSS.*, no. 16 J.

[8] See above, p. 324.

[9] Green, *Chronicles and Records of the Northern Coal Trade*, p. 269. The corf at Lumley at this time contained nearly 2 cwt. of coal (see Appendix C (iii)).

[10] See below, vol. ii, Appendix E, and pp. 74–5.

during and after the first decade of the seventeenth century, when landowners generally had come to realize that the returns from their leases were inadequate, and when the leases made before the phenomenal expansion in the market for coal were expiring, that the new method of adjusting royalties to output began to be adopted. In order to get an accurate picture of the differences between royalties at the end of Elizabeth's reign and at the Restoration, therefore, we should have to compare the returns per ton obtained by landowners who leased their coal for a fixed rent. Satisfactory materials for making this comparison are not, of course, available ; but such information as we have is sufficient to show that the payments in respect of royalties had much more than doubled ; it is not unlikely that they were from four to six times what they had been. During the thirty years from 1580 to 1610, when most mining lessees paid a fixed annual rent, calculated on an output much smaller than that which was actually obtained, it may be doubted whether landlords, at least along the Tyne and Wear, enjoyed on the average as much as three per cent of the pithead price of coal.[1] After 1660, it was not exceptional at highly productive mines in South Wales, Durham, Northumberland, or the Midlands, for the landlord to receive more than a fourth of the pithead price. It is true, no doubt, that the average royalty absorbed a smaller proportion, possibly an eighth. While it is possible to cite cases in which the lessee paid 8d. or 9d., or even 1s. per ton, the normal charge did not perhaps exceed 5d.[2] But even 5d. represented an enormous increase over the royalties commonly paid at the beginning of the seventeenth century. It is clear that the rise imposed a much larger burden upon the lessee of the coal, and at the same time added greatly to the income of the lessor.

If an exception be made of the rents sometimes obtained by the Bishops of Durham during the Middle Ages, royalties appear never to have been so high in British coal mining history as in the century following the Restoration. Sometime after the middle of the eighteenth century we find a tendency for the value of royalties to fall. One-fifth was a very high royalty, one-tenth the common royalty in Adam Smith's day.[3] Since that time the fall has been marked. In Durham and Northumberland between 1824 and 1834, according to the

---

[1] See figures of the rent paid to the Crown by its coal lessees in the north of England, and the " value " of these collieries to the lessees, as given above, pp. 146–7.

[2] When, in 1665, a group of London capitalists, attracted by false reports of the discovery of coal in Windsor Forest (see below, vol. ii, p. 73), proposed to set up a colliery there, they agreed to pay a royalty of 6d. for every London chaldron, i.e. 4½d. per ton. As the enterprise was an hypothetical one, undertaken by a mine manager with experience in the Newcastle district, it is possible that the proposed " rent " had been determined on the basis of sums then received by various landlords in that district (*Stowe MSS.*, 744, no. 45).

[3] *Wealth of Nations*, bk. i, chap. xi. At the time of Jars' visit to Newcastle-under-Lyme in 1765 (see *Voyages métallurgiques*, vol. i, p. 255), one citizen received 10d. in royalties on every 15 cwt. of coal, a quantity then selling for 3s. 6d. at the pithead. His share was therefore a shade under 25 per cent. In the parish of Kilwinning, in Ayrshire, three small collieries paid, in 1792, £250 in royalties on an output of perhaps 10,000 or 11,000 tons (Sinclair, *Statistical Account of Scotland*, vol. xi, p. 147), i.e. about 6d. a ton. The selling price had doubled during the previous fifty years.

comprehensive statistics of the Coal Trade Committee of 1836, the royalties averaged about one-fifteenth of the selling price ;[1] in 1889 the average for the whole country was about the same ; during the last two years of the Great War about one thirty-sixth.[2]

Besides this notable tendency for the average royalty per ton to diminish in terms of real purchasing power during the last two centuries, there is also a tendency for the tonnage royalties at all British mines to approach a common level. At the end of Elizabeth's reign, the " rents " charged probably ranged from one half per cent to as much as twenty per cent of the pithead price, some mine owners paying 5d. or 6d. a ton, others less than ¼d. During the seventeenth and eighteenth centuries this margin was narrowed considerably ; since 1800 it has been narrowed still further. In 1836 the maximum royalty in Durham and Northumberland amounted to 1s. 3d. per ton, in 1889 the maximum was 10d., to-day it is 9d.[3] All the representatives for the mineral owners, who testified before the Coal Commission of 1925, agreed that, at the overwhelming majority of mines throughout the country, royalties ranged between 5d. and 7d. a ton. Thus the average has tended more and more to become the usual charge.

In explanation it may be urged that the nature of what we call a " royalty " has been somewhat modified. Originally the landlord thought himself entitled to a share in the windfalls of the industry ; if a mine proved successful, either because of its geographical situation, the fertility of the seams, or as the result of a boom in the market, he attempted to claim a share in this success. Being bound by the terms of his lease, he could do this only after the lease had expired. During the seventeenth century, he succeeded in forcing up the " rent " more than in proportion to the increased output. But, since the eighteenth century, a royalty in Great Britain has become something in the nature of a fixed return levied on a given quantity of coal, regardless—to a large extent—of the conditions under which it is mined, the nature of the market, and the profits of the colliery owners. Practically all the differential or speculative element has disappeared.

This change in the nature of royalties helps to explain certain differences of opinion concerning them which have prevailed among economic theorists. Adam Smith and Ricardo both lived at a time when great discrepancies in the returns of landlords, who leased their

---

[1] *Report of Select Committee on the Coal Trade*, 1836, pp. xxii, 49 (" the average mine rent is between 5d. and 6d. per ton ") and xvi, 47 (" the highest pithead price per ton was 8s. 3d., the lowest 6s. to 6s. 6d.").

[2] *Royal Commission on Mining Royalties, Final Report*, 1893, app. no. 5 ; *Coal Industry Commission*, 1919, *Report*, vol. iii, app. nos. 8–11. See also Redmayne and Stone, *op. cit.*, pp. 72–3.

[3] *Report of the Roy. Com. on the Coal Industry*, 1926, vol. ii, pt. B, pp. 780–1. Ninepence per ton is the highest royalty paid to the Ecclesiastical Commissioners, who enjoy one-fifteenth of all royalties in Great Britain. Most of their holdings are in Durham and Northumberland. The highest royalty paid in South Wales is 1s. per ton. For a special case, in which 1s. 6d. per ton was offered in royalties, see *ibid.*, p. 789.

minerals, were still common.[1] Both spoke not of royalties, but of the
" rent " of mines, using " rent " in the same sense that it had been used
in seventeenth-century mineral leases. Neither probably distinguished
sharply between the portion of this " rent " paid for the coal itself,
and the portion paid for interference with the surface[2] rights of the
landlord. Both had before them the concept of a marginal mine,
which could be worked only by the landlord, because it could not bear
the charge of paying " rent ". " The return for capital", wrote
Ricardo, " from the poorest mine paying no rent, would regulate the
rent of all the other more productive mines. This mine is supposed to
yield the usual profits of stock. All that the other mines produce
more than this will necessarily be paid to the owners for rent".[3] It
was possible in Ricardo's day, owing to the wide margin between the
" rents " at different mines, to argue with some plausibility that the
landlord enjoyed the whole differential advantage which one mining
site possessed over another. But it is clear that in our time the major
portion of the additional return from the better situated and more
productive mines no longer goes to the landlord, who tends to receive
royalties at the same rate in all parts of Great Britain, but to the
investors in colliery enterprises. No doubt the tendency of royalty
rents to become an equal charge on every ton of coal produced
influenced Sorley and Marshall in their dissent from the theory of
the classical economists.[4] In stating the reasons for their view of
the matter, they insist upon the difference between the rent of land
and the rent of minerals, which are a wasting asset.[5] Whereas,

[1] For Smith's views, see *Wealth of Nations*, bk. i, chap. xi; for Ricardo's,
see *On the Principles of Political Economy*, chaps. iii and xxii. They differed
as to what regulates the price of coal at a number of mines in one neighbourhood ;
Smith holding that it is the most, Ricardo that it is the least, fertile mine. This
contradiction may be partly explained by the different conditions upon which
these two thinkers based their conclusions. Smith, who was born at Kirkcaldy,
probably had under consideration the collieries round the Firth of Forth. In
this district, until just before the publication of the *Wealth of Nations*, the demand
for coal was expanding much less rapidly than it had expanded in the sixteenth
and early seventeenth centuries, or than it was destined to expand at the end of
the eighteenth century (cf. below, vol. ii, p. 195). Many new mines were available ;
the supply of coal, on the whole, exceeded the demand ; cut-throat competition
raged to such an extent that the conditions of production at the more fertile
and better situated mines could at any rate be plausibly represented as tending
to settle the selling price (cf. Thorold Rogers' footnote in his edition of the
Wealth of Nations, vol. i, pp. 177–8). When Ricardo wrote some fifty years later, in
the midst of a great boom in the British coal industry, demand over-reached supply,
and price much more obviously tended to be regulated by the worst mine worked.

[2] Underground wayleaves, which now figure prominently in mineral rent
(Redmayne and Stone, *Ownership and Valuation of Mineral Property*, pp. 82–5),
were no doubt less important at the end of the eighteenth century.

[3] Ricardo, *op. cit.*, 1817 ed.

[4] W. R. Sorley, " Mining Royalties and their Effect on the Iron and Coal
Trades ", in *Journal of the Royal Statistical Society*, vol. lii, 1889, pp. 60–98.
John Stuart Mill, whose position is about midway between Ricardo and Marshall,
found difficulty in accepting the theory that royalties were precisely analogous
to land rents (*Principles of Political Economy*, bk. iii, chap. v, par. 3).

[5] The conception of mineral fuel as a wasting asset was not a new one,
as is proved by an early letter of the English Privy Council, written under the
guidance of James I (see below, vol. ii, p. 325). But the " rent " of a mine
had not, apparently, been thought of as strictly a payment for the exhaustion
of a natural resource.

Marshall pointed out, the farmer normally gives back the ground to its owner " as rich as he found it ", the mining company lowers its value by taking away riches which cannot be replaced by geological processes, except over a period of many centuries. Consequently, Marshall concluded, the " royalty on a ton of coal, when accurately adjusted, represents that diminution in the value of the mine, regarded as a source of wealth in the future, which is caused by taking the ton out of Nature's storehouse ". Marshall saw that royalties are now largely independent of the fertility of the seams, and held that they therefore enter into the selling price of coal, " in addition to the marginal expenses of working the mine ".[1]

Whatever may be thought of Marshall's distinction between royalty and rent, an investigation of the royalties actually paid on British coal suggests that the concept of a royalty as a differential return for the possession of the more fertile and better situated mines, can be made to correspond with the return which the British land-lord in fact receives, only if it is so qualified as to lose all its essential meaning. Although, in the past, the landlord sometimes succeeded in claiming a portion of the return which arose from advantages of situation or of fertility, the royalties paid at different mines have never been an accurate measure of the differences in natural advantages between these mines. Mining conditions and the state of the market for coal changed from year to year, and even from month to month, while colliery leases have always been drawn for much longer periods, during which the royalties have been fixed, in the sixteenth and seventeenth centuries most commonly at a definite sum per annum, and since the seventeenth century more frequently at a definite sum per unit of coal extracted. No method of assessing royalties to absorb everything above " the usual profits of stock " throughout the term which leases have to run has yet been devised.[2]

(ii) *Wayleaves*

In addition to royalties, considerable sums have been paid to landowners in Great Britain in respect of wayleaves. Wayleave rent may be defined as a payment for the right to make use of the land or subsoil for purposes connected with the mining and marketing of minerals. In connection with the coal industry in the sixteenth and seventeenth centuries such leave was needed for the use of ground taken up for sinking shafts and ventilation pits, driving adits, setting up machinery, stacking coal, and building colliers' houses. It was needed, finally, for making paths and wagonways for carrying away the produce of the mines, and, in case the coal was destined to be shipped by water, for building wharves (or " staiths " as they were called in the north of England) in order to store it until it could be loaded into keels.

Although one often finds the words *sufficiens chiminum*, or wayleave, in medieval leases of coal,[3] there are few instances of separate

---

[1] *Principles of Economics*, bk. iv, chap. iii ; bk. vi, chap. ii.
[2] Cf. Sorley, *op. cit.*, pp. 79–81.
[3] Garnier, *History of the Landed Interest*, vol. ii, p. 31.

payments for it.[1] Workings were shallow and took up little ground ;
coal was seldom marketed at a distance from the mine, except on
the Tyne, where the early pits were on the very edge of the river.
When the need arose for carrying an occasional load through a
neighbour's farm, it was thought scarcely more necessary to offer
" rent " for the privilege than to pay interest to-day for the loan
of a book from a friend. If in our time, when the barking dog and
the barbed wire fence are among the first enemies of our childhood,
it is difficult to picture a state of things when the hedge and the fence
had hardly been invented, we shall do well to remember that on the
medieval manor rights to property in land were not so awe-inspiring
as to-day.

" Le premier qui ayant enclos un terrain s'avisa de dire : *Ceci
est à moi*, et trouva des gens assez simples pour le croire, fut le vrai
fondateur de la société civile ", wrote Rousseau. Less than a century
after we first notice in England an extension of enclosures, symbols
of the tightening bonds of personal ownership, the coal industry
begins seriously to interfere with the surface ground of landholders.
A colliery enterprise comes to demand the use, the desecration, and
sometimes the subsidence, of considerable areas ; and, as a
consequence, it frequently provides a landholder with a just claim
to compensation. In 1609, in Elswick manor on the Tyne, several
houses and a barn collapsed owing to the underground passages dug
by the coal miners, and, in addition to the damage done to personal
property, the operations of the colliers caused the loss of land used for
pasture and farming, two acres having sunk down in the oxen pasture
and cornfield.[2] In Lamesley parish near Chester-le-Street, in 1634,
subsidence did much harm to the fields used for grazing, and damaged
the houses of numerous poor tenants.[3] Nor was such damage known
only in Durham and Northumberland. The provost and scholars of
King's College, Cambridge, complained, in 1686, that the colliery
operations of their tenants in Prescot manor, near Liverpool, destroyed
the grass and prevented them from holding their annual horse fair in
two acres of unenclosed land.[4] As Peter Banckes, a Newcastle yeoman,
observed in 1677, "wayleave and spoil" were worth more than the
yearly letting value of the land occupied, because the ways through
it " much spoiled " the ground for cultivation.[5]

Wherever there were coal mines, there began to be a clamour for
wayleave rents, until, by the time of the Civil War, it was as true
for concessionnaires throughout the country as for a tiny mining
partnership on the manor of Stratton in Somerset, that " without
. . . fines, yearly rents, and compositions with other lords for water-
courses, and trespasses for carrying away water "—and, one might
add in many cases, for digging shafts and moving coal—pits " can no

---

[1] In 1441 wayleave was granted to the Master of Gateshead Hospital to
carry coal from his land to the Bishop's "staith" on the Tyne (Brand,
History of Newcastle, vol. i, p. 472).

[2] *Exch. Spec. Com.*, no. 4355.

[3] *Palat. of Durham, Decrees and Orders*, vol. i, ff. 130–2.

[4] *Palat. of Lancs. Bills*, 38/107.

[5] *Exch. Deps. by Com.*, 29 Charles II, East. 20.

longer be wrought ".[1] Wayleaves had become, in fact, a common item in the expense account of every colliery enterprise.

Most mineral leases, by means of some such phrase as " free way, liberty of passage, egresse and regresse, for all manner of persons, carts, carriages, horses, oxen . . . etc.",[2] included all the rights needed for the mining enterprise, in so far as they concerned the land held by the owner of the coal. But, even when he was granted by the lord full power to mine in any part of the manor, the concessionnaire had often to make separate agreements with copyholders and other tenants before he could enter their grounds, or, if he entered without an agreement, he had usually to make reparation for damage actually done.[3] Further, mining adventurers were frequently obliged to encroach upon land to which they had no title whatever, for the purpose either of carrying away coal or of draining off water. In such cases they had to ask for what is called to-day " foreign " wayleave, or permission to make use of lands outside their original mining concession.

Wayleave rents, like royalties, were ultimately settled largely according to the bargaining power of the parties. Sometimes they were fixed by referees, chosen either by both sides, or, if the dispute had come before the courts, by a judicial authority. The legal power of landowners, unless there existed a manorial custom to the contrary, to refuse the concessionnaire access altogether was not denied,[4] and, although the courts sought to induce them to accept reasonable compensation,[5] they were not always willing to do so. Compensation might take the form of a lump sum, intended to cover mining operations for a term of years, as at Cowpen in 1598, or merely the sinking of a single pit, as on Leeds manor in 1588, and on Houghton manor in 1628.[6] It might be based directly upon the actual loss suffered by the landowner in harm done to his ground and in the foregoing of those parts used for coal mining. That was the usual method of assessment at Liége, where the surface owner received

---

[1] *Parl. Surveys, Somerset,* no. 39.

[2] Green, *op. cit.,* p. 197. This lease applied to the Bishop's mines in the manor of Evenwood in Durham. To cite two of many other examples, the lessee of the coal mines of Westerleigh in Kingswood Chase had liberty " to break the soyle . . . digg and worke levells ", convey the coal, etc., " with free ingresse, egresse, and regresse " (*Chanc. Proc.,* Series II, 371/4), while a charter to the coal of the barony of Carriden (Linlithgowshire) granted full power to break the ground, and erect necessary " Ingynes for wyning and transporting of the . . . coales to sea and Land . . . with frie ische [exit] and entrie of wayis and passages for winning, carieing and transporting . . . etc." (*Acts of the Parliaments of Scotland,* vol. vi, pt. i, p. 606). The words " egress and regress " were also commonly used in grants of pasture.

[3] Cf. above, pp. 312–15.

[4] This power was ordinarily used as a lever to raise the rents. It was only in a case like that of Broseley in Shropshire (see above, p. 308), where the landlord wished to prevent his rivals from selling coal, that wayleave was refused on any terms. In the reign of Henry VIII, Sir John Willoughby refused to allow his rival, Nicholas Strelley, to drain away water from his mine through Willoughby's land (*Star Chamb. Proc., Henry VIII,* 22/94).

[5] Cf. *P.C.R.,* vol. xliv, pp. 544–5.

[6] *Exch. Deps. by Com.,* 41 Eliz., East. 19 ; 3 and 4 Charles I, Hilary 2 ; *Duchy of Lancs., Decrees and Orders,* vol. xx, f. 427.

double damages, and on several manors in Somerset, where by an old custom he received triple damages.[1] In Somerset compensation was occasionally reckoned on another basis. On the manor of Benter, miners usually granted tenants a part interest in the coal in return for permission to work under their ground,[2] and on Stratton manor, during the reign of James I, a small mining partnership was committed to giving to a copyholder one-eighth, and to a copyholder in reversion one-sixteenth, of all coal raised, besides an allowance for trespass and loss of herbage.[3] The most common method throughout the country, however, was to lease a wayleave, like a royalty and a plot of land, in return for a fine and fixed annual rent, rated sometimes according to the number of acres granted. And, in the case of wayleaves, this method of payment proved more enduring than in the case of royalties. Until the nineteenth century wayleaves were rarely assessed by the ton, or by any other unit of output.

Whatever the method of assessment, it was customary in the seventeenth century, as it is to-day, to require the concessionnaire, when he ceased mining, to restore the soil to a proper condition for arable or pasture farming. At Stratton, for instance, he engaged to fill up the pits and cover them over " two foote deepe with good earth ".[4] During the term of his lease he had to take precautions for protection of the landholders' herds and soil : at Chester-le-Street to fence off all ventilation pits, as well as the ground about the shaft head ; at Whickham to ditch off the way used for the carriage of coal ; at Wigan to pave the " gutter " used for draining the mine, in order that the water should not flood the highway.[5]

In opposition to the view that wayleave rent should pay only " good and full recompense " for the damage actually sustained by the landholder, there grew up during the seventeenth century the view that this rent should be in proportion to " the benefitt " reaped by the concessionnaire from his colliery.[6] It was of some consequence which opinion gained the ascendancy, for, with the increasing importance of the coal industry, the figure arrived at on the second view became ever higher in comparison with that reached on the first. Notwithstanding opposition from the Privy Council before the Civil War, and from the House of Commons after it, rents came in practice to be settled in accordance with the second view. Every landholder fought to get as much as he could from the colliery company, and the number of claimants of wayleave rent was much greater than the number of claimants of royalties, for copyholders and leaseholders, as well as freeholders and manorial lords, could assert a right to compensation for damage done to the surface land, though not for the removal of the minerals underneath it. The ability of a tenant to exact high payments for wayleave depended on his economic strength,

---

[1] Louvrex, *Recueil contenant les édits . . . pour le pays de Liége*, p. 196, art. 5 ; *V.C.H. Somerset*, vol. ii, p. 381.
[2] *Chanc. Proc., James I*, B. 20/72.
[3] *V.C.H. Somerset*, vol. ii, p. 381.   [4] *Ibid.*, p. 381.
[5] *Palat. of Durham, Decrees and Orders*, vol. i, ff. 130–2, 228–30 ; Crofton, *Lancashire and Cheshire Coal Mining Records*, p. 55.
[6] *P.C.R.*, vol. xxxii, p. 499 (1624) ; vol. xxxiii, p. 66 (1625).

and—though probably to a lesser degree—on the custom of the manor. Some landholders were left without even adequate damages. By the Wakefield Enclosure Act, passed in the eighteenth century, tenants could not get any money from the concessionnaire when their houses or cellars caved in because of his mining operations.[1] Poor tenants found much difficulty in obtaining damages for injury done to their land or their movable property by the colliery owners, and this difficulty may have been increased by the fact—long ignored— that subsidence does not always take place vertically, or even approximately vertically, above the place where the mining occurs.[2]

Although many poor tenants had difficulty in getting compensation for damage actually sustained, others were able to exact from the colliery owners in advance a price beyond the value of any possible damage that they might sustain. In the Bishopric of Durham copyholders and leaseholders were usually successful in preventing the use of their soil for colliery operations, except on their own terms. A landholder with an undisputed title to land in a key position, by exercising his legal right to refuse entry, often held the colliery company at his mercy.[3] When it was a question of allowing wayleave for an adit to drain the seam, he held all the aces, for his refusal might result in flooding the mine, and in forcing the concessionnaire to abandon the whole enterprise, with all the capital invested in it.

Worst of all for the concessionnaire, a settlement once made might not be honoured by the landowner, if he saw an opportunity to improve his rent. Sir William Murray of Natown, " being heretour of the ground throw which the levell " ran from Major Biggar's colliery of Wolmet in Midlothian, " after much . . . good deid receaved by him [Murray] since . . . 1658, haveing sometyme . . . half of the free profite of the said coall, and after many solemne oaths . . . never to trouble . . . [Biggar] by troubleing the course of the water throw his ground, did by a bond of [March 9, 1665] . . . bind . . . him upon his honor . . . never to stop directly or indirectly any of the watergoing levells or runnes that runne throw his ground nor to cutt doune any stoupes, yet the said Sir William has of late [1667] threatened . . . [Biggar] that he will drowne his coall of Wolmet unlesse he will give him what soumes of money he pleases ".[4] Landowners seized on any pretext to wriggle out of what they regarded as a bad bargain. In the case of Player *versus* Hobbes, Player was able to prove by witnesses that on April 14th, 1627, Hobbes had signed a bond to allow the colliers " free egress and regress " to his land, with leave to " digge, land and carry coles " until Christmas, 1630, in return for a payment of £60 ; but that Hobbes had then taken advantage of the fact that the bond, whether by a slip of the pen or by malice, had been wrongly

---

[1] *Roy. Com. on Mining Royalties, 3rd Report,* 1891, Qs. 17327–17402.
[2] Redmayne and Stone, *Ownership and Valuation of Mineral Property,* pp. 25–6.
[3] When the land was owned by a public authority, the colliery owners might obtain liberal terms. In 1657 the Corporation of Newcastle waived its claim to a wayleave rent, upon a report that the " coleway is little hurtful to the ground " (*MS. Journals of the Common Council of Newcastle,* vol. i, ff. 462, 477).
[4] *Privy Council Register of Scotland,* 3rd Series, vol. ii, pp. 371–2.

dated " 1610 " instead of " 1630 ", to deny Player entry.[1] Again, in spite of an agreement still in being, the wayleave granted to a colliery company near the Tyne was contested in the Court of Exchequer by a greedy landowner, because rails of timber had been laid for the carriage of coal along the usual wagonway. He demanded a higher rent because the rails permitted his lessees to market a greater quantity of coal, notwithstanding that, as the Lord Chancellor remarked, the new device was less prejudicial to the soil than was the use of the common way.[2]

If he controlled the only route available for the carriage of coal, the landowner could force the colliery enterprise to settle with him on his own terms, or else prevent it from competing with its rivals. In 1632 the pit at Softly in south Durham was surrounded by a holding, the two lessees of which refused passage to the concessionnaires, stopped country boys bearing coal from the mine, and cut their sacks.[3] Violent interruptions of this sort were common throughout the country.[4] In the Tyne district, wagons loaded with coal were frequently stopped on their way to the waterside by claimants for wayleave rent.[5] A copyholder of Whickham alleged in 1678 that it was the custom in the neighbourhood for landowners, in default of payment for wayleave and spoil, to prevent (the assumption is, with force) the carriage of coal. At this time the copyholders demanded 5s. for every " rigg " of meadow and 6s. 8d. for every " rigg " of arable land spoiled, besides fire coal at cost price.[6]

Wayleave charges became extremely oppressive to colliery owners everywhere,[7] and especially to those along the Tyne and Wear rivers. And the great number of claimants for them led to a general movement for the acquisition of wayleaves by the State, a movement such as would probably have developed concerning royalties if the number of persons entitled to them had been larger than it was. As early as 1610, the colliery owners of Newcastle complained that " the Rent of wayleave and staithroomes is of late so far inhansed by the freholders . . . that the prices of coles is therby the rather occasioned to be raised ".[8] In 1674, the colliery owners of the same

[1] *Chanc. Deps.*, P. 18/H, and see above, p. 313.

[2] Bainbridge, *Law of Mines*, 1878 ed., p. 211.

[3] *Exch. Spec. Com.*, no. 5276.

[4] *Duchy of Lancs. Pleadings*, 131/T/4 ; *Duchy of Lancs. Deps.*, 75/13 ; *Palat. of Durham, Bills and Answers*, bdl. 2 (Selby v. Hedworth) ; *Privy Council Register of Scotland*, 2nd Series, vol. viii, p. 267 ; *Acts of the Privy Council*, N.S., vol. xxix, pp. 657–8.

[5] *Exch. Deps. by Com.*, 29 Charles II, East. 20 (testimony of Lancelot Cramlington, gent., of Newcastle). In 1617 the rector of Whickham refused wayleave to the partners in the Grand Lease colliery (*P.C.R.*, vol. xxix, p. 60). In the sixteenth century the Earl of Westmorland prevented passage from the Bishop of Durham's mine in Raby lordship, until the Bishop agreed to pay for wayleave (*Exch. Spec. Com.*, no. 752).

[6] *Exch. Deps. by Com.*, 29 Charles II, East. 20 ; 32 Charles II, Mich. 43.

[7] The undertakers of Bedworth colliery in Warwickshire (to cite an example from another mining district) were obliged to pay, as we learn from the testimony of a mercer of Coventry in 1684, " great damages to various tenants, mostly husbandmen, for passing through their farms with coal carts or pack horses " (*Exch. Deps. by Com.*, 36 Charles II, Mich. 43).

[8] Dendy, *Records of the Company of Hostmen*, p. 59.

town, "upon consideration of the great abuses and exaccons upon Collieries for their way-leave and Staith-roomes", determined to work for the passage of an Act of Parliament to restrict the amount of rent which a landowner might ask. The comments of impartial observers suggest that landowners were accustomed to drive an extremely hard bargain.[1] Roger North writes in 1676 that " the owner of a rood of ground will expect £20 per annum for this leave ", many times its value for farming.[2] In 1725, Lord Harley mentions the excessive wayleave rents charged to the owners of collieries near Chester-le-Street.[3] The resentment of the north of England colliery owners against the exactions of certain landowners resulted in an agitation, during the thirties of the eighteenth century, for the State to purchase all the land needed for wayleaves in the north of England at a fair valuation, and then to build the railed " wagon-ways " as they were required, charging the colliery owners rent for the use of them.[4] But nothing came of this suggestion ; and, when Jars visited the north of England some thirty years later, the wayleave rents were still fixed as the result of private bargaining, and the landowner often pocketed from twenty to forty times the value of the land for cultivation.[5]

Wayleave rents (reckoned by the ton mile) tended to reach a high point at the end of the seventeenth or the beginning of the eighteenth century, since which time they have absorbed a steadily diminishing portion of the selling price.[6] Their importance has been largely reduced, also, by the introduction of steam railroad transport, the right of way now being secured by the railroad, rather than by the colliery company. To this decline in the importance of wayleave rents, and to the elimination by enclosures of many claimants to them, we may perhaps attribute the decreasing evidence of a serious movement among colliery owners, after the middle of the eighteenth century, for the acquisition of wayleaves by the State. But this must not obscure the fact that, in the century and a half preceding the Industrial Revolution, wayleave rents imposed a serious handicap upon the development of colliery enterprise.

[1] *Ibid.*, pp. 136–8, 143, 146.
[2] Quoted Galloway, *History of Coal Mining*, p. 64. Already in 1628 Sir Richard Lumley had charged £40 annually for a short wayleave to the river Wear (*Palat. of Durham, Bills and Answers*, bdl. 29 : Lumley *v.* Chambers and others).
[3] *Journeys in England*, printed in *H.M.C., Report on the MSS. of the Duke of Portland*, vol. vi, p. 104.
[4] E. R. Turner, " The English Coal Industry in the Seventeenth and Eighteenth Centuries ", in *Amer. Hist. Rev.*, vol. xxvii, 1921, p. 9.
[5] Jars, *op. cit.*, vol. i, p. 179. An attempt had been made by the colliery owners in 1696 to get Parliament to pass a statute restricting the rights of land-holders to fix their own terms for wayleave rents, but although a bill was introduced in the Commons it failed to become law (Dendy, *op. cit.*, pp. 151–2 ; Brand, *History of Newcastle*, vol. ii, p. 301).
[6] The wayleave rent per ton mile in the north of England in the period between 1830 and 1836 varied from $\frac{1}{2}d.$ to $2\frac{1}{2}d.$ (*Select Committee on the Coal Trade*, 1836, p. xxii), as compared with $\frac{1}{4}d.$ to $1\frac{1}{2}d.$ at the close of the Great War (Redmayne and Stone, *op. cit.*, p. 81). About £5,800,000 annually is paid to-day in royalties as compared with about £200,000 in wayleave rents (*Report of Royal Com. on the Coal Industry*, 1926, vol. ii, p. 753 [73]).

CHAPTER V

CONCLUSION

IF, as seems probable, the right of landowners to demand whatever terms they liked for the use of their land for purposes of colliery enterprise had a tendency to retard the growth of the coal industry, what can be said of the influence upon this growth of the general method of settling disputes in the coal fields, on the one hand, and of the private ownership of royalties, on the other.

The struggles over the definition of the mineral rights which went with different kinds of tenure, and over the fixing of the sums due for wayleaves and royalties, were responsible for only a part of the cases concerning coal mines brought into court during the sixteenth and seventeenth centuries. In another category from the suits between tenant and manorial lord, or between tenant and concessionnaire, are those between rival colliery owners, whether single persons, partners, or companies. First of all, there are the disputes caused by an uncertain surface title, where each side advanced a claim to mine under the same plot of ground.[1] Secondly, there are the disputes arising from the fact that the same seam of coal was worked under two or more separate titles. Beneath the surface it was not possible to fence off one's own part of a seam as one would enclose an arable farm, and it was therefore often hard to determine the dividing line between two concessions.

The only way was to measure the surface distance from the pit-head to the hedge separating one holding from the other ; then to descend the shaft and work out the horizontal distance from the shaft bottom to the working face farthest towards the rival holding—an operation requiring, as underground galleries were rarely level, a skilful mathematical surveyor, especially where the seams slanted steeply. Commissioners appointed by the Court of Chancery of the Bishopric of Durham in 1663, to investigate the action brought by the farmers of the Grand Lease mines against the farmers of the mines under Brinkburne freehold in Whickham, were told to "view" the colliery in the usual manner, to "make use of the standerd rowler, roapes and other like instruments", and also to "carry with them spades, shovells, hacks, pickes, and other towles and instruments wherewith to remove . . . all lettes, obstructions and impediments above or under ground".[2] In one instance,

---

[1] The following are references to a few examples : *Court of Augm., Misc. Bks.*, vol cxvi, ff. 55–65 ; *Exch. Deps. by Com.*, 8 James I, Trin. 1, Hil. 1 ; *Star Chamb. Proc., Henry VIII*, vol. v, f. 138 ; *ibid., James I*, 154/12 ; *Duchy of Lancs. Pleadings*, 73/M/10, 78/A/7 ; *Chanc. Proc., Eliz.*, O. 2/46 ; *Privy Council Register of Scotland*, 1st Series, vol. x, pp. 5–6, 27.

[2] *Palat. of Durham, Decrees and Orders*, vol. ii, f. 131.

concessionnaires were said to have taken 20,000 tons of coal to which they had no title.[1]

Undermining of this sort might also endanger the life of an adjacent colliery. Careful technique made it indispensable to leave— and it was generally prescribed in mining leases that there should be left—a certain amount of barrier coal, known as " sinzies ", " sinzie walls ", or dykes in Scotland,[2] for the reason that they dammed out water from other workings. But, though prudence often demanded the sacrifice of a quantity of good coal, there were always some who preferred to take the risk in order to increase their " winning ", and others who, out of malice, knowing that a rival's mine was dry when their own was full of water, ordered their hewers to attack the barrier. Whatever the motive, the result of piercing such a coal wall might be disastrous. At the " heugh " of Bo'ness, in Linlithgowshire, where large sums had been spent on drainage, the wall was pierced by rivals in 1622, and " the watter, getting ane litle vent and passage . . . soone . . . overquhemled and distroyit the said wall and bulwark, and drowned nyne of the best headis ".[3]

The drainage problem was complicated in other ways where two or more colliery enterprises were carried on in close proximity. When a water-raising machine had been installed by a partnership, it tended to suck up water, not only from its own mine, but from all the pits and shafts of the neighbourhood. A pump erected on Whick-ham manor in 1618 drained not only the Grand Lease colliery, but also the important adjoining mine of Allerdeans, and the relative share of each colliery in the drainage expenses became a subject of dispute between the two partnerships.[4] Henry Liddell, a coal owner on the Wear, alleged in 1623 that a rival mining enterprise " received benefitt to the value of £1,000 a year by one water pitt within his ground ".[5] In 1645 the " maine cole at Harraton ", also on the Wear, " beinge hitherto a drowned mine ", was, as a result of " water drawinge " at the adjacent colliery in Lambton manor, " drayned . . . and water free and for like to be as long as Lambton shall drawe water and can work ".[6]

It might prove precarious to depend upon a neighbour's pump. In the Mendip region in 1617, when Richard Burke introduced a " mill wheel . . . to drawe the water " from his small diggings in Midsomer Norton, his competitors, James and Clement Huishe and others, " so worked underneathe the earth that they caused the water to runne into [Burke's mine] . . . whereby . . . the mill wheele might drawe the water of both ". Instead, they " utterly drowned " the works.[7] According to the story of Thomas Robinson and John Briggs, undertakers of the great Bedworth colliery in Warwickshire, told in answer to charges of having flooded the pits of their rivals at

[1] *Palat. of Durham, Bills and Answers*, bdl. 24 (Stevenson *v.* Liddell and others).

[2] *Privy Council Register of Scotland*, 3rd Series, vol. ii, pp. 563–4.

[3] *Ibid.*, 1st Series, vol. xiii, pp. 751–2.

[4] *Star Chamb. Proc., James I*, 245/6.

[5] *P.C.R.*, vol. xxxi, p. 718.

[6] *S.P.D., Charles I*, vol. Dvi, no. 59.

[7] *Chanc. Proc., James I*, B. 20/72.

Griff, the latter were themselves to blame, because they depended for drainage solely upon the two " engines " erected at Bedworth. They were without resources of their own, therefore, when the Bedworth partners, finding fresh coal nearer the surface, allowed the water to stand fourteen " ells " higher in the pits.[1]

Needless to say, disputes without number could arise where several mine owners depended surreptitiously, and without written agreement, upon the drainage system of a rival. That was a chief reason in favour of wide mineral concessions, which took in an entire drainage area. When Lowther estimated that, if all the collieries about Newcastle had been operated as a single enterprise, the expense for drainage would have been only one-sixth of what it was,[2] he may well have taken into account the cost of litigation engendered by multiple management.

In still another category come suits over the actual management of collieries. First, there were many disputes between the owner of the coal and his lessees, usually over an alleged failure to comply with some provision in the lease. For example, John Nicholson, owner of the coal of Lasswade in Midlothian, was in continual trouble with his concessionnaires over a clause providing that they should leave his " coill in als goode estait and als low a watter as scho wes the tyme of the conditioun maid betuix thame ".[3] Trouble was especially likely to occur when the landlord sold his title to minerals with a lease still in being, because the new owner was loath to be bound by terms made with his predecessor.[4] Secondly, there were disputes between colliery owners and contractors or managers, usually concerning the deliveries of coal.[5] Thirdly, there were disputes between partners in the enterprise, sometimes over their share in the payment of expenses,[6] sometimes over their share in the profits on sales.[7] Fourthly, there were disputes between colliery owners and their creditors over the repayment of loans, and the foreclosing of bonds signed as security when the loans were made.[8] As all these disputes relate to the operation and financing of collieries rather than to the mineral rights of lords of manors and their tenants, the issues raised are reserved for more detailed treatment in Part IV. We are here concerned only with some of the consequences of the methods adopted in settling these cases and those dealt with in earlier chapters.

---

[1] *V.C.H. Warwickshire*, vol. ii, p. 222.

[2] Birch, *History of the Royal Society*, vol. iii, p. 439.

[3] *Privy Council Register of Scotland*, 1st Series, vol. xiii, pp. 368-9.

[4] *Exch. Deps. by Com.*, 29 Eliz., East. 4 ; *Star Chamb. Proc., James I,* 50/23 ; *Chanc. Proc., James I,* A. 9/24.

[5] *Chanc. Proc., James I,* G. 1/76 ; Green, *op. cit.*, pp. 267-70 ; *Star Chamb. Proc., James I,* 155/5.

[6] *Chanc. Proc., James I,* D. 1/29 ; *V.C.H. Shropshire*, p. 458.

[7] *Palat. of Durham, Bills and Answers*, bdl. 24 (Carr v. Bartlett) ; *ibid., Decrees and Orders*, vol. ii, ff. 168 sqq. ; *Exch. Deps. by Com.*, 3 Charles I, East. 19 ; *Duchy of Lancs. Pleadings,* 171/A/12 ; *Chanc. Proc., James I,* J. 3/41 ; *Court of Requests Proc.*, 40/106.

[8] *Court of Requests Proc.*, 28/56 ; *Star Chamb. Proc., James I,* 69/18 ; *Palat. of Durham, Bills and Answers*, bdl. 24 (Parkinson v. Lee) ; *ibid., Decrees and Orders*, vol. i, f. 533.

Friction over tenants' rights, over mineral rights, over the management of collieries, or between one colliery owner and another, was not peculiar to the British coalfields. It was bound to occur in the early stages of the mining industry. What we want to know is whether in Great Britain the courts were prompt and just in settling these disputes, and effective in establishing precedents to limit their recurrence.

Behind all violence there is weakness. Just as the success of the Hague Court, of the League of Nations, or of any international tribunal, will depend ultimately upon its ability to settle conflicts between nations without recourse to war, so the strength of courts in dealing with mining disputes must be judged, first of all, by the willingness which they inspired in the two sides to seek a peaceful arrangement. We have seen that nearly every English and Scottish jurisdiction dealt with coal mining suits, and the amount of lawlessness seems to have been almost directly proportional to the number of law courts. When every allowance is made for the natural desire of each party to strengthen its case by a charge of *sabotage*, the charge was advanced too frequently, and was too often admitted, to be dismissed as a mere fabrication. Pitched battles between tenants and concessionnaires, some of which have been already referred to, do not show any greater marks of sour and premeditated ill-will than do the raids carried out against rival colliery owners. A certain George Morton, Esq., in 1608, assembled a large group of north-countrymen, who proceeded to the pits of his competitor, Ralph Orde, gent., where they cast down his " howeses and Covers ", broke down and cut into small pieces with axes and saws three " windows of the Colepits " and all the " timberworke, Ropes, frames and buildinges thereon made . . . for the . . . wynninge of Coles ", cast all these broken pieces into the shaft, together with forty " Chawder of Coles ", thus filling the shaft to the brim with coal and rubbish.[1] In 1622 a number of Scottish colliers were " sent to ward " in the tolbooth at Edinburgh for having gone to a neighbouring " heugh " in Lanarkshire, belonging to Margaret Hamilton, and " cuttit and hoghit the whole pillaris of the said . . . heugh, so that the rooff thairof is fallin to the sole, and hir heugh thairby maid unprofitable unto hir ".[2] John Ramsay, a coal mining adventurer, jealous of the success of Sir John Maxwell's " coalehewes " at Southside, in Midlothian, was said to have come with intent to murder William Wilson, one of Maxwell's colliers, but " missing him, he cutted the towes, brake the buckets and other instruments made for drawing away the water frome the coale, and hes thairby drowned the compleaners whole coale ".[3] At Llanelly in Carmarthenshire, in 1613, John Griffiths, who worked mines in his own lands, was charged with having entered the pits of his rival, William Vaughan, and having burned straw, " or some ill flavoured stuff ", which nearly suffocated two workmen, his aim being to smoke out all the colliers.[4] On the

---

[1] *Star Chamb. Proc., James I*, 224/19, 20.
[2] *Privy Council Register of Scotland*, 1st Series, vol. xiii, p. 47.
[3] *Ibid.*, 2nd Series, vol. iii, p. 563.
[4] *Star Chamb. Proc., James I*, 288/5.

manor of Colne in Lancashire, in 1574, James Hartley and a group of other yeomen are said to have entered the lands of Anne Townley, where there were many of her " servants " at work in a coal pit, assaulted them, broken the "wyskytts and basketts", hewed down the " Turne," and the "rope serving . . . for the wyndinge and drawinge upp of the coles ", thrown earth and timber into the shaft, where the servants were imprisoned, "not beinge able to wynd themselves owt ", and threatened with straw and gunpowder to set fire to the pit.[1]

Many examples could be offered of each special kind of violence. In some cases persons maliciously set fire to the coal and works. At Cliviger, in Lancashire, it was estimated in 1576 that more than £200 worth of coal had been wasted by such action,[2] and not many years later a Scottish collier, John Henry, was hanged, beheaded, and his head placed on a pole beside the mine as a warning to other colliers, for having set fire to the coal " heugh " of Fawside.[3]

For these conditions of lawlessness the method of administering justice was largely to blame. No common set of regulations for mining practice, such as prevailed in the seventeenth century in all the important continental fields—Mons, Liége, Aachen, and the Duchy of Magdeburg—existed in Great Britain. The rules were likely to be just as different at two adjacent mines in the same county, as they were between a mine in Wales and one in Durham. When there were so many standards, it was difficult to enforce any laws whatever, and it is small wonder that many should have acted as if none existed. To get a decision even in the most just cause was often an interminable process. There was seldom a lawsuit which did not give rise to a counter-suit, if the defendant was rich enough to bring one ; and, the longer the case dragged out, the greater was the cost, and the greater also the hardship imposed upon the party who needed an immediate judgment to save his colliery from destruction. In Germany, when a colliery owner benefited by his neighbour's drainage system, he was obliged to contribute his proportional share of the expense.[4] But in Great Britain it was rarely possible to obtain an unbiased and expert judgment as to how much his benefit was worth. Important questions, which had to be investigated by visits to the pits and measurements underground, could never be settled so satisfactorily in England or Scotland as at Liége, where experts made it their business regularly to enter the mines, and kept records and plans of the progress of each enterprise from year to year, even from month to month. Nowhere in Great Britain was there any similar procedure. Visits to the mines could be made only with the consent of the owners,[5]

---

[1] *Duchy of Lancs. Pleadings*, 102/T/10. [2] *Ibid.*, 100/G/4.
[3] Cochran-Patrick, *Records of Mining in Scotland*, p. xlvi. Local inhabitants long remembered this execution, because Henry was thought to have kindled the underground fire which burned in the coal at Fawside for nearly a century (G. Sinclair, *The Hydrostaticks*, 1672, p. 284).
[4] Jars, *Voyages métallurgiques*, vol. i, p. 285. The same obligation applied in the Forest of Dean after 1678 (Galloway, *Annals of Coal Mining*, p. 204).
[5] The lessor often reserved the privilege of viewing the coal workings of his lessee (*Chanc. Proc., James I*, D. 9/10) ; but neighbouring landlords, when they suspected a colliery company of undermining them, had to get an order from the courts before they could enter the mine.

or upon a court order, and there was no place in England as there was at Liége for depositing the reports of such visits, even when they took place. As a consequence, the weight of an opinion based on expert judgment and reliable data was lacking. Finally, except for the Scottish law of 1592, which made it a capital offence to set fire to a colliery,[1] there were practically no punishments prescribed by statute for overt attacks upon a rival's mine. At Mons it had been decreed as early as 1486 that coal barriers (*espontes*)—from fifteen to thirty fathoms thick—should be left between concessions, and early in the sixteenth century it was decreed that persons touching these *espontes* should be sent to prison.[2] Not until 1740 was an Act [3] passed in England against attempts to drown neighbouring collieries. It was only shortly before that date that it had been made felony to set collieries on fire.[4]

The legal machinery for settling disputes in British coal mining during the seventeenth century left much to be desired. Apart from the damage done to colliery plant by *sabotage*, which could have been largely prevented by an efficient administration of justice, the great expense of litigation imposed an unnecessarily heavy charge upon the colliery owners. The more we consider the state of mining law in Great Britain in the seventeenth century, the more we are likely to conclude that the coal industry expanded in spite of it. Nor is this impression offset by a consideration of the probable effects of the private ownership of minerals.

While the transfer of mineral property at the time of the Reformation undoubtedly contributed to the expansion of the coal industry in the Age of Elizabeth,[5] the private ownership of minerals which became a principle of English law at this time was hardly an advantage to the subsequent development of mining. We have seen that a great many of the innumerable disputes which fill most English and Scottish judicial proceedings between 1560 and 1700 arose out of the artificial boundaries between colliery concessions, made to correspond with the boundaries between surface holdings. Even in the seventeenth century, when the area required by a single mining enterprise was much smaller than to-day, the necessity of leaving barriers of coal, which was one of the principal consequences of the private ownership of minerals, added to the expense of litigation, and at the same time involved the loss of great quantities of coal, which could have been worked if the granting of mining leases had been taken out of private hands.

It has sometimes been argued that the undoubted disadvantages of private ownership have been offset, particularly in the early stages of the mining industry, because private ownership enlists the interest

---

[1] *Acts of the Parliaments of Scotland*, vol. iii, p. 575 (" For the better puneis-ment of the wickit cryme of setting of fyre in coalheuchis be sum ungodlie personis upoun privat revenge and despite ").

[2] Decamps, *Mémoire historique sur l'industrie houillère de Mons*, p. 155.

[3] 13 Geo. II, c. 21. This Act appears to have been passed in response to complaints of certain colliery owners in south Nottinghamshire, that their rivals had maliciously flooded their pits (see *The Case of the Petitioner John Fletcher and his Co-Partners . . .* , 1739).

[4] 10 Geo. II, c. 32.   [5] See above, pt. ii, ch. i (i).

of the landlord in the development of the mineral wealth of the country. In some cases, as in Cumberland,[1] the opportunity afforded private persons to acquire mineral-bearing property, as a result of the secularization of lands in the sixteenth century, undoubtedly stimulated the growth of the coal industry. But, in many cases, the greatest collieries were started, not in manors which had come into private hands, but in those held by the Crown.[2] Although the interest of great lay landowners in mining in the sixteenth and seventeenth centuries was encouraged by the facilities afforded before the Civil War to obtain mineral concessions on favourable terms, they were scarcely less eager to invest capital in mines owned by other persons than in those which formed a part of their own estates.[3] The private ownership of minerals was not indispensable in order to elicit a great deal of interest from this class. In France, where the mines were temporarily granted to the landowners in 1601, on the ground that they would be more effectively exploited under a *régime* of private ownership, the result was disappointing, and the first period of intensive development in French coal mining occurred after the passage of the decree of 1744, which once more took the ownership of minerals out of private hands. Monsieur Rouff's investigation does not suggest that the change in ownership discouraged the great landlords from investing their capital in the coal industry. It shows, on the contrary, that during the half century preceding the Revolution, the French landowning aristocracy, which held aloof from most industrial ventures, made an exception in the case of coal mines, and became heavily interested in colliery concessions granted by the Crown.[4]

In England the disadvantages of private ownership in the sixteenth and seventeenth centuries were not so great as in France. This was especially true during the reigns of Elizabeth and James I, when, as a result of the great increase in the demand for coal, the royalties exacted were low.[5] As royalties increased and landlords came to regard their minerals primarily as a source of revenue rather than as an opportunity for investment, the disadvantages of private ownership were augmented. The English were able to afford it, partly, perhaps, because the natural conditions were more favourable to the development of colliery enterprise than on the Continent,[6] partly because the decisions of the law courts, which deprived many seventeenth-century landholders of rights to coal under their holdings, minimized the evils which are an inevitable result of the multiple ownership of the subsoil.

While the private ownership of minerals hardly contributed to the expansion of the coal industry in the sixteenth and seventeenth centuries, and while the manner of settling disputes over mining was expensive and inefficient, the decisions of the courts in cases relating to the mineral rights of manorial tenants appear to have encouraged indirectly the rise of modern capitalism. Wherever coal mining became important, it stimulated the movements towards curtailing the rights of customary tenants and even of small freeholders,

---

[1] See above, pp. 143–4.
[2] See above, pp. 144 sqq.
[3] See below, vol. ii, pp. 9–17.
[4] Cf. below, vol. ii, pp. 4 n., 5.
[5] See above, pp. 322–3, 326.
[6] Cf. above, p. 133.

and towards the enclosure of portions of the wastes. And it is not unlikely that the claims of customary tenants to wayleave rents sometimes provided a motive for their ejection, and for the consolidation of their holdings by means of enclosures. When we consider that pits were being sunk, adits driven, and ways for carrying coal made through hundreds of manors in more than a score of counties, it seems probable that the expansion of the industry—for which, as we have seen, there was no parallel on the Continent—had a considerable bearing upon the decline of peasant proprietorship and the increase of the great landed estate. Moreover, the supply of cheap coal made possible, as has been pointed out, the expansion of other industries, such as salt-making, metallurgy, and even to some extent textiles—all of which required extensive tracts of countryside, to be used for purposes with which the rights of small freeholders and customary tenants interfered. While the relation of coal mining to the agrarian changes was doubtless less important than that of sheep farming, coal, like wool, should be regarded as a factor which helps to explain why the history of land tenure in sixteenth- and seventeenth-century England differed from that in France and other continental countries.

The enclosure movement is commonly associated with a more intensive cultivation of the land, and an increase in the proportion of the population engaged in industrial and commercial pursuits. In its later phase it is regarded as an important aspect of the Industrial Revolution of the late eighteenth and early nineteenth centuries. If there is, in fact, a connection between the decline of peasant proprietorship in Great Britain and the beginnings of modern capitalism, it follows that the rise of the coal industry helps indirectly to explain the latter. As we attempt to show in Part IV, it has also a direct and more important relation to the rise of capitalism.

8.

PART IV

COAL AND CAPITALISM

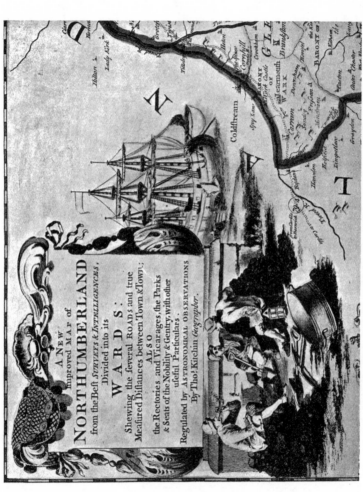

MINERS WITH THEIR EQUIPMENT, AND A COLLIER LOADING

(" A New Improved Map of Northumberland," in Bowen's *English Atlas*, 1769, f. 27)

PART IV

COAL AND CAPITALISM

ECONOMIC theorists have sometimes held that the gauge by which we may measure the growth of a capitalistic society, much as we may tell the age of a turtle by the number of marks on its shell, is the length of the chain separating the producer of goods from the consumer. Thus, the greater the differentiation of stages in the productive process, the more we may count ourselves capitalistic. With respect to domestic fuel, however, in spite of the manifold advances in capitalism since the time of Milton, the London householder, in point of the number of operations required to bring coal to his cellar, is to-day no farther from the producer than he was in the seventeenth century. To convince ourselves that this is true, we need only turn to Defoe (who, though he wrote early in the eighteenth century, painted a picture not unfaithful to conditions fifty years before), and follow with him " the Process of the Coal-Trade from the Mine to the Consumer " of Newcastle coals :—

" They are dug in the Pit a vast depth in the Ground, sometimes fifty, sixty, to a hundred fathoms, and being loaded, for so the Miners call it, into a great basket or Tub, are drawn up by a wheel and horse, or horses . . . to the top of the shaft or pit-mouth and there thrown out upon the great heap, to lye ready against the Ships come into the Port to demand them.

" They are then loaded again into a great machine, called a Waggon, which by the means of an artificial road, called a Waggon-way, goes with the help of but one horse, and carries two Chaldron or more at a time ; and this sometimes three or four miles to the nearest River or Water-carriage they come at ; and there they are either thrown into or from a great store-house, called a Stethe, made so artificially, with one part close to or hanging over the Water, that the Lighters or Keels can come close to or under it, and the coals at once shot out of the Waggon into the said Lighters, which carry them to the Ships ; which I call the first loading upon the Water.

" From the Lighters they are thrown by hand into the Ships, which is the second loading.

" From the Ships being brought to London they are delivered by the Coal-meeters into Coal-lighters or Vessels in the river, which is the third loading.

" From those lighters into the great West country Barges, suppose for Oxford, or Abbington ; which is the fourth loading.

" From those barges they are loaded into carts and waggons to be carried to the respective Country Town to the last consumer ; which is the fifth loading.

" But if you include the account of their digging and loading into the Waggons, this makes seven several removes, all which adds to the expences, and heightens the price of Coals ".[1]

To-day, when a railway wagon, loaded often at the pithead, brings coal direct to the warehouse of some large London merchant or co-operative society, the process is, in point of the number of " removes ", actually simpler than it was in Defoe's time.

But it is not possible to judge the extent to which an industry is capitalistic simply by the number of removes between producer and consumer. It is necessary to examine also to what extent labour is subdivided within the industry, and to what extent this labour is carried on entirely by wage-earners. In both respects we find that, by the middle of the seventeenth century, coal mining must be regarded as a " capitalistic " industry. Both in England and in the Low Countries, the miners were separated already into several groups or gangs, performing different functions. Some were occupied exclusively in sinking pits, others in digging adits for drainage, others in pumping out water, others in repairing machinery and equipment, still others in timbering and keeping clear the passages underground ; there were the hewers who severed the coal at the face, the " barrowmen " or " hurriers ", who dragged it to the " eye " or shaft bottom, the " winders " who raised it to the surface, the " weallers "[2] or " ridlers " who separated it from the stones and dirt and graded the different varieties according to their quality.[3] There were a number of other craftsmen, such as carpenters, wrights, coopers, and smiths, who supplied tools and machinery for the workmen at the pits. Thus a mine could with justice be called " a Jack of all Trades shop ".

To describe the organization at the mine is not, as Defoe's words should remind us, to describe the organization of the whole industry. Towards the end of the seventeenth century coal at the pit head fetched from two to four shillings per ton ; but it is doubtful whether consumers paid on an average less than ten shillings ; indeed, Londoners considered themselves fortunate when they could buy their supplies at the rate of fifteen shillings per ton, or twenty shillings per London chaldron.[4] Between £1,000,000 and £1,500,000 per annum was probably paid by the consumers of British coal by the time of the Restoration.[5] By the end of the seventeenth century the proceeds were probably somewhat larger, and, although an increasing proportion of them went to the Crown and to municipal governments in the

---

[1] *The Complete English Tradesman*, 1726, vol. ii, pt. ii, pp. 29–30.
[2] *S.P.D., James I*, vol. lxxxvii, no. 61.
[3] Owen, *Description of Pembrokeshire*, pt. i, pp. 91–2.
[4] On prices see below, vol. ii, pp. 79–83, and Appendix E.
[5] Assuming that the annual output between 1660 and 1700 approached three million tons. The coal carried by sea, which amounted to something like a million tons at the end of the seventeenth century, sold for 15s. per ton and often more, so that the proceeds on sea-borne coal alone must have exceeded £500,000 every year, and may have approached £1,000,000 in some years. It is impossible to estimate, with any feeling of assurance, the proceeds on the sale of river- and land-borne coal, but it is improbable that they fell short of £500,000 per annum after the Restoration. (Cf. above, Tables I and VI, and below, Appendix D and E.)

form of taxes,[1] the amount which the various middlemen paid for coal at the pits was much less than half the amount which they realized from its sale.   More labour was engaged and more capital invested in the business of transporting than in the business of mining coal. There were workmen who loaded the wagons at the mine, others who drove these wagons to the river, keelmen who manned the lighters, seamen who manned the ships, coal-heavers who unloaded the cargoes into barges, lightermen who manned the barges, porters who stored the fuel on the wharves or in coalyards or warehouses, carmen who drove it in carts to the consumer.[2]

The vast majority of all the workers engaged both in the mining and the transport of coal were hired for wages, and had come to depend for their living entirely upon the adventurers who employed them.[3]   There was no other British industry of equal importance which had advanced so far on the road to modern capitalism.   This observation leads naturally to the question : How far does the expansion of the coal industry in Great Britain at an earlier period than in any other part of the western world account for the fact that the new capitalistic order, which, before the reign of Elizabeth, had found more fruitful soil in Italy, Flanders, and southern Germany than in England, should have obtained, during the seventeenth and early eighteenth centuries, a tighter hold on the economic life of England than on that of any continental country ?   How far, in other words, is the growth of modern capitalism as the dominant form of economic organization related to the rise of the coal industry ?

---

[1] See below, vol. ii, pp. 308–12.
[2] This description is meant to fit the trade from the Tyne to London.   With a few alterations, however, it applies to all the trade in coal shipped by water.
[3] Cf. below, pt. iv, ch. ii.

CHAPTER I

A FERTILE FIELD FOR THE NEW ECONOMIC ORDER

(i) *Capital Costs of Mining Coal*

THE new industry provided a fertile field for the development of capitalistic forms of enterprise, because the amount of capital required to participate at all either in coal mining, or in the coal trade, was, in an increasingly large majority of cases, much greater than any group of manual workers could raise. Let us consider first the costs of starting a colliery.

" At Newcastle, as in Wales ", wrote an anonymous petitioner, who urged Queen Elizabeth to prevent the export of mineral fuel to foreign countries, " the coalworks . . . are so greatly wrought that they are grown so deep and drowned with water, as not to be recovered without extreme charges ".[1] The need for more capital was not limited to any one coalfield, but had become general throughout the country before 1600. There remained a few areas, like the hilly country west of Halifax, or the Ribbledale above Clitheroe,[2] where seams outcropped at many different places so that very shallow diggings could still meet all local demands,[3] and where the nature of the terrain made it possible to sink pits exclusively on high ground, and thus to postpone the drainage problem. Under these conditions a few odd pounds of stock might suffice, as they had sufficed during the Middle Ages,[4] to carry on a small mining venture. During the seventeenth century, however, owing to the increasing demand for large supplies of coal, the areas in which only the surface deposits were worked became relatively unimportant.

From every district comes evidence of the increasing depth of the pits.[5] A mine of twenty-four fathoms (counting six feet to a fathom)

---

[1] *H.M.C., Report on the MSS. of the Marquis of Salisbury*, vol. xiv, pp. 330–1.

[2] Lister, " Coal Mining in Halifax ", in Wm. Wheeler, *Old Yorkshire* (2nd Series), 1885, pp. 275–6 ; *Duchy of Lancs. Pleadings*, 96/T/15.

[3] Pits varying in depth from 4 to 10 fathoms were common during the reign of James I, e.g. in the Mendip Hills (*V.C.H. Somerset*, vol. ii, pp. 382–3), in Westerleigh manor, Kingswood Chase (*Chanc. Deps.*, P. 18/H), near Newcastle-under-Lyme (*Star Chamb. Proc., James I*, 218/13), and in the West Riding (*V.C.H. Yorkshire*, vol. ii, p. 356). Even at these shallow depths, the capital required might be beyond the means of any but the wealthy ; both in the Mendip Hills and in Kingswood Chase, as we know from other sources, drainage had become already a serious problem at a number of pits. Several hundreds of pounds had been invested in a colliery in Bradford manor, near Manchester, where one of the pits was only 11 fathoms deep (*Star Chamb. Proc., James I*, 106/7).

[4] See above, p. 8.

[5] Not all coal mines in the seventeenth century were exploited by means of vertical pits. Apart from open work mining, which doubtless still predominated in certain remote parts of the country, some of the more important seams, notably those along the banks of the Severn near Broseley and in Cannock

at Brandon in south Durham, and a thirty-six fathom mine at
Gateshead, are heard of shortly after the middle of the sixteenth
century ; [1] and in the first quarter of the seventeenth century mining
at such depths was commonly undertaken in the Tyne valley
by the owners of all the principal enterprises there.[2] In Scotland,
before the Civil War, Robert Bruce's famous colliery at Culross had
been dug to a depth of forty fathoms under the sea ; [3] and in the
important early coalfield of East and Mid-Lothian mining was seldom
carried on less than sixteen fathoms beneath the surface.[4] In Pembrokeshire, according to Owen, who wrote in 1603, it was common
to sink twelve, sixteen, or twenty fathoms before reaching the coal,
" whereas in ould tyme fowre fathome was counted a great laboure ".[5]
    Before 1700, mines at least twenty fathoms deep were common
in nearly all the British coalfields.   Descending, as they frequently
did, one hundred and twenty feet or more " from the grass " in Yorkshire and Lancashire, in the Midlands,[6] in the Lothians,[7] and on the
West Coast,[8] and three to four hundred feet near the Tyne and Wear,[9]

Chase, Staffordshire, could be reached by direct levels driven into the hillside,
sometimes for several hundred yards ; the coal being extracted as easily as
from a quarry, " without windlass, rope or corf " (Galloway, *Annals*, p. 203 ;
Plot, *Natural History of Staffordshire*, pp. 129–30 ; *V.C.H. Shropshire*, p. 451).
In Scotland such mines received the name " creeping heughs ", as distinguished
from " windlass heughs ".   David Earl of Wemyss drove a " level mine "
1,200 yards into the hills rising up from the Fifeshire coast (Sinclair, *Statistical
Account of Scotland*, vol. xv, p. 331 ; vol. xvi, p. 519). But the great majority
of mines in the seventeenth century were worked by vertical shafts.   And
even where levels could be driven, it might be impossible to escape heavy
capital charges.
    [1] *Augm. Partics. for Leases*, 39/78 ;   *Court of Augm. Proc.*, 14/76 ;   *Court
of Augm. Misc. Bks.*, vol. cxi, ff. 16–19.
    [2] In James I's reign thirty-five fathoms was by no means an unusual
depth for pits along both sides of the Tyne ; elsewhere in Durham and
Northumberland ten to twenty fathoms was the common depth (*Exch. Spec. Com.*,
no. 5037 ; *A History of Northumberland*, vol. iv, p. 173, note ; *Star Chamb. Proc.,
James I*, 53/10).
    [3] Birch, *History of the Royal Society*, vol. i, p. 339.
    [4] Sir William Brereton, " Travels in Holland, etc.", in *Publications of
Chetham Soc.*, vol. i, p. 112.
    [5] *Description of Pembrokeshire*, pt. i, p. 90.
    [6] Birch, *op. cit.*, vol. i, p. 133 ; *Roy. Soc. Philos. Trans.*, vol. viii, 1674,
p. 6181 (reference to finding certain white liquor in a Yorkshire coal pit, forty-
nine yards deep) ; Galloway, *op. cit.*, p. 185 ; *V.C.H. Derbyshire*, vol. ii, p. 353 ;
Plot, *Natural History of Staffordshire*, p. 125 ; Dud Dudley, *Mettallum Martis*,
1665, p. 36.
    [7] Sinclair, *Hydrostaticks*, p. 280.
    [8] *Royal Soc. Philos. Trans.*, vol. xii, 1677–8, p. 895.
    [9] J. C., *The Compleat Collier*, p. 32 ; G[uy] M[eige], *The New State of
England*, 1691, p. 164 ; Lord Harley, *Journeys in England*, in *H.M.C., Report
on the MSS. of the Duke of Portland*, vol. vi, p. 106. Galloway in his *Annals*
(p. 157) speaks of shafts near the Tyne and Wear 300 to 600 feet (50 to 100 fathoms
deep. He gives no authority for this statement, and I myself have been unable
to find any record of pits deeper than 60 fathoms (360 feet). In his *History of
Coal Mining* (p. 76) Galloway himself mentions 60 fathoms as the greatest depth
attained in the coal pits before 1700, and one is inclined to think that his " 600
feet " may have been a slip of the pen. According to Decamps (*Mémoire historique
sur l'industrie houillère de Mons*, p. 139), some pits near Mons were sunk at the
end of the seventeenth century to a depth of from 80 to 95 fathoms, "the
maximum before fire engines ".

the pits had reached depths not generally exceeded to-day in some of the bituminous fields of the United States. To the modern British or German mining engineer, accustomed to shafts ten times as long, up and down which cages, driven by electrical power, rush at a speed of a hundred feet a second, three or four hundred feet seems insignificant; but to the seventeenth-century Englishman this was a formidable distance to descend into the earth, and an unknown distance to rise above it. Though members of the Royal Society listened to a description of sal-gemine mines in Poland two hundred fathoms deep, though they heard of the profundity of certain silver or copper workings in Bohemia, the Tyrol, or the Harz, they found in the British coal pits, which in the matter of depth developed simultaneously with those of Liége and Mons,[1] their first native experience of a large industry carried on so far underground.[2] Like Guy Meige, they found it " a great Depth for workmen to go and rake a Livelyhood ".[3]

It was a depth at which the labour of " finding " the seams, unless they had been revealed by previous workings, required large capital reserves ; for the expense was uncertain, and out of the quest, as Crown Commissioners reported in 1589, " infinite charge may grow ".[4] A single shaft might be sunk twenty or twenty-five fathoms for a relatively small sum, in places where the strata were neither so hard as to hold up the sinkers for want of efficient methods of blasting,[5] nor so soft as to make successful timbering well-nigh impossible. But the first shaft rarely struck the coal. In sinking, the capitalist was guided in Elizabethan times " by the judgment of those that are skillful in choosing the ground for that purpose ".[6] Their skill depended largely upon the extent to which the mines in the district had been exploited ; for every new success in reaching coal added to

---

[1] Pirenne, *Histoire de Belgique*, vol. iii, pp. 248–9 ; Decamps, *op. cit.*, p. 139.

[2] Depths of 30 fathoms and more are heard of in the Scottish gold, the Devonshire silver, the Derbyshire lead, and the Welsh copper mines, and in the stannaries (Lewis, *op. cit.*, p. 9 and note, also *S.P.D., Eliz.*, vol. clxx, no. 37), but such depths were most common in the coal industry, which had far outstripped the other native mining industries in importance.

[3] G[uy] M[eige], *The New State of England*, p. 164.

[4] *Exch. Spec. Com.*, no. 966.

[5] Gunpowder for blasting a " stone head " (or bed of solid rock) was introduced into England soon after the Civil War (*Royal Soc., Philos. Trans.*, vol. i, pp. 82 sqq.), but although this method of blasting was practised at lead mines in the Mendip Hills, at copper mines in Staffordshire, and at tin mines in Cornwall (Granville Poole, *Historical Review of Coal Mining*, 1924, p. 83 ; Galloway, *History of Coal Mining*, p. 61), Galloway has found no evidence that explosives were turned to account in sinking coal pits until long after 1700. Sometimes, at least in the south Staffordshire colliery district, they made " great fires to soften the rocks to make them yield to the pickaxe " (Plot, *Natural History of Staffordshire*, p. 134), a process known as " annealing " and adopted before 1668 to reach lead mines in Somerset (*Royal Soc., op. cit.*, vol. iii, p. 767). After " annealing ", the miners found it dangerous to enter the mine again " before it be quite clear'd of smoak ; which hath killed some ". It was observed that the collieries in which these fires had been burnt were seldom free from damp ; and this fact may have prevented the general adoption of the process. In the Wear district, at the beginning of the eighteenth century, sinkers still depended on well-sharpened picks to pierce the rocky strata (J. C., *The Compleat Collier*, p. 22).

[6] *Duchy of Lancs. Pleadings*, 185/T/12.

the data available concerning the nature, the position, and the declivity of local seams. Even in the most extensively exploited district, the miners sank many useless shafts.[1] Prospecting charges were reduced somewhat at the beginning of James I's reign by the invention of boring rods,[2] which made it possible to try the strata without digging pits. While useful for revealing the absence or the presence of mineral fuel in the earth, the rods employed during the greater part of the seventeenth century were too crudely constructed to enable prospectors, without sinking trial shafts, to plan the lay-out of a colliery, or even to determine whether the nature and thickness of the seam were such as to make it desirable to mine at all. " I have known ", wrote George Sinclair, the seventeenth-century Scottish mineralogist and mathematician, " a Coal boared, which the Boarer . . . hath judged four foot in the thickness, yet . . . hath not proven one ".[3] And, although the experience gained during the seventeenth century concerning the position of the seams tended to diminish the amount of futile prospecting within the mining districts,[4] the greater depth of the mines increased the expense of such prospecting as remained indispensable. Adventurers had to be prepared to spend anything from £100 to £1,000 or more before reaching seams.[5]

Frenchmen at the beginning of the eighteenth century considered coal mining at depths of twenty-five to thirty fathoms so costly a venture that, as Monsieur Turgot, *intendant* at Tours, wrote to the *Contrôleur Général*,[6] no local capitalists would undertake to start a new colliery in the Anjou coalfield, although the existence of valuable seams at such depths had been established. These capitalists were concerned less with the expense of sinking pits, than with the expense of keeping the pits free from water once they had been sunk. " To describe the miners' battle with water underground," observes a well-informed student of the coal industry in Belgium, " is to write, in brief, the history of coal-mining exploitation before the introduction of steam pumping engines ".[7] For the British mine owner, too, drainage dwarfed all other problems in importance, and determined to a large extent the structure of the seventeenth-century colliery.

Mines were drained either by adits (narrow tunnels driven from a convenient point in the nearest valley on a slight incline up to the seam), or by various types of pumps,[8] operated usually by horses, less often by the force of a falling stream, occasionally (at the smallest

---

[1] *Exch. Deps. by Com.*, 29 Eliz., East. 4 (testimony of Thomas Elmer) ; ibid., 13 Charles I, East. 14 ; *Exch. Spec. Com.*, no. 1739.

[2] For a description of seventeenth-century boring technique see Appendix N.

[3] *The Hydrostaticks*, 1672, p. 295.

[4] On the searches made for coal in places where none was to be found, see below, vol. ii, pp. 72–3.

[5] J. C., *The Compleat Collier*, p. 22. See also *V.C.H. Salop*, p. 450 (the town of Shrewsbury decides to raise £100 " towardes the fyndinge of the colle pit ").

[6] Boislisle, *Correspondance des Contrôleurs Généraux*, vol. iii, no. 496.

[7] Decamps, *op. cit.*, p. 142. The wettest strata, met with first not far from the surface (but which the medieval coal miners avoided by sinking new pits), do not continue down beyond 300 to 600 feet (Galloway, *Annals*, p. 157).

[8] For a description of the various types of pumping machinery in use and the technique of driving adits in the seventeenth century, see Appendix O.

mines) by hand labour.[1] Investigators have not been very successful in tracing the existence in Great Britain of genuine adits, as distinguished from mere ditches open to the sky, farther back than than the early sixteenth century.[2] By the beginning of the seventeenth century, however, both adits, or " soughs " (the name most frequently given them in the Midlands),[3] and pumping machines, had come into common use in all the coalfields. They were to be found, not simply at large mines with access by water transport to wide markets, but at small inland enterprises like those at Morpeth in sparsely settled Northumberland and at Stratton-on-the-Fosse in Somerset.[4] Whoever compiles a Domesday Book for the reign of James I will have to count the adits and the pumps by hundreds, if not by thousands.

Adits could serve, of course, only such mines as were above the level of free drainage, having an elevation somewhat higher than that of the adjacent valley. With the increasing depth of mines,

---

[1] Windmills were also tried ; but the wind was found too capricious an element to supply the constant force indispensable for drainage (*The Compleat Collier*, p. 22 ; Galloway, *Annals*, p. 160).

[2] While the adit was introduced on the Continent before 1300, and was used in medieval times to drain coal pits round Liége, Lewis was unable to trace it farther back than the seventeenth century in connection with the tin mines of Devon and Cornwall (*Stannaries*, p. 11). It has been generally supposed that adits were driven to drain English coal pits during the Middle Ages —indeed, the Middle Ages have been called the " pit and adit stage " of coal mining (Galloway, *op. cit.*, pp. 56-7, 71)—but this view, in the cases with which I am familiar, is based on the assumption that the word " aquaeductus ", or " watergate ", implies a genuine tunnel ; whereas it means, if strictly rendered, simply a passage for water ; and this passage might have been merely an open ditch. Sinclair speaks of the adit in 1672 as a comparatively new invention in Scotland, where one could still see the deep open ditches, or " cuts ", formerly used for draining the collieries (*Hydrostaticks*, p. 298). There is, however, no doubt that genuine adits became common in English mining during the sixteenth century. Fitzherbert, one of the earliest authorities on husbandry, referred in 1539 to a means of drying marsh ground for farming, by making " a sough undernethe the erthe as men doo to gette cole, yron, stone, leade or tin " (*Surveyinge*, chap. xxxv) ; and Owen in 1603 defined the word " levell ", used by the coal miners in Pembrokeshire, as " a way digged undergrounde " (*Description of Pembrokeshire*, pt. i, p. 90).

[3] " Suffe ", a term used in Lancashire (*Duchy of Lancs. Pleadings*, 108/W/12), and " surfe ", a term used in the Forest of Dean (*Exch. Deps. by Com.*, 27 Charles II, Mich. 28), were apparently corruptions of " sough ". There were many other names for these passages. In the Mendip Hills an open ditch was apparently called a " gout ", an adit proper a " cutt pitt " (*V.C.H. Somerset*, vol. ii, p. 380 ; *Land Rev. Misc. Bks.*, vol. ccvii, ff. 138–46) ; in Kingswood Chase the words were " trench " and " levell " (*Exch. Deps. by Com.*, 27 Charles I, Mich. 29). A ditch in Staffordshire was called a " gutter ", in Carmarthenshire a " channel " or " cut ", in Lancashire a " great ditch " (*Chanc. Proc. Eliz.*, U. 1/7 ; *Star Chamb. Proc., James I*, 288/5 ; *Duchy of Lancs. Pleadings*, 156/B/5) ; while an adit in Scotland, as in Pembrokeshire and Kingswood Chase, was called a " level ", in Northumberland a " trencher pit ", in Flintshire a " witchet " (Sinclair, *Hydrostaticks*, p. 298 ; *P.C.R.*, vol. xxxi, p. 622 ; *Royal Soc., Philos. Trans.*, vol. xii, pp. 895–9).

[4] *Exch. Spec. Com.*, nos. 1760, 5037 ; *Chanc. Proc., James I*, B. 20/72. The earliest reference yet found to the use of a pump for drainage at an English coal pit occurs in a document of 1486-7 (Galloway, *op. cit.*, p. 71). Pumping machinery was rarely used in the Mons colliery district until the middle of the sixteenth century (Decamps, *Mémoire historique*, pp. 152-3), and a water pump erected at a coal mine near Aachen about 1616 was spoken of as if it were a novelty (Käntzeler and Michel, in *Aachener Echo der Gegenwart*, 1873, no. 126).

the number that could be drained by adits diminished. " Soughing "
persisted, nevertheless; for, until the invention of steam power, pumps
could not force up water more than fifteen or twenty fathoms at
a single flight, although various ingenious devices had been contrived
to accomplish greater heights in two or three stages.[1] Where, therefore, the pit heads, if not the seams, were above the level of free
drainage, operators sometimes used pumps to raise the water to the
point necessary to enable it to be drained off down an adit.[2]

A seventeenth-century colliery drainage system frequently
included an adit, and more than one type of pumping machine. Jeffrey
Foxe, a London merchant who helped to finance a colliery in Warwickshire, described with some precision the system which he and his
partners installed about 1600 in order to free from water certain
mines under the manors of Chilvers Coton and Temple, a few miles
to the north of Coventry. They " did sincke divers pitts as well
for the drayninge of the . . . coale mines as allsoe for the gettinge
of the . . . coale, and did make a Soughe to convey the water from
the . . . mynes and . . . a poolé or dame to hold water for the
continewall draininge . . . of the . . . mynes and did force a springe
forth of his course to maintaine the . . . poole with water and brought
the same into the . . . poole, and in the same place did builde an
House for a ginne house, and there did fixe and make both a water
. . . and a horse mylne with Ingens thereunto belonginge which
before were never invented [3] for the Continuall dreyninge of the
. . . Cole mynes ". From the pool the water was made to run down
a slanting wooden trough, carried on poles to a point above the water
mill—a large wheel, which rotated under the impact of the water
pouring against its paddles, and set the pump at the pit head in motion.[4]
The erection of Foxe's drainage system cost him £600.[5] Not many
years later £4,000 were required to make two " engines " and a new
" sough " for the neighbouring colliery at Bedworth.[6]

Where drainage could be accomplished either entirely or in part
by driving an adit, adventurers had to weigh the estimate of its
ultimate cost against that of exhausting the water entirely by pumping
machines. Their choice depended largely upon the nature of the
terrain, the probable presence or absence of rocky strata to be
pierced by tunnelling, and the presence or absence of soft, crumbling
strata which made it difficult to support the walls. If the hillside
sloped steeply, so that there was but a short distance between the

[1] See Appendix O.
[2] For a description of such a process see Houghton, *Husbandry and Trade*,
vol. i, p. 106.
[3] The owners of the coal, which Foxe worked on lease, denied that his
" Ingins " differed from the " usuall and comon workes " (cf. above. pp. 242–3).
[4] See the diagram in Galloway, *Annals*, p. 158. For descriptions of water
mills at small collieries in the Mendip Hills and in Kingswood Chase, see *Parl.
Surveys, Somerset*, no. 39 ; *Chanc. Proc., James I*, B. 20/72 ; *Exch. Deps. by
Com.*, 27 Charles II, Mich. 29. At Culross in Fife the water wheel was 24 feet in
diameter (Birch, *History of the Royal Society*, vol. i, p. 339). These " engines "
had been copied from German water wheels (see above, p. 242), which had had
the same diameter (cf. Agricola, *De Re Metallica*, ed. Hoover, pp. 184 sqq.).
[5] *Chanc. Proc., James I*, F. 4/53.
[6] *V.C.H. Warwickshire*, vol. ii, p. 222.

"head" of the projected adit and the point at which the water could issue into the valley, it might be possible, if conditions of the strata were favourable, to drive a "sough" at a cost of from £20 to £50,[1] which was almost certainly less than the charge of erecting satisfactory pumping machinery.[2] Indeed, drainage "engines", as they were called by contemporaries, might cost far more to set up than short adits, particularly when, as in the case just described, is was necessary to divert a stream from its course and build a reservoir.

In general, however, adits were the more expensive contrivances to construct and pumps the more expensive to operate. Two "soughs" driven during the sixteenth century for draining mines in the Trent and the Wear valleys, cost £1,000 and £2,000 respectively.[3] During the seventeenth century many adventurers invested far larger amounts in adits;[4] Sir William Blackett, a leading Newcastle merchant, is said to have lost £20,000 in an unsuccessful attempt to drain a seam several miles north of the Tyne, by tunnelling clear through a hill which intervened between his colliery and the river valley. Even where mining was undertaken to serve only a small local market, men often spent considerable sums in driving soughs. One driven during the reign of Charles II at Clayton-le-Moors, between Burnley and Blackburn, cost "£700 or upwards";[5] another at Smalley, near the town of Derby, cost from £900 to £1,000.[6] To get the equivalent of any of these sums in terms of modern currency, we must multiply them by six or seven at the least.

It must not be supposed that the expense ended when the adit had been completed. Men had to be employed thereafter in keeping the passage clear, for, notwithstanding the practice of supporting the walls with timber and stone, the adit frequently became choked with falling earth or with rubbish washed down from the mine.[7] In

---

[1] Owen, *Description of Pembrokeshire*, pt. i, p. 90; *P.C.R.*, vol. xxxi, p. 622.

[2] I have found no exact statement of the cost of purchasing and setting up a drainage machine at the English collieries. At Mons, where conditions were similar, such a machine might cost as much as £240 (Decamps, *Mémoire historique*, p. 184).

[3] *H.M.C.*, *Report on the MSS. of Lord Middleton*, p. 149; *Cal. S.P.D. Addenda*, 1580–1625, pp. 327–8.

[4] Along the Tyne and the Firth of Forth, and on the coasts of Ayrshire and Cumberland, capitalists undertook to drive tunnels two miles and more to drain seams in the hills behind (Galloway, *History of Coal Mining*, pp. 57 sqq.; Poole, *Hist. Rev. of Coal Mining*, p. 185; *V.C.H. Cumberland*, vol. ii, p. 359; Sinclair, *Statistical Account of Scotland*, vol. vii, pp. 10–11). When in 1623 Samuel Johnston, a prominent Scottish colliery owner, contemplated driving a new adit to drain his mine at Elphinston in East Lothian, he estimated that the passage "being begun and weill bakit [backed] and holdin foreward, it will necessarilie requyre siven yeares space to bring it to perfectioun, the expenssis quherof will be so great . . . that the work itself will not be able to beare the tenth part of the saide charges" (*Privy Council Register of Scotland*, 1st Series, vol. xiii, pp. 207–8).

[5] *Palat. of Lancs. Bills*, 33/29. £500 was the estimate made in 1651 for the probable cost of driving an adit at another mine not far from Clayton-le-Moors (J. G. Shaw, *History and Traditions of Darwen*, 1889, pp. 67–8).

[6] Galloway, *Annals*, p. 185.

[7] *V.C.H. Somerset*, vol. ii, p. 380; *H.M.C.*, *Report on the MSS. of Lord Middleton*, pp. 181–2; *Star Chamb. Proc.*, James I, 288/5; *Chanc. Deps.*, P./7/T; *Duchy of Lancs. Pleadings*, 156/B/5.

addition, precautions had to be taken to provide it with a continual supply of fresh air—sometimes by sinking ventilation shafts at regular intervals,[1] sometimes by artificial devices to force out or draw off the stagnant air [2]—and to keep a few inches of dead water standing all along the passage, experience having taught that mines drained too dry were more susceptible to damps than others.[3] At Bo'ness colliery, in Linlithgowshire, the charge of keeping an adit in proper condition is said to have " eaten up the whole free rent and commoditie of the . . . heugh ".[4]

Yet the cost of maintaining adits is not to be compared with that of operating most early pumping machinery. The difficulty of obtaining a constant supply of water from altitudes above the mining sites, and the danger that streams might dry up during hot summer months, or be diverted from their courses by rival mine owners before they reached the shafts,[5] made it precarious for colliery owners to depend on the fall of a stream to turn their drainage wheels ; and they almost invariably built " horsemilnes " as well as " watermilnes ".[6] Anyone who reads the manuscripts which mention the drainage of collieries in the seventeenth century will agree that the operators used many times as much horse-power as water-power. In order to transmit horse-power, a wheel smaller than the water wheel had to be erected to turn in a plane horizontal with the ground (rather than in a vertical plane as in the case of a water-mill), and teams of horses were harnessed to a horizontal bar, attached to the axle of the wheel, and were driven round this axle all day and often all night.[7]

The costs of drainage by horse-power were scarcely more certain than the costs of finding a coal seam ; but they were nearly always great.[8] When in 1676 water flooded Sir Ralph Delaval's colliery at

---

[1] See Sinclair, *Hydrostaticks*, pp. 291–2. The cost of these ventilation shafts, " in regard to their vast depth, hardness of the Rock, drawing of Water, etc., doth sometimes equal, yea, exceed the ordinary charge of the whole Adit " (*Royal Soc., Philos. Trans.*, vol. i, 1665–6, pp. 79–81).

[2] Galloway, *Annals*, p. 162 ; *Royal Soc., op. cit.*, vol. i, pp. 79–81.

[3] *Ibid.*, vol. xii, 1677–8, pp. 895–9 ; Sinclair, *Hydrostaticks*, pp. 285–6. A " Memorial concerning the Kinneil Coal Work ", written in 1717 (*Hamilton MSS.*, no. 592, ii, bdl. 2), states that " there must be special care in carrying up the Level by having 5 inches of dead water in the nether side of the Level roume ".

[4] *Privy Council Register of Scotland*, 1st Series, vol. xiii, pp. 751–2.

[5] *V.C.H. Somerset*, vol. ii, p. 383 ; *Palat. of Lancs. Bills*, 33/29. And cf. above, pt. iii, ch. v.

[6] *Privy Council Register of Scotland*, 1st Series, vol. xiii, pp. 668–9. See also below, pp. 375–6. John Gilbert, in seeking a patent for a pumping device, was careful to state that his " Engin . . . [might be] moved and driven either by some Current or Streame of Water, or, for want thereof, by strength of Horses ". (Rymer, *Foedera*, 1704–35 ed., vol. xvii, p. 102.) For driving the pump which drained his famous colliery at Culross, Robert Bruce had both a " wattermilne " and a horse-mill (*Privy Council Register of Scotland*, 1st Series, vol. vii, pp. 313–14 ; Taylor, *The Pennyless Pilgrimage*, 1618, in Hume Brown, *Early Travellers in Scotland*, pp. 116–17). At Ravensworth near the Tyne, Sinclair found the pumps driven by water (*Hydrostaticks*, pp. 298–9), but the colliery there was particularly well situated for obtaining a supply from the stream which flowed through the manor. We cannot be sure that horse-power was dispensed with at all seasons of the year.

[7] See diagram, Galloway, *op. cit.*, p. 168.

[8] Cf. Primatt, *The City and Country Purchaser and Builder*, 1680 ed., pp. 28–30.

Seaton Delaval, " £1,700 was spent upon engines and they could not sink [the flood] . . . an inch, yet £600 more emptied it, so that it had no more than ordinary springs ; and in about six weeks he raised coal again ".[1] At the " poorest " of all those Firth of Forth collieries whose owners sold part of their output for export, the " ordinarie chargeis for intertenying of thair watter ingynis ", in 1609, were said to have exceeded every week 300 Scottish marks, or about £17 sterling, while at four of the largest collieries—Airth, Alloa, Carriden, and Sauchy—these charges reached 500 or 600 Scottish marks.[2] Early in the reign of Charles I, the partners in Benwell colliery spent every week £26 " more or less, for the horses for draweing the water ".[3] Assuming that drainage was what one Lancashire colliery owner reported it to be at his mine near Manchester, " a dayly and Contynuall Chardge ",[4] it must have cost at least £1,000 per annum at several collieries in Scotland and the north of England. As a matter of fact, we have the word of the two chief undertakers of Bedworth colliery in Warwickshire, that they had spent from £1,000 to £2,500 a year during the last part of James I's reign, in their efforts to cope with the water in their pits.[5] From experience gained in the last decades of the seventeenth century as a colliery manager in the Sunderland district, an anonymous writer estimated the annual charge of maintaining all the drainage engines in the lower Wear valley at several thousands of pounds.[6] These maintenance charges were greatly reduced by the installation of steam engines during the eighteenth century.[7]

Since coal sold at the pithead for less than 4s., and often for about 2s., per ton,[8] it is obvious that the entire proceeds from a sale of 6,000 tons (perhaps the maximum annual output of a single mining enterprise in Great Britain during the Middle Ages,[9] and many times

---

[1] Roger North, *Life of Guilford*, as cited Galloway, *op. cit.*, p. 162.

[2] *Privy Council Register of Scotland*, 1st Series, vol. viii, p. 568.

[3] *Exch. Deps. by Com.*, 12 Charles I, East. 15.

[4] *Star Chamb. Proc., James I*, 106/7. The drainage problem was naturally most serious during the " watry winter months ", when the heavy rains soaked into the earth, causing colliery owners to fear lest their pumping machinery prove inadequate to cope with the additional water (*ibid., James I*, 245/6).

[5] *V.C.H. Warwickshire*, vol. ii, p. 222. The costs of running horse engines at Bedworth colliery in 1684 are said to have reached £40 per week (*Exch. Deps. by Com.*, 36 Charles II, Mich. 43).

[6] *The Compleat Collier*, p. 24.

[7] It was estimated in 1734 that the purchase of a fire engine for Bo'ness colliery, though it would cost £1,500, would reduce the annual expense of drainage from £500 to £200 (*Memorial Concerning the Coal and Salt Works at Burrowstones*, preserved among the *Hamilton MSS.*).

[8] See Appendix E.

[9] An output of almost 6,000 tons was obtained from the Bishop's colliery in south Durham in 1461 (see above, pp. 138–9) and 301 " lasts ", 3½ " weys ", or about 5,000 tons (cf. Appendix C), from a colliery within Gower manor, Glamorganshire, in 1400 (Appendix K (i)). The profits from Wollaton colliery in 1502 amounted to £283 16s. (*H.M.C., Report on the MSS. of Lord Middleton*, p. 302), so that one may assume perhaps that the annual output there had reached, if it had not exceeded, 5,000 tons. Possibly the colliery at Whickham had a larger output than any of these enterprises. But, except for the mines at Whickham, Raby, Wollaton, Strelley, and Gower, there were probably few medieval enterprises with an output of more than 2,000 or 3,000 tons (cf. above, pp. 8–9).

more than the output of most medieval diggings) would not pay even the overhead costs of operating pumping machinery at some seventeenth-century collieries. Although operating charges of such magnitude as those which we have been citing were exceptional, the sum required to drain a new seam in most of the coalfields had grown large enough, and withall unpredictable enough, to put a premium upon mining far greater quantities of coal than had been commonly obtained from one enterprise before Elizabeth's reign. Wherever the " charge of drawing water wilbe great ", it is " not worth undertaking for one pit only ", explained a Crown concessionnaire, who had obtained a grant in 1609 to mine in Kingswood Chase.[1] An " engine " shaft sunk, or an adit driven, to an appropriate part of the seam, could be used to clear the water from a large area of coal at only a slightly greater cost than from a small area.[2] Pumps at Ravensworth in the Tyne valley drained every pit within a radius of three miles ; and Sir John Lowther believed that five-sixths of the sum spent on drainage in the Newcastle district could have been saved by operating all the mines as a single enterprise.[3] The more coal raised within one drainage area, the less would be the weight of drainage charges upon each ton.

But conditions of transport generally prevented the development of genuine large-scale production wherever the coal seams lay at long distances from the sea or from navigable rivers. It would have been useless to attempt to supply fuel for the whole West Riding from one colliery, at—let us say—Elland, just south of Halifax ; for, no matter how much the overhead costs of mining at Elland might be reduced through the economies made possible by a large output, the price of carriage thence to Wakefield, Leeds, or Bradford—a distance of ten miles or more in each case—must have been double or treble the price of producing coal at smaller collieries near those three towns. It would have been equally futile to have attempted to supply the whole of Staffordshire from a mine at Beaudesert or Dudley ; or the whole of Lancashire from one at Wigan or St. Helens or Bolton. Accessible coal seams were found at any number of places scattered about these districts, and (unless there was little difference between the two mines as regarded distance) the consumer's choice was determined almost entirely by the distance from which his fuel had to be fetched, and scarcely at all by the pithead price. All through Yorkshire, Lancashire, and the Midlands, the typical colliery enterprise, of which one finds scores of examples, inevitably commanded only a small market, and had an annual output of from 2,000 to 5,000 tons.[4] Small as such

---

[1] S.P.D., James I, vol. xlviii, no. 118, I.

[2] By driving short tunnels water could be collected in the " sump " or reservoir, at the bottom of the engine pit, from all seams in the drainage basin having an altitude higher than that of the " sump ".

[3] Birch, History of the Royal Society, vol. iii, p. 439. Lowther's scheme probably envisaged the intensive working of the coal within a small area, until the easily accessible coal should be exhausted, after which colliery operations would be transferred to another drainage area.

[4] This estimate of the size of the typical " land-sale " enterprise is not based solely, or even chiefly, on Dud Dudley's well-known statement, made shortly after the Restoration, that there were, in the neighbourhood of Dudley, twelve

an output seems to us, it indicates a great increase in the scale of mining enterprise, when compared with medieval diggings in the same regions. Until the use of steam railroads for the transport of coal, the average production at an inland mine, if one may judge by statistics of 1810 for thirty-five " land-sale " collieries in Durham county, did not exceed 6,000 tons per year.[1]

Where population was concentrated within a small area, or where one mining site was closer than any other to a large consuming district in which no seams had been discovered, there might be an opportunity for a larger enterprise in spite of the heavy cost of transport over ground. Bradford colliery, conveniently placed to supply fuel for the growing town of Manchester, had an annual output of about 10,000 tons in James I's reign.[2] Bedworth colliery probably produced from about 20,000 to 30,000 tons, if not more, throughout the seventeenth century,[3] for its situation at the southern end of the southernmost of all midland fields and near the junction of Watling Street with the Fosse Way, the two most travelled of early English roads, gave the mine owners an advantage over all others in loading fuel to be carried south into the populous counties of Warwickshire, Northamptonshire, Oxfordshire, and Buckinghamshire.[4]

It is in mining districts bordering the sea or navigable rivers that the tendency towards large-scale production is most notable. Mine owners there had access to wider markets than those in " land-sale " districts. Customs commissioners reported in 1695 that the collection of a tax on coal shipped along the river Severn would be easy, because practically all the trade was with three " great collieries " at Broseley, Benthall, and Barr.[5] Since the annual coal traffic on the Severn approached 100,000 tons by this time, in addition to considerable quantities consumed locally in the Broseley district[6],

or fourteen collieries, each producing from 2,000 to 5,000 tons a year (see above, p. 66). All the evidence which I have obtained from manuscript sources tends to show that such came to be during the seventeenth century the usual output of collieries, not only in south Staffordshire but in many inland coalfields (see above, pp. 58–9, 63, 74).

[1] J. Bailey, *General View of the Agriculture of Durham*, 1810, pp. 23–6. The 35 collieries produced 147,080 London chaldrons; that is to say, the average output of a colliery amounted to about 4,200 London chaldrons, or about 5,600 tons. For the size of the London chaldron, see Appendix C.

[2] *Star Chamb. Proc.*, *James I*, 106/7. Seventeenth-century Bristol provided a much larger market than Manchester; and there is some reason to think (see below, vol. ii, p. 30) that, at least for a few years, most of the pits in Kingswood Chase (from whence came all the coal for the city) were operated as one enterprise.

[3] The annual output at Bedworth already amounted to 20,000 loads in 1622 (*V.C.H. Warwickshire*, vol. ii, p. 222). For the probable weight of the load in this case, see Appx. C. The falling off in the output at Bedworth, referred to by Fuller in 1662 (*V.C.H. Warwickshire*, vol. ii, p. 223), was probably temporary, for the colliery was frequently referred to after that time as being of greater importance than any other in supplying Coventry (see, e.g., *Cal. S.P.D.*, 1671–2, pp. 159, 181 ; *Exch. Deps. by Com.*, 36 Charles II, Mich. 43).

[4] Cf. above, pp. 101–2.

[5] *Treasury Board Papers*, 34/51. The same method of collection which the Commissioners propose for the Severn might—they say—be applied on the Trent as well, " if the principal places of shipping were observed ".

[6] See above, pp. 97 n., 230–31.

it may, perhaps, be assumed that each of these three collieries produced upwards of 30,000 tons in a year. A colliery at Winlaton produced about 20,000 tons in 1581,[1] and it was certainly not the most important enterprise in the Tyne valley. Along the south bank of the Tyne, in 1636, were twenty-two collieries producing probably more than 350,000 tons annually, an average of about 17,500 tons for each.[2] From the figures given in the assessment for ship money for that year concerning the " value " of these collieries, it is clear that a few of them were much more important than the others. The value of the four principal enterprises, the Grand Lease, Blaydon, Stella, and Ravensworth collieries, is set down as £9,600 per annum, compared with a total value of £16,250 for all the twenty-two collieries, and the Grand Lease colliery alone is valued at £4,500, or nearly 30 per cent of the total.[3] If it may be assumed that the estimated output of these twenty-two enterprises can be divided between them somewhat according to their value, then it would appear that the four chief collieries produced between them upwards of 200,000 tons per annum, and that the annual output of the Grand Lease colliery already amounted to from 75,000 to 100,000 tons.[4] This enterprise appears to have produced about 50,000 tons annually at the end of Elizabeth's reign.[5] Possibly an output of more than 100,000 tons a year was obtained from it after the Restoration.[6] Before the end of the seventeenth century there is also evidence of large mining ventures on the north bank of the Tyne, on the Wear, on the Trent, on the Cumberland and Northumberland coasts, on the Firth of

---

[1] 23,603 " fothers " (Appx. K (ii)). For the size of the fother, see Appx. C.

[2] See the list of collieries, with their " value ", given in Appendix I. In estimating their combined output at about 350,000 tons, I have been guided by the following considerations. On the eve of the Civil War, the annual shipments of coal from Newcastle amounted to about 450,000 tons (see above, table iv), while the salt works at Shields consumed about 80,000 tons more (see above, p. 208), so that the quantity produced in the Tyne valley must have approached 600,000 tons. By far the greater proportion of this—perhaps 400,000 or 450,000 tons—came from the south bank (cf. above, pt. i, ch. i (i), a). There were possibly a few small collieries on that side of the river which were not mentioned in the document printed in Appendix I, and I have assumed that they produced from 50,000 to 100,000 tons.

[3] See Appendix I.

[4] While the " value " cannot, of course, be regarded as an exact means of estimating output, it appears to provide a fairly satisfactory method of determining roughly the relative importance of the various collieries within a district (cf. Appendix B). We know from other sources that Blaydon, Stella, and Grand Lease collieries were much the most important enterprises in the Tyne valley during the first half of the seventeenth century. From the figures in the assessment, it is reasonable to assume that on the eve of the Civil War only nine of the twenty-two collieries had an annual output of more than 10,000 tons. The remaining thirteen were valued at £300 and £200 apiece.

[5] The output of the Grand Lease colliery is said to have accounted, during the last decade of the sixteenth century, for one-fifth of the total output of coal in the lower Tyne basin (*Harleian MSS.*, 6850, no. 39), then from 200,000 to 250,000 tons (cf. above, pp. 36, 207).

[6] The annual output of coal from the manor of Whickham undoubtedly exceeded 100,000 tons before the Civil War, for the assessment for 1636 shows that not only the Grand Lease colliery, but also at least seven of the smaller collieries, were within the boundaries of the manor.

Forth, and possibly in Glamorganshire.[1] A British census of production, had one been taken during the reign of William III, would doubtless have revealed more than a score of collieries with an annual output exceeding 25,000 tons.

To a modern mind, accustomed to thinking of production in terms of millions, an output of 25,000 tons will seem insignificant. But the historical student must divest himself of present-day habits of thought. He must measure early developments, not by the standards current in his age, but by those current in the age that he strives to recreate. For centuries prior to the reign of Elizabeth most changes had been extremely slow. New economic forms had grown up while others were fading away; yet so measured had been the coming and the going, so gradual the process of evolution, that a contemporary must have found it almost as difficult as does a modern scholar to connect the variation with any particular reign or even with any particular century. With his background of gradual change, an Englishman of the time of Sir Thomas More would have been hardly less astonished, had he lived to see mining enterprise conducted on the seventeenth-century scale, than would have been a contemporary of Adam Smith, had he lived to see the giant holding company or the international cartel of the present day. Before Charles I lost his throne, the annual output of one colliery on the Tyne probably exceeded the entire annual output of coal in Durham and Northumberland, if not in all England, in the reign of Henry VIII. In fact, the rate of growth in the size of a mining enterprise may well have been actually more rapid in the century following 1560 than in the succeeding century. In 1810, nearly fifty years after the Industrial Revolution is supposed to have begun, the average output of one of the thirty-four " water-sale " collieries of Durham, amounted to about 55,000 tons,[2] which is scarcely more than three times the average output of one of the twenty-two collieries on the south bank of the Tyne, as listed in the assessment for 1636. On the Continent, no mine appears to have produced as large an output as that already produced by the Grand Lease colliery before the Civil War, until the middle of the eighteenth century, when the recently discovered coalfield of northern France was first exploited by the Anzin Company on what then seemed a

---

[1] On the north bank of the Tyne, Denton and Benwell collieries, and probably Elswick colliery as well, appear to have produced 20,000 tons and upwards per annum before the end of James I's reign. In 1610 Denton colliery was said to have an annual " value " of £1,200, equal to that of the colliery at Ravensworth in 1636 (S.P.D., James I, vol. lviii, nos. 17–19). In 1631 the output of Benwell colliery was variously estimated at from 100,000 to 300,000 tons during a ten-year period (Exch. Deps. by Com., 11 Charles I, Mich. 46; Exch. Spec. Com., no. 5996). Before the end of the century there were other important collieries farther north, at Walker, Walbottle, Fenham, etc. (cf. above, pp. 26–7). On the river Wear, Harraton and Lambton collieries, and perhaps Lumley colliery as well, were in the same class as Blaydon, Stella, Ravensworth, and Elswick before the outbreak of the Civil War (cf. above, p. 30). For the size of the collieries in south Nottinghamshire, in Cumberland, in Northumberland, and in Scotland, see above, pp. 59–60, 71, 34–5, 46–9. The profit on a colliery near Neath in Glamorganshire is said to have amounted to £600 in the year 1660 (Phillips, Pioneers of the Welsh Coalfield, 1925, p. 15).

[2] Dunn, View of the Coal Trade, p. 29.

colossal scale.[1]   In 1768, the colliery at Vetine in the Duchy of Magdeburg, possibly the largest enterprise in Germany at this time, produced about 25,000 tons per annum.[2]   As late as 1812, there was no single colliery in the Liége district with an annual output exceeding 50,000 tons.[3]

The spread of large-scale mining in Great Britain greatly increased the amount of capital required to enter the industry.   Large overhead expenditures had to be incurred in digging and operating a number of separate pits.   Having sunk their shaft straight down to the seam, which was usually " founde to lye slope in the grounde and seldome downeright ",[4] the miners' natural objective was to work out the coal in all directions as far as possible.   A limit was imposed, however, on the distance from the shaft to which they found it profitable to push their workings.   One cause for this was the difficulty of getting air to circulate through the workings if the headways were dug far from the " eye ", or shaft bottom.   The danger of " damps " was already great in the seventeenth century, and contemporary descriptions of collieries show that the problem of ventilation, like the problem of drainage, added much to the cost of starting and maintaining a mine.   By means of " vent-pits ", which were connected underground with the main shaft, the mining experts sought to provide a better circulation of air than could be secured by a single opening at the " eye ".[5] The cause of all damps, wrote Robert Plot, in summing up the scientific opinion on the question in the time of Charles II, was to be found in lack of motion and most of the remedies " may plainly be reduced to motion, which I take to be the catholic remedy of all damps ".[6] A good circulation was indispensable to drive out the " fire-damp ", before enough could collect to cause explosions, which had become common in all the British coalfields in the seventeenth century.   Set on fire— frequently by the miners' candles—this " sulphurous matter . . . breaks out like a Thunderbolt, carries all away with it higher than the Pits-Mouth, and that with a dismal noise, as it were with a crack of Thunder. In this case one might compare the inflamed Sulphur to Gunpowder, the Coal pit to a great Gun, and what it brings up with it to Bullets, it comes up with such a force ".[7] Often it blew off the drum, or " turn beam, which hangs over the top of the shafts ", and wrecked other parts

---

[1] Rouff, *Les mines de charbon en France*, p. 424.   The annual output at Anzin exceeded 100,000 tons after 1756, 200,000 tons after 1772, and touched the 300,000 ton mark in 1790, by which time its output probably approached, if it did not exceed, that of any colliery in Great Britain.

[2] 2,400 " wispels " (Jars, *Voyages métallurgiques*, vol. i, p. 318).

[3] Thomassin, *Mémoire statistique*, pp. 416 sqq.

[4] Owen, *Description of Pembrokeshire*, pt. i, p. 90.   Seams approaching the vertical were worked, at least in Staffordshire, in the seventeenth century (Plot, *op. cit.*, p. 129), but such workings were undoubtedly rare and need not concern us here.   Steeply inclined measures are seldom mined in Great Britain even to-day, except in Staffordshire, Lancashire, Somerset, and certain parts of Scotland (cf. Granville Poole, *op. cit.*, p. 57).

[5] Birch, *History of the Royal Society*, vol. i, p. 134.

[6] Plot, *Natural History of Staffordshire*, pp. 139–40.

[7] G[uy] M[eige], *The New State of England*, pt. i, p. 165 ; see also Birch, *op. cit.*, vol. i, pp. 135–6.

of the plant.[1] Such material damage disposed the adventurers to
spend money on preventive measures more freely than had " choke-
damps ", which " kill insensibly " ; [2] for rare in all ages is the employer
who does not value his property more than the lives of his workmen.[3]

When he could get a circulation of fresh air by driving a " vent-
head ", or tunnel, to another working pit or to an old abandoned
shaft,[4] which could be reopened, the mine owner escaped the expense
of sinking a vent pit. But this was frequently impossible, and, if
many headways were driven far from the " eye ", the cost of vent-pits
soon exceeded that of sinking a new shaft.[5]

The problem of ventilation was not the only difficulty which caused
the mine owners to abandon going pits. Just as the high charge of
moving coal overland prevented the owners of a " land-sale " colliery
from selling their output far from the pit-head, so the high charge
of mine haulage, before the introduction of horse traction underground,[6] prevented them from mining coal far from the pit bottom.

Although the so-called " long way " or " long wall " system of

---

[1] *Ibid.*, p. 135 ; *Royal Soc.*, *Philosophical Transactions*, vol. x, 1675,
pp. 450–4.

[2] Roger North, *Life of Guilford*, as quoted R. Fynes, *The Miners of
Northumberland and Durham*, 1873, p. 10. On " choke-damp " see also Plot,
op. cit., pp. 133 sqq. ; Sinclair, *The Hydrostaticks*, pp. 284–93 ; *Court of Augm.
Misc. Bks.*, vol. cxii, ff. 17–24.

[3] It is not strictly true that no preventive measures were taken until after
the first explosions. A curious recipe " for the dampe " was proposed by one
William Poole in 1620. By putting unslacked lime in the pit it was thought
the moistness of the damp would slack the lime, thus depriving the damp of
its objectionable properties (*H.M.C.*, *Report on the MSS. of Lord Middleton*,
pp. 192–3). It was not until later that colliery owners began to spend large
sums in ventilating the pits.

[4] At a colliery within the Town Moor operated in 1656 by the Newcastle
Corporation, " the pit now working is troubled by styth for want of Ayre by
reason of the auld Pitts being filled up which were ounce open." The Corporation directed, therefore, that one of the old pits be " Ridded for Ayre " (*MS.
Journal of the Common Council of Newcastle*, 1650–6, f. 355).

[5] Attempts were made to stimulate the circulation of air by building fires
underground, or by lowering into the pit an iron cradle full of burning coal
(Birch, *op. cit.*, vol. i, pp. 133–6 ; Poole, *Historical Review of Coal Mining*,
pp. 128–9 ; Plot, *op. cit.*, p. 138). Plot considered the latter method a " secure
one . . . but very chargeable ".

[6] Professor Granville Poole (*op. cit.*, p. 90) is probably mistaken in thinking
that horses were used for underground haulage in 1667. The passage that he cites
in favour of such a view—" to find sufficient horses and drivers for drawing the
coals to bank "—obviously refers to the use of horses at the pithead to operate
the winding machinery by which baskets of coal were " drawn " up the shaft
(see below, p. 373). During the seventeenth century horses may have been
introduced for traction purposes within the mine, where, as in Shropshire, mining
was carried on by means of levels driven into the hillside, but I have found no
clear evidence that horses were ever lowered into the pits until after 1700.
During the eighteenth century, however, sledges with wheels, made to fit
railed wagonways laid along the underground " lanes ", were dragged from the
working face to the shaft bottom by horses or ponies, lowered into the mines
for that purpose (Jars, *Voyages métallurgiques*, vol. i, pp. 242–4). The use of
horse traction reduced the expense of underground haulage, and led undoubtedly
to a great increase in the distance which workings were carried from the " eye ",
and therefore to a great increase in the quantity of coal raised from and the length
of life of a single shaft. Horse traction helped, therefore, to make it practicable
to undertake the increased sinking charges involved in mining at greater depths.

mining was used, and perhaps had its origin, in the Shropshire coal-field,[1] the system of mining generally adopted in Great Britain was that known as " pillar and stall ".[2] A description of this system in its primitive form is given in a report made in 1610 concerning diggings in the parish of Clutton, in Somerset. " There be now three pits near widow Blacker's house, the highest about four fathoms, the middle six fathoms, the lowest eight fathoms deep. At these depths they cut out their lanes about four feet high and broad . . . The lane we crept through was a good quoit's cast in length, wherein we found but two cross lanes, whereby it may appear that the mine is yet but newly entered into . . . At the end of every lane a man worketh, and there maketh his bench, as they call it, and, according to the vent, they make more or fewer benches ".[3] When the thickness and the strength of the seam made it feasible, these lanes and cross lanes might be higher and wider, and often they were longer and more numerous. In Durham and Northumberland they were called " boards " and " headways ", in Scotland " roumes " and " throughers "; but everywhere they left a similar pattern. Cut into the vein at right angles to each other, these two sets of excavations formed a honey-comb between pillars, or " stoops ", of coal, square or oblong or irregular in shape, which were left standing to support the roof ".[4]

At the " bench ", or " room ", in which the miner hewed coal from the working face, it was loaded into baskets, or " corves ", made of interwoven twigs, and these baskets were then dragged or. pushed along the lanes in the darkness to the " door ", or " eye ", of the shaft, by colliers variously called " bearers ", " hurriers ", " coal-putters ", or " trotters ". In Durham, before the end of the seventeenth century, the baskets were loaded upon a sort of sledge, known as a " tram ",[5] which may or may not have had wheels ; [6] and these trams, which were moved by the strength of two " barrowmen " (one behind, the other in front), undoubtedly effected some reduction in the cost of haulage. Sometimes, at least in Pembrokeshire and

---

[1] Galloway, *Annals*, p. 203. It was also used at some Scottish collieries, for Jars found that no pillars were left in the mine at Carron (*Voyages métallurgiques*, vol. i, p. 268). It was a feature of the long wall system that the whole seam was removed, and the strata overhead supported by filling in the empty space with refuse (cf. *V.C.H. Shropshire*, p. 458).

[2] Galloway, *op. cit.*, pp. 176–7, 180–2 ; *Star Chamb. Proc., James I*, 161/17 (which contains a reference to " the pillers and props . . . made and left . . . for supporting of the . . . colemyne and the earth above the same " at Gateshead) and 228/13 ; *Duchy of Lancs. Pleadings*, 153/F/3 ; *Chanc. Proc., James I*, G. 1/76 ; Birch, *History of the Royal Society*, vol. i, p. 136 (a report concerning certain pits in Lancashire, with a diagram showing " walls or pillars " of coal, which walls " are always left ungotten and to support the roof ") ; Owen, *Description of Pembrokeshire*, pt. i, p. 90.

[3] *H.M.C., 12th Report*, appx. pt. i, p. 71. Owen (*loc. cit.*) describes the same process in his native Pembrokeshire. On reaching the coal seam " they worke sondrie holes one for every digger, some two, some three, or four, as the number of diggers are ".   [4] See diagrams in Galloway, *op. cit.*, pp. 181–2.

[5] J. C., *The Compleat Collier*, p. 39. The word " tram " was used as early as 1635 in connection with the equipment at Harraton colliery on the river Wear— " divers ingines, trammes, shovells, pitts, and Lodges " (*Palat. of Durham, Decrees and Orders*, vol. i, f. 212).

[6] T. J. Taylor, " Archæology of the Coal Trade ", in *Proc. Arch. Inst.*, vol. i, 1852, p. 180 ; Galloway, *op. cit.*, pp. 177–8.

in the north of England, the bearers worked in relays, each of them carrying the baskets a distance of from twelve to twenty yards, or more, and handing them over to another bearer.[1] Whether this haulage was accomplished in relays or by one bearer or by a tram, it is clear " that the more and further a pit is wrought . . . the dearer she lies in the charge of barrowmen ".[2]

Eventually this charge became so great that the mining entrepreneur found it to his advantage to abandon the working for another shaft dug to the seam at an appropriate distance from the old one, even though the problem of ventilation had not become serious. How far from the " eye " it proved economical to push the " rooms " depended largely on the costs of starting new pits. If the seam to be worked lay only six or seven fathoms (thirty-six to forty-two feet) from the surface, the cost of sinking was likely to be small ; both at Westerleigh, in Kingswood Chase, and at Griff, in Warwickshire, pits were sunk to these depths at a rate of 6s. per fathom.[3] At greater depths, owing to the probability of encountering water, gas, or rocky strata, and the increasing difficulty of supporting the walls with timber, every additional fathom was likely to be paid for at a higher rate than the last.[4] Yet, until the use of steam power for drainage, most of the workings were shallow enough so that the cost of sinking a single new pit seldom exceeded £50 to £60 in the north of England and in Scotland.[5] This did not cover, of course, the charge of unsuccessful prospecting. Although there was probably relatively little of this once a seam had been found and a colliery started,[6] new pits

---

[1] Owen, *op. cit.*, pt. i, p. 90 ; *Exch. Spec. Com.*, no. 5996.

[2] J. C., *The Compleat Collier*, p. 39.

[3] Richard Poole, a coalminer of Westerleigh, testified in 1630 that a certain coal pit was 7 fathoms deep, " which cometh at 6 shillings a fathom unto 42 shillings " (*Chanc. Deps.*, P. 18/H). At Griff in 1603 a pit sunk 25½ ells had been paid for at the rate of 3s. an ell for the first 11 ells, 6s. for the next 12½ ells, and 6s. 8d. for the last two ells (*V.C.H. Warwickshire*, vol. ii, p. 221). Three shallow pits sunk within Beaudesert Park in Staffordshire, in 1582, cost £5 14s., £5 12s., and £2 13s. respectively (*Exch. Q.R. Accounts*, 632/17). In the Halifax district, at the beginning of the eighteenth century, pits were sunk from 7 to 10 fathoms at a cost of from £1 10s. to £3 each (Lister, " Coal Mining in Halifax ", in Wheeler, *Old Yorkshire*, 2nd Series, pp. 280–2).

[4] During the forties of the sixteenth century, Thomas Walker, a " sinker ", and two of his fellows were hired by James Lawson, the owner of a colliery at Gateshead, to dig a new pit, their work to be paid for at the rate of 6s. 8d. per fathom. After they had sunk 12 fathoms, Walker testified " he and his felowes did leive worke bicause they were not able to performe ther covenante by that wage " (*Court of Augm. Misc. Bks.*, vol. cxii, ff. 17–24).

[5] The expense of sinking a new pit at Newcastle in 1709 was about £55 (R. Bald, *A General View of the Coal Trade of Scotland*, 1812, p. 8). In Elizabethan times a shaft of from 25 to 35 fathoms is said to have cost in sinking charges from 20 marks (£13 6s. 8d.) to £20 (*Court. of Augm. Misc. Bks.*, vol. cxii, ff. 17–20 ; *Exch. Spec. Com.*, no. 1739 ; *Exch. Deps. by Com.*, 29 Eliz., East. 4). A " new pit " sunk within the " Town Moor " in 1658 by the Corporation of Newcastle cost £21 (*MS. Journal of the Common Council of Newcastle*, 1650–6, f. 383). That was the estimated cost of a new " sink " for the colliery at Kinglassie in Fife at the beginning of the eighteenth century (*Hamilton MSS.*, no. 592, bdl. 2).

[6] When the author of *The Compleat Collier* wrote (p. 23) that £1,000 was sometimes spent in " sinking ", he had in mind the cost of " finding " the seam and of digging the first shaft, not the cost of digging supplementary shaft once the seam had been found.

occasionally produced nothing. In most cases the costs of sinking and of such prospecting as proved necessary were probably not the chief expenses in connection with starting a new pit. To them were added the costs of ventilation and of drainage. Besides the charge of digging one or more vent-pits, and of other protective measures against damps, the mine owner had usually to drive a drainage head to get the water from the new working to run either into the engine pit, or into the adit in case the mine was above the level of free drainage.

Although these charges added to the expense connected with starting a new pit, the increase in the costs of ventilation and underground haulage when workings were carried far from the " eye " forced most colliery owners to sink new pits frequently, which naturally added to the capital required to enter the industry. While the actual sinking of one pit was seldom very costly, even sinking charges might mount up beyond the means of small producers when a number of shafts were required. Before the eighteenth century it was seldom deemed advantageous, even in the north of England, where the mines were deepest, to haul coal more than about 200 yards underground.[1] Even for that distance the cost of " barrowmen " was nearly three times as great as it was for a shaft newly sunk. The workmen might protest if they thought that the headways were too long. At a small enterprise near Halifax, where coal had been worked 180 yards from the " eye ", the " hurriers " declared that the expenses attending " the bringing and hurrying of coals such an extraordinary . . . length of way underground" had diminished their earnings and had caused a " scarcity of coal in the country dependent on this pit ". Some of the hewers refused to go " into the far heads because of their great distance " ; and one " hurrier ", Henry Vicars, said " he would be at some part of the charge of sinking a new pit out of his own pocket rather than hurry so far in the old pit ".[2] One wonders what language these miners would have used, had they been informed that their descendants in the twentieth century would be compelled to walk two and three miles underground to reach the working face.

Exceptional conditions might make it desirable to drive longer headways. Such was the case at Culross, where Sir George Bruce worked seams under the sea by means of a shaft sunk in a reef, free from water at low tide and artificially protected at high tide by the building of " a round circular frame of stone . . . joined together with glutinous and bitumous matter, so high withall, that the sea . . . can neither dissolve the stones, so well compacted in the building, or yet overflow the height of it ". To connect this ocean shaft with their drainage pit, sunk on the shore, Bruce's labourers had driven a passage " more than an English mile under the sea, [so] that, when

---

[1] Galloway, *op. cit.*, p. 179. See also *Exch. Spec. Com.*, no. 5996 (1632), wherein reference is made to a pit at Benwell colliery which was wrought to six " barrowmen rancke "—i.e. 120 yards. For the length of headways in some other mining districts, where they were generally somewhat shorter than in the north of England, see Lister, *op. cit.*, p. 279 ; *V.C.H. Derbyshire*, vol. ii, p. 353 ; *Star Chamb. Proc.*, *James I*, 228/13.

[2] Lister, *op. cit.*, pp. 281–2.

men are at worke belowe, an hundred of the greatest shippes in Britaine may saile over their heads ". The mine, continues Taylor, in his description of " this unfellowed and unmatchable worke ", is " most artificially cut like an arch or a vault, all that great length, with many nookes and by-wayes ; and it is so made, that a man may walke upright in the most places both in and out ".[1] Had it not been for the fact that at high tide Bruce could load ships as they anchored beside the reef, without any charges such as other mine owners paid for carrying coal overland to a harbour, it is doubtful whether he could have afforded what must have been exceptionally heavy charges for haulage underground.

With a very few exceptions, shafts were abandoned in the seventeenth century with a recklessness that would scarcely have been possible had accessible minerals been less abundant. Modern experience in the bituminous fields of the United States suggests that, when plenty of coal is to be had, the tendency is to attack the best seams in quest of large immediate profits, without regard for the waste involved, and with little heed for the needs of future generations. In Great Britain in the seventeenth century the miners not only failed to drive headways beyond a few score yards, but, like some Illinois miners to-day, they left behind, both in Scotland and the north of England, more than half the seam in pillars—for " wante of good rooffe of stone ".[2] If the surrounding strata seemed shaky, they

---

[1] Taylor, *The Pennyless Pilgrimage*, 1618, in P. Hume Brown, *Early Travellers in Scotland*, pp. 115-17. We have records of other mines, besides Culross, with exceptionally long headways. At Gateshead, shortly after the middle of the sixteenth century, a working was carried 300 yards from the " eye " (*Court of Augm. Proc.*, 14/76). At Smalley in Derbyshire, in 1693, an underground passage had been driven nearly half a mile from the shaft (Houghton, *Husbandry and Trade*, vol. i, p. 105). In the Mons colliery district the longest headway driven before the eighteenth century was about 400 yards (Decamps, *Mémoire historique*, p. 161).

[2] The breadth and height of the pillars, and of the passages between them, depended on the thickness and the quality of the seam and on the nature of overhead strata. Where the seam was surmounted by a layer of solid rock, the danger of " thrusts " was lessened (Owen, *Description of Pembrokeshire*, pt. i, p. 90 ; *Exch. Spec. Com.*, no. 1739), and the problem of support simplified. " The breadth of the space which they could . . . trust without wooden props in the Newcastle coal, was not more than 3 yards ", Sir John Lowther reported to the Royal Society, " but the Scots coal, being a stronger and greater coal, would support at least 7 yards width " (Birch, *History of the Royal Society*, vol. iii, p. 439). His was too broad a generalization, for not all Scottish seams could support such wide passages. A " Memorial ", submitted in 1717 by the Duke of Hamilton's mine manager, " concerning the regular working " of the " main ", or chief, seam of coal at Kinneil colliery, Linlithgow-shire, recommended that the breadth of " Roumes " should not exceed four yards, the breadth of " Throughers " two and three-quarters yards, and that " Stoops ", or pillars, should be left five yards long and three ells thick. In mining the upper seams of " Burnt " coal, the manager added, these dimensions must be reduced (*Hamilton MSS.*, no. 592, ii, bdl. 2). If his recommendations were followed, it is clear that the pillars would occupy more space than the passages, and that more than half of the seam would be left underground. In Durham also, according to Galloway, at least half the seam was lost (*Annals*, p. 181). Sometimes thicker pillars were left than even seventeenth-century mining technicians considered necessary (see *Chanc. Proc.*, *James I*, B. 40/70), and attempts might be made to work through the passages a second time to get more of the pillars (*Parl. Surveys, Somerset*, no. 39 ; *Exch. Spec. Com.*, no. 5037).

often left, in addition, a thick layer of coal overhead and sometimes underfoot.[1] Notwithstanding this waste of coal in supports, few mining experts understood the technique of upholding walls underground ; and from all parts of the country came reports of pillars on the verge of collapse.[2] Henry Power, a Yorkshire scientist who carried out various experiments in coal mines, writes to the Royal Society that one experiment involving the firing of a gun at the shaft bottom is too risky to try, " by reason of the craziness of the roof of their works, which often falls of its own accord, without any concussion at all ".[3] Frequently these falls, or " thrusts ", forced the miners to abandon a shaft even before the costs of underground haulage were large enough to make this desirable.

The amount of coal which was raised before a shaft was abandoned varied considerably under different conditions and in different colliery districts. In the Tyne basin, there was a clear tendency for the quantity to increase as seams were worked at greater depths ; but the increase was far less rapid than the expansion in the total output of the district would lead one to expect. Two pits worked in the manor of Gateshead at the end of the reign of Henry VIII " cast " about 2,500 and 3,000 tons respectively before they were given over by the miners ; [4] twenty-two pits worked at the beginning of the reign of Charles I in the manor of Benwell, which lay across the river from Gateshead, produced quantities ranging from 4,000 to 10,000 tons each.[5] By the end of the seventeenth century it was not unusual in the north of England for entrepreneurs to get nearly 20,000 tons from one pit.[6] Elsewhere the output per pit was much smaller ; Staffordshire mine owners after the Restoration still considered 5,000 tons a large quantity to draw from a single shaft.[7]

---

[1] At Smalley colliery in Derbyshire in 1693, Houghton's friend " went underground . . . in a . . . vein, which was about 6 foot, where were coals overhead and under foot, the workmen knew not how thick " (Houghton, *op. cit.*, vol. i, p. 105).

[2] The " roof is badd and falls of itself ", Thomas Tueddy, a Northumberland hewer, told Crown Commissioners in 1611. These commissioners found the same conditions at most of the mines which they visited (*Exch. Spec. Com.*, no. 5037). See also *Exch. Deps. by Com.*, 12 Charles I, East. 15 ; Birch, *History of the Royal Society*, vol. i, pp. 177–8 ; Owen, *Description of Pembrokeshire*, pt. i, p. 91.

[3] Birch, *op. cit.*, vol. i, pp. 133–6.

[4] 3,057 " pytt " chaldrons, 6 corves and 3,703 " pytt " chaldrons, 10 corves (*Court of Augm. Misc. Bks.*, vol. cxii, ff. 17–24).

[5] *Exch. Spec. Com.*, no. 5996. A twenty-third pit was abandoned after 400 tons had been raised.

[6] The author of *The Compleat Collier* (pp. 32, 39) gives the daily output of a pit of 60 fathoms as about 21 score corves, and implies that the pit would be worked for 300 days or so. Assuming that each corf load weighed about 3 cwt. (see Appendix C), the daily output must have exceeded 60 tons and the total output approached 20,000 tons. Galloway (*Annals*, p. 178) estimates that the " corf " in this case weighed 4½ cwt.

[7] £500 worth of coal (Plot, *Natural History of Staffordshire*, p. 128).   I have assumed that each ton was worth about 2s. (cf. Appendix E (i)). It is probable that, at many collieries outside the north of England, the average yield per pit dug during the period from the accession of Elizabeth until 1700 did not much exceed 1,500 tons (see *Exch. Q.R. Accounts*, 632/17 ; *Chanc. Deps.*, P. 7/T ; Lister, *op. cit.*, p. 280).

During the first half of the seventeenth century, it was common to raise from 50 to 75 tons a week from each shaft at the more important collieries in Pembrokeshire,[1] and from 100 to 125 tons at collieries on the Tyne ; [2] and at this rate the headways were driven as far as most miners cared to " hurry ", or as most mine owners found profitable, within a year, or two years at the most. As most enterprises had several pits in operation at once, new shafts might have to be sunk nearly every month. Contemporary seventeenth-century descriptions bear witness to the feverish haste with which the miners deserted old workings for new. Parliamentary commissioners, sent by the Commonwealth to survey Kingswood Chase, counted by the hundred abandoned pits,[3] some filled in with stones, small coal, and rubbish, others dangerously open, and gaping to receive any ox or cow which grazed unwittingly about the pastures. Within one small Tyneside holding of twenty acres, scarcely larger than a modern cricket field or baseball park, commissioners reported in 1636 that forty separate pits had been worked and abandoned during the two previous decades, " as by the heape-roomes may appeare ".[4] According to the deposition made to the commissioners by Guy Smith, a freeholder in Benwell (of which this small holding was parcel), the mine owners who had leased in 1617 the royalty rights attaching to the manor had dug in all 148 pits and eleven " groves " [5]

[1] From 80 to 100 barrels (11 to 13 tons) a day at each pit (Owen, *op. cit.*, pt. i, p. 91).   From one pit within the manor of Wollaton, Notts., the miners " got coles at the benke to the value of 40 or 30 rookes [55 to 75 tons] a weeke " (*H.M.C., Report on the MSS. of Lord Middleton*, p. 164).

[2] At Winlaton colliery in 1582 the daily output at some pits is said to have averaged 19 tons per pit ; at other pits as many as 25 tons were produced each day (*Exch. Deps. by Com.*, 29 Eliz., East. 4).   In 1595 " full work " meant an output of 40 tons (20 score bowles) per day per pit, at collieries within the manors of Ravensworth and Lamesley, on the Durham side of the Tyne (*Chanc. Proc., James I.*, A. 9/24) ; but one may doubt whether so large a quantity was regularly obtained for any considerable period.   According to testimony given in 1635 by Nicholas Hodgson, one of the owners of the colliery in Benwell manor, a pit sunk to the " stone ", or 27 inch, coal might " cast " per week 4 " tens " (80 to 100 tons) ; a pit sunk to the " crawe ", or 3 foot coal, 5 " tens " (100 to 125 tons) ; a pit sunk to the 5 foot coal, 6 " tens " ; a pit sunk to the 8 foot coal, 8 " tens " (*Exch. Deps. by Com.*, 11 Charles I, Mich. 46).   In fact, however, the output per pit over a period of months seldom exceeded on the average 20 tons per working day, or from 4,000 to 5,000 tons per year, even at the most productive pits.   During the second half of the seventeenth century, to judge from statements in *The Compleat Collier*, the output per pit in the north of England had greatly increased, but the life of each shaft was still very short (see above, p. 369, note).

It is interesting to compare these conditions with those on the Continent at the beginning of the nineteenth century.   Statistics for 1838, which include all the coal mines in France, Belgium, and Prussia (M. Dunn, " Remarks on the State of Belgium and France in respect to the Production of Coal ", in *Proc. North of Eng. Inst. of Mining Engineers*, vol. iv, 1855–6, pp. 289, sqq.), show that the average annual yield per pit amounted to about 20,000 tons for France, about 7,000 tons for Belgium, and about 3,700 tons for Prussia.   In the Ruhr district in 1791 the annual output is said to have been 231,788 tons, produced from 158 pits ; i.e. an output of 1,466 tons per pit (Hue, *Die Bergarbeiter*, vol. i, p. 347).

[3] *Exch. Deps. by Com.*, 27 Charles II, Mich. 29.

[4] *Exch. Deps. by Com.*, 12 Charles I, East. 25.

[5] Probably open works (cf. Galloway, *Annals*, pp. 17, 39, 67).

during a period of twenty years.[1]   Rarely, even in the case of the smallest enterprise in remote mining districts, did the miners work in only one pit at a time.   Usually they kept two, three, or four pits " casting " coal.[2]   Where the market warranted it, there might be seven or more working pits.   At the Grand Lease colliery there were fourteen in 1642.[3]   A survey made of the coal mines about Newcastle in 1622, when there were probably between twenty and thirty sea-sale collieries in the Tyne valley, shows that approximately 130 pits had to be kept going to supply the demand, or an average of about five pits to each enterprise.[4]

A large colliery in the seventeenth century required an extensive surface plant, and presented a scene of busy activity above as well as below ground.   Let us try to visualize its equipment, as it may have appeared to a contemporary observer.[5]   Taking up our position before the " engine shaft " at the lower end of a gently sloping field, we see a team of four horses [6] circling about the pit to the crack of their driver's whip and harnessed to a cross-piece descending from one end of a horizontal wooden beam, about eight yards long and raised ten feet above the earth.   The other end of this beam is fastened to the vertical axle rod of an iron wheel, furnished with a number of teeth so set as to catch the rungs of an iron drum, or " turn ", suspended horizontally across the shaft from two wooden frames, one on either side.   As the horses are driven in a continual circle about the axle rod, their motion is transmitted to the wheel, which causes the " turn " to rotate in its wooden frames.   Another set of teeth, or cogs, round the centre of the turn, are fashioned to fit the links of an endless iron chain, and as the turn rotates, this chain rises out of the shaft on one

[1] *Exch. Spec. Com.*, no. 5996.   Cf. map, facing p. 305.

[2] At Wednesbury in Staffordshire four pits were worked at once in 1584 (*Exch. Q.R. Accts.*, 632/17).   At Strelley colliery, Notts., in 1610, there were four working pits (*H.M.C., Report on the MSS. of Lord Middleton*, p. 176).   Although engaged by the terms of their lease of minerals within the manor of Stratton, Somerset, to work only two pits at once, William Long and Hercules Horler had, according to the testimony of one of their miners, " landed cole at 3 pitts at a tyme in one day and did keep another pitt there going att the same tyme for to find out more cole " (*V.C.H. Somerset*, vol. ii, p. 380).   Sir Thomas Mostyn kept three pits in operation in 1617 at his colliery near the Flintshire coast (*Exch. Spec. Com.*, no. 3848).   In starting a new colliery at Sutton in Lancashire, an engine pit was finished, three old pits " scoured up " and four new pits sunk between September, 1717, and May, 1718, when the mine began to yield coal (*Palat. of Lancs., Bills*, 61/11).   At Bedworth colliery in Warwickshire in 1684 at least six pits were being operated simultaneously (*Exch. Deps. by Com.*, 36 Charles II, Mich. 43).

[3] *MS. Account Books of the Newcastle Corporation.*

[4] *Rawlinson MSS.* (cf. below, vol. ii, pp. 241 n., 246).

[5] The description which follows is fictitious in that it is not based on information concerning a specific mining enterprise.   For parts of it I am particularly indebted to the illustrations in Galloway's *Annals* (pp. 158, 168) ; but the whole is a composite interpretation of a number of seventeenth-century manuscripts.   The type of machinery and equipment in use differed at different collieries, owing to special conditions of mining and to peculiarities of mining practice in the various districts.   I have tried to indicate in footnotes the principal alternative types of machinery and equipment.

[6] There were collieries where the pumps were operated by water-power or by human labour (Appendix O).   For the number of horses employed at the pits, see below, p. 375.

side and descends into it on the other.[1] At intervals of a few seconds, large oxhide dippers, fastened to the chain, emerge full of water.[2] As the dippers rise with the chain over the turn, they empty themselves into a slanting wooden trough ; and down this trough the water pours into a ditch or gutter,[3] which has been dug downhill towards the valley to join the nearest stream. Planks are laid to permit the horses to cross the gutter as they plod round their circular path. Over the shaft has been built a great shed or lodge with a roof, nearly twenty yards from side to side, which projects far enough to cover the path of the horses.[4] This lodge protects the iron work from rain, which brings rust, and the whole apparatus from damage by storms. Though now open on two sides, the lodge is furnished with great doors ; and, in case the water in the workings should be temporarily drained off, these doors can be closed under padlock and key, to shut out rogues, rivals, or other ill-wishers bent on theft or *sabotage*.

As we look farther up the sloping field, we see seven or eight similar lodges, at distances of a few score yards the one from the other. About them there appears to be some considerable movement of men, horses, and wagons, while under them are great heaps, or, as they are called in Scotland, " hills " of loose coal. Each lodge indicates a working coal pit. At other points, as may be seen from the piles of cut timber, colliers are engaged in sinking and supporting new shafts ; at still other points, where we see mounds of stone and black dust, more colliers are engaged in filling up shafts already abandoned. Not far from the engine pit is a cluster of wooden buildings. One is a " counting house " occupied by the " viewer ", or supervisor, and by the " clerk of the works " who pays the labourers ;[5] there the account books are kept [6] and all clerical business pertaining to the

---

[1] Other drainage devices were sometimes employed (see Appendix O).

[2] At Culross in Fife the chain had 36 buckets, 18 descending empty while the other 18 were rising full (Galloway, *History*, p. 56). At another Scottish colliery by the seaside, the owner ordered " fourtie bucketis weill dressit and wrocht in all things necessar for drawing of watter " (*Privy Council Register of Scotland*, 1st Series, vol. xi, p. 301).

[3] This description does not apply to mines where drainage was accomplished altogether or in part by means of an adit.

[4] Pit lodges were commonly built in all districts, even at small " land-sale " collieries (*Star Chamb. Proc., James I*, 161/17 ; 224/20 ; 227/3).

[5] Cf. *Exch. Deps. by Com.*, 36 Charles II, Mich. 43 (testimony of Samuel Troughton, clothier).

[6] Accounting had already become the subject of treatises in the sixteenth and seventeenth centuries (see R. D. Richards, " Early History of the Term Capital ", in *Quart. Journ. of Econ.*, vol. xl, 1926, pp. 329–38). At all the principal collieries regular accounts were kept in writing before the end of the sixteenth century, and not a few of these accounts have been preserved. At Wollaton in Nottinghamshire a rough attempt was made to separate capital charges from current costs by keeping a " synkyng booke " for all payments for digging, repairing, or timbering pits and adits, and a " colpytt booke " for payments for pumping water and other incidentals. More frequently all expenses and receipts seem to have been kept in one ledger. Though the methods adopted may seem crude to an age in which accounting is almost a science, these early ledgers have the merit of detail, giving, as they do, full particulars of the output of each hewer, the names of all the colliers, the nature of the work performed, the wages paid, the prices at which coal was sold. Great pains must have been

colliery transacted.   The remaining buildings—long sheds or barns—
are used for storing materials necessary for the maintenance of the
enterprise ;[1] in one is a smith, who spends his time resharpening
pickaxes, repairing driving instruments—" weadges ", " fleadges ",
" mandrels ", " mattocks ", " hammers ", " hacks ", " gavelocks "—
and all the other so-called " furniture " belonging to the miners ;[2]
in another a chandler heats his tallow and prepares the candles for
the workmen to light their way underground ; in another a " corver "
weaves hazel rods into the " corves ", or baskets, used for carrying
coal.   Finally, beside this cluster of buildings, several acres are fenced
off for pasture ; and in the pasture we see grazing one or two oxen
and a number of horses, some of which will soon relieve those at present
engaged in work at the shafts.

If we now proceed from the engine pit to one of the working
coal pits, we shall find that the " winding " machinery differs from the
machinery for raising water.   Instead of the heavy iron chain, there
is a light rope of stout hemp, wound round the turn, with a noose at
the two ends which hang down on either side.[3]   The force of the
horses,[4] transmitted to the turn, causes one noose to descend as the

---

taken with the entries ; for the books which I have seen are not only easily
intelligible, but remarkably legible, after nearly three hundred years.   Their
compilation involved no small labour, to judge by the fact that when, in 1586,
an abstract was wanted of the accounts at Winlaton colliery in connection with
a case to be tried before the Court of Exchequer, it took several persons six or
seven days, working continuously " from morning to night ", to supply it (*Exch.
Deps. by Com.*, 29 Eliz., East. 4).   Merely to transcribe the *Register of the
Tulliallan Coal Works*, 1643-7, as it is preserved at Edinburgh, would require
more than a month of a record searcher's time.   References to the whereabouts
of seventeenth-century colliery account books are given in Appendix A (i).
For examples of colliery accounts, see Appendix K.

[1] Sometimes the entrepreneur also built cottages for his miners (cf. below,
vol. ii, pp. 147, 187).

[2] *Star Chamb. Proc., James I*, 224/19, 20 ; 245/6 ; *Palat. of Durham,
Bills and Answers*, bdl. 6 (Tomlinson *v.* Cole) ; Owen, *Description of Pembroke-
shire*, pt. i, p. 91 ; *V.C.H. Warwickshire*, vol. ii, p. 221 ; Lister, *op. cit.*, p. 280.
These tools were made altogether of wood, except for the cutting or striking face
(*Star Chamb. Proc.*, 218/13 ; *V.C.H. Shropshire*, p. 458 ; cf. above, p. 191).

[3] Decamps (*Mémoire historique*, p. 182) refers to the use of iron chains
for winding in the Mons colliery district ; but I am inclined to think he may
have confused a winding with a drainage " engine ".   The lighter the cable
the less, of course, the horse power required to pull it (J. C., *The Compleat Collier*,
p. 34).

[4] At the less important collieries, where smaller quantities of coal were
raised at a time, the turn, or windlass, was generally operated not by horses
but by hand, the labourers being known as " winders ", or—in Scotland—as
" gatemen " (*Exch. Spec. Com.*, no. 5037 ; *V.C.H. Somerset*, vol. ii, p. 384 ;
Owen, *Description of Pembrokeshire*, pt. i, p. 90 ; Decamps, *op. cit.*, pp. 182-3 ;
Lister, *op. cit.*, p. 281).   The use of a crude windlass for raising coal is said to
have been common even during the Middle Ages (Salzman, *English Industries
of the Middle Ages*, p. 12).   In parts of Scotland coal was still raised in the
seventeenth century without the aid of any machinery.   Stooping women
bearers carried the baskets, known as " skiffies " or " hutches ", up ladders
placed against the walls of the shaft (Galloway, *History of Coal Mining*, pp. 26-7 ;
Cunningham, *Mining in the Kingdom of Fife*, pp. 12-13).   I doubt, however,
whether this method was as prevalent in seventeenth-century Scotland as
Galloway supposed.   At a colliery at Gorballs belonging to the town of Glasgow,
all raising and lowering of coal and men was apparently accomplished by means
of a rope (*Acts of the Parliaments of Scotland*, vol. vii, appx. p. 31).

other rises from the pit ; consequently the team is driven round its circle first in one direction, then in the opposite, according to which end of the rope is to be lowered.[1] To go down the mine, a hazardous and sometimes fatal passage, the collier gets " a Leg and Knee " into the noose " as far as the very Hip. Thus hugging the Rope with one Arm, his life wrapt up with it ", he gives the driver a signal to set the team in motion, and " down he goes ", scraping against the sides of the narrow shaft.[2] At the " eye ", or bottom, is a miner, sometimes called a " filler ", who receives the baskets which the bearers or barrowmen have dragged along the headways, and hooks them on to the same noose ; for one rope serves to raise both men and coal.[3] At the pit head several workmen are employed. One removes the baskets as they come to the ·surface, and places them on a sort of sledge, pulled by an old horse and driven by another labourer, who empties the coal upon the " heaps " and returns the baskets to be sent back down the pit.[4] Another culls over the coal to rid it of extraneous matter,[5] and helps to load it by hand into the wagons that come to fetch it to market. Still another drives the team which moves the winding machinery.

Should we pass to the six or seven other working pits, belonging to our hypothetical colliery, we should find essentially the same apparatus set up and similar labour done at all of them. It is evident that, quite apart from the work of sinking shafts, draining and ventilating the mine, the expense of merely providing the materials and animals required for such a colliery must have been considerable.

[1] The device here described is generally called a " cog and rung gin " (see illustration in Galloway, *Annals*, p. 168). In the case of the more primitive " turns ", the windlass was rotated in one direction only, and only one rope end descended into the shaft (see Appendix O). A device known as a " whim gin " (see illustration, Galloway, *op. cit.*, p. 178) tended to supersede the " cog and rung gin " for " winding " coal. In the " whim gin " the " turn " set over the shaft was replaced by a drum, or barrel, of larger diameter set up upon a vertical axis several feet away from the shaft. The two ends of the rope, which wound round this drum, were carried down to the " eye " through pullies hung from a frame over the pit. This new gin is said to have worked in a manner very like the seventeenth-century malt mill (Galloway, *op. cit.*, p. 185) ; and indeed the windlass was sometimes called a " mill " (*Palat. of Durham, Decrees and Orders*, vol. ii, f. 439) and also a " loome " (*Star Chamb. Proc., Edward VI*, 6/99) and a " frame " (*Court of Augm. Misc. Bks.*, vol. cxvi, ff. 55–65).

[2] G[uy] M[eige], *The New State of England*, 1691, pt. i, p. 165. For another description see *Star Chamb. Proc., James I*, 53/10. At the less important collieries, the shafts, which were usually dug square or rectangular and seldom round, might be less than 4 feet across (*Duchy of Lancs. Pleadings*, 174/N/5). As deeper seams were attacked, the sinkers increased the breadth of their shafts. In the Wear valley, at the end of the seventeenth century, shafts were at least 6 feet broad, according to the author of *The Compleat Collier*, p. 20; cf. Galloway, *Annals*, p. 176. In the Tyne valley, where the shafts " are all square ", the sinkers dug them 7 or 8 feet broad (M[eige], *op. cit.*, pt .i, p. 164), and by 1765, when Jars visited Newcastle, the sinkers there dug circular shafts from 10 to 12 feet in diameter (*Voyages métallurgiques*, vol. i, p. 191).

[3] *Star Chamb. Proc., Henry VIII*, 23/46; *ibid., James I*, 109/9 ; *Palat. of Durham, Decrees and Orders*, vol. ii, f. 439 ; *Duchy of Lancs. Pleadings*, 102/T/10.

[4] J. C., *The Compleat Collier*, pp. 34, 36–7.

[5] See above, p. 348 ; and below, vol. ii, pp. 242–3.

The Whim Gin

(Morand, *L'art d'exploiter les mines de charbon de terre*, 1768-77, Plate XXXIV, No. 2)

The following explanation is based on *ibid.*, vol. i, pp. 696-7. A. Drum fixed to axle B. C. Chief beam in which the pivot of the axle turns. D. Transversal beam.
E.E. Cross-pieces holding up the framework of the pit pulleys. F.F. Rollers on which roll the ropes of the drum. G.G. The great pulleys on which the same ropes
roll as they descend into the pit H. I.J.J. Cross-pieces holding up the pulleys. K. The swing-bar to which are attached as many as eight horses. L. Piece containing the hole in which the axle rotates. The hole is lined with metal. M. Air shaft, 38 feet high, built of brick and communicating with the pits by the covered galley N.,
in order to extract the damp. O. Heater suspended on a rope by means of which it can be pulled up or lowered into the air shaft. P. Device to prevent the
wind from blowing into the air shaft.

Fig. 2. A piece of a boring rod used at Newcastle-on-Tyne.

[*To face page* 374, *Vol.* I

Vast supplies of timber were no less essential for coal mining than for iron smelting or for ship building in the seventeenth century. " In all partes of [Kingswood] Forest ", writes Norden, " coal mines . . . devoure the principall hollies for the supportation of their pitts ".[1] Wood was the chief material used for putting up surface lodges and sheds, for making tools, machinery, and equipment of all kinds. Every shaft, every drainage or ventilation head, must be carefully lined with deal boards, held in place by pointed oaken or fir bars, to support pressure from the walls and from above.[2] The rapidly rising price of wood added, therefore, to the expense of producing coal. Mine owners in the Tyne district began to grumble as early as 1610 over " the charge of tymber, dailes [deals] and corvinge,[3] where of there is great scant, and yet of necessitie must be used in great quantitye ".[4] At Coventry, in 1684, the price of timber and firewood had been forced higher than ever, " by the spending of them about the [coal] works ".[5] More than a thousand trees and shrubs were purchased in a single year for the colliery of Tulliallan in Fife.[6] Before Elizabeth's reign the lessees of coal mines were usually entitled by a clause in their lease to provide themselves with " sufficient timber " for their work, by cutting down trees within the manor.[7] But in the seventeenth century, though this clause persisted in most leases, the supply of wood that could be obtained on the manor was hardly sufficient for building the pit lodges, and the mine owners sometimes had to spend several hundred pounds in buying timber,[8] before they had begun to sell any coal.

They had to lay out nearly as much again in the purchase and maintenance of horses to run the pumping and the winding machinery. Bedworth colliery maintained, in 1622, a force of sixty horses, " employed in drawing water out of the pits " ;[9] and some collieries in the North of England must have maintained an even greater number. " For a train of artillery . . . this place [Newcastle-on-Tyne]

---

[1] *S.P.D., James I*, vol. lxxxiv, no. 46, 1615 (?), cited *V.C.H. Gloucestershire*, vol. ii, p. 236.

[2] *The Compleat Collier*, p. 15.

[3] i.e. branches for making corves in which to draw coal up the shafts (see above, p. 373).

[4] Dendy, *Records of the Company of Hostmen*, p. 58.

[5] *Exch. Deps. by Com.*, 36 Charles II, Mich. 43.

[6] *Accounts of the Tulliallan Coal and Salt Works*, 1679–80.

[7] See above, p. 319.

[8] In 1600 Jeffrey Foxe (see above, p. 355) purchased £150 worth of timber, preparatory to starting a colliery at Chilvers Coton in Warwickshire—a colliery which could not have ranked among the principal enterprises of the time (*Chanc. Proc., James I*, F. 4/53).

[9] *V.C.H. Warwickshire*, vol. ii, pp. 222. At Griff colliery at the beginning of the eighteenth century fifty horses were employed in raising water (Galloway, *Annals*, p. 241). Several horses might be harnessed at a time to operate a drainage engine (Taylor, *The Pennyless Pilgrimage*, 1618, in Hume Brown, *Early Travellers in Scotland*, p. 117 ; *The Compleat Collier*, pp. 32–3). Even in the Belgian provinces, where there were presumably no collieries as large as Bedworth (see above, pp. 360–63), from 15 to 20 horses were frequently employed at a single colliery in the Mons coalfield (Decamps, *Mémoire historique*, pp. 152–3), and from 32 to 40 horses at a single colliery in the Liége coalfield (Pirenne, *Histoire de Belgique*, vol. v, p. 358).

affords many horses, which they use in [*sic*][1] their coal mines ",
wrote Sir Jacob Astley, who had been sent thither in 1639 to prepare
for the threatened invasion of the Scottish Covenanters.[2] The
expanding mining industry appears to have created a great demand
for beasts of burden, to turn the machinery at the pithead and to
haul away the coal. This demand the breeders could not easily supply,
and they raised their prices accordingly. Mine owners of the Wear
valley had to pay from £6 to £12 for each horse purchased at the
end of the seventeenth century,[3] and to provide, besides, sufficient
hay and fodder for feeding them.

What with supplies of tallow for candles, iron for forging tools
and machinery, ropes, timber, beasts of burden, and heaps of unsold
coal, the movable property of a mine owner at the surface might
represent a very substantial investment. When various creditors
confiscated Bedworth colliery in 1637 for non-payment of debts,
the debtors declared that the stock on the ground " in coale, horses,
wood, and hay, with other materials " was worth £1,400, or about
£10,000 in terms of modern currency.[4]

Just as the modern shopkeeper finds it necessary to crowd his
windows with tempting wares, so the seventeenth-century mine
owner was obliged to display large stocks of coal at his pits ; otherwise
the traders or consumers might desert him for a competitor who
could give immediate service. The amount of these stocks varied
from a few hundred tons at a small " land-sale " colliery, like that
within the manor of Eckington [5] in the heart of Derbyshire, to several
thousand tons at large midland collieries like Bedworth or Strelley,[6]
and to tens of thousands of tons at the largest collieries, in the north
of England. On the Wear a mine owner had to be prepared to advance
from £2,000 to £3,000 towards the provision of such a stock, in
addition to all the other expenses connected with starting a colliery.[7]
Throughout the British Isles summer was the best season to market
coal ; for then the roads were more easily passable and the sailor
less exposed to tempestuous seas.[8] Winter was on the whole the
best season to mine it, although the pits in the Tyne valley are said
to have been idle sometimes for two months at that time of year.[9]
In the cold weather the damps were less prevalent in most districts ;[10]
notwithstanding that in the Mendips the collieries, which were " apt
to take Fire all the year ", were said to do so " most, and with most

[1] This should probably read " *about* their coal mines ". There is no reason
to believe that horses were as yet used *underground* (see above, p. 364).
[2] *Cal. S.P.D.*, 1638–9, pp. 481–3.
[3] *The Compleat Collier*, p. 33.
[4] *Chanc. Proc.*, Series II, 425/47.
[5] *Exch. Deps. by Com.*, 11 James I, Trin. 6. See also *Duchy of Lancs.
Pleadings*, 73/M/10.
[6] When Strelley colliery was seized in 1618 for non-payment of debt (see
below, vol. ii, p. 34), the debtors had, according to their own testimony, 8000
" rookes " (about 12,000 tons) of coal " ready stacked upp above the ground and
ready for sale " (*Chanc. Proc.*, James I, B. 17/61).
[7] *The Compleat Collier*, pp. 46–7.
[8] See below, p. 396.
[9] Galloway, *Annals*, p. 387.
[10] Plot, *Natural History of Staffordshire*, p. 140.

violence in the Winter, and chiefly in a Black Frost, when the air runs best ".[1]

If coal getting was abandoned for some weeks or months of the year,[2] the mine owners had to carry stocks for longer periods than would otherwise have been necessary. And the current operating expenses of the mine did not cease entirely during the dead season.[3] Pumping engines had to be kept at work. Wagon horses or cattle could not be dismissed like superfluous labourers ; they must be fed, for, if they starved, they could not be replaced without the expenditure of a substantial sum. And, since most of the coal was sold at the height of summer, the mine owners had the delicate task of estimating the demand in advance. An under-estimate meant loss of trade ; an over-estimate forced them to carry their stocks an additional twelve months.

Few mine owners could know even approximately how much capital they would be called on to invest before their colliery began to yield a profit. Mining has always been ranked among the speculative business enterprises ; in the seventeenth century, when the value of the mine, the future of the market, and the costs of drainage were unpredictable, it was, as we attempt to show in a later chapter, speculative in the highest degree. Although there are some examples in Elizabethan times of adventurers who were able to start a colliery

---

[1] *Royal Soc.*, *Philosophical Transactions*, abridged edition, vol. ii, 1716, p. 382. In 1675 a great explosion at Mostyn, in Flintshire, occurred in February, " being a Season when other Damps are scarce felt or heard of " (*ibid.*, unabridged edition, vol. xii, 1677–8, p. 899).

[2] The season for abandoning mining operations was not the same in all the coal fields, but almost nowhere, it appears, were the hewers kept busy steadily throughout the year (cf. below, vol. ii, pp. 183–4).

[3] Some idea of the current charges at a working colliery can be gained from the following list of a few of the " oncosts " at Tulliallan between August 28th, 1643, and December 29th, 1644. (The price is given in pounds Scots, each worth approximately a twelfth of an English pound.) " For Beiging [?] up the walles of 4 Collhedlars houses ", £26 13s. 4d. ; " 2 hundred firr deal ", £80 ; " Horses and mens charges at mending of coal gate and sea gate ", £85 4s. ; " For laying steaves " (timbering ?), £14 17s. 6d. ; " Cost of cutting the dam and Reding of Milne " (diverting a stream from its normal course to operate a drainage pump), £13 10s. Similar expenditures were also incurred for the following work :—" Salteris [salters] bringing up a levell " ; " For cutting the netherdyke to drye the nether wall " (refers no doubt to digging a ditch in one of the underground roadways to drain off the water into the pool of the engine pit) ; " New sink chairges " (cost of digging a new pit) ; " Wrights wages " ; " Levell cutting " ; " For an watter wheill and a turne wheill with airmes " (evidently to complete the mill run by water power) ; " Shering of 27 fadom of levell att 43/- the fadom, and of coll myne 12 fadom att 30/- the fadom " ; " For virleing of tua [two] fadom of the coll myne " (evidently another pit with underground trench has been dug and timbered) ; " Extraordinair expenses for wineing and leding of stone " ; " For shering of the turne [windlass] of the trap heads of the new sinke [pit] " ; " Extraordinar expense of the mill sink [engine pit] ". The charge of digging " levells " and " sinks " and erecting pumps occurs in the account almost every week (*Register of the Tulliallan Coal Works*). These accounts show that a considerable part of the current expense of operating a colliery was not directly connected with hewing, bearing, and winding, but was of such a nature that much of it had to be carried on to protect the mine from flooding and to prepare for a future demand, even when there was no sale for the coal.

with an outlay from £100 to £200,[1] or even less, the opportunities for entering the industry with as little capital as that diminished during the course of the seventeenth century. Larger investments became the rule. £400 was invested as early as 1599 in Aspull colliery, one of the two principal cannel coal mines in the Wigan district ; and by 1626 the capital sunk in this colliery is said to have reached £3,000.[2] In 1607, a few years after the opening of a new colliery at Broseley, in Shropshire, the costs in " wyning and fynding " the seams and in " makeing of devises and engynes to avoyde the water " amounted to £2,000 more than the undertaker had " yet received commodity ".[3] In 1623 a Scottish mine owner found himself above 20,000 marks (more than £1,000 sterling) out of pocket ; and another, in 1634, declared that he had spent above 36,000 marks on the working of his mine.[4] During the reign of James I, more than £6,500 was spent in an attempt to develop the coal industry at Cowpen on the Northumberland coast.[5] The capital invested in a mining enterprise near Bristol ran into thousands of pounds.[6] Between 1621 and 1626, the expense of operating Benwell colliery is said to have added £4,500 to the original investment.[7] Between 1660 and 1667 more than £8,000 sterling was sunk in opening up a new mine at Methil in Fife.[8] The net losses in connection with an attempt made during the 'seventies to revive Bedworth colliery were variously estimated at £14,000, £15,000, £16,000, and £17,000, not counting the losses sustained by the same undertakers at the adjacent Hawkesbury colliery.[9]

There is no reason to suppose that investments such as these were the exception. Had we information concerning the largest enterprises near Newcastle and Sunderland, we should undoubtedly find even larger sums tied up in them. And where, as at Cullercoates or Hartley on the Northumberland coast,[10] a harbour had to be dredged or a

[1] *Duchy of Lancs. Pleadings*, 108/W/12, 153/F/3 ; *Chanc. Proc., James I*, B. 20/72, B. 16/50 ; Galloway, *Annals*, p. 104.

[2] *Duchy of Lancs. Pleadings*, 211/G/6 ; *Duchy of Lancs. Deps.*, 75/13.

[3] *Star Chamb. Proc., James I*, 109/8.

[4] *Privy Council Register of Scotland*, 1st Series, vol. xiii, p. 207 ; 2nd Series, vol. v, p. 278.

[5] *Chanc. Proc., James I*, B. 17/61 ; and see below, vol. ii, p. 16.

[6] *Chanc. Deps.*, P. 7/T.

[7] *Exch. Deps. by Com.*, 11 Charles I, Mich. 46 (see testimony of Thomas Crome of Newcastle, merchant—that for his one-ninth part in Benwell colliery he laid out more than £500 between July, 1621, and March, 1626).

[8] £100,000, Scots, was spent on the " Happy Mine ", the harbour, and the salt pans (A. S. Cunningham, *Mining in the Kingdom of Fife*, p. 30).

[9] *Exch. Deps. by Com.*, 36 Charles II, Mich. 43.

[10] *A History of Northumberland*, vol. viii, pp. 281–3 ; Jars, *Voyages métallurgiques*, vol. i, pp. 207–8. Defoe has left us a description of Sir Ralph Delaval's successful attempt to scour the haven of Hartley and create Seaton Sluice. " In the construction of this small Harbour ", he writes, " he found enough to exercise his Skill and Patience ; the Stone-pier which covered it from the North-east Wind being carried away by the Sea more than once ; and when he had overcome this Difficulty by using Timber as well as Stone, he felt a new Inconvenience, by his Ports filling up with Mud and Sand, though a pretty sharp Rill ran through it, which had so hollowed the Rock, as to produce that very Bason which Sir Ralph would convert into an Haven. In order to

breakwater built before shipmasters would venture in to take a lading of coal, as much as £10,000 might be spent by the owners, in addition to all their charges about the colliery itself. Before the end of the seventeenth century, reserves of from £15,000 to £20,000, and often more, were required for undertaking any of the great sea-sale colliery enterprises, whether in Scotland or the north of England, or along the West Coast, nearly as much for some of the river-sale collieries in the Midlands, and from £3,000 to £5,000 for the more important land-sale collieries.

To understand the significance of these sums in seventeenth-century economic life, it is not enough to multiply them by six or seven, though that will give us their approximate equivalent in terms of our own currency. It is necessary to consider them in relation, not to the private accumulations of wealth common to-day, but to those common in Stuart England. To-day the holdings of the richest capitalists are—to put it at a low estimate—from five hundred to a thousand times the real property value of the holdings of the richest men in the Elizabethan Age. Compared with Ford or Rockefeller or Sir Basil Zaharoff, Thomas Sutton was miserably poor. There were probably not more than fifty men in Elizabethan England with sufficient wealth to finance single-handed the largest colliery of the day, even had they been able to realize all their assets. Although £100,000 to £250,000 (the equivalent in our own currency of the sum invested in a few seventeenth-century coal mines) is a small capitalization for a modern industrial enterprise, it was an enormous capitalization for a seventeenth-century one. If we consider the state of joint stock enterprise, the method of financing some of the principal companies in Stuart times (though not that of financing collieries),[1] we find that, out of 140 companies whose capitalization could be estimated for the year 1695, only five—the East India Company, the Royal African Company, the New River Company, the White Paper Makers, and the Bank of England—had capital conspicuously in excess of that probably invested in the largest colliery on the Tyne. After excepting the twenty-seven chief companies, Professor Scott has estimated the average capital of a joint-stock enterprise in 1695 at £5,000 for England and £3,000 for Scotland, no more than the sum then invested in a number of " land-sale " collieries in the Midlands.[2] Most of the joint-stock companies were engaged either wholly or mainly in commercial rather than industrial enterprise. If we were able to compare the average capital required to start a colliery with that required to set up in other industries of corresponding importance in the national economy, we would probably find that the former

remove this Mischief, he placed a new strong Sluice with Flood-gates upon his Brook ; and these being shut by the Coming-in of the Tide, the Backwater collected itself into a Body, and forcing a Passage at the Ebb, carried all before it, and twice in twenty-four Hours scoured the Bed of the Haven clean, . . . It admits small Vessels, yet larger Vessels may lie safe and receive their Lading in the Road " (*Tour*, vol. iii, p. 240). See also *Privy Council Register of Scotland*, 3rd Series, vol. vi, pp. 383–5, for mention of the dredging of the port of Easter Cockenzie in Midlothian.

  [1] See below, pt. iv, ch. iii (ii).
  [2] W. R. Scott, *Joint-Stock Companies*, vol. i, pp. 325, 334–6.

greatly exceeded the latter.[1] For the first time in western Europe, in connection with an industry employing a considerable portion of a country's population,[2] large capitals had become the rule.

(ii) *Capital Costs of Distributing Coal*

Had the demand for coal been limited, during the sixteenth and seventeenth centuries, to a few country husbandmen, all living within a mile or two of the pits, there would have been no more place for the middleman than there had been during the Middle Ages. It was the growth of markets in populous towns and at great distances from the mining districts which led to the investment of substantial capitals in the coal trade, just as the increasing depth of the seams worked was responsible for the large sums invested in mining. Many manorial tenants or cottagers in sparsely settled areas continued to fetch their own coal. Townsmen, however, were not disposed to do so, even if they could. A smith, a brewer, or a merchant of Bristol did not care to spend his own time, or the time of his servants, in fitting out a gang of packhorses and driving to the pits in Kingswood Chase to bring home sacks of coal by the dozen. He preferred to purchase his supplies from a dealer. The presence of scores of like-minded smiths, brewers, merchants, and other citizens within a town like Bristol, Edinburgh, Leicester, Nottingham, or Aachen,[3] tempted some persons to seek their fortunes as dealers. While the citizens of these towns lived near enough to the mines to fetch their own coal if they wished, there was an increasing market in places, such as Cambridge, London, Oxford, Bridgwater, Aberdeen, Dublin, Gainsborough, and Gloucester, with their surrounding villages, which were situated so far from the pits that some days or even weeks might be required to bring coal over land and water to the consumer.[4]

---

[1] While it seems certain that, in the seventeenth century, more capital was required, on the average, to start a coal mine than to set up a manufacturing plant, this is obviously no longer the case. An explanation may perhaps be found in the fact that coal mining, being an extractive industry, is even more governed as to the size of the unit of investment by the situation of natural resources than are manufactures. The depth, quality, and thickness of seams in the earth fix definite limits between the amount of capital indispensable for mining, and the maximum amount which can be profitably employed within a given area. These limits are narrower than the limits imposed upon the amount that could be profitably employed, in an area of the same size, in the production of iron wares, textiles, or other manufactured goods. During the period with which we are especially concerned in this book, industrial units were still generally small, and the development of coal mining therefore necessitated the investment of larger amounts of capital than had been commonly employed in the past. But, with the evolution of modern industry during the eighteenth and nineteenth centuries, the upper limit for the size of investments in a single enterprise, the point at which, in the economists' language, diminishing returns on each added increment of capital set in, is reached sooner in coal mining than in most manufactures. [2] See below, vol. ii, pp. 135–45.

[3] Cf. Beck, *Geschichte des Eisen*, vol. i, pp. 769–71.

[4] As a matter of fact some London citizens did send for their own supplies. Under date of September 16th, 1661, Pepys relates: " This morning I was busy at home to take in my part of our freight of coles, which Sir G. Carteret, Sir R. Slingsby, and myself sent for" (*Diary*, ed. H. B. Wheatley, vol. ii, p. 97). But

Except for a few men and women carriers who bore the fuel short distances from certain Scottish collieries [1] to ships and salt pans, horses, mules, ponies, or oxen [2] came to be generally employed for surface haulage in the seventeenth century. At the pithead, pieces of coal were filled into sacks not unlike those of a modern city carter, and made to contain slightly more than a hundred-weight. The coal " drivers ", or " carriers ", then strapped these sacks on the backs of their packhorses, a pair to each horse,[3] and drove to market. Small two-wheeled carts, pulled by one or two beasts of burden, and capable of holding from seven hundredweight to a ton or more of coal, loose or in sacks, provided drivers with a cheaper means of transport than did packhorses ; the cost of " horse-carriage being in proportion to wheel-carriage as three to two ".[4]

Such carts could be used only where the condition of the roads permitted. Money provided during the seventeenth century by the Crown or by local authorities for the upkeep of the public highways did not suffice to repair the damage done by the vastly increased traffic in all kinds of commodities, particularly in coal.[5] Continual processions of cattle, sheep, and fowl (destined for slaughter in London),

---

few citizens could have been in a position to make their purchases in this way ; and Pepys himself, some years later, got his coal through a London trader. Many of the wealthier citizens, who bought from traders, arranged to have coal brought home by their own servants (*The Case of the Coal Trade, c.* 1702).

[1] Such as Tulliallan colliery in Fife (*Register of the Tulliallan Coal Works*, 1643–7). These particular carriers were called " leaders " ; some were women. Coal from the mines in the Mons district was borne the short distance to barges on the river Haine on the backs of women, picturesquely called " cars-à-fesses ", or " boraines " (Decamps, *Mémoire historique*, pp. 195–6 ; Arnould, *Mémoire sur le bassin houillier du Couchant de Mons*, p. 87). Sometimes men or women carriers bore the sacks, or " creels ", of coal from the pits to the houses of consumers (M'Neill, *Tranent*, p. 17). In 1634 Patrick Edmonstone charged James Rait with sending his bastard son to the common highway to wait upon the " coale-carriers " coming from Patrick's " coal pot " to Edinburgh, and to pull " the creels aff the poore creatures backes " (*Privy Council Register of Scotland*, 2nd Series, vol. v, pp. 276–8) ; yet we know from other sources that coal had been brought to Edinburgh from East- and Mid-Lothian by the horseload some decades earlier (*ibid.*, 1st Series, vol. v, p. 227 ; Chambers, *Domestic Annals of Scotland*, vol. i, p. 24).

[2] Apparently ponies were sometimes used in Scotland (H. G. Graham, *The Social Life of Scotland in the Eighteenth Century*, 1906, p. 531), mules in South Wales (Wilkins, *South Wales Coal Trade*, pp. 66–7), and oxen in the Tyne valley (*Cal. S.P.D.*, 1689–90, p. 527) ; but horses most commonly hauled coal in the towns and through the country generally. Macaulay (*History of England*, 1849 ed., vol. i, p. 330) speaks of the use of dogs for the haulage of heavy goods through the streets of Bristol, but I have found no evidence that dogs were used to haul coal carts either in the country or in the towns.

[3] Lord Harley, " Journeys in England ", 1725, in *H.M.C., Report on the MSS. of the Duke of Portland*, vol. vi, p. 100. According to Lord Harley, each sack contained " something more than a bushel " (cf. Appendix C). By an ordinance passed in 1606, the Corporation of Bristol declared that the sacks used to bring coal into the town should contain either one or two bushels (*Egerton MSS.*, 2044, f. 12).

[4] Birch, *History of the Royal Society*, vol. iii, pp. 207–10 ; *Royal Soc., Philosophical Transactions*, vol. xiv, 1683–4, pp. 666–7 (" Experiments relating to Land Carriage " by Sir William Petty).

[5] For the administration and condition of the roads see Sidney and Beatrice Webb, *The Story of the King's Highway*, 1913, chaps. ii–v.

of packhorses and other beasts of burden, proceeding often in a drizzling rain, turned both main thoroughfares and byways into appalling rivers of mud ; while wheeled wagons (now numerous for the first time) sank deep into these rivers, leaving impressions which the winter frosts solidified into ruts. In 1634 the highways of Brislington parish, near Bristol, are reported to be " of late yeares . . . very founderous and in decay by means of the greate resorte of colliers with their horses to certain cole pitts there of late yeares found out ".[1] In 1667 the streets of Liverpool are spoken of as much decayed " by the frequent driving of carts laden with coal ".[2] At Newcastle in 1656 the " coleway " had been so torn up by the driving of carts and wains that it was practically impassable.[3] We have Defoe's word that Watling Street, since Roman times perhaps the most travelled of English highways, " is not passable but just in the Middle of Summer, after the Coal Carriages have beaten the Way ; for, as the Ground is a stiff Clay, so, after Rain, the Water stands as in a Dish, and Horses sink into it up to their Bellies ".[4] Out of a bog such as this, drivers found it easier to extricate a packhorse than a cart and pair of oxen. A driver could not keep both wheels of his cart upon the narrow " causeways " of stones and round pebbles, which were now built along the roads in order that the animals might keep out of the mud ; [5] nor could carts be driven as easily as packhorses where, as for stretches on the route to many small, newly-started collieries, there were no roads at all.[6] It was inevitable that, while carts or " wains " for carrying coal were introduced in nearly every mining district at the end of the sixteenth century,[7] horses, mules, and oxen should remain, because of their greater mobility, the principal means of conveyance,[8] except in areas such as the Tyne and Wear valleys

[1] E. H. B. Harbin, *Quarter Sessions Records for the County of Somerset*, vol. ii, 1908, p. 203. See also vol. i, 1907, p. 227.

[2] J. A. Picton, *Memorials of Liverpool*, 1873, vol. i, p. 131.

[3] *MS. Journal of the Newcastle Common Council*, 1650–6, ff. 375–6.

[4] Defoe, *Tour*, 1748 ed., vol. ii, p. 425.

[5] Webb, *op. cit.*, pp. 65–6.

[6] Often the mine owners themselves built a pathway from their pits to the nearest thoroughfare. In 1632 the " railing and repairing " of such a " way " from a mine at Softly in south Durham is said to have cost £7 or £8 (*Exch. Spec. Com.*, no. 5276). The word " railing " in this case probably indicates some sort of paving ; it cannot mean the laying of wooden rails, for coal was carried from the pits by packhorses. Lessees of a colliery at Orrell (Lancs.) specify in 1711 that the lessor shall provide four separate " cartways " for carrying coal to the " Queen's highway " (*Palat. of Lancs., Bills*, 55/32).

[7] There are definite references to the use of carts or wains for carrying coal in south Durham in 1580 (*Exch. Deps. by Com.*, 22 and 23 Eliz., Mich. 5) ; in Eckington manor, Derbyshire, in 1614 (*ibid.*, 11 James I, Trin. 6) ; in Wakefield manor, Yorks., in 1593 (*Duchy of Lancs. Pleadings*, 162/A/35) ; in Orrell manor, Lancs., in 1600 (*ibid.*, 195/S/10) ; in Warwickshire, Staffordshire, and Leicestershire, in 1626 (*Records of the Borough of Leicester*, vol. iv, pp. 240–1).

[8] It would be unsafe to assume from the frequent mention of wains and carts that they provided the principal means of conveying coal in the inland districts during the seventeenth, or even during the eighteenth, century. Although wains are mentioned in connection with Raby colliery as early as 1580 (see preceding footnote), almost all the coal carrying in south Durham was done by packhorses at the beginning of the eighteenth century (Lord Harley,

or the Severn basin in the neighbourhood of Broseley, where the
vast coal traffic, all converging upon a few wharves at the river,
made it worth while for mine owners themselves to build and maintain
roads,[1] and in London and some other towns and villages, where
the local authorities managed to keep the streets in tolerable repair.[2]

The beasts of burden engaged during the seventeenth century in
moving coal could be counted by the tens of thousands. For supplying
the town of Bristol with fuel in 1675, the " coal drivers " of Kingswood
Chase " kept over 500 workeing horses and over 100 cows ".[3] All
summer long a traveller journeying through Somerset, Lancashire,
Staffordshire, the Tees basin, or almost any part of Great Britain
within ten miles of a working colliery, met every mile or so gangs
of packhorses laden with sacks of coal.[4] A single carrier, with the
help of one or two boy-apprentices might hitch together a string
of packhorses, as many as " thirty and forty in a gang ", and, mounting
the first horse (equipped with " a bell to give warning to travellers
coming in the opposite direction, by any sharp turn or narrow pass "),[5]
proceed from the pit with a load of from three to six tons [6] to the
nearest market town. Only a fairly well-to-do yeoman could afford
to purchase and provide fodder for so large a " gang " or " parcel "
of horses or cattle.[7] Yet it was possible to enter the inland carrying

op. cit., p. 100). In Kingswood Chase coal carriers used only " horses and other
cattell " in 1624 (Chanc. Proc., Series II, 371/4). When we hear of carts in
connection with inland collieries, we usually hear of packhorses also. It is
possible that the use of wheeled vehicles actually decreased in certain districts,
owing to the deterioration of the roads. One cannot be certain, I think, that the
term " coal carriages ", employed in the seventeenth and eighteenth centuries
(Defoe, in passage cited above ; Exch. Spec. Com., no. 752), means wheeled
vehicles. It may mean simply a train of packhorses laden with coal.

[1] Yet at " sea-sale " collieries in Cumberland all carriage from the pits to
the wharves on the coast was done by packhorses until the very end of the
seventeenth century (V.C.H. Cumberland, vol. ii, p. 360). The Assembly of the
Etats du Hainaut considered, in 1723, a project for building a road some forty
miles long from Mons to Tournai to enable coal owners to send their coal in
wagons, instead of by boat, and thus avoid paying the French tariff at Condé
(L. Devillers, Inventaire des archives de la ville de Mons, vol. ii, p. 363). There was
not, as far as I know, any project for constructing so long a " coal way " in
Great Britain ; but neither was there any considerable tax on the passage of
coal from one inland county to another (cf. below, vol. ii, pp. 313–4).

[2] Carts were used to carry coal even in small Scottish burghs in the early
seventeenth century (Privy Council Register of Scotland, 1st Series, vol. xi, p. 117 ;
2nd Series, vol. v, p. 456). On the other hand, the town authorities prohibited,
in 1667, the passage through the streets of Liverpool of carts laden with coal
destined for shipment by sea (Picton, loc. cit.).

[3] Exch. Deps. by Com., 27 Charles II, Mich. 29.

[4] See, besides the passage in Lord Harley's Journeys already referred to
(above, n. 8), Smollett, Humphry Clinker, 1771, vol. i, p. 132 (Matt Bramble
to Dr. Lewis, from Bath, May 8th) ; Webb, op. cit., pp. 64–5, 70–1, 76–7 ;
V.C.H. Somerset, vol. ii, p. 382 ; Mantoux, The Industrial Revolution, pp. 113–4.

[5] P. Whittle, The History of the Borough of Preston, 1821–37, vol. ii, p. 61.

[6] Since each horse carried from 2 to 3 cwt. (see above, p. 381, n. 3)

[7] Early in the reign of James I, the mine owners in the Tyne valley attribute
the high price of coal partly to the increased cost of fodder for the beasts of
burden which drag the carts and wains from the pits to the river bank. " The
sharpnes of the [late] winter ", they declare, " being such as little or no . . .
cariage could be hadd, and the great want of haye and fother and the death of

trade with a small number of cheap mules ; [1] and the capital required even for the most expensive outfit of a provincial coal carrier in south Durham or in Mid-Lothian was insignificant in comparison with that needed in order to operate a " land-sale " colliery in the same districts.

A London carmen's cart, or " carr ", together with its carthorse and sacks to carry about 14 cwt. of coal, was said to be worth £30, when one was seized in 1608 because the driver had run over and killed a six-year-old boy playing in Creed Lane.[2] Presumably most of the carts and " wains " used in other towns, or in the Tyne valley and other colliery districts, did not cost much more to make and fit out than did this city " carr ".

It was the introduction of railed " wagonways " in the colliery districts that involved the first considerable investments in the transport of coal overland.[3]   James Clifford, the Shropshire mining adventurer, spent 200 marks (£133 13s. 4d.) in 1606 in setting up his " very artificiall engine of timber ", to convey coal through the demesne lands of the manor of Broseley to the banks of the Severn.[4] A complete wagonway consisted of two sets of wooden rails,[5] each with its groove for the wheels ; in descending one track the driver hitched his horse or ox to the rear of the wagon, unless the descent

many cattle and horses was so great that we were enforced to give greater prices . . . [and] whereas by this time we expected the prices for loadinge of Coles should have fallen to a lower rate, we find that by reason of the dearnes of cattle, the unseasonableness of the yeare, the raisinge of the price of pasture gates, which all the groundes lyinge conveniently for that purpose is not able to supplie, there is no likelyhood that the price of the cariage of Coles can so fall as Coles maye be afforded here at lesse prices then nowe they are " (Dendy, *Records of the Company of Hostmen*, p. 59).   On the one hand, the new demand for beasts of burden had resulted in a great increase in the purchase price (cf. above, p. 376), and even in a failure of the supply.   " There is little or no profit to be made of the collieries now ", writes in 1646 a former mayor of Newcastle, " by reason there are not cattle in the country to convey the coals from the pits to the waterside " (Welford, *Records of the Committees for Compounding*, 1905, p. 164). On the other hand, the rise in rents is said to have forced " the gentlemen " of the Tyne valley in 1595 to pay for their pastures " more . . . than heretofore by threfold " (*Harleian MSS.*, 6850, no. 40).

[1] " Gangs " as large as the one described in the text were probably the exception in the seventeenth century.

[2] *Star Chamb. Proc., James I*, 1/38.   A tragic accident of the same kind occurred in 1635 in the street before a small house in the little Scottish village of Culross.   John Clarke, coal carter of the burgh, " come by the foote of the staire with a horse loadned with coalls within a great coale cairt ", one wheel of which ran over and killed a small child playing at the bottom of the steps (*Privy Council Register of Scotland*, 2nd Series, vol. v, p. 456).   Although the Court exonerated Clarke, local opinion was so outraged that he was banished from the burgh and forbidden to return under pain of death ; for street accidents were not in the seventeenth century the common occurrence that they are to-day.

[3] Although, even when no rails were laid, considerable sums were spent in making " watergates, trenches, and ditches " to keep dry certain of the most important ways from large collieries near Newcastle to the river Tyne (*Palat. of Durham, Decrees and Orders*, vol. i, f. 228).

[4] *Star Chamb. Proc., James I*, 310/16 ; for proof that this " engine " actually ran on rails, see above, pp. 244, 309.

[5] See the description of the two " way leaves " running from the colliery at Chester-le-Street to the Wear river (Lord Harley, *op. cit.*, p. 104).   The rails were of oak, ash, or birchwood.

A Coal Wagon and its Wagonway

(Morand, *L'art d'exploiter les mines de charbon de terre*, 1768-77. Plate XXXIV, No. 3)

As the drawing was made somewhat after the middle of the eighteenth century, the wagon and the wagonway are probably improvements on those common in the seventeenth century, but the principle upon which the earlier ones were built was the same. The reader will find it interesting to compare this plate with Jars, *Voyages métal-lurgiques*, vol. i, plate v, and pp. 201sqq.

The following explanation is based on Morand, *op. cit.*, vol. ii, p. 698 : A, is the wagon coming from the mine and descending the hill along a sloping wagonway on its way to the staith. B, Wheel which, like its fellow, has a groove of cast iron, 30 inches in diameter. C, A similar wheel made of wood. D, Long piece of wood which the wagoner uses as a brake. E, Road-bed along which the wagon runs from the top of the hill to the river; the road is planked with strong pieces of wood in order that it shall hold together, and that the weight of the loaded wagon shall not sink the rails into the ground. F, Staith, seen from a distance, showing the manner in which the coal is emptied from the wagon into the keel.

was too gradual to put it in motion of itself ; in ascending the other track the animal pulled back the empty vehicle. Presumably special sidings were laid from the various pits to connect with the main line of rails. In order to secure a steady and easy descent, it might be necessary, as it is in laying a modern railroad, to span ravines by bridges and to drive cuts through rising ground.[1] This added enormously to the expense, which might run into hundreds, if not thousands, of pounds before the wagonway could be opened for the passage of wagons. The new road, by relieving the carriers from dependence upon the muddy, and often almost impassable, cartways,[2] enabled every driver with his horse or ox to pull a large four-wheeled wagon, usually holding about two tons,[3] and to move from five to eight or more times as much coal from the pits to the water as had been possible with the old carts and wains ; [4] but the wagonway represented so heavy an investment that it was worth while to lay one only if the conditions of the market made certain an abundant traffic. These conditions prevailed only in the case of some of the largest collieries producing coal for transport by water from a single wharf.[5] Wagonways were financed, therefore, not by

[1] See the description of the arch belonging to Blackburn colliery in John Loveday, *Diary of a Tour in 1732 (Publications of Roxburgh Club*, vol. cvii, 1890), p. 172. " It is the highest Arch I ever saw, perhaps has not its Equal . . . 'Tis a little stream that flows under it ; the occasion of making it so high was that the coal-carts might go on plain ground . . . All the way is contrived to be upon a gradual descent that the Carriages may go without horses . . . The proprietors have been at a great charge in levelling hills, raising valleys, etc."

[2] Undoubtedly the deplorable condition of the ordinary roads and paths encouraged, and may even have been the principal cause of the building of wagonways. In 1610 a midland coal dealer, who had contracted to market the output of Sir Percival Willoughby's colliery at Wollaton near the Trent, asks Willoughby to get the permission of Sir Thomas Beaumont, one of the owners of the adjoining colliery at Strelley, " to bring coales downe the rayles . . . , for Strelley cartway is so fowle as few cariadges can passe " (*H.M.C., Report on the MSS. of Lord Middleton*, pp. 176–7).

[3] Green, "Chronicles and Records of the Northern Coal Trade", in *Trans. North of Eng. Inst. of Mining Engineers*, vol. xv, 1865–6, pp. 187–8 ; and see below, Appendix C (iii). Not all the coal wagons which ran on rails could carry as much as 2 tons. The wagons used at Crawcrook colliery on the Tyne held 15 bowls, which Galloway thinks were equivalent to 33½ cwt. (*Annals*, p. 156, but see below, Appendix C). Some wagons, on the other hand, carried much more. The two front wheels were made larger than the rear wheels, so that, in descending with a load of coal, the wagon remained level, and the loss of coal in spillage along the tracks was thus reduced to a minimum (*A History of Northumberland*, vol. viii, pp. 20–1).

[4] Where a pit was situated about a mile from the river, the carrier, if he drove one of the new wagons which ran on rails, could carry every day from 25 to 40 tons of coal to the wharf—making 15 or 20 trips. But if the mine owners still depended on the old wains and carts, the carrier could seldom move move than 5 or 6 tons to the river in a day (Green, *op. cit.*, pp. 187–8). Railed wagonways also made it possible for mine owners in the Tyne valley to transport some coal during the winter months (Jars, *Voyages métallurgiques*, vol. i, pp. 199–200), when the ordinary roads were impassable.

[5] Railed wagonways were first built in Nottinghamshire and Shropshire to bring coal from certain pits to the Trent and Severn rivers ; they were introduced in the north of England during the reign of James I (see above, pp. 244–5) ; but, even at the end of the seventeenth century, there were still some important collieries in the Tyne valley served exclusively by the old wains and carts

independent adventurers, but by the mine owners themselves; the rails, wagons, stables, and horses became a part of their movable property, like the pit lodges, " engine " horses, " turns ", and cut timber.

They also built the wharves, or " staiths ", at the water's edge. A staith might involve an additional capital outlay of hundreds of pounds, for these structures served not simply as gangways for the loading, but also as storing places for the " safe keeping of the coals ". At small ports, where the trade was casual, there was, it is true, no need for storage. We have a description of the arrival at Irvine, in Ayrshire, of the " birlings " which carried coal to Ireland. One of the sailors " blew a large horn which was fixed to a post at the quay by an iron chain; and, upon this signal, the country people loaded their coal ponies or small horses, and carried down what quantities were wanted ".[1] At those ports where coal shipments were of some importance, the mine owners found it desirable to place the larger portion of their stock of fuel on their staiths, ready to load at a moment's notice, rather than on the " heaps " beside their pits; because shipmasters, who usually arrived in great fleets, did not relish delaying their sailings while wain drivers collected the cargoes. During the summer season there might be from 100,000 to 200,000 tons waiting sale at the six or eight principal staiths on the Tyne; [2] so that each staith had to be built to hold upwards of 25,000 tons. Great care had to be taken to prevent fires; [3] and the storage space had to be large. The wharf erected in 1716 for the Duchess of Hamilton on the Clyde between Glasgow and " Broomielaw " occupied a space along the water front three yards in breadth and more than a quarter of a mile in length, and cost more than £500 sterling to build.[4] It was entirely surrounded by a high limestone wall to prevent " pickeries " of fuel—then very frequent. At Newcastle some staiths

(Rawlinson MSS., a. 241, ff. 79–82). One great colliery made use of 600 such vehicles (Dunn, View of the Coal Trade, p. 19). In Cumberland the first " coalway " was laid in 1682 (V.C.H. Cumberland, vol. ii, p. 360), and in Glamorgan at the very end of the seventeenth century (Galloway, History of Coal Mining, p. 65; E. Phillips, Pioneers of the Welsh Coalfield, p. 10; V.C.H. Salop, p. 465). The date at which wagonways were introduced in Scotland is uncertain. Possibly enterprising Londoners attempted to lay rails along some of the city streets to facilitate the carriage of coal; for we hear in 1623 of " divers carrmen " who work with " engine carrs newly invented " (Journal of the Common Council, vol. xxxii, ff. 159–63).

[1] Sinclair, Statistical Account of Scotland, vol. vii, pp. 173–5.

[2] In response to a request from the Privy Council, the Mayor of Newcastle reported, on August 19th, 1640, at the time of year when stocks neared depletion (cf. above, p. 376), that there were on the staiths along the Tyne 100,000 tons of coal, " a fifth or a sixth fewer than in former years " (Cal. S.P.D., 1640, p. 606). We know from the MS. Account Books of the Newcastle Corporation, that there were not more than six or eight staiths on the Tyne in the years immediately preceding the Civil War. The active life of a staith was limited. " When the mines . . . are wrought elsewhere, they must have other staiths in places fittest or nearest to those mines or pits " (Welford, History of Newcastle and Gateshead, vol. iii, p. 354).

[3] Cal. S.P.D., 1639, p. 157.

[4] Hamilton MSS. (Account of Wm. Lawson, mason in Glasgow, for building the wharf, 1716).

were covered with a roof of timber to keep the coal dry.[1] Labourers with wheelbarrows ordinarily collected the coal from the stacks, and emptied it into the keels which drew alongside the staiths.   But, before the end of the seventeenth century, Tyneside mine owners probably commenced their practice of building gangways from the wharves, on wooden piles running out into the water : so that by means of a trap door in the floor, coal could be dropped directly into the waiting boats.[2]

At some ports coal was loaded from the staiths directly into sea-going vessels.[3]   Rivers like the Tyne and the Wear were too shallow, however, for the hoys to draw near the banks.   During the seventeenth century shipmasters entering the Tyne dropped anchor farther and farther downstream ; for, on the one hand, the size of ships increased,[4] while, on the other, the river bed became choked with ballast cast carelessly overboard, and with loose coal washed off the staiths.   Shippers depended almost entirely on " keels ", or " lighters ", to load their hoys.[5]

These small boats changed little either in size or appearance from the fourteenth century, when they are first heard of in connection with the coal trade,[6] until the nineteenth, when they were superseded

---

[1] Covered staiths were known as " trunks " (Lord Harley, *op. cit.*, p. 104). If long exposed to the elements, coal might depreciate substantially in value, for water made it heavy, " which doth not relish well with the Shipmaster or first Buyer, because he buys . . . by weight, and therefore, being heavy, . . . a less quantity of . . . Wet . . . then of dry bright Coals, which he finds in his making out in Sale at London or elsewhere, where he sells by Measure, and not by weight " (*The Compleat Collier*, pp. 29–30).   According to Sir John Lowther, the soft coal mined near Newcastle was a grade especially subject to depreciation if left standing in the rain (Birch, *History of the Royal Society*, vol. iii, p. 439).

[2] The original gangways, or wharves, for loading coal appear to have been far more primitive.   They were nothing but narrow banks extending a few yards from the shore, made of earth, clay, rubbish, and ballast cast off from the sea-going vessels.   It was inevitable that the current should continually wash away a portion of this sandy gangway, clogging up the river basin beside the wharf, so that the keelmen found great difficulty in drawing their barges alongside. And, if this forced the staithmen to load coal into keels across a yard or two of open water, between wharf and barge, a considerable quantity of dust and small pieces fell into the river, further clogging up the basin.   In fact, refuse washed from the bases of these early wharves, or spilled in the process of loading keels, must have been partly responsible for the increase in the size and number of bars which choked the whole Tyne basin and caused the citizens of Newcastle great concern for the continued prosperity of their shipping trade during the seventeenth century (see above, p. 255 n., and below, vol. ii, p. 127).   The surface of the old earthern wharf was raised so little above the water level that a river flood might overflow it, and damage the stock of coal.   Perhaps the disastrous floods in December, 1672, which swept " a vast quantity " of fuel into the river (*Cal. S.P.D.*, 1672–3, pp. 299, 310–11), convinced colliery owners and coal traders that it was imprudent to begrudge any expense in improving their wharves.   It was probably soon after this that they began to build the gangways on wooden piles.   Before the accession of George III all the wharves on the Tyne were of this type (Jars, *Voyages métallurgiques*, vol. i, pp. 201–2).

[3] See, e.g. *Privy Council Register of Scotland*, 2nd Series, vol. vi, p. 126.

[4] See below, pp. 390–91.

[5] On the Tyne a small quantity of coal was still loaded from the staiths directly into sea-going ships even at the end of the seventeenth century (*Rawlinson MSS.*, a. 241, ff. 79–82).

[6] Galloway, *History of Coal Mining*, pp. 15–16.

by other loading devices.[1] Pennant made a note concerning them
in 1769 as he crossed the Tyne on his journey to Scotland.
" These boats ", he wrote, " are strong, clumsy, and round, will carry
about 25 tuns each ; sometimes are navigated with a square sail, but
generally are pushed along with large poles ".[2] As a matter of fact,
they seem to have been slightly oval in shape,[3] measuring perhaps six
or eight paces from prow to stern, and as flat-bottomed as possible
in order to displace little water.   Keels were manned invariably by
four " keelmen " : [4] one, the captain or " skipper ", took charge ;
another, the " haddock ", placed himself in the stern and steered with
a kind of oar, large in size and called a " swope " ;   the last two,
" hillies ", stood on either side, equipped with long poles (" puys "),
which they dug into the muddy river bed to propel themselves
along,[5] in much the same way as to this day bulky skows, full of
merchandise, are pushed along the shores of the Italian lakes.   When
they had made fast to the hoy, the keelmen fell to with their bare
hands and heaved coal over the sides into the hold—taking a shovel
for the small pieces and dust.

Keels usually carried about 21 tons,[6] not 25 tons as Pennant
stated.   When fully equipped with the necessary " furniture ", a
" cole lighter " of the type used at Sunderland in 1618 was said to be
worth £100.[7] But building costs increased considerably during the
seventeenth century, owing to the shortage of timber and the rise
in carpenters' charges ;   and the expenses of repairing keels grew
heavier.   " Whereas the Collieries ", complain the mine owners
of the Tyne valley in 1656, " are much more burthened with the
nomber and charge of Keeles then formerly, when onely shippes of
smale Burthen frequented this harbour, who tooke in all theire ladinge
of coles before the Towne, and above the Owesburne, a place neere
adioyninge to the same,[8] whereby a smaller nomber of keeles would
supplie a great collierie, with little charges in repayringe : which Keeles
also would continue fiftie or Threescore yeeres, never beinge put to the
danger of any stresse of weather belowe the Towne and Owesburne.
And whereas at this time the most shippes that come into and frequent
this River are great shippes, in whom the greatest part of the coles are
laden aboard at Sheeles or thereabouts, so as thereby everie cole
Owners worke requires a double nomber of Keeles, which keeles also
cost double as much in repayringe, besides treble hazard of the losse
of keeles and goods, yet they continue and last not above a fourth
part of the time as formerly ".[9] At the beginning of the eighteenth

---

[1] Cf. Dendy, *Records of the Company of Hostmen*, p. 1. At the time of Jars'
visit (1765), the merchants still loaded nearly all coal by means of keels (*Voyages
métallurgiques*, vol. i, pp. 205–7).

[2] Pennant, *Tour in Scotland*, p. 29. For another description of a Newcastle
keel see Westerfield, *Middlemen in English Business*, pp. 223–4.

[3] Galloway, *op. cit.*, p. 15.

[4] *Exch. Decrees*, Series iv, vol. iii, f. 106 (1626) ; Lord Harley, *op. cit.*,
p. 105.   [5] *MS. Bell Collection*, vol. xiii, pp. 504 sqq.

[6] Eight Newcastle chaldrons (see Appendix C).

[7] *Palat. of Durham, Bills and Answers*, bdl. 24 (Parkinson *v.* Lee).

[8] The " Owesburne " is a small brook which flows into the Tyne from the
north, just below Newcastle.

[9] Dendy, *op. cit.*, pp. 107–8.

Unloading a Collier and a Coal Lighter in London
(From Ambrose Heal's Collection of Tradesmen's Cards)

Reproduced through the courtesy of Mr. Heal

century the coal traffic on the Tyne employed about 400 keels,[1] representing, no doubt, an investment of well over £40,000.[2] There were probably between one and two hundred more on the Wear.

Small boats, called " lighters " or " barges ", of a nature very similar to the keels of the North of England, served to unload most of the colliers entering the Thames. Until Elizabeth's reign, the greater number of coal cargoes for London had been unloaded from the sea-going ships without the aid of lighters. But, during the reign of James I, if not before, the use of lighters in bringing coal to the capital became almost universal.[3] A letter written by the Lord Mayor to Chief Justice Coke in 1616 explains the reason for the change. Most of the sea-going hoys, being " built of greater bulk than formerly ", drew too much water to anchor above Tower Wharf. As long as the " cars " were permitted to pass through the bulwark of the Tower to reach the wharf, many shipmasters continued to unload their cargoes near the Tower. But recently, writes the Lord Mayor, the Lieutenant of the Tower, affecting causes of difference with the City, forbad the passage of cars through the bulwark. If cars are now to reach the wharf at all, they must be driven by a roundabout route, which doubles the cost of carriage. Shipmasters are obliged, therefore, to sell their coal to " woodmongers and other engrossers ", who have lighters to fetch it.[4]

It was not only on the Thames that the expansion of the coal industry involved substantial new investments in river craft. Provincial traders with their lighters waited, at Hull or King's Lynn or Bridgwater, for the arrival of colliers from Newcastle or Sunderland or Swansea, to receive a portion of the cargo, which they then ferried up the Humber, the Ouse, or the Parrett, for consumption at inland towns and villages.[5] Somewhat more capital was required to own and operate one of these provincial lighters, than to purchase and equip a string of packhorses, such as made the journey from the pits at Brislington to Bath, or from those in Kingswood Chase to Bristol.

---

[1] Lord Harley speaks in 1725 of 800 keels (*loc. cit.*), but this would seem to be an overstatement. The number of Tyneside keelmen in the first decade of the eighteenth century probably did not exceed 1,600 (see below, vol. ii, p. 142), so that, allowing four keelmen to each keel, the keels would have numbered about 400 at that date. They had increased in number with great rapidity ever since the accession of Elizabeth. In 1617 the keelmen are spoken of as " having been a company formerly consisting of one hundred and sixty " (Welford, *History of Newcastle and Gateshead*, vol. iii, p. 217), which would mean, again allowing four keelmen to a keel, that there had been only about forty keels. " Formerly " may well have referred to a time within the memory of men living in 1617. In 1626 the number of keels actively employed on the Tyne was estimated at 300, " all which are to goe every tide three or fower miles to fetch coales to the shippes " (*Exch. Decrees*, Series IV, vol. iii, f. 106). Another estimate, however, made during the reign of Charles I, puts the number of Tyne keels at 200 (*S.P.D., Charles I*, vol. clxxx, no. 58). Ralph Gardiner's estimate for 1655 was 320 (*England's Grievance Discovered*, 1655, p. 98).

[2] Assuming that each keel cost £100 or more to build (see above, p. 388).

[3] *City of London Repertories*, vol. xiii, f. 130 ; vol. xxxii, f. 131*b* ; vol. xxxix, f. 279.

[4] *Remembrancia MSS.*, vol. iv, no. 12.

[5] *S.P.D., James I*, vol. cxxxviii, no. 120 ; *Star Chamb. Proc., Henry VIII*, vol. v, f. 21 ; *Exch. Deps. by Com.*, 23 and 24 Charles II, Hil. 18.

To build a sea-going collier, or even a large river " trow " of the
type that carried coal from Broseley down the Severn, required a still
larger amount of capital.   The average size and carrying capacity of
ships engaged in the coal trade increased notably during the seventeenth century.   Until late in the reign of Elizabeth, coal had been
shipped from all British ports in ordinary merchantmen designed for
the trade in corn, cloth, beer, or French wine ; and English merchantmen, narrow and light of keel, were, as one anonymous authority
observed, ill-adapted for holding so " grosse [a] commodity " as
coal.[1] The ideal " collier " had to be wide and heavy of keel, and
so built as to hold the maximum quantity of coal and to be navigated
with the minimum number of seamen—for every seaman added to the
overhead costs of each voyage.   By the time of the Restoration nearly
all the old English merchantmen had been superseded, at least in the
trade between Newcastle and London, by a new type of vessel especially
designed for the transport of coal.   Colliers " are deep ships and unfit
to carry men ", wrote in 1672 a government agent, charged with the
task of assembling transports to bring Scottish soldiers to the Dutch
coast to make war on Holland.[2]   The average size of a cargo of coal
imported at London increased from 56 tons in 1592 to 73 tons in 1606,
83 tons in 1615, 139 tons in 1638, and 248 tons in 1701.[3]   During this
period the labour needed to sail a collier was reduced at least by half.
Ten seamen had been hired in the eighties of the sixteenth century to

---

[1] *S.P.D.*, *James I*, vol. clxxx, no. 77. Foreign ships were much better adapted
for carrying coal than were English merchantmen (*S.P.D.*, *Charles I*, vol. xiv,
no. 9 ; and cf. below, vol. ii, p. 24).
[2] *Cal. S.P.D.*, 1672, p. 346.   For all that, eighteenth-century travellers
sometimes took passage in a collier (see Smollett, *Roderick Random*, chap. viii ;
Peregrine Pickle, chap. lxxxi ;   *Humphry Clinker*, 1771 ed., vol. ii, p. 154
(J. Melford to Sir Watkin Phillips, from Newcastle, July 10th)).
[3] These figures have been worked out from *Exch. K.R. Port Bks.*, 9/3, 13/4,
18/1, 41/6, and *Additional MSS.*, 30504, f. 18, which give a complete record of all
London coal imports for the years in question.   Coal imports are recorded in
these manuscripts in London chaldrons, which I have reduced to tons, counting
1 chaldron equal to 1⅓ tons (see Appendix C).   There is no evidence of any
similar increase in the size of the coal cargoes shipped from Newcastle during
the fourteenth, fifteenth, and sixteenth centuries (*Exch. K.R. Customs Accounts*,
106/4, 107/57).
When seventeenth-century traders and customs officers spoke of the
" burden " of a collier, they meant the weight of her full cargo of coal.   Thus
in 1625, when a Newcastle chaldron was commonly supposed to weigh about
2 tons (though in fact it probably weighed more, see Appendix C), mention was
made of " a ship of burthen of 100 chaldrons, which is 200 tons " (*S.P.D.*,
Charles I, vol. xiv, no. 9 ; cf. *Exch. K.R. Port Bks.*, bdl. 187, no. 5, etc.).   The
average cargo of coal, as given in the text, may be considered, therefore, as
equivalent to the average tonnage burden of a collier engaged in bringing coal
to London.
It was the largest colliers that came to London.   An examination of the
imports at Hull, King's Lynn, Yarmouth, and Southampton (*Exch. K.R. Port Bks.*,
354/2, 439/4, 449/11, 500/16, 821/5) shows that the average coal cargo unloaded
in these towns towards the end of the seventeenth century did not exceed
50 tons.   (One must remember, however, that not all the shipmasters unloaded
their entire cargo when they dropped anchor in these ports.)   While London
was supplied almost entirely with coal produced in the Tyne valley, an increasing
proportion of all the coal imported at small towns north of the Thames came
from Sunderland, where the harbour was accessible only to ships of small burden.

handle a 100-ton coal-carrying bark ;[1] in 1665 the same number could manage a 220-ton collier [2] of the type employed in the trade between Newcastle and London, although from ten to fifteen seamen were probably needed to navigate an ordinary merchant vessel, half as large.[3] Many of the Newcastle colliers were built after 1625 to transport from 200 to 300 tons of coal ; [4] and the average cargo carried coastwise from the Tyne in 1634 was about 150 tons.[5] Later in the century the number of very small colliers diminished, and a few ships of 400 tons and more entered the trade, so that the average cargo of coal must have increased considerably.[6] Apparently shipbuilders believed that the maximum efficiency was obtained with a vessel of 500 tons, for

---

[1] See table in Welford (*History of Newcastle and Gateshead*, vol. iii, p. 45) giving the number of men employed on thirteen ships owned by Newcastle merchants in 1587. The total tonnage of these ships is 1,105 tons, the total number of men 115 ; more than ten men to a hundred tons of shipping.

[2] See warrant granted April 21st, 1665, for ships and seamen to continue to serve in the Newcastle coal trade during the Dutch War (*P.C.R.*, vol. lviii, pp. 111–12). There are 71 "colliers", with a total tonnage of 12,765 tons, and with 523 men (" besides boys ") ; that is, 10 men to every 244 tons of shipping. Another warrant was granted May 2nd, 1665 (*ibid.*, p. 160), for 28 "colliers", with a total tonnage of 6,270 tons, and with 254 men (" besides boys ") ; that is, 10 men to every 247 tons of shipping. A royal proclamation of 1691 provides that, for every 100 tons of shipping, four men shall be freed from conscription to man the collier fleet during the French war (Steele, *Bibliog. of Tudor and Stuart Proclamations*, no. 4069). In 1703 the masters of Trinity House, London, report (M. Dunn, *View of the Coal Trade*, p. 21) that 4,500 seamen are required to sail 600 colliers, with an average carrying capacity per ship of 80 Newcastle chaldrons, or 222 tons (Appendix C) ; that is, 10 men for every 296 tons of shipping. These numbers are the minimum considered necessary for manning the colliers, for all these proposals were drawn up in war-time, with the object of supplying the navy with sailors. In proposals submitted to Charles I for bringing the coal shipping trade under royal control, it was estimated that 3,000 sailors would be required to bring (in one voyage) 50,000 London chaldrons, or 66,666 tons, of coal from the north of England to the south-eastern counties (*Sloane MSS.*, 2902, no. 9), that is, 10 men for every 222 tons of shipping.

[3] There is no doubt that it was the expansion of the coal industry in the seventeenth century which led to the changes in shipbuilding which increased the hold space and reduced the number of seamen (cf. below, vol. ii, pp. 90–1). In the eighteenth century other vessels were still more costly to navigate than colliers. Seamen on voyages to the south seas were glad if they could make use of ships built for the coal trade, because they were " more roomy . . . and might be navigated by fewer men than other vessels of the same burden " (John Hawkesworth, *An Account of the Voyages undertaken . . . for making Discoveries in the Southern Hemisphere*, 1773, pp. iii–iv).

[4] In 1626 a certain Edward Bennett suggested to Buckingham the passage of an Act of Parliament to prescribe that " noe shipp should be built hense forward " for carrying coal unless it be nearly 300 or 400 tons in burden and fashioned according to the best advice of sea carpenters (*S.P.D., Charles I*, vol. xxv, no. 1).

[5] 58·4 Newcastle chaldrons, according to the records for nine months, from March 1st to November 30th (*Exch. K.R. Port Bks.*, 190/9). The average was pulled down considerably by a number of very small cargoes—10 chaldrons and even less—which were shipped especially during the summer months. A large proportion of all the cargoes carried were of from 200 to 300 tons, and as the century proceeded the proportion increased.

[6] This is a reasonable conclusion to draw from the evidence which we have of an increase in the size of coal cargoes at London (see above, p. 390). See also the preceding note.

they made no attempt to launch larger colliers,[1] although a few 1,000-ton merchantmen had been launched for use in other carrying trades.[2]

Newcastle colliers ranked, nevertheless, among the large trading vessels of the seventeenth century ; coal shipped from other British ports was carried for the most part in ships of less than a hundred tons burden. Until the harbour was deepened in the eighteenth century, it was impossible for very large vessels to enter the Wear ; the average size of ships loading coal at Sunderland increased from about 20 to about 60 tons during the seventeenth century.[3] On the west coast, the colliers were also much smaller than those engaged in the Newcastle trade. As the Lord Deputy for Ireland explained in 1636 to the English Privy Council, the coal freight was cheaper in proportion from Newcastle to London than from Chester to Dublin (a voyage half the distance, and across quieter seas), " in regard of the greatness of the ships managed with far less charge and fewer men ".[4] For the trade between Cumberland and Dublin some merchants built between 1660 and 1690 a new fleet of sixty colliers, ranging in burden from 70 to 150 tons ;[5] but the average coal cargo shipped at the end of the seventeenth century from west-coast ports, other than Whitehaven or Workington, did not exceed 30 or 40 tons.[6]

---

[1] Even in the eighteenth century a collier never carried much more than 500 tons of coal, but the number of very small colliers diminished steadily (*Exch. K.R. Port Bks.*, bdl. 215 ; *City of London MS. Accounts of Orphans Duty* for 1732–3).

[2] Cf. R. G. Marsden, " English Ships in the Reign of James I ", in *Trans. Roy. Hist. Soc.*, N.S., vol. xix, 1905, pp. 309–42.

[3] Between 1594 and 1634, the average cargo of coal shipped at Sunderland increased from 10 Newcastle chaldrons to 18 Newcastle chaldrons (*Exch. K.R. Port Bks.*, 185/6, 188/6, 8 ; 190/9). For the average cargo at the end of the seventeenth century, see above, p. 31.

[4] *Cal. S.P. Irish*, 1633–47, p. 130.

[5] According to testimony said to have been made before the Privy Council in the 'nineties by citizens of Whitehaven and Workington (Walter Harris, *Remarks on the Affairs and Trade of England and Ireland*, 1691, p. 19). The average coal cargo per ship sailing from Whitehaven to Ireland in 1708 amounted to more than 100 tons (*Exch. K.R. Port Bks.*, 1449/6, 11).

[6] From a scrutiny of the Port Books, it appears that the average cargo at Liverpool, Chester, and Milford increased from some 12 or 14 tons at the beginning of the seventeenth century to about 30 tons at the end. At Swansea the average cargo was somewhat larger (see Appendix D(ix)). Colliers from Scottish west-coast ports bound for Ireland were probably no larger than the colliers sailing from South Wales. Tucker, who furnished the Protector with a survey of the state of Scottish shipping, found half a dozen 100-ton vessels belonging to the port of Glasgow ; but most of the " smiddy coales " were still carried to Ireland in tiny open boats smaller than river keels (Tucker, *Report upon . . . Excise and Customs in Scotland*, pp. 38, 40). Nor were the Scottish colliers, which took their ladings on the Firth of Forth, much larger, to judge from a warrant granted April 28th, 1665, for 18 Scottish ships to be freed from conscription for war purposes, upon an offer made by their masters to bring coal from Newcastle for the relief of London (*P.C.R.*, vol. lviii, p. 121). All these vessels must have been under 75 tons in burden, for eight were allowed only one seaman per ship besides the master, six others were allowed two seamen, and the remaining four were allowed three seamen (only three vessels employing boys in addition to seamen). Most of the Scotch coal exported was probably carried by foreign shippers, and most of their vessels were much larger than the Scottish vessels. In general, colliers engaged in the foreign trade from the north of England were somewhat smaller in size than those engaged in the coastwise trade between Newcastle and London, but larger than those engaged in the west-coast trade (*Exch. K.R. Port Bks., passim*).

Customs officers appear to have discouraged the construction of large colliers, at least for the South Wales coal trade, because the master of a large vessel, which could easily be sailed across the open sea, was likely to defraud them by slipping over to France with a coal cargo entered in their books to pay coastwise duty only.[1] Customers stationed at Swansea brought an action in 1636 against the shipowners of Bridgwater, on the ground that the owners had adapted their colliers for " foreign transportation ". Several seamen of Swansea testified that " the trowes or vessels of Bridgwater are of far greater burden than they were 11 years since . . . and . . . have been fitted with close decks, masts, sails and tacklings, fit for sail beyond seas ". For the defendants, William Meyricke, a sailor of Bridgwater, protested " that he would not hazard his life in any of the trows for Foreign parts (being now master of one) ".[2]

These boats of Bridgwater—if they were designed merely to cross the Bristol channel with fuel for the counties of Somerset, Devon, and Cornwall—may well have been simple river " trows ", similar to those which carried coal from the wharves at Broseley to the towns and villages along the Severn.[3] River " trows ", and even the smaller " frigates ", were sailing vessels. Trows " are from 40 to 80 tons burthen ; . . . have a main and top mast, about 80 feet high, with square sails, and some have mizen masts ; they are generally from 16 to 20 feet wide and 60 in length ", i.e. nearly as wide and almost three times as long as a keel ; " they are mostly navigated with three or four men, who being generally robust and resolute, may be esteemed a valuable nursery of seamen ".[4]

Trows probably differed from sea-going colliers mainly in that they were built with shallow hulls to avoid the river shoals.[5] They provided, therefore, comparatively little space for storing coal ; and, being somewhat top-heavy, were bound to be tossed unmercifully in a North Sea or an Atlantic gale. A Newcastle collier, on the other hand, must have been built with a heavy iron keel, so that the greater part of her hull was submerged. Without being, perhaps, very much longer or broader than a river trow, the large collier gained so much space in depth and breadth of hold, that she could carry from five to

[1] See below, vol. ii, pp. 236-7.

[2] *Exch. Deps. by Com.*, 12 Charles I, Mich. 45.

[3] " Trows " were sailed on the Severn at least as early as 1606 (*Star Chamb. Proc., James I*, 310/16).

[4] *Gents Magazine*, vol. xxviii, 1758, pp. 277–8. This description probably fits conditions at the end of the seventeenth century, for the traffic on the Severn was then already of considerable importance (see above, pp. 96–7). The smaller Severn " frigates " were from 20 to 40 tons in burden ; they had a single mast, and a square sail. On nearly all rivers except the Severn, simple barges, propelled by oars, were used for moving coal. (Cf. *H.M.C., Report on the MSS. of Lord Middleton*, pp. 172–3.) As late as the eighteenth century, it was impossible to carry coal down the Meuse from Liége, except in flat-bottomed boats (*A State of the Coal-Trade to Foreign Parts*, 1745, p. 5). The Severn iron barges were drawn along by men on the banks until the end of the eighteenth century (Ashton, *Iron and Steel in the Industrial Revolution*, pp. 242–3).

[5] The river barges used to bring coal down the Trent from Nottingham, in 1605, drew four feet of water ; and the owners of these barges were much concerned about " sandbead[s] or gravell [which] shall heppen to growe " in the river bed (*H.M.C., Report on the MSS. of Lord Middleton*, pp. 171–2).

ten times as much coal. A collier might be equipped with three tall
masts ; and (except at the prow, where there was a high forecastle,
or at the stern, where there was cabin space for a few sailors), the
side came so near the water that, when a keel drew abreast, the
keelmen had no difficulty in shoveling coal into the hold through
a port-hole. To protect the cargo from rain water and sea spray,
the hold was probably covered by a thin wooden deck, with hatches
which could be closed down as the ship sailed out into the open
sea.[1] She was manned by common seamen, who rarely concerned
themselves with handling coal, unless, in a tempest, they were obliged
to heave overboard a part of the cargo, in order to allow their ship
more lightly to ride the waves.[2] Keelmen did all the loading ; and
when a collier reached her destination, a gang of " coal heavers "
came aboard with shovels and with standard-sized wooden " vats ",
or " colemetts ", which they proceeded to fill and to empty over the
ship's side into a waiting lighter.[3]

Under normal sea conditions, it took about a fortnight for the
collier to make her 350-mile voyage from Newcastle to London,[4] and
less than a week for the west-coast trow to make her 50-mile voyage
from Swansea to Bridgwater.[5] Newcastle coal hoys had a reputation
for mobility—their "nimbleness", it was said, making them a difficult
catch for a heavy man-of-war, provided that they could take to flight
while still out of range of its gunfire.[6] Apart from the danger of
capture by men-of-war or by pirates, there was little risk of the actual

---

[1] See statement made in 1672 by a boatman of Bridgwater to the effect
that a ship master, who had arrived with coal from Wales, " shut up his decks "
and would not expose his cargo to the view of the city officers (*Exch. Deps.
by Com.*, 24 Charles II, East. 24).

[2] *Exch. K.R. Port Bks.*, 195/5 ; 197/1 ; 492/3 ; 792/3 ; *passim ; Cal. Treas.
Bks.*, vol. v, p. 224 ; *Sackville MSS.*: Certificate of customers at Newcastle,
dated November 14th, 1622, that the whole lading of a ship of Woodbridge had
to be cast overboard, and perished at sea. Other accidents, such as fire at sea,
might sometimes occur.

[3] *Additional MSS.*, 4459, no. 8 ; Dale, *Fellowship of Woodmongers*, p. 35.
Actually the coal heavers filled the coal into baskets, which were passed up from
the hold to a foreman, who turned them into the vat. For the quantity of coal
contained in a vat and a coalmett, see Appendix C (ii).

[4] *Exch. K.R. Port Bks.*, bdl. 198/3, 4, 16 ; *A Letter from a Master of a Collier
. . . shewing the Causes Coals are so dear*. There being four months during which
sailings were practically suspended (see below, p. 396), a collier averaged from 5
to 9 round trip voyages between London and Newcastle during a year of normal
trade (*S.P.D., Charles I*, vol. xiv, no. 9 ; *Cal. S.P.D.*, 1637–8, p. 47 ; *Sloane
MSS.*, 2902, no. 9, f. 52). The statement in *The Present State of the Coal Trade*,
1703, that some colliers made from 12 to 14 trips each year can have applied only
to a very few. Petty estimated in 1677 that, in the coal trade from Whitehaven
to Dublin, ships averaged ten round trips a year, " which is more than hath
hapned in any other sort of Trade ". Some of these west-coast colliers had made
as many as sixteen round trips, but the voyage was much shorter, of course,
than that from Newcastle to London (*Petty-Southwell Correspondence*, ed.
Marquis of Lansdowne, pp. 26–7).

[5] The round trip took from 8 days to 2 weeks, according to the testimony
of Thomas James, a Somerset husbandman, who had sometimes served as
" pursur " aboard such a trow during the reign of Charles II (*Exch. Deps. by
Com.*, 24 Charles II, East. 24).

[6] *Acts of the Privy Council*, N.S., vol. xxvi, p. 61.

loss of a vessel.[1] What added most persistently to the costs of the voyage were delays in loading and unloading, in clearing for tolls at the custom houses, in taking ballast at London, or in fighting " crosse windes or tempestuous weather ".[2] " The freight of shipping ", as Sir Robert Southwell expressed it, " swells chiefly from their attendances and waiting on their trade, not on the very value of the labour ".[3]

To build and equip a seventeenth-century collier for the coastwise or foreign trade, cost from £300 to £3,000, according to the size of the ship.[4] A Severn river trow is said to have been worth in the eighteenth century about £300, "when new and completely rigged ".[5] In addition, there might be frequent maintenance charges for repairing the hull or the anchors, and for the purchase of new ropes, sails, and rigging. Shipmasters must have on hand a stock of £50 or more, before setting out on the voyage north from London, Ipswich, or Harwich, in order to provide food, ballast, and other provisions for the trip,[6] and must also have means to pay for the coal at Newcastle, Sunderland, or Blyth. For the export trade from Newcastle to France or the Low Countries, a " 40 pound stock " was required in 1624, to " freight " a coal ship of 160 tons.[7]

The capital needed to buy coal for future sale might be far larger than those needed to purchase a string of packhorses, a wagon, a keel,

---

[1] But shipwrecks were not unknown (*P.C.R.*, vol. xxxv, p. 402 ; *The Compleat Collier*, p. 53). During the great flood of 1672, when the Tyne overflowed its banks, nine colliers sank in the river (*Cal. S.P.D.*, 1672–3, pp. 310–11).

[2] *P.C.R.*, vol. lii, p. 679; vol. lix, pp. 364–5; *Cal. S.P.D.*, 1638–9, pp. 91–2; *Lansdowne MSS.*, 156, no. 105, f. 419; *Star Chamb. Proc., James I*, 10/13; *Chanc. Proc., James I*, C. 20/74. Plagues were another cause for delay. During the Great Plague of 1665 in London the inhabitants of the Tyne valley avoided contact with the sailors of colliers from the metropolis, who were forbidden to come on shore. The farmers of the excise at Newcastle asked for a defalcation of £1,000 from their rent on this account (*Exch. Decrees*, Series IV, vol. x, f. 104*d* ; cf. Defoe, *Journal of the Plague Year*, in *Works*, ed. Hazlitt, 1840, vol. ii, pp. 74–5). Similarly the plagues in the Netherlands in 1635 hindered the Dutch shippers in taking their cargoes in Scotland. " Owners of the coal hewes [are allowed] to furnishe coales to the Hollanders presentlie in the firth and to lay the coales at the ship side, provyding that they nor the caryers of the coales doe not enter within the shippes nor have no handling nor medling, brocking, changing nor wissiling with anie of the companie of the saide ships " (*Privy Council Register of Scotland*, 2nd Series, vol. vi, p. 126 ; see also pp. 118–19, 123).

[3] Birch, *History of the Royal Society*, vol. iii, p. 208.

[4] Mention is made in 1653 of the capture by the Dutch of two English colliers valued at £300 and £800 respectively (*Cal. S.P.D.*, 1653–4, p. 542). In 1649 an eighth " part " (see below, pt. iv, ch. iii (ii)) in a small coal hoy was valued at £50 (Welford, *Records of the Committees for Compounding*, p. 163). In 1700 Charles Povey, himself a coal merchant, refers to some owners who have spent from £2,000 to £3,000 in fitting out a ship for the Newcastle trade (*A Discovery of Indirect Practices in the Coal-Trade*, p. 43). The smallest collier employed in the Newcastle trade must have cost at least as much as a fishing-smack, which, according to an estimate made in 1580, " will coste two hundreth pound the Shippe, with the furniture, if it be readie furnished to the Sea, in all thynges necessarie " (Tawney and Power, *Tudor Economic Documents*, vol. iii, p. 241). [5] *Gents Magazine*, vol. xxviii, 1758, pp. 277–8.

[6] *S.P.D., Charles I*, vol. xiv, no. 9. £50 was the stock required to fit out a 200-ton ship for the export trade to the Continent. I have assumed that roughly the same stock was required in the coastwise trade.

[7] *S.P.D., James I*, vol. clxxx, no. 77.

a trow, or even a collier. Summer was the season for moving coal, whether by land or sea. Most carriers deemed it inadvisable to pass along country roads or paths after the first frost.[1] And, although winter sailings could be made regularly on the Irish Sea, the danger of storms on the North Sea reduced the trading season for east-coast colliers to a bare eight months. " I have heard good Saylers say ", wrote a mine manager in 1709, " they had rather run the Hazard of an East-India Voyage, then be obliged to sail all the Winter between London and Newcastle ".[2] Not one shipmaster in three attempted regular sailings between November 1st and March 1st, a time of year when most of them discharged their crews and laid up their hoys in dry dock.[3] Winter, on the other hand, was the principal season for selling coal, because a greater quantity was consumed by house-holders, who wanted most of their supplies during the cold weather, than by artisans or manufacturers. A shipload, mined in January and loaded into keels in May or June, might not find its way into the cellar of a London citizen until the succeeding January or February. Mean-while someone had to pay for the labour involved in mining and moving this coal. Money might change hands several times—when the mine owner sold to a " fitter ", when the " fitter " sold to a shipper, when the shipper sold to a London dealer [4]—and, with each successive change, the sum to carry, and therefore the capital, in the form of cash or trade credits or both, required to carry it, grew in amount. Whether money changed hands once or many times, all the accumulated charges had eventually to be borne (not only in London but in most large towns) by wholesale traders, who bought from the shipmasters, the lightermen, or the drivers of carts and packhorses. These traders had either to remain out of pocket or to obtain credit until they could find buyers.

With the object of relieving the city poor from depending for coal supplies upon traders, whose motive was their personal gain, many municipal corporations undertook, during the sixteenth and seventeenth centuries, to perform without profit the operations of traders.[5] In London the Corporation kept a coal fund of £4,000 to supply 4,000 London chaldrons (about 5,333 tons) every year.[6] A

---

[1] *Star Chamb. Proc., James I*, 106/7 ; *Harleian MSS.*, 6850, no. 39, f. 163 ; *Exch. Deps. by Com.*, 22 and 23 Eliz., Mich. 5. " During hard winters it was impossible to supply coal from the more distant mines for the town of Coventry " (*Exch. Deps. by Com.*, 36 Charles II, Mich. 43). For the description of an attempt made in the reign of James I to carry coal to Wigan from Aspull colliery, when " the lane . . . was very deep with snow ", see *Duchy of Lancs. Deps.*, 1 Charles I, 75/13.   [2] *The Compleat Collier*, p. 46.

[3] The figures of coastwise shipments from Newcastle in the years 1612, 1634, and 1655, show that only about 20 per cent or less was carried during the five winter months. The exports overseas from Newcastle (though these are more irregular), and the coastwise shipments from Sunderland, follow the same course (Appendix D (x), tables i–iii). It was not until the eighteenth century that shipments during the winter months began to increase in relative importance (Appendix D (x), table iv ; Westerfield, *Middlemen in English Business*, p. 235).

[4] For the operations performed by " fitters " and " crimps ", and the manner of financing the middleman's trade see below, vol. ii, pp. 58–9, 85–6, and pt. iv, ch. iii (iv) generally.

[5] See below, vol. ii, pp. 260–1.

[6] *City of London Repertories*, vol. xiii, ff. 108, 111, 133b ; vol. xv, f. 240 ; *Alchin MSS.*, bdl. 31, no. 12 ; and see below, pt. v, ch. iii (iii).

private trader had a similar need for capital. In fact, he had to have even more capital in proportion to his annual turnover than had the Corporation, for he often sold on credit to his wealthy customers, and could not count on receiving payment from them before the summer buying season.[1] Thomas Bagshawe, who had given up the profitable profession of a London fishmonger for what he considered the still more profitable one of a London coal trader, declared in 1607 that he always kept " a great stock of money " for the purpose of buying coal.[2]

Besides this stock, the prospective wholesale trader had to provide a place to store his coal pending sale. It had become common by the end of the sixteenth century for traders, not only in the metropolis, but in most towns, and at all points where coal was landed in considerable quantities, to fence off, by high brick or limestone walls, conveniently situated plots of land as " Coale Yards ".[3] In London these yards had been built originally at or near the wharves where the ships and lighters unloaded.[4] They cost anything from a few score to a few hundred, or even a thousand, pounds. As the trader came to depend more and more for his profits on deceiving the consumers as to the true quantity of fuel which he had on hand, he disclosed only a small portion of his supply to the public scrutiny, which focused on his wharf or his yard, and hid the bulk, sometimes in his own cellar, if it was large, sometimes in a supplementary subterranean vault.[5] So clandestine and secure a hiding-place did these vaults provide,

[1] Westerfield, *Middlemen in English Business*, p. 236. London coal merchants in the seventeenth century had to provide for a certain number of bad debts (Povey, *A Discovery of Indirect Practices*, 1700). When the debtor was a court favourite, the merchant could not force him to pay, as is shown by the well-known attempt of Sir Edmund Godfrey, one of the most powerful London " woodmongers ", to arrest King Charles II's physician (Pepys, *Diary*, ed. Wheatley, vol. viii, p. 310 : May 26th, 1669 ; Dale, *Fellowship of Woodmongers*, p. 60). Most of the coal merchants were so notorious for their dishonest dealing that they could hardly look for sympathy from the populace in their efforts to collect a debt, even if, like Godfrey, they pleaded that they were acting on behalf of the poor. (Cf. below, vol. ii, pp. 102 sqq).

[2] *Star Chamb. Proc., James I*, 10/13 ; 13/2.

[3] *City of London Repertories*, vol. xiii, f. 131*b* ; Bateson, *Records of the Borough of Leicester*, vol. iii, p. 153 ; *P.C.R.*, vol. liii, pp. 27–8 (reference to a " Coale Yard " in Maldon). In 1558 the Lord Mayor of London applied to the Lord Treasurer for a portion of the " Church Yards ", once the property of " the late " Augustine Friars, to be utilized for storing the City's stock of coal for the poor (*Repertories*, vol. xiv, f. 168*b*). In Edinburgh coal dealers sometimes stored their stocks in the back yards of their own houses (*Cal. of Charters* (General Register House, Edinburgh), 1587–91, no. 2976). To store 2,000 London chaldrons, or slightly more than 2,700 tons, of coal, one city merchant had a yard 126 feet long by 49 feet broad (Dale, *Fellowship of Woodmongers*, p. 75).

[4] *Repertories*, vol. xiv, f. 167 ; *Order of the Company of Woodmongers*, 1657. When a certain William Archebolde set up in 1594 as a London coal dealer, he took a lease of a convenient dwelling-house and wharves in and above Milford Lane, St. Clements Danes (*Star Chamb. Proc., James I*, 41/12). The space reserved for storing coal at a wharf was sometimes called a " warehouse ", or a " magazine ", rather than a yard (*Court of Requests Proc.*, 91/33 ; *P.C.R.*, vol. xlix, pp. 499–500).

[5] *The Two Grand Ingrossers of Coles*, 1653 (cf. below, vol. ii, p. 103). In 1675 the London Court of Aldermen referred to the city coal vaults lately made on each side of the new channel from Holborn Bridge to the Thames (*Repertories*, vol. lxxx, f. 403).

that in one of them were assembled the materials for the Gunpowder Plot [1]—a fact which may reveal to some of the less historically-informed readers of Dickens why it was natural for so many Chuzzlewits, who took pride in their descent from Guy Fawkes, to " set up as coal merchants ".

Operating charges connected with the business of wholesale trade increased during the seventeenth century. The trader might hire an assistant or two to help him in dealing with his customers and in keeping his somewhat sketchy accounts ; [2] he might run lighters on the river and coal " carrs " on the city streets ; he might even spend a small sum on advertising his business.[3] In 1700 Povey, himself a member of the trade, estimated the ordinary overhead charges of a London coal merchant at from £350 to £400 a year.[4] The weight of these charges was, of course, less heavy in proportion if the annual turnover was large. And shipmasters gave substantial rebates to the dealer who could take whole cargoes off their hands. Generally the city middleman, like the mine owner, had a better chance to succeed if his capital reserves were larger than those of his rivals, so that he might conduct his enterprise on a larger scale. The small man might be forced to borrow more heavily than his resources justified, and thus be driven from the trade because of inability to pay his debts.

Whoever else carried stocks of coal—whether it were the mine owner, the " fitter ", the shipowner, the merchant who undertook to export to foreign countries, or the dealer who entered the trade on the Trent as intermediary between the mine owner and the town retailer in Newark or Gainsborough—also needed funds to bridge the gap in time between his purchases and his sales. In 1612 Robert Fosbrook of Nottingham (yeoman and " poore servant " of Sir Percival Willoughby, whose output from Wollaton colliery he has contracted to market) sends his " master " a letter, showing that lack of capital is preventing him from out-trading a rival dealer named Hentworth, who probably sold the output from the neighbouring mine at Strelley. " Towching our sale of coles ", runs the letter, " we have solde more by many than Hentworth hathe. Butt Hentworth maie well overgoe us in cariadg of more coles to Newark, bycause he nether payeth for coles nor cariadg till he have sold them and wee paie beforehand, so that we are nott able to have great stackes standing by us for wante of stock, as he maie. Butt yf ytt pleased your worship to afford us a competent somme of money upon sufficient securitie and

---

[1] Humpherus, *History of the Company of Watermen and Lightermen*, vol. i, p. 162.

[2] Thomas Bagshawe (*supra*) testified in 1607 that he kept a book for the coal he sold on credit, some notes of the coal he sold for ready money, " and of other soms he doth not keepe anie note at all " (*Star Chamb. Proc., James I*, 10/13). Bookkeeping of a sort has been a general, if not a universal, practice among London coal merchants at least from the beginning of the seventeenth century (*ibid.*, 13/2).

[3] Povey, *A Discovery of Indirect Practices in the Coal-Trade*, pp. 26–7. The first business advertisements were printed in London about the middle of the seventeenth century (*The Times*, July 10th and 25th, 1924).

[4] Povey's own wharf represented an investment of from £2,000 to £3,000, but much of this sum was tied up in a new " engine ", which made it possible to unload the incoming colliers without the use of lighters (see below, vol. ii, p. 101 n.).

for interest, or coles upon securities to paie for them when we have
sold them, then we wold cary more than he can ".[1] Some fifty years
later, it was estimated that to market at Newark an annual output
of about 16,000 tons, " from the pitts neere Nottingham ", would require
£500 in " standinge stocke . . . for horses, waines, boats ", and
£700 a year for other charges in transport, besides the ready money
needed to buy coal on the " heap ".[2]

Large-scale enterprise for marketing coal was practicable only
where, as on the Trent, or the Tyne, or in the port of London, a large
traffic passed through a single channel. Where, as in Derbyshire or the
West Riding, the chief centres of population did not absorb annually
more than five or six thousand tons, brought by carriers along the half-
dozen different country roads which converged upon the town, there
could be no employment in the coal trade for such great capital
reserves as were kept by some London traders, by some shipowners
in the Newcastle trade, and by some dealers on the Trent and Tyne.
It is necessary to remark, moreover, that, while the city middleman, like the mine owner, stood a better chance of success when
his capital was large, a stock of £3,000 was by no means indispensable
for starting business as a London retail trader, or for entering the
trade at some other stages between the mine and the consumer,
although such a stock was very nearly indispensable for setting up a new
colliery in all the important mining districts. " Footmen, pawn-
brokers, and such like ", runs a complaint addressed in 1745 to an
alderman of London, adopted the more respectable title of " Coal
Merchant ".[3] Among these new adventurers the servants of great
families predominated ; some of them even retained their old positions
for wages.

While there are many examples during the seventeenth century
of large accumulations of capital used to finance the trade in, as
well as the mining of, coal, large accumulations for trading purposes were by no means so novel a feature of European economic
life as were those employed for industrial ventures such as a colliery,
a brewery, a salt work, or a glass furnace. Commercial capitalism
had thrived in the medieval towns of Italy, Flanders, and southern
Germany, and had developed to some extent in England before there
were signs of a genuine industrial capitalism in the western world.

For those interested in the causes behind the spread of this
new capitalism during the seventeenth century, the cost of mining
is perhaps more important than the cost of distributing coal. Yet
we must not overlook the fact that the coal trade was likely to
encourage the accumulation of larger stocks of money than the trade
in wood which it superseded. Timber grew, or could have been made
to grow, almost everywhere in Great Britain, hence, if coal had never
replaced wood as fuel, the average distance which fuel had to be
transported to reach consumers would have been shorter, and the

---

[1] *H.M.C., Report on the MSS. of Lord Middleton*, pp. 172–3, 176.
[2] *Additional MSS.*, 33509, f. 10b. It is uncertain in this case whether
or not the dealer had to pay for the coal which he purchased before he had
sold it.
[3] *The Frauds and Abuses of the Coal-Dealers*, 1745.

concentration of traffic over any single route less. And it was long-distance transport and the heavy concentration of traffic that made necessary the large sums of capital employed by some middlemen in the coal trade during the seventeenth century. Had Londoners after 1550 depended upon the planting of trees, rather than upon the exploitation of mines, no doubt they would have received supplies in part by land from the neighbouring forests of Kent, Surrey, Essex, and Hertfordshire, in part by river from counties farther up the Thames, in part by sea from Norway and the Baltic provinces. Since fuel would have come from many sources instead of from one, there would have been less opportunity for a single middleman to handle large quantities ; since the consumer would have been generally at fewer " removes " from the producer, there would have been fewer functions for middlemen to perform. While notice must be taken of small men who set up as retailers in London, the control over the trade between the north of England and the Thames came more and more to rest with a few principal merchants, whose interest in the traffic of coal certainly ran into thousands of pounds.[1]

(iii)   The Freedom from Medieval Economic Traditions

The high costs of working a mine, and of setting up business as a trader in coal, made inevitable the development of capitalist forms of organization within the industry during the seventeenth century. This development was facilitated by the fact that the great demand for coal arose in an age when the regulations associated with the economy of medieval towns had ceased to serve the purposes for which they were originally designed. Coal mining grew up largely outside the sphere of influence of gild and other medieval regulations which sometimes helped the manual workmen to remain independent of outside capital. The coal trade, on the other hand, was brought within the framework of gild organization in Newcastle and in London, but the gild, in these cases, proved so flexible an institution that the principal coal merchants were able to use it to further their ends as capitalist traders. In town life, the position of the manual worker had become very different from his position in the thirteenth, the fourteenth, or even the fifteenth century. It was no longer common for him to be at once workman, foreman, employer, merchant and shopkeeper. He was tending to become simply a wage-earner.[2] Although the power of the handicraftsman, as a member of his gild, to deal in the product of his own labour had by no means disappeared in the seventeenth century, the merchants had become dominant in the control of both municipal politics and economic life.[3] At Newcastle an attempt by the artisans to challenge the right of the commercial gilds to dominate the town government ended in complete failure in 1516.[4]

[1] Cf. below, pt. iv, ch. iii (iv).
[2] Cf. Unwin, *Industrial Organization in the Sixteenth and Seventeenth Centuries*, chaps. i–iii.
[3] *Ibid.*, pp. 16–17.
[4] Leadam, *Select Cases in the Star Chamber*, vol. ii, pp. xcvi, 75 sqq. See also Unwin, *op. cit.*, pp. 75 sqq.

Coal was not one of the commodities which had been of importance in the town economy of the Middle Ages, and it had therefore largely escaped regulation until after the commercial gilds had gained their ascendancy. Except at Newcastle, where the " colliers " appear to have been organized as a craft in 1516,[1] and where the trade in coal had long been restricted to a group of hostmen, there is no evidence of the existence of gilds of mine workers or of coal dealers in any English town before the beginning of the seventeenth century. We have unfortunately no information concerning this early craft of colliers. It probably included only such coal workers and " carriagemen " as lived within the town of Newcastle.[2] It was certainly never expanded to include the thousands of pitmen who came to labour in the Tyne valley during the late sixteenth and the seventeenth centuries.[3] There is no evidence that these pitmen, whose dwellings were seldom within the town of Newcastle itself, ever sought incorporation like their fellows, the keelmen, who lived for the most part within the poverty-ridden section of the town known as Sandgate.[4] Coal mining in Great Britain was from the beginning a country industry ; there was no need to move away from old centres of production in order to escape the restrictions imposed by gild regulations. And the expansion of the coal industry occurred at so late a period, that its organization was little affected either by ancient mining jurisdictions or by manorial custom.[5]

It is not suggested that, if miners' privileges, like those enjoyed by colliers in the Forest of Dean, had been general in England, or if the English colliers generally had been members of gilds like the gild of *houilleurs* at Liége, this would have stemmed the rising tide of capitalist enterprise in sixteenth- and seventeenth-century mining. In the Forest of Dean, as has been shown above, many free miners sold their mineral holdings to outside adventurers, and they sometimes acquired a holding simply for the purpose of selling it to an adventurer, who alone could provide enough capital for a more intensive exploitation of the seams.[6] At Liége the *métier des houilleurs* may once have admitted all working colliers in the district (we know that there were two thousand members at the beginning of the fifteenth century),[7] but before the end of the seventeenth century

---

[1] Welford, *History of Newcastle and Gateshead*, vol. ii, p. 46 ; Leadam, *Select Cases in the Star Chamber*, vol. ii, pp. 75 sqq. I am satisfied that this was in fact a craft of workers engaged in digging and carrying mineral coal. But it should be pointed out here that in most parts of England the word " collier " was used to mean charcoal burner until near the end of the sixteenth century. It is difficult to determine at what period the meaning changed. The more modern usage was already to be found in the Tyne valley in 1582, in northern Staffordshire in 1609, in Lancashire in 1620, and in the Mendip Hills in 1650 (*Exch. Deps. by Com.*, 29 Eliz., East. 4 ; *Star Chamb. Proc., James I*, 228/13 ; Crofton, *Lancashire and Cheshire Coal Mining Records*, pp. 56-7 ; *Parl. Surveys, Somerset*, no. 39).

[2] Cf. Welford, *op. cit.*, vol. iii, p. 160.   [3] Cf. below, vol. ii, pp. 137-8.

[4] The keelmen, like the colliers, were organized as a craft in 1515, but they appear to have lost the privilege of ordering their own labour early in the seventeenth century, if not before (cf. below, vol. ii, pp. 177 sqq.).

[5] Cf. above, pp. 298-301.   [6] See above, pp. 279-80.

[7] R. Malherbe, *Historique de l'exploitation de la houille dans le Pays de Liége*, pp. 333-7.

all the power resided in—if indeed membership was not limited to —adventurers who furnished the money and foremen who managed the pits. The organization at all Liége collieries had become definitely capitalistic in character.[1]

Nevertheless, the ancient privileges of the free miners of the Forest of Dean, and the ancient regulations of the colliers' gild in the province of Liége, imposed restraints upon the rise of capitalist enterprise. In both districts some effect was given to the theory, which appears so strangely misguided to most modern captains of industry, that participation in the actual labour in and about the pits carries with it a special claim to ownership and control of the mine. Custom prescribed that only natives born in the Hundred of St. Briavels, men who had worked a year and a day with pick and shovel, could obtain, in the first instance, a coal or an iron concession in the Forest of Dean.[2] By the charter granted to the Liége *houilleurs* in 1593, and still in force in the middle of the eighteenth century, only persons who had exercised the *métier* (either as technicians or as foremen) were eligible for membership.[3] As we shall see, the London gilds, in which the traders in coal associated, readily lent themselves to monopoly and exploitation. Doubtless this was true to some extent of the Liége coal gild. But there were important differences between the organizations of the London coal traders and the organization of the Liégeois. The gild at Liége, unlike the London gilds, dealt with the production of coal. It had a history and traditions reaching back into the Middle Ages, and its regulations had been framed when the market for coal was still local, and the amount of capital required to enter the mining industry slight. But in London coal was brought within the framework of the gild economy only in the late sixteenth and seventeenth centuries, when the principal wholesale coal merchants utilized the machinery of the Woodmongers and Lightermens Companies —neither of which had been originally formed to deal in coal—to serve their ends. There were, in the case of the Liége gild, old regulations which hampered the flow of capital into the mining industry, and afforded some protection to the small investor, if not to the small working partnership of miners. Originally, all the partners in the Liége mining *sociétés* were probably members of the gild ; and, according to the regulations, no partner might sell his interest to a " foreigner " until the other partners (as individuals and collectively) had refused it.[4] Every partner was entitled to an equal voice in the management of the colliery, whether his financial interest in it was large or small.[5] In the principal English coal fields, on the other hand, the interest of the small partner was frequently ignored, and the sale of shares in collieries was much freer.[6] This gave the adventurer

[1] Pirenne, *Histoire de Belgique*, vol. iv, 1911, p. 427.

[2] *Royal Commission on Mining Royalties*, 2nd Report, 1891, Appendix lii, pp. 416–19. See also above, pp. 277–8.

[3] M. G. de Louvrex, *Recueil contenant les édits et réglements faits pour le pays de Liége et comté de Looz*, vol. ii, pp. 208–15 ; Jars, *Voyages métallurgiques*, vol. i, p. 381.   [4] *Ibid.*, pp. 380–1 ; Malherbe, *op. cit.*, pp. 333–7.

[5] See below, vol. ii, p. 57. This was true also of collieries in the Mons district (Jan St. Lewinski, *L'évolution industrièlle de la Belgique*, 1911, pp. 35–6).

[6] See below, vol. ii, pp. 56–7, 61.

whose resources were considerable an added advantage, and facilitated the flow of capital into the industry, making it easier to develop mining enterprise on a large scale. In the Forest of Dean, the collieries have always been smaller than in the midland fields. At Liége, in spite of the facilities provided by the river Meuse for marketing large quantities of coal, and in spite of the fact that pits were sometimes sunk to as great depths as in the north of England, the normal output of a colliery after the sixteenth century appears to have been less than in the British fields which had access to water transportation.[1]

It was natural, as historians have often pointed out, that a capitalist form of organization (during the sixteenth and seventeenth centuries still in its infancy) should impose itself more easily upon industries set up outside those places which had been the centres of such industrial life as existed during the Middle Ages. In the case of the textile industry, it is in the villages of Gloucestershire and Wiltshire rather than in York or Coventry, which had formerly ranked among the most important seats of the English cloth trade, that the business of the capitalist clothier develops on the largest scale. Municipal and gild regulations could not of course prevent the spread of capitalism in the textile industry when other powerful factors were working in its favour, as in the Flemish towns during the fourteenth century, or in Norwich and Colchester during our period. Such regulations frequently provided, nevertheless, something of a brake upon capitalistic development. The striking growth during the fifteenth and sixteenth centuries of capitalism in connection with the textile industry of rural Belgium has been attributed partly to the freedom from gild and municipal ordinances.[2] It was partly because of the fetters imposed by town regulations during the same period that the woollen industry of Yorkshire declined at York and Beverley and expanded in the West Riding.[3] An especially fertile soil was provided in the rural districts for the growth of capitalism in connection with other industries besides textiles. Without the advantage of any craft regulations—such as gave temporary protection to many town artisans—the small independent salt manufacturer of the Elizabethan Age found himself caught in the net of some rich adventurer, who had offered him bait in the form of ready money, advanced on condition that the salter's output should become the property of his creditor. At Droitwich, writes Harrison, " though the commoditie . . . [of salt] be singular great, yet the burgesses be poore generallie, bicause gentlemen have generallye for the most part gotten the great gaine of it into their hands, whilest the poore burgesses yeeld unto all the labour ".[4] Along the Northumberland coast, as John Mount informs Cecil in 1566, " there liethe shepes all wayes waytinge for salt and there comethe merchaunts which dothe give a hundred pounds before hande to them that maketh salt which are but haglayers ".[5] Before 1600 most of the latter

---

[1] See above, p. 363.
[2] Pirenne, *Histoire de Belgique*, 1922–3, vol. ii, p. 427 ; vol. iii, pp. 236 sqq.
[3] Heaton, *Yorkshire Woollen and Worsted Industries*, pp. 50 sqq.
[4] Harrison, *Description of Britain*, 1577, bk. iii, chap. xvii, quoting Leland.
[5] *S.P.D., Eliz.*, vol. xli, no. 13.

were forced to labour for wages as employees of capitalists, who had now installed sets of larger and more expensive iron pans.[1] Speaking of Belgium, Professor Pirenne tells us that industrial capitalism was far less advanced at the end of the seventeenth century in the ancient municipality of Liége than in the recently created town of Verviers, an offspring of the new competitive spirit.[2] Given the need for large capitals, which became technically essential in British coal mining in the late sixteenth and seventeenth centuries, corporate organization could not of course have prevented the sharp cleavage between capital and labour which is discussed in the next chapter. But the progress of capitalism in connection with coal mining was facilitated by the absence of a medieval system of regulation.

The coal trade, which unlike mining centred in the towns, could not escape municipal and gild regulations altogether. Medieval traditions, however, had always imposed less of a handicap upon commercial than upon industrial capitalism. Moreover, the fact that coal became important in the national economy only in the late sixteenth century put it in a somewhat different place from older commodities so far as regulation was concerned. During Elizabeth's reign the Merchant Adventurers' Company sought to keep new adventurers and their capital out of the foreign trade in cloth, in wine, and in other commodities.[3] But there is no evidence that they ever established a monopoly over the foreign shipments of coal, comparable to their monopoly over the foreign shipments of cloth in Elizabeth's reign. Their lack of interest in the new trade may perhaps be ascribed to the fact that the great expansion in the coal traffic by sea was largely confined to shipments from one English port to another ; for the coastwise trade, hitherto of small consequence, seems to have been neglected both by the ancient regulated companies and by the new joint-stock companies. Their members, tempted by the glitter of the precious metals and the flavour of the spices of the east, ignored the prosaic, but growing, commerce in a bulky commodity carried up and down their native coasts. Without interference from these companies, any native merchant might rig out a ship to carry coal coastwise. Capital was the only passport required. The Society of Coal Merchants, which enjoyed a brief existence during the last years of Charles I's reign, was an organization of owners and masters of ships carrying coal from Newcastle and Sunderland to other English ports ; but it was not in a true sense a gild or regulated company. It resembled the Society of Soapmakers, and the Society of Saltmakers, in being one of the new capitalist monopolies for fostering which Charles I brought upon himself such general unpopularity.[4]

---

[1] In 1589 one adventurer, Robert Bowes, who had invested £4,000 in salt pans, is said to have employed 300 people in his works (*Lansdowne MSS.*, 59, no. 69).

[2] *Histoire de Belgique*, vol. v, p. 356.

[3] Unwin, " The Merchant Adventurers' Company in the Reign of Elizabeth ", in *Studies in Economic History*, 1927, pp. 133 sqq.

[4] For its history, see below, vol. ii, pp. 281–2.

It is true that the Newcastle coal merchants, as Hostmen, and the London merchants, first as Woodmongers, and later as Lightermen, formed themselves into incorporated companies, which were directly derived from the earlier gild forms. But, in each of these cases, the merchants used these associations to monopolize the trade in coal, and thus strengthened their positions as capitalist traders.

The Hostmen's Company, which monopolized the sale of all coal raised in the Tyne valley for shipment by sea, had both a different origin and a different status from the other trading associations of Newcastle.[1] There had been established in the town during the reign of King John a gild merchant of local artisans, retailers, and other inhabitants, all recruited from the twelve " mysteries " which controlled the town government, and which included the merchants of woollen cloth, the mercers, skinners, tailors, sadlers, bakers, tanners, cordwainers, butchers, smiths, and corn dealers. Members of the gild merchant, while retaining their affiliation with their old fraternities, devoted themselves more and more to the pursuit of foreign trade. By an early agreement, the nature of which remains obscure, certain of them came to deal, to the exclusion of the rest, in the sale of coal to the shipmasters. They called themselves " free hosts ", or " hostmen ", words which might have been applied to all citizen traders, but which belonged in fact only to those who sold coal and grindstones for shipment.[2] Until 1600 these hostmen had no official status. Under the charter granted to the town of Newcastle in that year they secured incorporation from the Crown, and the monopoly which they had already obtained over the shipment of coal was legalized. Although by incorporation the Hostmen's Company became a town gild, its members could be at the same time members of any of the twelve companies which elected the municipal officers, and this enabled them to control the local government in the interest of their coal monopoly. They maintained the gild form because it guaranteed their monopoly a certain immunity from the attacks of rival coal owners, traders, and consumers ; but in fact the operations of the chief merchants greatly resembled those of modern financiers.[3]

Under the municipal regulations no London gild ever had a complete monopoly of the coal imports, like the hostmen's monopoly of coal exports from the Tyne. When Elizabeth came to the throne in 1558, the business of providing coal in London is said to have been in the hands of fourteen citizens—fishmongers, haberdashers,

---

[1] A number of writers have already investigated the Hostmen's Company. More detailed references to their work will be found below (vol. ii, p. 111 n.), where the nature of the hostmen's agreements is discussed at greater length. For what follows in the text, see Cunningham, *Growth of English Industry and Commerce, Modern Times*, pt. i, pp. 247–8, 527–8 ; Dendy, *Records of the Company of Hostmen*, Introduction ; Boyle and Dendy, *Records of the Merchant Adventurers of Newcastle*, pp. xxxii–iii ; Welford, *History of Newcastle and Gateshead*, vol. ii, p. 56.

[2] The practice of " hosting " merchant strangers, which obliged the merchant to buy and sell his goods through a citizen, originally bore no special relation to the trade in coal, and existed in many English ports besides Newcastle (cf. Dendy, *op. cit.*, pp. xiii–xx).

[3] On the whole subject, see below, vol. ii, pp. 110–15, 119 sqq., and cf. *City of London Repertories*, vol. xiv, f. 172b.

leathersellers, grocers, and others—" who do moche occupy and use the trade of byinge and sellynge of See Cols ".[1] They found in the city coal trade a new opening for their capital, similar to that which the goldsmiths and the scriveners were finding in making loans and in the " colouring of other men's money ".[2] Neither banking nor coal trading had any regularly appointed place in the municipal economic system of the sixteenth century. At first the dealings of various citizens in coal were secondary to their dealings in the commodities of their own gilds. But, before many years had passed, some of them began to devote themselves entirely to the rapidly expanding business of importing coal from Newcastle. On February 15th, 1607, the Attorney-General brought charges against Thomas Bagshawe, fishmonger, Thomas Careles, wharfinger, Francis Clarke, ironmonger, Thomas Bence, Thomas Morley, and " others unknown", who were alleged to have combined to enrich themselves as wholesale coal dealers. Some of them were merchants, some vintners, some victualers, some woodmongers, who, " having bene brought upp and instructed in other trades, have now lefte theire former Courses and imployed theire stockes and travell wholy for the buyinge and ingrosseing of . . . Coales ".[3]

Before the accession of James I, London citizens, who had begun to occupy themselves with the new trade, saw the need of obtaining special privileges which would enable them to resist the attacks on their monopoly made by a public growing every year more indignant over the high cost of fuel. Like the spectacle-makers, the soap-makers, and the tobacco-pipe makers,[4] the coal traders might have sought direct incorporation by the Crown as a city company dealing with a commodity which was now important for the first time. But would the citizens of London have tolerated a chartered company of coal merchants ? It is true that without endangering their existence, these other corporate monopolies raised the price of their commodities ; but the public was much more concerned about the price of coal, which was a necessity, than about the prices of pipes, spectacles, and soap, which were seldom indispensable, and the purchase of which in any case involved the expenditure of a much smaller portion of its income.

Under the circumstances, it is not surprising that the more wealthy coal traders should have sought to monopolize the wholesale traffic under cover of an old gild, and by means of a privilege the true purpose of which did not disclose itself at first sight. They found in the ancient Fellowship of Woodmongers an organization well suited to their designs. For several centuries [5] the woodmongers had possessed the right to bring into the City all wood and charcoal

---

[1] *Ibid.*, vol. xii, f. 363 ; vol. xiii, f. 130 ; vol. xiv, ff. 172*b*, 436, 455*b* ; vol. xv, ff. 107, 138 ; vol. xxvi, pt. i, f. 195*b*.

[2] Cf. R. H. Tawney's Introduction to Thomas Wilson, *Discourse upon Usury*, 1925, pp. 86–104.

[3] *Star Chamber Proc., James I*, 13/2 ; see also 10/13.

[4] Unwin, *Gilds and Companies of London*, p. 302.

[5] The earliest recorded appearance of the woodmongers as a craft is in 1376 (Dale, *Fellowship of Woodmongers*, p. 1).

for the inhabitants.   Citizens of London had been accustomed to purchase their fuel from woodmongers ; consequently it did no violence to their habits to go to that company for their sacks of coal, as they had gone to it for their bundles of faggots.[1]   Coal traders, formerly members of other gilds, joined the ancient fellowship during the reigns of Elizabeth and James I, and set about strengthening its powers.   Francis Dodd, a haberdasher who " has taken up the trade of carman ", was permitted in 1606 by the Court of Aldermen to transfer from the Haberdashers to the Woodmongers Company.[2] Thomas Morley, one of the defendants in the action brought by the Attorney-General, was not a woodmonger in 1605,[3] but became " warden " of the company at the beginning of the reign of Charles I.[4] On August 29th, 1605, the woodmongers obtained from the Crown a charter changing the ancient fellowship into an incorporated company.   There is nothing in the words of the charter [5] to suggest that the new company has any special rights over the trade in mineral fuel.   Coal is not once mentioned.   What the charter gives the wood-mongers is control of all the " cars ", or carts, permitted under the muncipal laws to carry goods within the City.   It was upon this control that their monopoly over the coal trade rested.

In order to limit the number of carts, and thus avoid traffic congestion along the narrow streets, the right to lease out " car-rooms "— that is, to grant licences to drive carts—had been vested by the City early in the sixteenth century in Christ's Hospital, which was presumed to be a disinterested institution as far as street traffic was concerned.   Every carman paid a fine and a fixed annual rent for his " room ", which reverted to the Hospital upon his death.   The coal merchants, in their disguise as woodmongers, wished to supplant the governor of Christ's Hospital as the authority for granting licences. They prevailed on the Court of Aldermen in 1580 to pass an order (justified on the ground that the woodmongers, as the chief dealers in fuel, had more use for carts than any other gild) giving them this power, upon the understanding that they should pay the Hospital a fixed sum of £150 each year.[6]   The charter of 1605 confirmed the woodmongers in this privilege, and, in addition, put the carmen, heretofore organized as an independent fellowship, under the government of the newly incorporated company, the Master, Wardens, and Assistants of which had full power to regulate " all matters and things touchinge or in any wise concerning the . . . trade or misterie of Woodmongers and Carremen or either of them ".[7]   Henceforward, when a carman died, the vacancy reverted to the woodmongers, who might retain possession, or might profit by leasing it out to one of their

---

[1] The chief traffic of the woodmongers apparently changed from wood to coal between 1580 and 1603.   (See indexes to the *Journals of the Common Council* in the London Guildhall.)

[2] *City of London Repertories*, vol. xvxii, f. 161.

[3] Dale, *op. cit.*, p. 11.

[4] *S.P.D., Charles I*, vol. lxviii, no. 48.

[5] Printed Dale, *op. cit.*, pp. 102–11.

[6] Unwin, *Gilds and Companies*, pp. 355–6 ; Dale, *op. cit.*, pp. 5, 9–11.

[7] Quoted Dale, *op. cit.*, p. 105.

own nominees.[1] Thomas Morley is said to have owned nine " rooms " on his own account in 1620.[2] Through their control over the carmen, the woodmongers could refuse the hire of carts to customers who patronized rival wholesalers. This gave them an indirect monopoly over the coal trade, for without carts it was impossible to make deliveries to the great majority of consumers.[3] By refusing " carrooms " to wharfingers " (who owned wharves), the woodmongers forced a number of them to abandon their wharves for want of means to transport the fuel which they purchased.[4] Woodmongers, or their nominees, alone had an interest in acquiring the abandoned wharves, for they alone possessed the means to make a profit out of the acquisition. Woodmongers like William Cory, who bore a name destined to become famous in the annals of the London coal trade, engrossed the landing places for fuel, as well as the conveyances for moving it, and hired labourers to keep their wharves and to drive their carts.[5] In their hands, the regulations of the Woodmongers Company served, not to maintain the status of the manual worker in London, but to widen the gulf between him and the citizen whose chief means of livelihood was his control of capital.

Neither the wharfingers nor the independent carmen submitted willingly to domination by the woodmongers.[6] Nor could the public be forever hoodwinked into believing that there was more than a shadowy difference between the ends attained by the newly incorporated company and those which might have been attained by a complete coal merchants' monopoly. Indignation over the continued rise in fuel prices led every year to renewed demands that the woodmongers' practices should be investigated by an impartial commission, and that their special privileges should be revoked. These privileges did not survive the second Dutch War, when the coal merchants profiteered without mercy towards their customers or consideration for their own future.[7] As a result, the carmen succeeded in breaking loose from their subservient position in 1665, and in gaining recognition as an independent fellowship three years later ; while the woodmongers were forced to surrender their charter in 1667.[8]

---

[1] Although the woodmongers were forbidden by the terms of their charter to take more than 17s. 4d. per annum in rent for a car room (*Journals of the Common Council*, vol. xxxii, ff. 159–63), they must have got round this restriction by charging heavy fines, for their profits from leasing rooms are said to have been large enough to pay for the many lawsuits in which the company became involved (*Lansdowne MSS.*, 162, no. 28). See also *Harleian MSS.*, 6842, no. 67. The company even made illegal gains between 1645 and 1651 by an abuse of its trusteeship over the sixteen rooms which had been reserved to be disposed of by officers of the Green Cloth. These rooms were all sold by the woodmongers for £549 19s., but they paid over to the Crown only £254 10s. (*Cal. S.P.D.*, 1660–1, p. 377).

[2] *Lansdowne MSS.*, 162, no. 28.

[3] See *P.C.R.*, vol. xlix, pp. 532–4 ; cf. *The Case of the Coal Trade, c.* 1702.

[4] *Journals of the Common Council*, vol. xxxii, ff. 159–163b ; *Harleian MSS.*, 6842, no. 67.   [5] *Court of Requests Proc.*, 64/67.

[6] Dale, *op. cit.*, pp. 13, 31–2 ; *Journals of the Common Council*, vol. xxxii, ff. 159–63b ; vol. xlvi, ff. 22b–24b (printed Dale, pp. 37 sqq.) ; *Repertories*, vol. xlii, f. 170 ; *Lansdowne MSS.*, 162, no. 67 ; *Harleian MSS.*, 6842, no. 28.

[7] See below, vol. ii, pp. 104–5.

[8] Unwin, *Gilds and Companies*, pp. 356–7 ; Dale, *op. cit.*, pp. 45–55, 65.

Coal merchants had found the disguise of woodmongers easy to assume when it served their ends ; now that it was exposed, they found it no less easy to discard. Even after the privileges of the incorporated company had been fully recognized, the retail traders who usually purchased their coal from the wholesalers to sell in small sacks containing a bushel, were mostly chandlers, leather-sellers, brewers, sugar-bakers, goldsmiths, tanners, ironmongers, or lighter-men,[1] not woodmongers ; and even the large wholesalers, who purchased fuel from the shipmasters, had not always traded as " wood-mongers ",[2] although nearly all of them were members of the company. As " traders in seacoles ", the latter lodged a complaint with the London aldermen in 1633 against " those that carrie seacoles upp and downe the streets ".[3] After the proceedings instituted by the Crown had made the dishonest dealings of the woodmongers notorious, few coal merchants, even if they remained members of the fellowship, cared to have their customers identify them with it. No astute merchant tries to market his goods under a discredited trademark. " The Woodmongers, . . . for reasons easy to be guessed at, thought fit now to style themselves Persons keeping Wharves ".[4]

At the end of the seventeenth century some of the principal coal merchants hit on another disguise, under cover of which they could make secret agreements to monopolize the wholesale trade in fuel. As we have seen, an increasing proportion of all the coal brought to London was unloaded into lighters.[5] A control over these boats became of as great strategic importance to the wholesalers who wished to monopolize the trade in fuel, as a control over the street carts. Before the Civil War, many woodmongers became owners of lighters as well as of carts and wharves ; and persons setting up as coal traders after the Restoration began to join the Fellowship of Lightermen, just as their predecessors had joined the Fellowship of Woodmongers. At the same time, some of the lighter-men, formerly employed by woodmongers to carry coal, saw that the ownership of lighters afforded them an opportunity to control the new trade. " From Carriers, they at once became Traders of a superior Class, and found themselves in a Capacity to treat those who had been their Masters,[6] as their Customers ; . . . having taken

---

[1] *The Two Grand Ingrossers of Coles*, 1653 ; *The True State of the Businesse of Glasse*, n.d. ; *S.P.D., Charles I*, vol. ccccii, nos. 4, 20 ; vol. ccccxii, no. 84 ; *H.M.C., Report on the MSS. of the House of Lords*, vol. v, p. 238.

[2] *Repertories*, vol. xl, f. 209b ; vol. xli, ff. 325b, 334.

[3] *Ibid.*, vol. xlvii, f. 155b. A certificate of " Traders in Newcastle coal," written about 1627, is signed by the warden and thirty-four other members of the Woodmongers Company (*S.P.D., Charles I*, vol. lxviii, no. 48).

[4] Dale, *op. cit.*, p. 71. Eventually the special odium attaching to " wood-mongers " was forgotten, and coal dealers were able to trade freely under that name. But the company never so completely dominated the city coal trade as it had during the first sixty-five years of the seventeenth century. For the attempts made by the woodmongers to regain their ancient power, see *ibid., passim*.

[5] See above, p. 389.

[6] By an order of the Court of Aldermen passed in 1627, lightermen who owned no wharves (and at that time few persons could retain the ownership of wharves except on sufferance of the woodmongers) were forbidden to trade in coal (*Repertories*, vol. xli, f. 325b).

Possession of the Hive, they resolved to keep all the Sweets to themselves ".[1] In 1700, the lightermen, who ferried goods, and the watermen, who ferried passengers, were incorporated by Act of Parliament in a single company.[2] Henceforth no one might operate a barge within the metropolitan area unless he was a member of this company ; unless, that is to say, he had permission from a few rich and powerful members, who turned out to be, as earlier in the case of the woodmongers, coal merchants in disguise.[3]

We shall have more to say below on the subject of these coal rings. For the moment, it is enough to point out that the association of the merchants in chartered companies, both at London and at Newcastle, actually helped to keep small dealers out of the coal trade. Thus a few powerful merchants used the gild forms to gain an exclusive control over the traffic between the north of England and the Thames. The gilds in so far as they touched the British coal trade contributed to the growth of capitalism.

[1] *The Frauds and Abuses of the Coal Dealers*, 1745.
[2] 11 and 12 William III, c. 12 ; Dale, *op. cit.*, p. 69.
[3] *H.M.C., Report on the MSS. of the House of Lords*, vol. v, pp. 235–8.

CHAPTER II

THE CLEAVAGE BETWEEN CAPITAL AND LABOUR

(i) *In Coal Mining*

THE modern organization of the coal industry, involving, as it does the employment of a large industrial proletariat by absentee capitalists, makes it easy to assume that the miner has always been a wage-earner. When a labour leader in our day refers to the miners as the aristocrats of the labour movement, he may have in mind the special power which they have exercised in British trade unionism, or their strong, and almost clannish, sense of solidarity. What he almost certainly has not in mind—though it might appeal to his audiences—is the position of the miner in the Middle Ages.

In medieval Germany, and in certain districts of medieval England, labourers in the metal mines were often granted special privileges and immunities of a kind rarely enjoyed by other manual workers.[1] There is reason to suppose that at one time most of the mines had been worked by serfs, but this arrangement seems to have become the exception at a comparatively early date. Kings, princes, and bishops depended to a large extent upon the produce of mines within their own domains for the supplies of gold, silver, and other metals. Perhaps it was to encourage resourcefulness in the management of these mines, and thereby to increase production, that feudal lords in some cases turned over their minerals, in return for a share in the output, to small autonomous associations of working miners, granting the associations protection from outsiders who were not members of the mining communities. Despite this protection, the original democratic associations, in which the members provided their own equipment and shared the physical labour, gave way to various quasi-capitalistic forms of industrial organization,[2] because the workmen could not supply enough money to cover the expenses of deeper and more intensive exploitation. But the old working partnership was by no means extinct in the fifteenth century.[3]

Its influence upon the organization of coal mining in the Belgian provinces seems to be unmistakable.[4] During the fourteenth century,

[1] For the following discussion of the industrial organization among the medieval metal miners, I am indebted to the excellent summary of the results of German scholarship contained in Lewis, *The Stannaries*, chaps. iii, vi, vii.

[2] The " cost agreement ", the " tribute system ", and the lease (see Lewis, *op. cit.*, pp. 177–99). For the chief differences which distinguish the industrial history of the medieval metal miners in England from that of the German miners, see *ibid.*, pp. 185–6.

[3] *Ibid.*, pp. 181, 198–9.

[4] But the similarity of the industrial organization in the Belgian coal mines to that in the German metal mines may be the result of similar working conditions rather than of a direct effort to copy the organization of the metal miners.

in the colliery district near Mons, many pits were worked by the
so-called *associations charbonnières*, or *bandes*—groups of half a dozen
or so local peasants who got permission from the *seigneur haut justicier*
to dig coal from a certain seam, and supplied their own tools and
materials.[1] No generalization is possible concerning the industrial
organization of the medieval coal miners in Great Britain. We hear
of the employment of forced labour at pits in the Bishopric of Durham,
and it is unlikely, though not impossible, that the workmen ever had
a considerable share either in financing or in managing any of the
Bishop's mines.[2] Small associations of colliers, not altogether unlike
those at Mons, appear in other English coalfields. Until the reign
of Elizabeth, the mine within the Crown manor of Newcastle-under-
Lyme was demised from time to time, by copy of court roll, to such
tenants as would agree to work it and to maintain at their own charge
the levels (or " gutters ", as they were called in north Staffordshire)
necessary for drainage.[3] Under this system the pits were " of long
tyme . . . preserved and maynteyned to the greatt benyfytt and
comoditie of the . . . Cuntrey in there provision of fewell ". Coal
within the Crown manor of Chesterfield, in Derbyshire, was leased
on similar terms in 1499 by Thomas Leeke, bailiff for the Crown,
to four tenants—Richard Rawlyn, Robert Mackaw, Saunder Wodde,
and John Oxle ; they " taking their advantage of the mine, paying
to the Bailie each of them 1*d.* a day " for sixty-seven working days.
But at Chesterfield the mine was not always in the hands of such
a partnership of co-labourers. On another occasion Leeke, apparently
without the knowledge or approval of superior Crown officials, hired
tenants to dig for day wages, and sold their produce " to his most
profit ". Again, in 1500, John Oxle obtained a lease for himself and
" his servants " ; he having no doubt assumed a dominant position
among his fellow workmen.[4]

At every period in the history of coal mining it is possible to
observe a number of different types of industrial organization, some
survivals from an earlier age, others precocious forerunners of an age
to come. But in the period between the Reformation and the
Revolution there was a general and rapid movement towards more
modern forms of industrial organization in the British coalfields.
We have seen how the new demand for mineral fuel led to mining
at greater depths, and how deeper mines, if they were to yield a
profit, had to produce a larger output, which could be obtained only

---

[1] Decamps, *Mémoire historique sur l'industrie houillère de Mons*, 1880,
pp. 138, 192 sqq.

[2] Cf. above, pp. 137–9.

[3] *Duchy of Lancs. Pleadings*, 90/B/41 ; 144/B/11. This method of leasing
coal to manorial tenants, who divided their interest into " parts ", is said to
have been in vogue before the reign of Henry VI. In 1572 the old method was
abandoned. All the minerals within the manor were leased by the Crown to
a Protestant courtier, Sir Ralph Bagnall, a prominent citizen who had already
represented the town of Newcastle-under-Lyme in Parliament, and who was
apparently engaged in accumulating grants and leases of Crown lands and
minerals in Staffordshire (*Dict. Nat. Biog.*, 1st Supplement, vol. i, pp. 96–7 ;
Pollard, *Political History of England, 1547–1603*, p. 128).

[4] *Duchy of Lancs. Deps.*, 4/R/1.

by the investment of much more considerable sums than had found their way into the medieval coal industry. The expansion in the scale of enterprise—involving as it did successive increases in the number of colliers employed, and further subdivisions of the labour in and about the pits—led inevitably to new forms of industrial organization.[1] A closer analysis reveals a dual process by means of which these changes in organization were brought about. On the one hand, increases in the number of colliers employed, or in the division of labour, made necessary more administrative and technical work ; and the managers and technicians who undertook this work tended to assume a status superior to that of the colliers, and thereby to add rungs to the economic ladder. On the other hand, unexpected increases in mining costs obliged the investors to borrow funds, frequently in return for contracting for deliveries of coal to their creditors ; and these contracts frequently proved their undoing. Finding themselves unable to meet in full the deliveries required of' them, they were forced to give up their share in the colliery plant, and sometimes to become the employees of their creditors.

Let us first observe how this dual process affected the autonomous working partnership. It was natural in a small association of manorial tenants, like those who mined coal in the manor of Chesterfield, for one of the partners, because he was richer or more skilful than the others, or because he possessed a natural capacity for leadership, to assume a dominant position among his fellows. During the reign of James I, in the manor of Tunstall (adjacent to Newcastle-under-Lyme), John Colclough and John Podmore each secured control of one colliery, and hired for wages various of their former brother tenants and squatters, who had now become " workmen and Labourers about Colemynes and for the getting of coles ".[2] The rough equality of economic conditions which had often existed among tenants had been based on a common subservience to the manorial authority, and, as that authority weakened, new inequalities developed. At the same time petty merchants, who sold coal in the neighbouring towns and villages but who had never been miners, might use the profits which they had accumulated in their trade to gain control of some local colliery. In the coalfield west of Mons, the members of the primitive *associations charbonnières* frequently placed themselves at the mercy of *marchands de charbon*, or middlemen, who first put in an appearance during the fourteenth century. To begin with, the middlemen contracted with the miners for monthly or annual deliveries of fuel. Soon they advanced capital to the *associations charbonnières*, reimbursing themselves out of the produce, until, little by little, they gained control of the concessions, and forced the original partners into a position of dependence which was frequently made perpetual in the sixteenth century ; when many miners became wage earners

---

[1] In France, where the small colliery enterprise was still the rule during the seventeenth century (see Des Cilleuls, *Histoire et régime de la grande industrie*, p. 38), almost all the mines remained in the hands of small peasant proprietors, whose stock of money consisted in savings out of their profits from agriculture (Rouff, *Les mines de charbon en France*, pp. 112–58).

[2] *Star Chamb. Proc., James I*, 228/13 ; and also 92/6.

pure and simple, hired by the coal merchants who had become the entrepreneurs.[1]

In the English coalfields, other than the Forest of Dean, the partnership of working miners had nothing like the same importance as at Mons, partly no doubt because the English coal mines had been less exploited than the Belgian during the Middle Ages. When the enormous new demand for mineral fuel burst upon the Elizabethan world, it was the great landlords, the rich merchants, and the courtiers who obtained most of the concessions.[2] Few peasants formed working partnerships to open pits without the support of outside capital. Where they did, they were doomed to fail. An instructive instance is afforded by the history of a tiny enterprise at Halton, in the sparsely populated and economically primitive country up the valley of the Wharfe in Yorkshire, where, if anywhere, one would have expected a successful outcome. Three brothers, Judson by name, joined hands about the year 1583 as working partners, William and John being occupied in getting the coal, Roger in selling it. He " received several times payment in advance before delivery ", which he shared with his brothers, but they were unable to repay their creditors before " the mine was taken from them ".[3] Attempts of this kind seem to have been rare after the accession of Elizabeth. In fact this is the only record that I have found, after the middle of the sixteenth century, of a colliery partnership in which all the capital was unmistakably supplied by the miners themselves.[4]

The influence of the working men's partnership can be traced, nevertheless, in the organization of a number of seventeenth-century colliery enterprises. Besides their tools, the miners sometimes furnished other capital. The workmen hired to operate the pumping machinery at Kincardine colliery in Fifeshire, in 1679, apparently provided their own horses for turning the beam at the shaft head.[5] Sometimes at small inland coal mines, the colliers joined together in a sort of partnership which contracted with the owner of the colliery to work his mine for a stipulated period.[6] There is no uniformity between these contracts, either as to the extent to which the colliers supplied their own tools and materials, or as to the manner in which they were paid for such equipment as they did supply, for their share in managing the enterprise, and for their manual labour.

John Collyer and eight other colliers agreed in 1593 with John

[1] Decamps, *op. cit.*, pp. 191 sqq.
[2] Cf. below, ch. iii (i). The cost of reaching the deeper seams was so great that few peasants could afford to undertake a colliery enterprise without the support of outside capital (*Duchy of Lancs. Spec. Com.*, No. 648).
[3] *Court of Requests Proc.*, 28/56.
[4] There remained, of course, sparsely settled districts in which the country people dug their own fuel from the outcrops.
[5] *Accounts of the Kincardine and Tulliallan Coal and Salt Works.* When horses were introduced for underground haulage at collieries in Cumberland in the eighteenth century, they often were the property not of the colliery owner but of local peasants who were hired to perform the work of haulage for daily wages (Jars, *Voyages métallurgiques*, vol. i, p. 241).
[6] Similar contracts with working miners were often made by the peasant proprietors in France as late as the eighteenth century (Rouff, *op cit.*, p. 120).

Freston, Esq., of York, who held the minerals within the manor of Kippax, near Leeds, by lease of the Crown, to work for him a pit already sunk ; he to furnish their ropes and picks and in return to have a third of all coal they mined, they to sell the other two-thirds for their own profit.[1] In 1637 the Bishop of Chester contracted with four " colers " to dig coal and cannel for a year in his pits at Farnworth, in Lancashire ; they to supply the necessary candles and bellows, to sharpen their own tools, and to find men to wind the coal up the shaft, the Bishop to provide them with baskets and ropes, " to build them upp a Hovell this summer ", and to allow them 8*d.* for every quarter (equivalent to about one-third of a ton) of coal dug.[2] In 1683 Isaac Blackburne, a yeoman, contracted with three colliers to be his " workmen and hired servants for getting " coals in his coal pit in a common called Land Moor, in Lancashire. The colliers were to sell their output on his behalf, and to hand over the proceeds of their sales after deducting their " wages "—a shilling each man for every working day.[3] In 1706 Sir Francis Blake, of Ford Castle in Northumberland, drew up and signed an agreement with seven " hughers ". They undertook to work the " stoney ", or thin surface, coal at Gatherick, Sir Francis' colliery, and to put into his banks-man's, or manager's, hands every fourth bowl they dug ; the other three bowls to be theirs to dispose of. Only four of the contractors were to hew coal. The others promised " to work at the drift until it is finished, according to a former agreement ".[4]

It is clear that, under all four contracts, the colliers were left to mine on their own initiative,[5] and that at Kippax and Gatherick they might sell for their own profit the major portion of the coal which they dug. Although the colliers at Land Moor contracted to labour for a daily wage, the owner, who lived fifteen miles away, was unable to enforce the terms which he had made—indeed he must have been both naïve and inexperienced in business to suppose that he could enforce them—and his three employees did not hesitate to pocket the entire

---

[1] *Duchy of Lancs. Pleadings,* 116/F/2.

[2] Crofton, *Lancashire and Cheshire Coal Mining Records,* pp. 60-1. A somewhat similar arrangement was made during James I's reign by a certain Francis Plot, who held a title to all the " Mynes and vaynes of Stone coale or Seacole " in the wastes and commons of the manor of Macclesfield. In 1612 he was the proprietor of two pits. At one of these were four workmen : two " getters ", a " drawer ", and a " winder ". The two getters " have one-half of all the coales gotten for their paynes and they pay the drawer ". Plot has the other half, in return for which he " findeth candles, . . . roapes and basketts, . . . and payeth the winder 3*s.* weekly " (*Land Rev. Misc. Bks.,* vol. cc, ff. 195 sqq.).

[3] *Palat. of Lancs. Bills,* 40/24.

[4] J. Robinson, *The Delaval Papers,* pp. 159-60. This mine was apparently being worked partly by open works, partly by a drift, which had evidently not been completed.

[5] Three of the contracts contain clauses as to the method of mining. At Kippax the colliers agree to yield up the pit free of water ; at Farnworth " to worke the worke substantially, fairly, justly and honestly as may bee best for the safetie and upholding of the Mines and most for the profitt of the . . . bishop and his assignees " ; at Gatherick " to drive her no wider than the Colliery will beare, and as to the Dip Room, . . . [the hewers] agree to [support] it very strongly as they go on " (*Duchy of Lancs. Pleadings,* 116/F/2 ; Crofton, *loc. cit.,* Robinson, *loc. cit.*).

proceeds from their sales, which are said to have averaged twelve shillings a day. Working partners under contracts of this nature often did not get so good a living ; sometimes they did well to earn as much as the miners who laboured for wages at neighbouring collieries.[1] And while the contracting workmen usually enjoyed more independence than simple wage-earners, their economic position was not incompatible with the lowest social status. Many of the Scottish landlords who owned salt pans as well as collieries habitually operated their pans by the labour of contracting workmen, to whom they paid " noe certaine salary or wages ". " Upon receiving such a quantity of coale ", the " makers ", as these workmen were called, undertook " to make and return the master such a quantity of salt as shall be agreed betwixt them . . . in lieu of coals and the use of their pans ; . . . the overplus that is made remayneing to themselves for theyre paynes ".[2] But these " makers ", like the Scottish coal miners, had to bind themselves to serve their masters for life ; and, as is well known, they actually lived in a quasi-servile condition until near the end of the eighteenth century.[3]

Whatever may have been the social status of contracting miners like John Collyer, it is unlikely that during the seventeenth century any large proportion of the mine labourers contracted in groups with the owners, and worked under the so-called " charter system ". As the number of pits operated by one enterprise and the number of miners employed in and about a single pit increased, the need for managerial skill to co-ordinate their work and to negotiate the sales of coal grew greater. As the depth and complexity of the diggings increased, the need for technical skill to determine the methods best suited for getting the coal grew greater also.

The larger the enterprise, however, the less was the likelihood that the persons who financed it would be either willing or able to undertake, on their own account, its administrative and technical management. Yeomen or husbandmen who, like Colclough or Podmore in north Staffordshire, had raised themselves to a place of dominance among their fellow colliers, might possess capital enough to finance one or two small pits producing some hundreds of tons in a year ; but they never possessed sufficient to finance large collieries of the type which began to appear at the end of the sixteenth century wherever there was a market for a large output. In the Tyne valley peasant mine owners, described as " persons not much superior to labourers ", continued to exploit seams at shallow depths as long as they could load their coal into sea-going hoys at the wharf side ; but when, towards the close of Elizabeth's reign, the hoys ceased to come up to the wharves, these mine owners were without " means to sell their

---

[1] At the pit in Macclesfield (referred to above), the contracting " getters " probably earned about 4s. each per week. The pit was said to yield 3 quarters of coal a day, each quarter selling for 16d., so that the weekly returns could not have exceeded 24s. even if the colliers laboured every week day. Out of this sum the hewers had 12s. and must pay the " drawer's " wages—perhaps 3s. a week. Plot paid the " winder " 3s. and spent 10d. for candles in a week.

[2] Tucker, *Report upon the Settlement of the Revenues of Excise and Customs in Scotland*, 1656, pp. 6–8. See also below, Appendix K (iv).

[3] See below, vol. ii, pp. 157–64.

coals for want of means to lade [them] ".[1] They could not afford the £100 or so needed to build a keel.[2] Nor could they make use of the keels already built, for the latter all belonged to Newcastle merchants, who refused to freight any coal except such as came from their own collieries.[3] These merchants, moreover, had been favoured by the Crown when they applied for mining concessions ; partly, no doubt, because the Crown officers, who disposed of so many valuable mineral holdings once the property of the Church,[4] were impressed by reports of the high costs of reaching new seams, and sought as concessionnaires rich traders—like Richard Hodgson of Newcastle— with large funds to " bestow . . . in serching and winning the mines ".[5] The new, wealthier adventurers, who took an ever greater share in financing collieries, not only in the Tyne valley but throughout the country, were seldom experts in the science of mining. What probability was there that a clothier or a fishmonger, a merchant adventurer who shipped cargoes to the Baltic, or a landlord who had hitherto drawn his revenue from surface rents, would know the external signs by which could be guessed the whereabouts of seams, or the proper place to drive an adit ?

Naturally the new adventurers took the same course as was pursued by John David Griffith, a country yeoman of Llanelly in South Wales, who had coal in his lands ; they turned for help to " one that dealt in such business and had skill therein ", asked his opinion whether there was hope of profit in the proposed colliery, and instructed him to sink shafts and order the work.[6] There grew up between the collier who laboured at the mine, and the adventurer who invested his money in it, an intermediate class of men " skilful and expert in the trade of gettinge Coales ".[7] Until Elizabeth's reign, there was not much division of labour, or much differentiation of status, within this intermediate class ; one expert (known variously as a " viewer ", a " banksman " a " worker of coal mines ",[8] or simply as a " collier " or " coalminer ") often performed all the functions connected with the technical and administrative management of collieries. Even at the end of the seventeenth century such was the practice at a great many small collieries. Meantime specialization had become the rule at the larger enterprises ; we find a wide gap in social station and in earning power between a humble foreman, whose job it was to oversee the labour of a tiny band of hewers, and a drainage engineer like Thomas Surtees—" a gentleman expert and skilful in Mineralls "—who undertook, during the reign of James I, to pump dry of water the mines of several colliery companies in Whickham, and earned, according to a report of some of the colliery owners, what was then the very large reward of £16 a week " clear profit ".[9]

---

[1] Welford, *History of Newcastle and Gateshead*, vol. iii, p. 164.
[2] See above, p. 388.   [3] See below, vol. ii, pp. 21–2, 125 sqq.
[4] See above, pt. ii, ch. i (i).   [5] *Augm. Partics. for Leases*, 111/15.
[6] *Star Chamb. Proc., James I*, 155/5.
[7] *Duchy of Lancs. Spec. Com.*, no. 433.
[8] See *Exch. Spec. Com.*, no. 2617.
[9] *Star Chamb. Proc., James I*, 245/6. Surtees was also a partner in various colliery enterprises in the Tyne valley (see below, vol. ii, pp. 21–2).

English and Scottish coal mining history during the sixteenth and seventeenth centuries is a history of infinite variations both in mineral law and industrial organization. It is unsafe to assume that the functions performed by the managers at one colliery, or the terms upon which they were employed, were ever duplicated exactly at any other colliery. Two officials, doing essentially the same work, might be called by different names at two mines in the same district ; again, two officials called by the same name might do entirely different work. All we can safely say is that, the larger the enterprise, the greater the number of officials and the more complex the system of management was likely to be.

At the very small mine, worked by less than a dozen colliers —a type of venture still to be found at the end of the seventeenth century all through the Midlands, Lancashire, the West Riding, and all those mining districts which had no access to navigable water— one manager generally sufficed. He appears under a great many different names in different parts of England, but his most common title in the Midlands, as in Durham, is " bankman ", or " banksman ", and in Scotland he is almost invariably called a " grieve ". During working hours he usually took up his post at the shaft head, called in Scotland the " hill ", in England the " bank ", and less frequently the " brow ". Besides deciding upon new sinkings and regulating all work above and below the surface, he often hired the colliers, paid them their wages, purchased the materials needed for the mine, sold " loads of coales . . . to such as came with their carriages to the . . . pitts ",[1] and kept some sort of account of his expenses and his receipts.

At collieries where the adventurers operated more than one pit at a time, it was common to have at least one official for each pit.[2] In 1582, when the owners of Winlaton colliery on the Tyne kept open four shafts, all " casting " coal, they employed four " overmen " whose duties—according to the testimony of a working collier, an overman, and a partner in the enterprise—were to see " the just number of corffs [as they] . . . come upp at everye pitt . . . well filled for the profitt of the owners ; [and] . . . to arrange for the carriage " to the river.[3] It is probable that the pit " overman ", who appeared

---

[1] Such is the description of a bankman's duties at a small mine within Eckington manor, Derbyshire (*Exch. Deps. by Com.*, 11 James I, Trin. 6).

[2] This cannot be said of a colliery worked by the Crown in Beaudesert Park, Staffordshire, during Elizabeth's reign. Coal was mined from several diggings, each employing a gang of three or four labourers, and although every gang had a nominal leader, he was usually paid the same wages as his " fellowes " (*Exch. Q.R. Accts.*, bdl. 632/17), and his work must have been almost entirely manual, so that it would hardly be possible to include him in the managerial class. Of course a distinction between the labourer and the foreman must sometimes be arbitrary, for the lowest grade foreman usually performed some manual labour. It may be difficult to see any difference between the leader of a mining gang at the colliery in Beaudesert Park and William Symonds, a " coleminer " of Westerleigh manor in Gloucestershire, who joined with two " fellows " to dig a pit within the manor on behalf of William Player, who was attempting to monopolize the coal industry of Kingswood Chase. Yet Symonds gives himself the high-sounding title of " director of Player's Coleworks in Westerley ", and a fellow miner refers to him as Player's " bayleif " (*Chanc. Deps.*, P. 18/H).

[3] *Exch. Deps. by Com.*, 29 Eliz., East. 4.

under various names at English collieries,[1] like his Scottish colleague, the " grieve ", and like the modern checkweighman, checked the output of each hewer, and thereby determined what wages he should receive.[2] Sometimes this official, like the bankman at a smaller colliery, sold all the coal mined at his pit, and rendered an account of his sales directly to the mine owners.   Indeed it was not uncommon for the overman to retain a considerable measure of autonomy within the enterprise : to be almost independent of the " viewer ",[3] or general supervisor, and responsible directly to the colliery owners or their representative, who was variously called an agent, a head clerk, a comptroller, an overseer, or a steward.[4]

Owing to the increase in the number of miners who worked in and about a single pit, its management came to require more than one official.   The duties of the foreman above ground might occupy so much time that he could no longer supervise the miners beneath the surface.   Already in 1582, at Winlaton colliery, it had become the function of the " viewer ", or general supervisor, " to see [the] workemen gotten to worke . . . and to see the ground upholden and

---

[1] In the Tyne valley he was first called a " bankman " (*Court of Augm Misc. Bks.*, vol. cxii, ff. 17–24), like the foremen at small collieries.

[2] See below, vol. ii, p. 183 n.

[3] Officials called " viewers " were associated with coal mining in Durham at least as early as the fifteenth century.   Originally the name applied to the representatives of large royalty owners like the King or the Bishop of Durham, it being the " viewer's " function to see that the lessee did not mine contrary to the terms of his lease (*Augm. Partics. for Leases*, 112/30 ; and cf. above, vol. i, p. 145).   The " viewer ", as inspector for the mineral owner, corresponds to the *Kohlwieger*, who remained the most important official in the Aachen coal field throughout the seventeenth and eighteenth centuries.   At Aachen the *Kohlwieger* had nothing to do with managing the colliery.   That was the function of a *Schichtmeister*, who represented the mine owner (*Aachener Echo der Gegenwart*, 1873, no. 207).   Mineral owners in England do not appear to have employed " viewers " to any great extent after 1600.   The name " viewer " came to be applied in the late sixteenth century to inspectors or surveyors (*Exch. Deps. by Com.*, 29 Eliz., East. 4) on behalf of the colliery owners ; it was still used in that sense in the Halifax district at the beginning of the eighteenth century (Lister, *op. cit.*, p. 280).   Eventually, in the north of England, " viewer " became a synonym for mine manager.   That is the meaning which has been attached to the word until recent times (Redmayne and Stone, *Ownership and Valuation of Mineral Property*, p. 88).   In the seventeenth century, however, there was not always a clear distinction between the use made in the Tyne valley of the words " viewer " and " overman ") ; sometimes the same official was called by both names (*Exch. Spec. Com.*, no. 5996 ; *Exch. Deps. by Com.*, 11 Charles I, Mich. 46 ; *Palat. of Durham, Decrees and Orders*, vol. i, f. 130, and vol. iv, ff. 193 sqq.), this confusion possibly arising out of the fact that some overmen were responsible directly to the colliery owners and not to the viewer.   For the most part the titles given to colliery managers in the seventeenth century were not derived from mining practice, but were merely carried over from some other employment, especially from farming (for the management of a large estate often engaged several officials).   It is probable that similar titles were adopted by the managers of the more important seventeenth-century textile establishments.   A petition presented to the Scottish Privy Council in 1641 is signed by John Haslie, " overman of the clothiers, for himself and on behalf of other manufacturers " (*Calendar of Supplementary Parliamentary Papers* (General Register House, Edinburgh), vol. ii, no. 107).

[4] In Scotland the word " overseer " referred to the general manager, but in Lancashire it apparently referred to the pit foreman (*Duchy of Lancs. Deps.*, Series II, 75/13).

truely wroughte underground ".[1] With the sinking of deeper shafts
and the driving of longer and more numerous headways, the task of
supervising all the miners in one pit became a full-time job, and we
begin to hear of a special underground foreman at each pit. In the
Tyne valley he was sometimes called an " under overman ",[2] in
Nottinghamshire an " underman ",[3] and in Scotland an " oversman ".[4]
At collieries in the Tyne valley, he showed the hewers where to drive
the headways and how much coal to sacrifice in pillars ; he " hurried "
the barrowmen along the tunnel between the shaft bottom and the
working face.[5] He was usually the principal official at the pit, being
of superior rank to the foreman above ground,[6] who now occupied
himself exclusively with checking the baskets of coal as they came up
the shafts, and arranging them on the heaps at the surface.

Besides the pit foremen, there were at all the large seventeenth-
century collieries special foremen, or " overmen " (called in Scotland
" oncost grieves "), who supervised the sinking of shafts, the driving of
adits, and all kinds of deadwork. In addition to the " viewer ", or
technical supervisor, there might be other officials connected with the
administration of the enterprise as a whole. At the beginning of the
eighteenth century, the pithead foremen at most of the great collieries
in Durham are said to have submitted weekly reports to a " clerk of the
works ",[7] and the miners no longer looked to their own " overmen " for
their wages, but to the general " paymaster ".[8] Before 1700, the Grand
Lease colliery, which probably employed about 500 colliers, not
counting the labourers who drove the wagons loaded with coal

---

[1] *Exch. Deps. by Com.*, 29 Eliz., East. 4.
[2] Lord Harley, " Journeys in England ", in *H.M.C., Report on the MSS. of
the Duke of Portland*, vol. vi, p. 106. The " viewer ", or general manager,
was sometimes called the " Head Under-over-Man " (J. C., *The Compleat Collier*,
p. 31).
[3] *H.M.C., Report on the MSS. of Lord Middleton*, p. 169.
[4] *Register of Deeds* (General Register House, Edinburgh), vol. x, p. 339. The
" oversman " is not to be confounded with the " overseer ", who was a general
manager.
[5] *Exch. Deps. by Com.*, 3 Charles I, East. 19 ; *Exch. Spec. Com.*, no. 4355 ;
Palat. of Durham, Bills and Answers, 2/24 (Stevenson v. Liddell). In the Aachen
colliery district the work of the underground foreman, the *Steiger*, has been
described as follows. " He saw to it that the hewers and other labourers underground worked faithfully and diligently, left sufficient pillars, sent up baskets of
coal filled properly and without deceit ; and took all possible care to provide
timbers to keep the headways from settling and doing damage to the mine "
(*Aachener Echo der Gegenwart*, no. 210).
[6] In Durham the foreman above ground was called the " upper overman ",
the " banksman ", the " merman of the tree ", or the " overman of the tree "
(*The Compleat Collier*, pp. 37–8 ; *Exch. Spec. Com.*, no. 5996). " Tree " may
have been a name for the " turn " at the pit head. In Scotland the pit foreman
was sometimes called a " hill clerk " and in the Aachen coalfield a *Kohlschreiber*,
because he kept accounts (*Accounts of the Kincardine Coal and Salt Works*
(General Register House, Edinburgh) ; *Aachener Echo der Gegenwart*, nos. 208–9).
[7] At Wollaton colliery, near Nottingham, there was a book-keeper at the
beginning of the seventeenth century (*H.M.C., Report on the MSS. of Lord
Middleton*, p. 164), at the colliery of Arthur Player near Bristol, a " clerk "
and an " underclerk " (*Chanc. Deps.*, P. 7/T.), and at Bedworth colliery near
Coventry a " clerk " (*Exch. Deps. by Com.*, 36 Charles II, Mich. 43).
[8] *The Compleat Collier*, p. 38.

to the river,[1] may have had a staff of about fifty managers, under-managers, and foremen of different ranks.

There were other specialists in mining operations, who did not attach themselves permanently to the staff of any particular enterprise, but put themselves at the disposal of all adventurers who wanted advice, and of some who did not. Mining promoters spent much of their time prospecting for coal and seeking to induce wealthy landlords or merchants to risk their money in starting new collieries.[2] Sometimes these promoters undertook to manage the enterprise. More often perhaps, as was the case with the " Prospecting Under-takers " in Durham, they terminated their connection with it as soon as they had convinced the adventurer that there was coal to mine, and had been appropriately rewarded for furnishing the information. Experts in drainage, who often pretended to possess the secret of some new pumping engine, also offered their services to the colliery owners.[3] In the eighteenth century we hear of a " master borer ", who lived at Newcastle, and who was called in by local prospectors to submit them reports concerning the whereabouts, nature, thickness, and declivity of new coal seams.[4] Another expert, a sort of inspector-general, was available to colliery owners all over England and Scotland for consultations concerning mining problems, whatever their nature.[5] Probably no such consulting expert had been available in the seventeenth century. At some collieries in the Tyne valley various overmen from the district had been called upon " to view " the workings.[6] When Scottish mine owners had been faced with an unusually knotty problem of mining technique, they had adopted the practice of calling upon the successful managers, or " overseers ", of neighbouring mines to inspect their pits, and to submit them written reports on the feasibility of driving an adit for drainage, the proper means of supporting overhead strata, and other problems.[7] The increase in the number of mine officials and mining experts during the period between 1550 and 1700 justifies us in speaking of the formation of a new social class within the coal industry ; a class which we may call professional because its members gained their living principally by brain work rather than by manual labour.

Whence came the recruits to form this new stratum in industry ? Sometimes they were miners, like Peter Pigg, of Newcastle, " overman " at Greenlaw colliery in 1627,[8] formerly a workman in these same mines ; or like Bryan Cole, of Tenby, in Pembrokeshire, and others who were " leaders " at Jeffreston colliery in 1576, after having been employed

---

[1] This estimate is based on the output of the colliery, see above, p. 361 ; and cf. below, vol. ii, p. 140 n.

[2] *The Compleat Collier*, pp. 10–31 ; and see below, vol. ii, pp. 47, 67, 72–3.

[3] *Star Chamb. Proc., James I*, 245/6 ; *Acts of the Parliaments of Scotland*, vol. iv, p. 276 ; *Privy Council Register of Scotland*, 1st Series, vol. xii, pp. 258, 277.

[4] Jars, *Voyages métallurgiques*, vol. i, pp. 182 sqq.

[5] *Ibid.*, pp. 186–7.

[6] *Exch. Deps. by Com.*, 3 Charles I, East. 19.

[7] Such a report was supplied the Duke of Hamilton by Lord Elphinston's " overseer " (*Hamilton MSS.*).

[8] *Exch. Deps. by Com.*, 3 Charles I, East. 19.

as miners in the pits most of their lives.[1] Sometimes they were ruined mining adventurers like Thomas Robinson, for many years one of the principal undertakers of the great colliery at Bedworth, but reduced in 1640 to the position of manager of the Nuneaton " coal works ", then in the possession of Edward Stratford, Robinson's " very good friend . . . under whome he gets that small meanes of livelyhood that he hath ".[2] Yeomen and husbandmen, who had some experience with coal works in their native villages, saw a chance to make more money than they could get by husbandry and offered their services to concessionnaires or royalty owners.[3] Even the merchant class might be represented. In the Tyne valley some overmen, like Thomas Wakefield, a yeoman of Whickham, stressed their humble origin,[4] but others, like the Hedworths, who managed pits in the same manor, described themselves as " gentlemen " and " merchants ".[5] There is no doubt that in some cases the managers came from the same stock as the adventurers. Richard Jackson, overman in 1634 at a mine in the Wear valley, was a poor relation of John Jackson, one of the owners of Harraton, the greatest colliery in that district.[6] William Liddell, who undertook in 1635 to manage the latter enterprise, belonged to the same family as Sir Thomas Liddell, one of the wealthiest colliery owners in Durham.[7]

All, or nearly all, the officials appear to have had some first-hand knowledge of mining technique. In the sixteenth and early seventeenth centuries most of them, except those in the very lowest grades, possessed some capital. John Bainbrigg, overman in 1606 of a pit at Whickham—a man who, according to one of his enemies, neglected his job to follow his " private pleasures ", and " to playe at dice and Cards " until he had " by his lewd courses and play decayed his estate utterly " [8]—was one of a number of officials who used their savings to help finance the collieries which they managed. Their financial participation took various forms. We find scarcely less diversity in the terms under which different seventeenth-century managers worked than in the functions which they performed. The

---

[1] *Exch. Spec. Com.*, no. 3493.

[2] *Chanc. Proc.*, Series 11, 425/47. It is very probable that some of the small mine owners, who were forced out of business in the Tyne valley at the end of the sixteenth century (see above, pp. 416–17), became overmen and viewers at the large collieries in the neighbourhood.

[3] For example, at Prescot and Wigan the collieries were managed by local husbandmen (*Duchy of Lancs. Pleadings*, 183/L/5 ; *Duchy of Lancs. Deps.*, Series 11, 75/13).

[4] Wakefield described himself as " a simple unlearned man " (*Star Chamb. Proc.*, *James I*, 53/10).

[5] *Ibid.*, 163/18 ; *Exch. Deps. by Com.*, 14 Charles I, Mich. 29. These Hedworths are not to be confounded with the family of that name which owned the manor of Harraton. The Hedworths with whom we are for the moment concerned had been, like the Wakefields, tenants of Whickham manor, and had first come into prominence at Newcastle in 1597, upon the appointment as town sheriff of Adrian Hedworth (see below, vol. ii, p. 40).

[6] *Palat. of Durham, Decrees and Orders*, vol. i, f. 130 ; *Exch. Deps. by Com.*, 9 Charles I, Mich. 11 ; Welford, *Records of the Committees for Compounding*, p. 254. John allowed Richard an annuity of £10 per annum.

[7] *Chanc. Proc.*, Series 11, bdl. 419/38 ; Welford, *op. cit.*, p. 269.

[8] *Star Chamb. Proc.*, *James I*, 53/10.

capital for certain mines in the Mendip district and in the Midlands was supplied by small partnerships in which one or more of the partners was by profession a " coal-miner ", who took charge of the administration of the enterprise.

Sometimes he himself had inspired the undertaking, and, finding his own fortune inadequate, had induced a few neighbours to help him provide the necessary funds. Thus in 1658 a collier from Rochdale, Arthur Hallowes by name, joined in copartnership with John Whitworth, a butcher, and Edward Atkinson, a tailor, to raise the fine of £100 needed to obtain a lease of mines of coal and cannel in Spotland manor ; the three partners agreeing to divide equally all charges for " chains, turns ", and other equipment, for rent and workmen's wages, and to share equally the prospective profits.¹ Under somewhat similar conditions, James and Clement Huish—both colliers —induced two local yeomen in 1615 to become their partners in a " Coalwork " within the manor of Midsomer Norton. Soon after, they took in four additional partners ; but their receipts from the enterprise, they said, no more than paid the wages of the " laborers and workemen which did digge . . . and were otherwise imployed ", and " the chardges of light and ropes and other necessaries imployed aboute the [pits] ".² It is certain that the Huishes managed the mine. More than twenty years later James Huish, then sixty years of age, still called himself a " coleminer ".³

Again, in the case of these small partnerships, it might not be a " collier " who inspired the enterprise, but some local adventurers, who then employed a " collier " as their manager, and offered him a share in the venture as a partial reward for his pains. Hercules Horler, of Stratton-on-the-Fosse, a yeoman active in financing several coal works in the Mendip district before the Civil War, testified in 1639 that he and his partner, William Long, Esq., owned a mine in Stratton, and had granted an eighth " part " to John Salmon, " yeoman ", on condition that he work the pits continuously for their benefit.⁴ In North Wales a husbandman testified in 1616 that he had been, " for divers yeares paste, . . . a digger and getter of sea coales " at a mine near the Flintshire coast, but had recently been made " a parte owner " by Mr. " Sallasburie ", the owner of the pits.⁵

Foremen or managers rarely, if ever, participated in this way as partners except in very small colliery enterprises. But, as farmers and contractors, they had a share in financing a number of the chief English collieries during the seventeenth century. The farmer was a sublessee who was granted by the mine owners a

---

¹ *Palat. of Lancs. Bills*, 32/44. For similar partnerships in Lancashire see *ibid.*, 30/108, 55/7, 65/23.
² *Chanc. Proc., James I*, B. 20/72.
³ *Land Rev. Misc. Bks.*, vol. ccvii, ff. 138–46.
⁴ *Ibid.* Salmon testified, however, that he had received this eighth part by conveyance from his father, and that he had paid £20 for two-fourteenths in the mine, assigned him as a reward for ordering the works.
⁵ *Exch. Spec. Com.*, no. 3648.

right to dig coal for a short term, seldom exceeding three years.[1] He might, like James Osborne—in 1609 the farmer of two pits belonging to the great colliery at Elswick—pay the mine owners an annual rent in money or in a specified quantity of coal ; [2] or he might pay a fixed sum per unit of coal that he sold, as did Thomas Colling, a Lancashire collier who, together with John Heyworth, yeoman, farmed the so-called Heyworth pit, " parcel of a coal mine in the Forest of Rossendale held on lease of the Duke of Albemarle by Joshua Nutthall, gent.", and paid Nutthall 1s. 6d. on every 20 " horse loades " sold.[3] According to the terms of the covenant under which Osborne farmed the two pits at Elswick, the mine owners bore the costs of sinking shafts, winning the " water-gaytt ", and putting everything in readiness " to sett on workmen to work Coales ". Osborne had to finance only the current costs of mining, and to bear the risk that he might not find a market for his output. At the end of a farmer's term, all the capital in drainage or winding equipment, originally supplied by the owners, reverted to them.

Contracting-managers, like farmers, worked under short-term agreements, which could in some cases be terminated by the mine owners without notice.[4] Like farmers, they did not supply the fixed capital. They met only the costs of workmen's wages and other current expenses. But, instead of themselves marketing the coal which they raised, they agreed to deliver it to the mine owners at a settled price, somewhat lower than that for which the latter hoped to sell it to middlemen or consumers. In the Tyne valley, where, during the early part of the seventeenth century, the chief colliery owners had frequently one or more of their pits worked under contract, the contracting overmen might receive anything from 19s. to 37s. for every ten (roughly from 10d. to 1s. 8d. per ton), stacked at the head of the shaft.[5] Their profits depended entirely, therefore, on keeping low their working costs, the chief among which was wages.[6]

[1] S.P.D., James I, vol. clxxi, no. 67 ; Green, " The Chronicles and Records of the Northern Coal Trade", in Trans. No. of Eng. Inst. of Mining Engineers, vol. xv, 1865-6, pp. 269–70.
[2] Exch. Spec. Com., no. 4355. Osborne paid in annual rent £100 for one of the pits he farmed, 60 " tens " of coal (valued at about £120) for another.
[3] Palat. of Lancs. Bills, 36/128. Nutthall allowed Colling and Heyworth to farm another mine of his in Spotland manor for one year at a rent of £5 10s.
[4] Palat. of Durham, Bills and Answers, bdl. 9 (Sir George Selby v. Clough and Wilson, contractors). On one occasion, at least, a contractor was ejected by the colliery owners because, in their opinion, his contract had proved too profitable to him (Star Chamb. Proc., James I, 163/18).
[5] Selby's contract with Clough made in 1608 for working a pit, parcel of Winlaton colliery, called for payments of 19s. 6d. per " ten " ; his contract with Wilson called for payments of 23s. per " ten ". Two contracts made in 1607 for working pits of the Grand Lease colliery called for payments to the overmen of 22s. and 25s. per " ten " (Star Chamb. Proc., James I, 53/10). A contract of 1615 for working a pit in the Brinkburne freehold in Whickham manor called for payments of 28s. (ibid., 245/6), another of 1622 for working a pit of the Grand Lease colliery for 37s. (ibid., 163/18), another of 1618 for working a pit of Benwell colliery for 4 nobles per " ten " (Exch. Spec. Com., no. 5996). These figures suggest that the first quarter of the seventeenth century was a period of rapidly rising mining costs at collieries in the Tyne valley. It is not surprising that the margin of profit should have been much reduced by 1625 (cf. below, vol. ii, p. 75).
[6] Cf. below, vol. ii, p. 186. Positions as overmen, like licences or patents,

Besides contracting to get their pits worked, mine owners contracted with prospectors or " sinkers " to have their shafts sunk at a fixed rate of so much per fathom,[1] and with drainage experts to have their mines drained ; the contractor in both cases undertaking all the expense connected with his job. Thomas Surtees, who undertook early in the reign of James I to drain the mines on behalf of the Grand Lease colliery partnership and other mining partnerships in Whickham, was said to receive four shillings for every " ten " of coal sold by any of these companies.[2] We have no records to show whether drainage contractors were generally paid on the basis of the owners' output.

Contracts for sinking appear to have been almost the rule at English coal mines during the late sixteenth and the seventeenth centuries, contracts for drainage the exception.[3] It is more difficult to generalize concerning the extent to which mine managers and pit foremen participated in financing the actual mining of coal—whether as partners, farmers, or contractors—and the extent to which they were merely salaried officials. But the trend was undoubtedly away from the former towards the latter status. Two pits belonging to the same enterprise might be worked on totally different terms. At Benwell colliery, early in the reign of Charles I, one of the overmen, John Osborne, worked under contract, while another, Richard Richardson, a man of the same age and social standing as Osborne, was paid a salary of ten shillings a week and relieved from every obligation to risk his own money in the pit which he managed.[4] The same manager might at one time work for wages and at another participate in financing a mine. Thomas Hulme, a Lancashire " yeoman ", hired himself out shortly after the Restoration as " banksman " to his neighbour, John Rigby, who worked a coal mine in his own lands within Shevington manor, near Wigan. Although Rigby paid Hulme a salary of four shillings per week for this work, he took Hulme into partnership as manager of a second mine in the lands of one Richard Blackburne.[5] In the Firth of Forth district at the beginning of the seventeenth century, mine owners frequently granted the working of their collieries, or of certain pits, to " tacksmen ", who apparently worked on terms similar to those of contracting overmen in the Tyne valley, and who usually contracted for a period of one year, but sometimes for as long a period as nineteen years.[6] After the Civil War, this

---

were sometimes bought and sold. In 1609 Thomas Gambeskie paid Matthew George £2 on condition that he should make Gambeskie overman of a " colepit " at Whickham in his stead, with the full consent of the " coleowners " and masters of the pits (*Palat. of Durham, Bills and Answers*, bdl. 6).

[1] Cf. above, p. 366.   [2] But see *Star Chamb. Proc., James I*, 245/6.

[3] But see *H.M.C., Report on the MSS. of Lord Middleton*, pp. 173–5 ; *Acts of the Parliaments of Scotland*, vol. iv, p. 176 ; *Privy Council Register of Scotland*, 1st Series, vol. xii, pp. 258, 277. Contracts for drainage probably became more common after the introduction of the steam engine (Lord Harley, *Journeys in England* in *H.M.C., Report on the MSS. of the Duke of Portland*, vol. vi, pp. 103–4).

[4] *Exch. Spec. Com.*, no. 5996 ; *Exch. Deps. by Com.*, 11 Charles I, Mich. 46.

[5] *Palat. of Lancs. Bills*, 30/108.

[6] *Privy Council Register of Scotland*, 2nd Series, vol. viii, pp. 22–5, 96–7 ; *Register of Deeds* (General Register House, Edinburgh), vol. vii, p. 881. The word " tack " sometimes referred to a long-term coal lease.

system appears to have become much less common in Scotland. The working of the colliery as a whole was given into the hands of a general overseer, who received a " commission ", like that granted in 1715 by the Duke of Hamilton to an expert named John Mann, " for overseeing the Collworks of Kinglass " in Linlithgowshire. Mann hired the workmen, purchased the equipment, and negotiated the sales of fuel ; but all this he did merely as agent for the Duke, who paid him " a competent salary " of about £20 sterling per annum.[1] Under the overseer it was common to employ a number of grieves, who received their salaries by the week, and who took charge of the details of mining and selling coal.[2] In England most of the large collieries came to be administered on similar terms.[3] At the smaller English mines the manager often remained, like Cuthbert Hartley, who had charge of pits at Helme Park in south Durham in 1611, both " farmer and bankman ",[4] for the working capital needed to run such a mine might be far less than that needed to run a single pit at one of the Tyneside collieries.

Every increase in the scale of mining tended to eliminate the manager or foreman who participated in colliery financing. Before the end of the seventeenth century the capitalist-official was probably in the minority in all the mining districts which had access to navigable water.   There remained, until the advent of railways widened enormously the market for coal from inland mines, a number of small collieries all through the Midlands (particularly, it is said, in Staffordshire and Derbyshire), worked by farmers (or contractors) who came to be called " butties ".[5] It would be a mistake, however, to suppose that the " butty system " was the rule at inland coal mines even during the late sixteenth and seventeenth centuries.   We find a number of cases in which the bankmen were paid wages by the colliery owners.[6] Everywhere the operating costs were increasing beyond the means of the small farmer or contractor, who, if tempted by the prospect of marketing a larger quantity of coal to hire additional workmen, often found that he ran short of cash to pay their weekly wages pending the summer selling season.   An interesting case in point is that of Sam Bure, a husbandman from Sutton, in Lancashire, who became in 1710 " Steward or Banksman " for a partnership owning a colliery in Huyton, a tiny village a few miles inland from

[1] Hamilton MSS., 592 (2), bdl. 3.

[2] Register of the Tulliallan Coal Works, 1643–7 ; Accounts of the Torry Coal and Salt Works, 1679–80 ; Accounts of the Kincardine and Tulliallan Coal and Salt Works, 1679–80 (all in General Register House, Edinburgh) ; Hamilton MSS., 592 (2), bdl. 5.

[3] See, for instance, Chanc. Proc., Series II, 419/38 (1637), in which the three partners who owned Harraton colliery brought charges against William Liddell, who worked the coal " as their agent ", the partners supplying all the capital.   See also Chanc. Deps., P/7/4 ; Exch. Deps. by Com. 36 Charles II, Mich 43 ; Court of Augm. Misc. Bks., vol. cxii, ff. 17–24.

[4] Exch. Spec. Com., no. 5037.

[5] R. N. Boyd, Coal Pits and Pitmen, p. 15 ; J. L. and B. Hammond, The Town Labourer, pp. 9 n., 174.

[6] Exch. Q.R. Accts., 632/17 ; Palat. of Lancs. Bills, 33/29 ; Duchy of Lancs. Deps., Series II, 75/13 ; Star Chamb. Proc., James I, 106/7 ; Chanc. Proc., Eliz., O.o. 2/46 ; Crofton, Lancashire and Cheshire Coal Mining Records, pp. 63–6.

the growing port of Liverpool. In the first two years of his steward-ship, Bure had to borrow £22 5s. 6d. to meet his obligations to the workmen he employed. As farmer of the pits, he had himself financed the carriage of coal to Liverpool, where he had sold it on credit to various west-coast shipmasters. Some of his creditors having disappointed him, he found it impossible to pay his first debt, and was forced to borrow, upon the security of the partners who owned the colliery, an additional £20 at interest, giving the partners notes against his failure to liquidate this new debt.[1] Just as borrowing was leading to the downfall of the independent working partnership, so it was leading to the downfall of the independent farmer-manager.

It had led early in the seventeenth century to the dis-appearance of many contracting-overman at collieries in the Tyne valley. To meet their current wages bills, these overmen came to depend upon money advanced them by the mining partners. In return the partner, whose profit depended on having supplies of fuel more than adequate to meet all demands from the shipmasters,[2] contracted with his overman for the delivery of a specified quantity each week, and required him to give his bond, or failing that, to get sureties from his neighbours equal to twice the amount of the desired bond, against failure to deliver the specified quantity.[3] Records of proceedings in the Chancery Court of the Palatinate of Durham show the partners in the larger collieries constantly bringing actions to have such bonds or sureties declared forfeit, on the ground that their overmen had fallen behind in their deliveries. Successive forfeitures must have deprived the contractors of their small capital, and must have diminished their chances of inducing their neighbours to go bond for them. The odium attached to an overman who had let down his guarantors is reflected in the bad reputation of John Bainbrigg, who had contracted to work a pit owned by the partners in the Grand Lease colliery, and had failed to fulfil the terms of his contract. The " sureties " confiscated by the partners amounted to £100, although they had advanced Bainbrigg little more than £55.[4] An impoverished contractor could hope for nothing better than to find employment as a salaried official.[5]

The process whereby the small producer is forced, for want of capital, into dependence upon outside capitalists, is repeated over and over

[1] *Palat. of Lancs. Bills*, 56/49. Bure brought an action against the partners on the ground that they had refused to deliver up his notes after he had paid his debts.

[2] See above, p. 376.

[3] *Star Chamb. Proc., James I*, 53/10 ; *Palat. of Durham, Decrees and Orders*, vol. ii, ff. 19 sqq.

[4] *Star Chamb. Proc., James I*, 53/10 ; and cf. above, p. 422.

[5] Overmen at the large collieries along the Tyne were already paid a salary of 10s. per week, while the chief grieve at Tulliallan colliery in 1643 and the " comptroller " at Kincardine colliery in 1679 were paid about 10s. and 8s. (English) per week. The yearly pay for an official of this sort at the larger collieries was apparently from £15 to £25 or £30 (*Waterford MSS.*, no. 43 (o) ; *Register of Tulliallan Coal Works ; Accounts of Kincardine and Tulliallan Coal and Salt Works ; Hamilton MSS.*, 592 (2), bdl. 3).

again in the course of the development of modern industrial organization. At the end of the seventeenth century this process was still
in its infancy in some of the backward coalfields. Where, as in the
district round Stratton-on-the-Fosse, the ordinary mining enterprise
was still small, the line separating adventurer from manager, even the
line separating adventurer from labouring collier, was blurred. Both
the Horlers and the Salmons—who, together with the Longs, controlled the colliery in Stratton before the Civil War—were simple
yeomen, some of whom thought it no shame to labour with picks.
There is a touch of derision, not unlike that which miners to-day
express for one of their number who chooses to become a
satellite of the employers rather than remain a trade unionist, in
the title " Gentleman " Salmon, assigned to a member of that family
who added " gent." after his name. His son John, though a share-
holder in the enterprise, was content to remain a yeoman, and,
in 1624 or 1625, set out " to work and order the works of the mines
taking such wages as other workmen had ".[1] But in the neighbouring
colliery district of Kingswood Chase, where all through the seventeenth
century mining enterprise was conducted on a larger scale, the labour
of coal digging had come to be regarded as the lot of a strange race
of " black men ", whose lives and manners were supposed to differ
radically from those of ordinary human beings. There is no doubt
that the general trend of the mining industry throughout the country
was towards the social and economic conditions which prevailed
about Bristol. Even where there was only a " land-sale " market,
the clusters of small collieries were, in fact, the nests from which a
capitalistic industrial organization tended to spread into the
surrounding regions.

At most of the larger collieries the separation between the functions
of financing and of management, as well as the separation between
the functions of financing and of labour, was clearly marked before
1700. Occasionally an important mine owner, like the second Earl
of Weymss, undertook the administration of his own enterprise ; [2]
but more often, as we have seen, those adventurers who supplied
the capital appointed a " clerk " or " agent ", to purchase supplies,
to deal with the viewers, the overmen and other under-officials,
and to negotiate the sales of coal. Even an adventurer like
Sir John Lowther, whose knowledge of the inventive achievements of his age enabled him to understand the problems of
mining much better than most technical experts, chose to leave
the routine management of his collieries in Cumberland to a
steward.[3] As early as 1587, Henry Killinghall, of Middleton St.
George, in Durham, a part owner of Winlaton colliery, declared that
he always " put [an]other in truste " to examine the account for his
" part " in the colliery.[4] Some of the partners in the great mining
enterprises along the Tyne paid almost as little attention to the
administration of the establishments from whence they drew their

<hr>

[1] V.C.H. Somerset, vol. ii, pp. 380–1 ; Land Rev. Misc. Bks., vol. ccvii,
ff. 138–46.   [2] A. S. Cunningham, Coal Mining in Fife, p. 47.
[3] P.C.R., vol. lxiv, p. 480.
[4] Exch. Deps. by Com., 29 Eliz., East. 4 ; and see below, Appendix K (ii).

dividends as do most modern royalty owners or shareholders. Henry Chapman, one of the principal mining adventurers among the citizens of Newcastle, told commissioners appointed to investigate the profits made from Benwell colliery during the decade which ended in 1631, that he had been a partner in that enterprise, " but does not know to what extent, because his servant, Charles Clarke, now dead, kept all the reckonings ".[1]

It is clear that, before the end of the seventeenth century, the need for the investment of considerable sums of money had led to an almost complete divorce between labour and capital in British coal mining. Even in many of the more backward colliery districts, the cost of starting a mine was beyond the means of a partnership of working miners. Workmen and foremen alike put themselves, to an increasing extent, into positions of complete dependence upon capitalists, by entering into contracts for the delivery of coal. In the more developed colliery districts the separation between capital and labour had already proceeded so far, that there had grown up between the two classes an intermediate class of managers and technicians, who undertook the management of the enterprise on behalf of the employers.

(ii) *In the Coal Trade*

To move coal from the bank beside a mining shaft to the cellar of a seventeenth-century householder involved three essentially different types of work : the handling of the coal—loading it at the pit and at the port of shipment, unloading it at the port of destination—the driving of packhorses or wagons, and the rowing or sailing of lighters, trows, and hoys. How far did the labouring men who moved the coal share in the ownership of the conveyances, the containers, and the tools with which they worked ?

Taken by itself, the handling of the coal required very little fixed capital—a shovel and perhaps a cloth sack or wooden vat. But it was usually undertaken in conjunction with other duties. Inland carters filled their sacks themselves at midland pits, sometimes with the banksmen's aid. Along the Tyne and Wear only a few special workmen were employed in emptying coal from the staiths into waiting keels, and the keelmen themselves tossed and shovelled it aboard the sea-going colliers. At King's Lynn in 1608 ships' " boys " took part in unloading a collier that had arrived from Newcastle.[2] Only at London (and possibly at a few other of the chief importing towns) was there a large special group of labourers occupied exclusively in delivering cargoes out of coal hoys.[3]

From very early times there had been within the port of London a lesser gild of unskilled, but independent, workmen—the Billingsgate or Fellowship porters—whose business it was to assist the city

---

[1] *Exch. Spec. Com.*, no. 5996.
[2] *Star Chamb. Proc., James I*, 145/18.
[3] The seamen rarely managed the unloading of the coal (see above, p. 394), although they frequently helped the workmen who performed this labour.

meters appointed by the municipal government to certify the exact
quantity of all measurable commodities (such as corn, coal, or salt)
brought in by sea.[1] To prevent disputes between sellers and buyers,
the meters were equipped with standard-sized vats. Under their
scrutiny, the porters filled these vats on board ship and then
carried them ashore. After the introduction of coal into the
port of London, and the appointment, probably towards the end
of the thirteenth century, of special " seacoal meters ", Billingsgate
porters took upon themselves the handling of this new commodity.
At least as late as 1600, ships bringing coal were still assigned to them
to unload.[2]

With rapidly increasing imports of coal, the problem of unloading
colliers changes. No longer does coal arrive as ballast in ships laden
with other commodities, which it is the porters' business also to
unload. Colliers laden almost exclusively with coal begin to arrive
in great fleets. While a few score porters had sufficed to handle all
the measurable goods imported before the reign of Henry VII, the
coal trade alone in the seventeenth century comes to require hundreds
of labourers,[3] who unite in gangs of sixteen to deliver each cargo.[4]
" These gangs take turns on board every ship laden with coales (as
the Coale-Meters doe) whereby every poore man will have his due
proportion of work ".[5] No longer do the colliers unload beside the
wharves, as all ships formerly did. They ride at anchor out in the
river ; so the coal meters must go to them in lighters. Not all the
coal now brought ashore by the lightermen is loaded directly into
carts as formerly. Often the larger portion is moved short distances
to the merchants' yards or vaults for storage. Instead of two possible
handling operations connected with the arrival of a cargo in the
Thames, there are now four. In the past, coal was carried directly
out of the ship on to a wharf and then loaded into a city cart. Now
it is first poured from the collier into a lighter, next it is heaved,
or carried (if it has already been put into sacks), out of the lighter
on to a wharf, and from the wharf it is either loaded into carts as
in the past, or else borne away upon men's backs.

Of these four operations, the last alone appears to have been
undertaken by the Fellowship porters during the seventeenth century.[6]
When, in the eighteenth century, they attempted also to undertake
the loading of carts at the wharfs, their original share in handling
the incoming coal cargoes had been so far forgotten that the London
coal merchants filed at the Guildhall in 1733 a petition protesting

---

[1] Cf. Unwin, *Gilds and Companies of London*, pp. 352–3, 358–9.

[2] On April 15th, 1600, 112 porters, freemen of the City, were sworn to
help the meters in the measuring of seacoal, at such times as they should need
help (*City of London Repertories*, vol. xxv, ff. 73–7). See also *ibid.*, vol. xxiii,
f. 100b (September 4th, 1593), " Ships of sea coal to be appointed to the gentlemen
porters of the Tower ".

[3] At the end of Charles II's reign the number of London " coal-heavers "
was estimated at 600 (*Additional MSS.*, 4459, no. 8). In 1696 a petition signed
by 614 coal-heavers was presented to the city aldermen (*Alchin MSS.*, bdl. 82).

[4] Dale, *Fellowship of Woodmongers*, p. 35.

[5] *Additional MSS.*, 4459, no. 8.

[6] Dale, *op. cit.*, pp. 142–3.

that " it hath been the custom and usage from time immemorial for
the Traders in Coals keeping Wharfs . . . to employ their own servants
to land and load Sea Coals, Billets, and Faggotts into their own carts
without any lett or interruption ".[1]   A tract printed in 1764 explains
how the Fellowship porters had ceased long since (probably soon after
Elizabeth's reign) to assist the coal meters.   Being " much employed
in several other Branches of Labour, which were more agreeable to them
than the slavish and dirty Part of Work, such as unloading Coal Ships
and Vessels ", they, " by degrees, forbore to have any further concern
therein ; and, at length, gave it up ".[2]

In their place the masters of colliers hired such " poore Labourers
as they could get to unlade the shipps ",[3]—most of them " persons
of the meanest sort, of little worth and no setled Habitation ",[4]
members of that growing section of the London population having no
status in the ancient order of city trades.   They came to be called
" coal heavers ", or simply " coal-labourers ", and later " coal-
whippers ".   Equipped with shovels and clad in ragged black, they
hung about a place called " Roomland " in Billingsgate, where the
masters of colliers came on shore to meet with the London buyers
" in the nature of an exchange ". [5]   After reaching an agreement
with some of the merchant-buyers for the sale and delivery of his
coal, the master applied at the neighbouring coal meters' office for
a meter to measure it, and then contracted with a gang of heavers
to return with him on board his vessel to unload it into lighters at
a piece-rate of from one to three shillings per London chaldron.[6]

Soon after the Restoration, the coal heavers lost their power to
bargain directly with shipmasters for the price of their own labour.
The constantly increasing imports of coal made room for a new type
of dealer, known as a " crimp ",[7] who acted as agent, or broker, for
the master in his dealings with the city buyers.[8]   These crimps are

---

[1] Dale, *op. cit.*, p. 142.

[2] *The Coal Heavers' Case*, 1764.

[3] Dale, *op. cit.*, p. 136.

[4] *Alchin MSS.*, bdl. 31:   *John Tyzack's Proposal about the Coal Heavers.*
It appears that soldiers of the Grenadier Guards were granted by Charles II
the right to act as coal heavers in their spare time (cf. Dale, *op. cit.*, p. 35).

[5] *Alchin MSS.*, bdl. 64 (Petition of various citizens of London to the Court
of Aldermen concerning the " Coal Traders ", June 13th, 1682).   For further
information concerning the coal exchange see R. Westerfield, *Middlemen in
English Business*, pp. 232–4 ; and below, pp. 443–4 ; vol. ii, pp. 96–7.

[6] *Alchin MSS.*, bdl. 31 (Petition of March 28th, 1706).   This price included
the unloading of an extra chaldron, for every score of chaldrons, free of charge.
(Cf. Westerfield, *op. cit.*, p. 231.)

[7] The earliest references that I have found to " crimps " occur in 1675
(*Repertories*, vol. lxxx, ff. 162b, 295b ; *Cal. Treas. Bks.*, vol. v, p. 147).   In
1676 a bill was introduced in the Common Council of the City for regulation
of " Coal Brokers and Crimps " (*Journals of the Common Council*, vol. xlviii,
f. 236).   The first complaint of the coal heavers concerning abuses received at
the hands of crimps is written upon the back of a bookseller's advertisement,
dated 1678 (*Additional MSS.*, 4459, no. 8).

[8] I doubt whether, as Professor Westerfield suggests (*op. cit.*, p. 232), the
" crimps " originally confined their operations to organizing gangs of coal heavers,
and became brokers, or factors, only after 1700.   It seems to me more likely
that they sometimes acted in the latter capacity before the eighteenth century.

said to have been mostly wharfingers or lightermen.[1] They " sett up places for Labourers to come in and be hired " to unload the ships, retaining " men when a fleet came in att 8*d*. and 10*d*. per chaldron and [making] the Owner pay 2*s*., 2*s*. 6*d*., and 3*s*. the chaldron ".[2]

It is significant that the London coal heavers should have been forced so early to depend on middlemen for their wages, and to accept a position comparable to that of the lowest class of city labourers, such as the stokers in glassworks,[3] when some other groups of carriers within the City preserved their independence down almost to the present day. The Billingsgate porters, who continued to handle all imports of corn or salt after they had abandoned the handling of coal, were organized as a fellowship which elected its own " rulers " and paid them regular salaries out of the common earnings. Through their rulers, the members retained the right to bargain collectively for the disposal of their labour. They became " co-operative capitalists ", the rulers being instructed to stop a certain percentage of all earnings to raise a fund out of which the society maintained the old and sick, and advanced money on tasks undertaken by the porters but not yet completed.[4] The rulers had " full authority to send as many of the company to any work (be it much or little) as by their discretion they shall think meet, having regard that they show favour as much as may be to ease the old and ancient and weaker persons ".[5]

Now the coal heavers desired to organize themselves into a fellowship modelled upon that of the Billingsgate porters. Their agitation to have this demand conceded by the municipal government probably started as soon as the crimps attempted " to beat downe " their wages.[6] It gathered additional momentum during the last decade of the seventeenth century, when the coal heavers, encouraged perhaps by the political revolution,[7] drew up and presented at the Guildhall a multitude of petitions [8]—some signed by scores, others by hundreds

---

[1] Westerfield, *op. cit.*, p. 232. Some crimps also acted as wholesale buyers of coal (see below, vol. ii, p. 86), appearing on the London exchange both as buyers and as sellers (*H.M.C., Report on the MSS. of the House of Lords*, vol. v, pp. 236–7).

[2] *Petition of the Masters of Coal Ships*, August 13th, 1696 (printed Dale, *op. cit.*, p. 136).

[3] M. Dorothy George, *London Life in the Eighteenth Century*, pp. 155–8.

[4] Cf. Unwin, *Gilds and Companies of London*, pp. 358–63.

[5] *Ibid.*, p. 362.

[6] During the late seventies of the seventeenth century (*Repertories*, vol. lxxxv, f. 71*b*; *Additional MSS.*, 4459, no. 8). On one occasion, the coal heavers complain that the " buyers and importers of coal ", as well as the " Crimps or Brokers ", " have grievously vexed your Petitioners in their wages . . ." (*Alchin MSS.*, bdl. Y : *Petition of Coalheavers*, 1695).

[7] The coal heavers first addressed a petition to King William III, who referred the matter to the Court of Aldermen (*Alchin MSS.*, bdl. Y : *Repertories*, vol. xcv, f. 52*b*).

[8] A large number of these petitions have been preserved at the Guildhall among the *Alchin MSS.*, bdls. 31, 64, 82, Y. See also *Miscellaneous MSS.*, bdl. 505 : *Petition of Coal Labourers*, 1699. These manuscripts have provided me with most of my information concerning the status of the coal heavers at the end of the seventeenth century. Extracts from certain of their petitions, preserved among the *Alchin MSS.*, have been printed by Dale (*op. cit., passim*). On one occasion (December 9th, 1695) a petition for incorporation seems to have

of their number. They disclaimed any intent themselves to deal in coal, " the proper business of coal merchants, lightermen, and coal traders ". All they asked was "to manage their own hard labour ", the fruits of which were being wrung from them by the crimps.[1] The readers of such documents were not always unsympathetic to the coal heavers' request, for there were substantial freemen of the City who begrudged the crimps their extortionate gains.[2] A commission appointed by the Court of Aldermen, after whittling down the coal heavers' demands,[3] reported on July 2nd, 1696, in favour of a " Fellowship or Fraternity of Coal-labourers ", which was to resemble, in many of its features, the Fellowship of Billingsgate porters, although it was to differ in the all-important respect that the twelve rulers were to be elected not by the coal heavers, nor even from among their number, but by the Aldermen and from among the city householders who were freemen of London.[4] Finding in this

been engineered by the crimps, with the object, doubtless, of securing control of the proposed organization. " We were drawn to petition incorporation ", write thirteen of the coal heavers who repudiate their signatures, " by the false suggestions and sly insinuations of some Crimps and others, who are not only the enemies of the coal dealers but of the coal heavers " (*Alchin MSS.*, bdl. 31). A few years later, the keelmen of Newcastle repudiated a petition in their name under similar circumstances (see below, vol. ii, p. 178).

[1] *Alchin MSS.*, bdl. 31 : *Reasons why the Coalheavers Corporation ought not to be governed by Coal Traders, Lytermen, and Crimps*, n.d.

[2] Among the petitions advocating a coal heavers' fellowship is one from " various London Citizens " (*Alchin MSS.*, bdl. Y), and another from one John Tyzack (bdl. 31), who bears a name famous in the history of English glass manufacture (cf. above, p. 180 n.).

[3] The commission reported that the abuses in unloading were only in part the fault of the crimps, and that the coal heavers themselves " have sometimes exacted what they pleased ".

[4] *Alchin MSS.*, bdl. Y : *Report of the Commission of the Court of Aldermen*, July 2nd, 1696. Tyzack's proposal (*supra*) had been to allow the " foremen of the cole-laborers " nine out of twelve places as rulers. Since the price of coal heavers' tasks (like the price of tasks done by the Billingsgate porters) was to be fixed by the municipality, it would be difficult to argue that the objections of the municipal government to a democratic fellowship were based on a genuine fear that the coal heavers would exact an exorbitant price for their labour.

According to the recommendations of the Aldermen's Commission, the fellowship was to be incorporated for three years under the control of a governor and deputy governor besides the twelve rulers (art. 1). They were to have thirty-six assistants to oversee the performance by the labourers of their respective duties (art. 3). An office was to be erected at Billingsgate, where one of the rulers should constantly attend and supply the shipmasters with labourers to unload their vessels. The master was to provide lighters and coal-meters, and to pay the rulers a fixed charge of 16*d.* for every London chaldron delivered out of the collier. Neither freemen of the City nor seamen of the ship were to be excluded from the labour of coal heaving, but those who worked constantly as heavers were to be employed by the rulers impartially, each in his turn (art. 5). The rulers were to pay 14*d.* on every chaldron to the workmen employed in unloading, and the remaining 2*d.* into the hands of such persons as the majority of the fellowship should designate (art. 6). No one except workmen appointed by the rulers, or seamen of the collier, were to be permitted to unload the coal into lighters (art. 7). Rulers were to hold regular meetings to decide, among other things, upon their own " wages " (art. 8). Books of the coal imported and the ships delivered were to be kept at an office in Billingsgate (art. 9). The Fellowship was to pay £500 annually to the City for charitable uses (art. 15), and was to give bonds not exceeding £7,500 against any damage or neglect arising through the fault of the rulers (art. 16). Finally, the aldermen reserved the right to alter the rules at any time (art. 17).

last provision a means of resisting the abuses of the crimps, without delegating any real power to those whom they regarded as the unruly coal heavers, the Aldermen adopted the report of their commission in favour of the fellowship.[1] But the crimps defied the ruling of the court, " gave out base language to the Fellowship, . . . beat some of their men, . . . tore down [and] stamp[ed] under foote my Lord Mayor's Order, and were encouraged by some to pull the [Coal] Meters' house down ".[2] They persuaded the shipmasters so effectively to boycott the new fellowship that the order creating it had to be withdrawn, apparently almost at once. A year later—in July, 1697—the Court of Aldermen, after heated debate over the projected re-enactment of the " Rules and Orders for the Coal heavers ", decided that " no such fellowship should be established ".[3] The Lord Mayor proceeded to wash his hands of the whole affair, on the ground that the coal heavers, being mostly " unfreemen ", could not be subject to his jurisdiction.[4] His attitude led them to bring their case before Parliament in 1699.[5] A recommendation in favour of the proposed fellowship was included in the Brokers' Bill of 1704 ; but the Bill did not pass,[6] and, in spite of repeated agitation for nearly a century, the coal heavers never obtained effective legal protection from the exactions of greedy middlemen.[7] Although the crimps, who came to be called factors, are said to have confined themselves soon after 1700 to brokerage operations on the London coal exchange,[8] their places as masters of gangs of coal heavers were immediately filled by " coal undertakers ". The latter set up employment agencies in certain public houses, and, by securing the exclusive right to purchase coal heavers' shovels from the makers, forced the heavers to come to the public houses in order to hire, at an exorbitant rate, the tools without which they could not gain their livelihood.[9]

The failure of the coal heavers to obtain recognition as a fellowship, or even to retain ownership of the inexpensive equipment requisite for their labour, may be ascribed in part to their appearance upon the

---

[1] *Alchin MSS.*, bdl. Y.

[2] This information is contained in a petition of the masters and owners of ships, dated August 13th, 1696, six weeks after the Court of Aldermen had created the fellowship (printed by Dale, *op. cit.*, pp. 135–8).

[3] *Alchin MSS.*, bdl. Y. Three principal reasons are given against incorporation : (1) It will limit the labour of coal heaving to specified persons, and prevent many seamen, lately employed in that labour, from taking part. (2) The number of regular coal heavers will be insufficient if a great many colliers arrive at once. (3) The coal heavers refuse to give security for quick dispatch.

[4] *Alchin MSS.*, bdl. 31 : *Petition of Coal Heavers to Queen Anne* (which contains a recital of events that occurred during the previous reign).

[5] *Alchin MSS.*, bdl. 31 : *Reasons for Making the Coal-Labourers on the River Thames a Fellowship, offered to Parliament* (printed broadside), 1699.

[6] *Alchin MSS.*, bdl. 31.

[7] Unwin, *Gilds and Companies of London*, pp. 363–4, note.

[8] Westerfield, *Middlemen in English Business*, p. 232. Later in the eighteenth century they organized themselves into the Coal Factors Society. The records of this society since 1761 have been preserved. (See Hooper, *The London Coal Exchange*, Appendix 2).

[9] M. Dorothy George, *London Life in the Eighteenth Century*, pp. 294–6, 397 ; Dale, *Fellowship of Woodmongers*, pp. 79–80.

London scene at a time when gild regulations were breaking down.[1] But that is less than half the story. As the late Professor Unwin pointed out, " the gilds of transport ", among which he included the porters, " entered upon the most active period of their existence at a time when the gilds of handicraft were becoming obsolete ".[2] Why did not the coal heavers receive the same consideration as the other porters ? As they themselves urged in one of their numerous petitions to Parliament,[3] they laboured as carriers just as did the Watermen, the Packer's Porters (who undertook to load and unload the goods of foreign merchants) [4] or the Hackney Coachmen, all of whom had been incorporated, although the latter two included non-freemen.

What disqualified the coal heavers was the nature of the commodity which they carried. Not by accident is the expression " to carry coals " the equivalent of to demean oneself. Notwithstanding its widespread use, coal was associated in the mind of the seventeenth-century Englishman with foul smells, bad tastes, and dirty clothing. Shunned by the Billingsgate porters, it was thought more filthy to handle than almost any other commodity, and he who earned his living by heaving it from the ships into lighters was regarded as something of a vagabond. This attitude cast a dark shadow over all who spent their lives handling or digging coal. Of its influence upon the social position of the coal miner we shall speak later on. At present it is enough to suggest that it played a part in depriving the coal heavers of the protection given to other porters in the City.

To the extent that those who transported fuel from the mine to the consumer performed functions other than that of handling coal, they escaped, no doubt, this special stigma. Their advantage in that respect might be counterbalanced, however, by the fact that more capital was required to own a vehicle or a vessel than to own a shovel and vat.

Next to the work of coal heaving, it was the carriage of coal overland—by packhorse, by cart, or by wagon—that required the least capital. In many colliery districts local tenant-farmers, who owned a beast of burden or two and perhaps an old cart, frequently did business as coal carriers in their off-hours from husbandry. Sometimes they were required to fetch fuel, as part of their obligations to the lord of the manor. Thus tenants of the Duke of Hamilton, living on his estate of Bo'ness, in Linlithgowshire, had to bring from his colliery at Corbie Hall enough coal to boil down all the salt made in his pans there.[5] The farmers of the granges within the bounds of Leicester agreed with the freemen of the town that " every farmer [should] bringe the Maior for the time beinge a loade of Coles after the rate of a loade for everie 3 yard land, Mr. Maior payinge for them at the Pitts ".[6]

---

[1] Cf. above, ch. i (iii). 　　　　　[2] Unwin, op. cit., p. 353.
[3] Alchin MSS., bdl. 31 : Reasons for Making the Coal-Labourers . . . a Fellowship, 1699.
[4] Unwin, op. cit., p. 360.
[5] Hamilton MSS. : Accounts of the Coal Works at Kinneil.
[6] Stocks, Records of the Borough of Leicester, pp. 214–15.

But the part-time labour of these farmers could not long satisfy the increasing demand for coal in Leicester. As early as 1617 " divers poore husbandmen, that dwell in the hundreds of Sparkenhoe and West Goscote neire to the Cole pitts, . . . weeklie bring Coles to Leicester Markett to sell . . . And have all waies . . . on the fridaie nights brought the Coles into Leicester forest, . . . and neere . . . the King's heighe way have sett downe there Carts and fettered there horses, . . . and on the Saturdaie morninge by breake of the daie have taken upp there horses and soe presentlie for Leicester to sell there Coles ".[1] Wherever, as at Leicester, there develops during the sixteenth and seventeenth centuries a considerable market for coal at a distance of more than a mile or two from the pits,[2] persons begin to specialize in the haulage of this new, bulky commodity. In Somersetshire there are " dealers ", who " goe to Mendip for coles ".[3] In all the shires bordering the Firth of Forth are " bass fellowis, the caryaris of coillis ", who wait beside the coal " hills " to receive a lading with which to serve the citizens of Edinburgh, Haddington, Linlithgow, and other smaller towns and villages.[4] Among the mining population living in Kingswood Chase in 1675 are many " coledrivers ", each with his tiny cottage, his garden, and his pasture for two or three horses, which he drives regularly into Bristol laden with coal.[5] Henry White, a blacksmith of Coventry, testifies in 1684 that great numbers of the poor are employed in carrying coal, " both themselves and with horses ", from the pits at Bedworth and Hawkesbury to the town.[6]

During the sixteenth and seventeenth centuries, a considerable proportion of all the coal drivers in British mining districts owned the conveyances which they drove and the horses or cattle to draw them. In the Tyne valley in 1595 many " Carriers of Colles [still bought] ther cattell necessarie thereunto ", although, " beinge for the moste parte pore men ", they could ill afford to pay the high purchase prices or the rents for pasture lands.[7] Occasionally these carriers contracted with colliery owners, after the manner of pit overmen in the district, to deliver at one of the staiths a specified quantity of coal within a given period.[8] As late as the reign of Charles II the copyholders of Elswick retained a monopoly of the transport of all coal from pits within the manor to the river. They kept wains or carts, which they

---

[1] *Ibid.*, p. 168.

[2] At Newcastle the " Common Pit " was so close to the dwellings of the citizens that many sent their own servants with carts or wains to bring home the fuel (*MS. Journal of the Common Council of Newcastle*, 1650–6, ff. 375–6).

[3] *V.C.H. Somerset*, vol. ii, p. 381. In 1632 John Dennison, a yeoman of Stainton (a small village to the south of Stockton), described to commissioners how he went the fifteen miles to the pits in south Durham " to fetch coals " for sale (*Exch. Spec. Com.*, no. 5276). See also *Exch. Deps. by Com.*, 11 James I, Trin. 6, and *Star Chamb. Proc., James I*, 154/12, for references to a similar practice on the part of yeomen in northern Northumberland and in Derbyshire.

[4] *Privy Council Register of Scotland*, 1st Series, vol. xii, p. 434.

[5] *Exch. Deps. by Com.*, 27 Charles II, Mich. 29.

[6] *Ibid.*, 36 Charles II, Mich. 43.

[7] *Harleian MSS.*, 6850, no. 39.

[8] *Palat. of Durham, Bills and Answers*, bdl. 6 : Rawling *v.* Harrison. And cf. below, vol. ii, p. 59.

commonly leased out to local drivers at an annual rent of 50s. per wain and 25s. per cart. An attempt on the part of the proprietors of Elswick colliery in 1678 to introduce conveyences of their own for carriage to the waterside was sharply and, for the time being at least, successfully resisted by the copyholders.[1]

These copyholders were protected by manorial customs peculiar to Elswick.[2] Long before the reign of Charles II, many colliery proprietors or the middlemen who purchased coal from them, in the Tyne valley and in other coalfields, began to keep wagons, carts, cattle, or horses of their own, and to hire labourers to drive for weekly or monthly wages.[3] In 1605 the " Grand Lessees " referred to their " servants and carriagemen ".[4] Not long afterwards they built special stables near the mine to hold their " wagon horses ".[5] Sir Thomas Liddell, who, as proprietor of Blackburn colliery, compounded with the copyholders of Whickham manor in 1635 for the right to carry coal through their lands, must have been in possession of a number of vehicles, for he agreed to pay an extra charge to the copyholders for all wains or carriages that he should " hire, gett, or imploy ", other than his own or those of his children.[6] In 1633 John Shepherdson, one of the partners in the great colliery at Harraton on the Wear, complained of the heavy taxes levied by the Bishop of Durham's Commissioners for Sewers upon his oxen, which he used to haul coal to the river.[7] Round the Firth of Forth, while coal and salt for local consumption were sold at the pits and pans to independent carriers,[8] all coal for export was moved to the wharves by " leaders " employed by the colliery owners on terms similar to those on which workmen were employed in the mines.[9] The evidence suggests, in short, that, wherever large quantities of coal were carried from the collieries over a single route, the supply of conveyances provided by the independent wagoners proved insufficient, and had to be supplemented by additional wagons and beasts of burden which were not the property of the drivers. These workmen were tending to become simple wage-earners like their fellows in the mines.

Even before the London Fellowship of Carmen succeeded in breaking loose from its subservience to the Company of Woodmongers,[10] most of the holders of " car-rooms ", or driving licences, did not drive the carts. Owing partly to the woodmongers' monopoly of these conveyances, the price of a "room" had been pushed in

[1] *Exch. Deps. by Com.*, 32 Charles II, Mich. 30.

[2] See above, p. 300.

[3] But as late as 1764 many of the horses used for hauling coal in the Tyne valley are said to have been owned by the "wagoners" (Jars, *Voyages métallurgiques*, vol. i, p. 204).

[4] *Palat. of Durham, Bills and Answers*, bdl. 2: Selby *v.* Hedworth. Apparently the labourers who carried coal in the town of Newcastle were originally members of the craft of colliers (see above, p. 401).

[5] *Star Chamb. Proc., James I*, 163/18.

[6] *Palat. of Durham, Decrees and Orders*, vol. i, f. 228.

[7] *Ibid.*, f. 19.

[8] Cf. above, p. 108; Tucker, *Report upon . . . Excise and Customs*, pp. 6–8.

[9] *Register of the Tulliallan Coal Works*, 1643–7; *Accounts of Kincardine and Tulliallan Coal and Salt Works*, 1679–80; Hamilton MSS., 592 (2), bdls. 4–8.

[10] See above, p. 408.

1620 as high as £40, £50, and £55, nearly twice the cost of a coal cart and horse.[1] Only a fairly well-to-do citizen could afford to purchase a "room". "The licenses of Cars", complain the wharfingers, "being become inheritable are engrossed into the hands of Brokers, Chandlers, Tapsters, Ostlers, Scriveners, and such like, who use no cars themselves but let them out by the week to poore Freemen and Wharfingers at 3s. or 4s. a week for every license, which is but to pass through the Kings highway or street with fuel for the service of the City".[2] In fact, "car rooms", like coal meters' "rooms",[3] yielded their holders steady incomes, not unlike those derived from stock in the more conservative modern corporations. By 1717 the value of a "room" had actually risen to £150.[4] These holdings were accounted assets, and were passed on—it is stated in a petition addressed to Cromwell by the "free carmen" of London in 1656—to wives and children.[5] The cart which ran over a little boy playing in Creed Lane in 1608[6] was driven not by its owner, but by one John Rubback, an "inholder". Thomas Morley, who as a woodmonger "using the trade of Carryinge Seacoles" owned this and other carts, told the Court of Star Chamber that his custom had been "to Comitt the government or direction of [this] . . . carr . . . to . . . Rubback". It was the holders of "rooms", like Morley, who made up the Fellowship of Carmen. The drivers, like Rubback, were not members. Although some of them may have owned the carts which they drove, the testimony of Morley suggests that many did not. The proportion of those who did probably diminished, for, in the eighteenth century, the drivers were reputed "an ill-mannered sort", being ranked socially along with the chairmen and workers at livery stables.[7]

Of all labourers in the London coal trade, the men who rowed the coal-laden lighters from the colliers to the wharves along the river Thames undoubtedly preserved the highest social status. We have referred already to the trading ring established by a few leading lightermen, after the incorporation of the Company of Lightermen and Watermen in 1700. The members of this ring were not labourers, but wholesale traders and brokers who purchased cargoes from the shipmasters on the coal exchange, just as the woodmongers did.[8] As coal merchants, they apparently organized themselves about 1730 into the so-called "Society of Owners of Coal Craft", while retaining their membership in the Lightermen's

---

[1] *Lansdowne MSS.*, 162, no. 28 ; and see above, p. 384.
[2] *Harleian MSS.*, 6842, no. 67, f. 256.
[3] See below, vol. ii, pp. 256–9.
[4] Unwin, *Gilds and Companies of London*, p. 357.
[5] *Cal. S.P.D.*, 1655–6, pp. 114–15.
[6] See above, p. 384.
[7] M. Dorothy George, *London Life in the Eighteenth Century*, p. 158. In 1661 the Privy Council asked the woodmongers to admit Bernard Dunce to membership in their company, as a reward for having chosen "rather to serve as a carman at a coal wharf " for twelve years "than to take arms for the late tyrants " (*Cal. S.P.D.*, 1660–1, p. 607).
[8] *H.M.C., Report on the MSS. of the House of Lords*, vol. v, pp. 235–8 ; Westerfield, *Middlemen in English Business*, p. 232 ; and see below, vol. ii, p. 86.

Company.[1] They hired workmen from among their lesser brethren to row their barges, and, if these barges did not suffice to unload all the coal, they hired those of other brethren to unload the remainder.[2] What is important is that this division into barge owners dealing in coal, simple barge owners, and rowers who had no share in the ownership of barges, all took place within the Lightermen's Company. Barge owners, even though they sometimes used abusive methods " of extracting money from the meaner members ",[3] could not follow the carmen in hiring labour outside their fraternity ; every rower of a barge had to be at least an apprentice of the Lightermen's Company, and membership in the Company was still limited, if the written rules were enforced, to men who had served seven years at the oar.[4] The different interests of the members as employers and employees were offset, to some extent, by their common interest in the maintenance of their power as a fellowship.

The working bargemen of London are said to have owed their strength partly to the government policy of resisting restrictions upon membership in the London Company of Lightermen, which provided during the seventeenth and eighteenth centuries a specially convenient recruiting agency for the royal navy.[5] It does not appear that the bargemen who rowed coal craft in other ports, or up and down navigable rivers other than the Thames, became members of such democratic organizations as this London company. During the sixteenth and seventeenth centuries many keels and lighters on the navigable rivers of England and France became the property of " merchants ",[6] and it is probable that the latter hired labourers to perform the hard work at the oar, without taking them into any sort of partnership. A Somerset family of the name of Bobbett, residents of the parish of Creech St. Michael, near Taunton, set up early in the seventeenth century as " dealers in sea coles " to supply the surrounding country towns and villages. One, Richard Bobbett, took a lease of a piece of moor ground, on the river Parrett near Ham Mills, to be used for storing fuel.[7] He and his brother Robert also owned packhorses to carry coal to neighbouring towns, lighters to row it up the river from Bridgwater, and even " trows " to bring it from South Wales. A number of husbandmen and boatmen, employed by the Bobbetts in driving their carts and packhorses, and in rowing

---

[1] Dale, *Fellowship of Woodmongers*, p. 80. The Society of Owners of Coal Craft was only one of several societies which grew up within the Lightermen's Company (Unwin, *op. cit.*, p. 364).

[2] *MSS. of the House of Lords*, as cited above ; London Guildhall, *Miscellaneous MSS.*, bdl. 505 : *Petition of divers Masters of Lighters.* According to the Webbs (*History of Trade Unionism*, p. 11, note), the administration of this company had been in the hands of master lightermen as early as the sixteenth century.

[3] *MSS. of the House of Lords*, as cited above.

[4] Dale, *op. cit.*, p. 80.

[5] Unwin, *Gilds and Companies of London*, pp. 364–5.

[6] *S.P.D., James I*, vol. cxxxviii, no. 20 (*Petition of the Mayor and Citizens of York to the Privy Council*, 1623) ; Bibliothèque de l'Arsenal, Paris, MSS. no. 4018, vol. i, f. 109 (*Mémoire sur le commerce des charbons de terre venant de Mons pour Tournay*) ; Boislisle, *Correspondance des Contrôleurs Généraux*, vol. iii, no. 1703.

[7] *H.M.C., Report on the MSS. of the Dean and Chapter of Wells*, vol. ii, p. 456.

and sailing their vessels, were among those who testified on their behalf in 1672 in an action brought against them by the town of Bridgwater, at the instigation—the Bobbetts alleged—of a "combynacon . . . using the same trade ".[1] Many of the barges used in carrying coal down the river Trent in 1612, were owned by Sir Percival Willoughby, proprietor of the colliery at Wollaton, who had turned them over to Robert Fosbrook, the dealer under contract to market Willoughby's coal as his agent.[2]

On the Tyne, the keels for loading the great collier fleets were built and owned by the same Newcastle merchants who became the principal colliery proprietors of the district during the last half of the sixteenth century.[3] Until the Civil War, most of these merchants and their successors engaged in a double business connected with the coal trade. They used their capital partly to finance mining enterprise, partly to supply keels for the enormously increased traffic on the river. At first they dealt directly with the keelmen, and entrusted the skipper, or chief keelman, with the work of fittage, that is with the work of providing the necessary keels to lade their coal.[4] As long as the skipper retained some initiative in dealing with the ship-masters for the disposal of the merchants' coal, the keelmen had a chance to set aside substantial savings, and to live in comfort and good social standing among the Newcastle citizens.[5]

By the beginning of the seventeenth century, however, there appeared on the scene a new intermediary called a fitter, known also as an agent, a bookkeeper, or a factor.[6] Originally the fitters acted simply as agents for the coal owners, who still owned the keels. As agents they arranged for the sales to the shipmasters, paid the Crown and municipal taxes at Newcastle, filed bonds with the customs officers against possible fraud by the shipmaster in carrying overseas a cargo entered in the books for delivery at an English port,[7] hired the keelmen, and determined the order in which the owners' keels could be most efficiently used in taking ladings at the several staiths. Before the end of the reign of James I, some of the north country merchants, while continuing to invest part of their money in mines, began to make an independent business out of fittage, buying coal outright from other colliery proprietors on their staiths and selling it to the shipmasters.[8] Eventually all fittage

---

[1] *Exch. Deps. by Com.*, 23 and 24 Charles II, Hil. 18 ; 24 Charles II, East. 24.

[2] *H.M.C.*, *Report on the MSS. of Lord Middleton*, pp. 175–6 ; and see above, p. 398.

[3] Welford, *History of Newcastle and Gateshead*, vol. iii, pp. 72, 79–80 ; Dendy, *Records of the Company of Hostmen*, pp. 47, 54, 69. On the Wear river, also, the keels were owned by the colliery proprietors (*S.P.D.*, *Charles I*, vol. Dii, no. 77).

[4] *Exch. Deps. by Com.*, 29 Eliz., East. 4.

[5] In 1566 George Kitchen, keelman, willed five tenements, besides other lands and goods, to his wife, appointing as his executors John Kaye, merchant, and Henry Temple, carpenter (Welford, *op. cit.*, vol. ii, p. 405).

[6] Dendy, *op. cit.*, pp. xlvii and *passim*.

[7] Cf. below, vol. ii, p. 59 ; and Westerfield, *op. cit.*, pp. 223, 231.

[8] *Exch. Spec. Com.*, no. 5996 ; *Exch. Deps. by Com.*, 9 Charles I, Mich. 11 ; 11 Charles I, Mich. 46. Cf. below, vol. ii, p. 132 n.

on the Tyne and the Wear came to be done by independent middlemen who owned keels, and who, while often retaining an interest in collieries, kept that interest separate from their interest as fitters. With the advent of the fitter, the keelmen lost all initiative in ordering the keels, and were simply hired as oarsmen to make such trips as the fitters designated. The immigration of men from other districts into the Tyne valley, in order to supply the rapidly increasing demand for keelmen,[1] undoubtedly had an influence in breaking down the protection once afforded these workers by their craft gild. When, at the beginning of the eighteenth century, the keelmen became suitors to Parliament for incorporation,[2] they did not speak of themselves as a craft, and it is therefore reasonable to assume that their old organization, which can be traced back to the early sixteenth century, no longer existed.

Sailors aboard the sea-going colliers had no more initiative than the Tyneside keelmen. Some of them might aspire, however, to the desirable post of shipmaster. All business connected with the managemen of a collier was usually confided to the master. He hired the seamen, paid their wages, provided food and other necessary supplies for the voyage, bought coal from the mine owners or the fitters and sold it to the town merchants ; he was at once captain, super-cargo, and agent for the shipowner in marketing the coal, and his " capacity for driving a bargain was as important to his position as his knowledge of seamanship ".[3] Except in rare instances, when he was paid a salary by the owner,[4] the shipmaster held a " part " in the vessel and shared in the profits, if, indeed, he was not himself the sole proprietor. No mariner could become a shareholding master unless he had saved a small stock of money, for he must purchase his share, though he was generally allowed to buy it at a rebate. In 1595 Richard Harregate, of Gateshead, paid £26 to acquire from George Anderson, a merchant of Newcastle, a quarter " part " in one of Anderson's light colliers,[5] Harregate " to be master of the ship and to have the government thereof ".[6] The difference between the price asked of Harregate and the actual value of the quarter " part "[7] represented a reward for the services he would render as master. His interest in the pursuit of profit was enlisted no less securely than that of other partners, who contributed nothing but money towards building or purchasing the ship ; for he apparently received no salary, but only the dividends to which his share entitled him.

By the middle of the sixteenth century, it had become the custom among members of the Company of Merchant Adventurers at

---

[1] See below, vol. ii, p. 148.
[2] See below, vol. ii, pp. 177–9.
[3] Hooper, *The London Coal Exchange*, p. 10.
[4] In 1595, Thomas Ireland, mariner, of Yarmouth, contracted to make, as master of a ship, a voyage from Yarmouth to Newcastle and from thence back by way of Grimsby for 40s. " wages " (*Court of Requests Proc.*, 48 /6).
[5] Carrying 28 chaldrons, or about 56 tons, of coal.
[6] *Palat. of Durham, Bills and Answers*, bdl. 5.
[7] Even so small a collier must have been worth a great deal more than £104 (cf. above, p. 395).

Newcastle for those who owned ships to take a mariner into partnership,[1] and it may have been from this source that owners of colliers derived the practice of enlisting their masters' interest in the pursuit of profit. Since the owners could not be on hand to supervise the transactions at both ends of the voyage,[2] they found it expedient to make the master a partner in order to encourage him to drive the best bargain he could get. When the shipowners did not deal as merchants in their cargoes, but merely carried them as freight, it was of less importance, of course, to enlist the master's interest.

To what extent, it may be asked, were the shipowners and other conveyors of coal themselves traders in the coal which they carried, and to what extent did they simply receive a fixed freightage payment, similar to that paid to the modern railway company by the factor, the city merchant, or the mine owner, whose coal the company undertakes to haul ? Much depended upon the importance of the market. At ports like Kinsale, on the southern coast of Ireland, where there was as yet no steady demand for considerable quantities of coal, a purchaser, like the Earl of Orrery, who wanted limestone and fuel to build a fort in the harbour during the reign of Charles II, had himself to hire the ships and send them to South Wales for the coal,[3] in much the same manner as a Crown official had secured supplies from the Tyne in 1366 for construction work at Windsor Castle,[4] or as consumers in the village of Broughty, on the Firth of Tay, had secured supplies from pits in the Lothians as late as the middle of the sixteenth century.[5] When, in 1608, the Bishop of Llandaff, lord of the manor of Mathern on the Monmouthshire bank of the Severn, desired " sea cole amounting to Twenty Tonne . . . for the necessary provision of his house ", he " caused [it] to be brought from Bristowe " and unloaded on the river bank, there to remain until he came with " conveyance " to fetch it home.[6] No doubt he had hired specially the trow which brought it across the Severn, or had at least guaranteed the master a certain price for the lading. Such guarantees were sometimes made by the Crown when the exigencies of state policy made necessary large deliveries at places not ordinarily important consuming centres. For building the fortifications undertaken during the reigns of Henry VIII and Edward VI at Boulogne,[7] the Privy Council promised the merchants of Newcastle, before they loaded ships on the Tyne, that the officials in charge of the fortification would pay a fixed sum for every chaldron delivered. Similarly, on

[1] Boyle and Dendy, *Records of the Merchant Adventurers of Newcastle-upon-Tyne*, p. 41.
[2] The owner could demand of the master a copy of the contract under which he had purchased the coal from the fitter or colliery proprietor, and of the contract under which he had sold his cargo to the city merchant or crimp (Westerfield, *Middlemen in English Business*, p. 231).
[3] *H.M.C., Report on the Ormond MSS.*, vol. iv, p. 98.
[4] See above, p. 11.
[5] *Accounts of the Lord High Treasurer of Scotland*, vol. ix, p. 376.
[6] *Star Chamb. Proc., James I*, 160/39.
[7] See above, p. 11 n.

October 26th, 1627, his Majesties' service requiring that "some good quantitie of Sea Coales . . . be forthwith sent to the Isle of Retz " for the provision of the army engaged under Buckingham in an attempt to storm the fortress of St. Martin, the Privy Council asked the mayor and aldermen of Bristol to " advise and conferr with the coale merchants and masters and owners of coale shipps belonging to that porte and members of the same, and in particular to those of Milford Haven, and invite and encourage them all you may to transport good quantities of Coales to the said Isle, assuring them that they shall receive good payment . . . either in readie money, salte, or other commodities there at their own election ".[1] This assurance was not enough to induce cautious traders to send cargoes to a place where the English army had so insecure a foothold.[2] A few months later one of Buckingham's lieutenants asked for a warrant to " press " barks to carry coal to La Rochelle.[3] But to places like Tangier, or Dunkirk, where the English occupation achieved at least an appearance of stability, native merchants undertook to send coal at their own risk, paying freight to the shipmaster and building storage yards at the terminal ports. Samuel Atkins, a London merchant, tells the Privy Council in 1665 that he usually trades with Newcastle for coal to supply the garrison of Tangier, " where hee now hath a Coleyard, and against the Vintage sends some to Boardeaux and other places ".[4] It is probable that many of the early shipments of coal to the Continent [5] and to the American Colonies were financed by native merchants in this manner.

To towns like Amsterdam, Rotterdam, Hamburg, Dublin, Bridgwater, Yarmouth, King's Lynn, and—above all—London, where the annual imports could be counted by tens of thousands of tons before the end of the seventeenth century, the shipowners in most cases transported coal ostensibly as independent traders. Upon the arrival of colliers in the Thames, the city buyers were accustomed to meet with the shipmasters, or their agents or brokers, every morning between eight and one in the open-air market or " change " at Billingsgate, while coal heavers and porters hung about smoking their pipes. This meeting was said in 1730 to have existed " time out of mind ".[6] Presumably regular meetings began to be held during the reign of Elizabeth, when the coal trade first became important. By the reign of James I, at any rate, it had become an

---

[1] *P.C.R.*, vol. xxxvi, pp. 184–5.

[2] Cf. F. C. Montague, *Political History of England, 1603–60*, pp. 144–5, 158.

[3] *Cal. S.P.D.*, 1628–9, p. 39.

[4] *P.C.R.*, vol. lviii, p. 115.

[5] *S.P.D., James I*, vol. clvii, no. 32.

[6] *Alchin MSS.*, bdl. 64 : Regulations proposed by coal traders and other inhabitants of London (1730). Cf. Westerfield, *Middlemen in English Business*, p. 233 ; and above, p. 431. After the Great Fire, the traders attempted to change their place of meeting from Billingsgate to Tower Hill, but an order of the Common Council in 1670 required them to return to Billingsgate (*MS. Journal of the Common Council of London*, vol. xlvii, f. 27). Again in the 'eighties they removed to Monument Yard on Fish Street Hill, only to be again ordered to return (*City of London Repertories*, vol. lxxxvii, ff. 184, 187). Not until 1768 was a building erected for the coal exchange (Hooper, *The London Coal Exchange*, pp. 11–12).

accepted practice for the wholesale buyers to meet the shipmasters as soon as a fleet was in, and the Lord Mayor had granted the masters licence to sell.[1]

As early as 1607 Thomas Bagshawe describes how he and other buyers bargain with the shippers for the sale of their coal.[2] At first sight, none but ordinary competitive forces appear to be at work. Yet, if we examine all his testimony with care, we discover that there were often commitments on the part of both parties which must have interfered with the perfect freedom of the market, and we are left doubting whether the " coastmen ", or shippers, can properly be described as trading on their own account. The burden of the charges brought on February 15th, 1607, in the Court of Star Chamber by Attorney-General Hobart against Bagshawe and other London coal merchants,[3] is that they send " factors " to Newcastle " or thereabouts ", where " from the very pittes, without bearinge the adventure by sea, [they] doe buy or Contract for the . . . Coales before ever [these] . . . come on Water, or els give earnest money . . . that they may have the Refusall of the . . . Coales when they come att or neere London att a price agreed upon ". Bagshawe denies these charges, stating that, when the ships arrive in the port of London, the cargoes belong to the " Merchant Marryner or other persons the owners of such Coales ". But from his answer to a different bill of complaint, filed by the Attorney-General on December 13th, it is clear that Bagshawe's assertion that the mariner always owns the coal is one of those superficial truths which are meant to conceal more than they reveal.[4] For years, Bagshawe admits, he has been, not only a London buyer, but also a " part owner " of certain vessels and hoys, in which he or his servants and " factors " go divers times every year to Newcastle for coals and to other places thereabouts, taking the usual risks of this business. He also deals with other shipowners mentioned in the bill of complaint, by adventuring with them, that is by advancing them money to buy coal on the Tyne, and by sharing in " victelling, hiring and furnishing " the hoys. Although he reserves the right to " view " the cargo in London before agreeing to accept it, this reservation is in many cases a mere formality. Often he " hath taken the words of the . . . Coastmen in what Place or Pitts the . . . Coales were gotten . . . the said Places or pittes beinge knowne unto [him] ". Quite accurately, he describes his " trade " as one " of buying, bringeing in, and dealing with . . . Seacoles ".

There are reasons to suppose that such participation in the business of bringing fuel from Newcastle was not exceptional for the London coal dealers, nor for the coal dealers of other ports along the east coast of England. Anthony Yound, a Yarmouth coal merchant, advanced money in 1596 to George Pym, the master of a collier

---

[1] According to Westerfield (loc. cit.), the first purchases on each market day involved considerable higgling as to price, but once a price had been established, it prevailed for the rest of the day.
[2] Star Chamb. Proc., James I, 10 /13.
[3] Ibid., 13 /2.
[4] Ibid., 10 /13.

called the " Blessing of God ", in return for which Pym agreed to deliver his cargo at Grimsby, and, as was usual in such cases, entered into bonds against failure to fulfill the written agreement.[1] At the time of another investigation into the practices of the London coal merchants, undertaken by officers of the Crown just before the Civil War, Giles Bagg, a woodmonger of Queenshithe, confessed that he had " agreed with Edward Peach, a shipmaster, before he left the river for all the coals he should bring back ".[2] A certificate, written in 1627, is signed by thirty-five " traders in Newcastle coal in and about London ", who say they are in the habit of putting " ready money in [the] . . . hands " of the shipmasters before they sail to the Tyne.[3] We find as early as 1598 an example of a contract for delivery entered into by a coastman, in return for money advanced him by a trading merchant who proposes to sell the cargo to consumers in Southwark. Richard Jentleman, a shipowner of Southwold, agrees to deliver before the coming Christmas, at one of two wharves in Southwark, eighty London chaldrons of Newcastle coal for the benefit of Thomas Holbeck, formerly a neighbour of Jentleman's in Southwold, now active as a coal merchant in the growing metropolis. Holbeck is to be given notice of the arrival of Jentleman's collier, and he is allowed three days for his " workmen and laborers " to unload the cargo. Jentleman deposits with Holbeck, as security for failure to deliver, a bond for 200 marks, or £133 13s. 4d., more than twice the value of the coal. In case war prevents a safe passage to Newcastle, Jentleman is liable to pay only 13s. 4d. for every chaldron he is short in his deliveries.[4]

The London dealers who signed the certificate of 1627 felt justified in stating that they were " not traders with the shipmasters at all ", inasmuch as the masters alone bore the risks of the voyage, unless the dealers were—as sometimes happened—part owners in the vessels.[5] But the coastmen who accepted money advanced on a promise to deliver coal were no more independent of the town dealers who advanced the money than contracting overmen in the Tyne valley were independent of the colliery proprietors for whom they worked the pits. Although a great number of coastmen undoubtedly traded without contracting in advance to deliver their cargoes to particular middlemen, the impression that there was little integration in the coal trade appears to be misleading,[6] for the business of shipping coal was coming increasingly under the control of the dealers. Merchants in London and other importing centres began to extend their financial interests over the whole process of distributing coal from the time it was loaded on board a collier. Just as contracts for delivery of coal at the pits were instruments which helped to reduce the workmen

---

[1] *Court of Requests Proc.*, 48 /6.
[2] *S.P.D., Charles I*, vol. ccccii, no. 20.
[3] *Ibid.*, vol. lxviii, no. 48.
[4] *Court of Requests Proc.*, 91 /33.
[5] Cf. below, vol. ii, p. 27.
[6] Professor Westerfield has been led to conclude that " almost total lack of integration is easily the most salient characteristic of the [coal] trade " (*op. cit.*, p. 238).

and foremen to positions of dependence upon wealthy mining adventurers, so contracts for delivery exacted by powerful merchants were helping to reduce the less powerful middlemen and even the mining adventurers to positions of dependence upon the merchants.[1]

At the collieries producing for shipment by sea, the proprietors, or their agents or fitters, controlled not only the mining but also the loading of coal. Drivers, or wagoners, even if they owned the vehicles and beasts of burden which they drove, were never traders in the fuel which they carried from the " sea-sale " mines. Round Newcastle, they received fixed " leading " charges, ranging from 4d. to 2s. 6d. per ton according to the distance the pits stood from the river.[2] Around Whitehaven or Workington, the carters who loaded their packhorses with produce from the Cumberland collieries received from an agent stationed by the colliery proprietors at their " depots ", or storage yards, beside the wharves, a " token ", or check, for every load delivered. Carters could then exchange these " tokens " for cash at the colliery on paynight.[3] In 1600 the Mayor of Liverpool refers to " the agent for the Coles, [who] is nowe present here in Towne ".[4] He was there presumably to arrange with shipmasters for the export to Ireland of the produce from the pits about Prescot. It is clear that, when the colliery proprietors, or the fitters, in these " sea-sale " districts did not themselves own all the conveyances required to load their fuel aboard the hoys, they hired extra conveyances for fixed charges, in the same way as London wholesale dealers hired the extra lighters, the carts, and the labour which they needed to unload their coal on the Thames, and to haul it to consumers' houses or to storage yards.[5] Occasionally proprietors and fitters also participated in building or

---

[1] For details, see below, esp. ch. iii (iv).

[2] From Winlaton colliery in 1582 the cost of carriage was sometimes 4d., sometimes 5d. per " fother " (Exch. Deps. by Com., 29 Eliz., East. 4). From Benwell colliery between 1620 and 1630 the cost was about 5d. per " fother " (ibid., 11 Charles I, Mich. 46). From Blackburn colliery in 1646 the cost was from 14d. to 16d. per " carriage " (probably a " wain "), having recently been raised from 9d. (Welford, Records of the Committees for Compounding, p. 164). From Elswick colliery in 1681 the cost varied from 3d. to 4d., 5d., and 6d. per " fother " (Exch. Deps. by Com., 32 Charles II, Mich. 30). For the size of a " fother " and a " wain " see Appendix C. During the reign of Charles I, the colliery proprietors of the Tyne valley say that the cost of " leading " a Newcastle chaldron of coal from the pits to the river varies from 16d. to 5s. (S.P.D., Charles I, vol. clxxx, no. 59). If we assume the average cost to have been 2s., and assume that approximately 180,000 chaldrons were carried annually, then the proprietors must have paid altogether nearly £20,000 a year in " leading " charges alone.

[3] W. Jackson, " The Colliery . . . Tokens of West Cumberland ", in Trans. Cumberland Archæological Soc., vol. xv, 1899, pp. 392–3. Perhaps a similar system of payment prevailed in the Tyne valley. We hear of " tallies " in connection with the deliveries from Winlaton colliery in 1582 (Exch. Deps. by Com., 29 Eliz., East. 4) ; and cf. below, vol. ii, p. 59.

[4] Lansdowne MSS., 156, no. 106.

[5] " Lighterage, Wharfage, and Carriage . . . are the Woodmonger's Profits " (The Frauds and Abuses of the Coal Dealers, 1745). See also City of London Repertories, vol. lxxvii, f. 140. A few of the more wealthy citizens were accustomed to hire their own carts for bringing home their supplies of coal (P.C.R., vol. xlix, pp. 532–4). But the woodmongers appear to have been so successful in monopolizing the " car rooms ", that any citizen who wanted such a conveyance had to apply to them (see above, pp. 407–8).

freighting colliers.[1]   If certain proposals (probably those of the romantic midland mining adventurer—Huntingdon Beaumont)[2] submitted in 1605 to the Lord Mayor had been successfully carried out, they would have meant that the proprietor of a colliery near Nottingham himself would have undertaken—without the intervention of any independent trader—to freight coal by barges down the Trent and by ships from Hull directly to the city store-yards on the Thames.[3] While the town merchant extended his interest in the coal trade towards the producer, the mine owner was attempting with less success to extend his interest towards the consumer.[4] The tendency was for the whole process of mining and delivering coal to come under the control of a few principal adventurers.

At first sight this tendency is not apparent in connection with the trade of inland districts, where coal is carted from the so-called " landsale " pits overground to neighbouring country hamlets, villages, or towns. Reference has been made already to the local " carriers " or " drivers " in the Lothians, in Leicestershire, Somerset, Gloucestershire, and south Durham, who transported coal, like some of the coastmen, at their own adventure.[5] A closer view suggests that the history of the inland trade resembles in miniature the history of the coastwise or the oversea trade. In places where there was no regular market the consumers themselves had to arrange to bring coal by land, just as the Earl of Orrery and the Bishop of Llandaff had to arrange to bring it by sea. Towards the end of Elizabeth's reign, Robert Horseman, of the hamlet of Kyme in Lincolnshire, " havinge occasion to provyde . . . Seacoles for the proper use and provision of his howse and household ", agreed with a neighbour jointly to purchase from Robert Falkingham, a beer-brewer of Boston, twenty London chaldrons " good and merchantable " ; Horseman himself to bring " carriages " to receive the coal at a place called " Skirby Gate " (probably Skirbeck), " nigh unto Boston " and about fifteen miles by land from Kyme.[6] When, in 1628, Joseph Giffard set up at Tetbury, in the Cotswolds, to manufacture saltpetre under warrant of the Privy Council, he charged the constables of Gloucestershire to provide him with carts for bringing coal more than twenty miles overland to Tetbury from a pit at Westerleigh, in Kingswood Chase.[7]

At towns such as Edinburgh, Nottingham, and Bristol, where the market for fuel was of considerable importance before the end of the seventeenth century, there were—as at King's Lynn, Bridgwater, and Maldon—coal merchants with " yards " in which to store their

[1] See below, vol. ii, pp. 24–6.
[2] We know that a few years later—in 1609—Beaumont was engaged, with the backing of a group of city merchants (see below, ch. iii (i)), in plans for supplying the London poor with coal from mines on the Northumberland coast (*Repertories*, vol. xxix, f. 29).
[3] H.M.C., *Report on the MSS. of Lord Middleton*, pp. 171–2.
[4] See below, vol. ii, p. 17.
[5] See above, p. 436.
[6] *Chanc. Proc., James I*, H. 22/15.
[7] *S.P.D., Charles I*, vol. cxxi, no. 10.

supplies.[1] It might be expected that carriers coming to these towns from the neighbouring pits would find ready sale with these merchants when they could not sell directly to consumers, and that the merchants would attempt—like the buyers in London—to get exclusive control over the incoming " carriages ". It might be expected, further, that the town merchants would advance the carriers money to make their purchases at the pits. There is some reason to believe that this process was taking place, at least in Bristol. As early as 1624 Francis Popham, an aged citizen of that town, told commissioners, appointed to take evidence in a Chancery suit involving the proprietors of the colliery of Kingswood Chase, that he was " ymployed in uttering a greate part of [these proprietors'] . : . Cole ". Since the mine owners were all merchants of Bristol, and since their leader, Arthur Player, was accused of having obtained a monopoly of the sale to consumers in the town, we are perhaps safe in assuming that a single group owned the fuel from the time it was mined until it was delivered into the cellars of the local householders, and that we have—at least temporarily—in the small Kingswood coal industry a case of complete integration, both vertical and horizontal.[2]

Before the end of the seventeenth century, therefore, the cleavage between capital and labour appears to have been scarcely less general in the case of the coal trade than in the case of coal mining. To an increasing extent the workmen who handle the new fuel, and who drive the conveyances in which it is transported from the mine to the consumer, become simple wage-earners like their fellows in the mines. To an increasing extent the tools, wagons, horses, boats, and storage places necessary for the carrying on of the coal trade become the property of capitalists, who hire wage workers to use and man them. Even in the few cases in which transport workers remain the owners of the tools and conveyances which they use, they cease to deal in the coal which they carry. The function of trading in coal is becoming more and more the monopoly of a few rich adventurers.

The whole of the great capital outlay which the expansion of the coal industry has been shown to involve, with the exception of the small sums supplied by working miners and mine managers, who frequently lost their investments altogether by entering into contracts for the delivery of coal, came from persons who took no part in manual labour. The financial history of the industry is largely the story of the control which the great town merchants obtained over the mining and the transport of coal. If the growth of coal mining during the century following the accession of Elizabeth is connected with the development of a capitalistic industrial organization, it appears also to be not less closely related to the increase in the economic and political power of the rising merchant class, which is perhaps the most significant social phenomenon in seventeenth-century England.

---

[1] See above, p. 397 n.; Stevenson, *Records of the Borough of Nottingham*, vol. v, pp. 400–1.

[2] *Chanc. Deps.*, P. 7/T.